PRINTED CIRCUIT BOARDS

Design, Fabrication, Assembly and Testing

PRINTED CIRCUIT BOARDS

Design, Fabrication, Assembly and Testing

Dr R S Khandpur
Director General,
Pushpa Gujral Science City, Kapurthala, Punjab

Formerly
Director General,
Centre for Electronics Design and Technology of India (CEDTI)
Dept. of Information Technology, New Delhi
and
Director
CEDTI, Mohali (Chandigarh) Punjab

McGraw-Hill

New York   Chicago   San Francisco   Lisbon   London
Madrid   Mexico City   Milan   New Delhi   San Juan
Seoul   Singapore   Sydney   Toronto

The McGraw·Hill Companies

Cataloging-in-Publication Data is on file with the Library of Congress

Copyright © 2006 by The McGraw-Hill Companies, Inc. All rights reserved. Printed in the United States of America. Except as permitted under the United States Copyright Act of 1976, no part of this publication may be reproduced or distributed in any form or by any means, or stored in a database or retrieval system, without the prior written permission of the publisher.

2 3 4 5 6 7 8 9 0   DOC/DOC   0 1 0 9 8

ISBN 0-07-146420-4

Printed and bound by RR Donnelley.

This book was previously published by Tata McGraw-Hill Publishing Company Limited.

McGraw-Hill books are available at special quantity discounts to use as premiums and sales promotions, or for use in corporate training programs. For more information, please write to the Director of Special Sales, McGraw-Hill Professional, Two Penn Plaza, New York, NY 10121-2298. Or contact your local bookstore.

Contents

| 2. Electronic Components | 25 |

3. Layout Planning and Design — 104

5. Artwork Generation — 193

6. Copper Clad Laminates

7. Image Transfer Techniques   283

8. Plating Processes — 310

10. Mechanical Operations   384

11. Multi-layer Boards — 414

12. Flexible Printed Circuit Boards — 427

13. Soldering, Assembly and Re-working Techniques — 453

Preface

Printed circuit boards are the most frequently used interconnection technology for components in electronic products. PCB requirements today have developed with the increase in the packaging density of modern electronic and mechanical components. They now include finer conductor tracks and thinner laminates, present in an ever-increasing number of layers. Integrated circuits have become dramatically sophisticated especially in the last decade. This has in turn created new design requirements for mounting them on the boards. While insertion was common with DIP (dual in-line-package) technology in the 1970s, surface mount technology in now being increasingly employed. In addition, line and space dimensions are diminishing; the number of conductors between through-holes is increasing; and hole diameters are rapidly decreasing. These requirements have lead to a rising trend in the implementation of microvias as blind-vias or through-holes. In the future, PCBs will have higher functionality/density, improved reliability and lower cost through better and more tightly controlled/cost-effective processing. The industry will also move towards more environmental friendly PCBs. Furthermore, the advantages in contract manufacturing at the global level will effectively ensure that the design and manufacture of PCBs are of internationally accepted quality.

Most of the books currently available on this subject, do not address several of the above mentioned (important) aspects. This book is a single-source reference covering these vital areas of PCB technology. This includes design, fabrication, assembly and testing, including their reliability and quality aspects. The book therefore, addresses not only the design considerations but also provides a general understanding on all the processes needed in the physical construction and testing of the printed circuit boards. Despite the several highly specialized disciplines in this field, such as, electronics, mechanical engineering, fluid dynamics, thermodynamics, chemistry, physics, metallurgy and optics, the attempt in this book has been to keep the text lucid and to explain the salient aspects of PCBs without indulging in an exhaustive theoretical approach. Extensive bibliographical references are provided to represent such specialized extensions of the subject which lie outside the domain of this book.

The book is divided into fifteen chapters. Each chapter is comprehensive in its coverage and can be read and well understood as an independent chapter. However, the chapters are so arranged that they represent the processes as they progress in actual practice. The book has been written keeping in mind professionals in the field for whom there is much practical information, coupled with information from manufacturers of various machines and materials.

Chapter 1 is an introduction to the field of printed circuit boards. From their historical developments, it progresses to the description of the types of PCBs and the sequence of their manufacture from design to assembly. The major drivers for modern PCB technology, particularly high density interconnects, are also illustrated. The field of printed circuits is now largely governed by a high degree of standardization, with IPC (Institute of Electronics Circuit Packaging) taking the central role, and accordingly, a brief about this aspect is also given in the first chapter.

In many design offices, the persons responsible for layout design and artwork generation are draughts-men trained in mechanical engineering who need to understand the basics of all electronic components including integrated circuits (ICs) and surface mount devices (SMDs). To fulfill this requirement, Chapter 2 is devoted to the fundamentals and characteristics of a wide variety of electronic components.

Chapter 3 deals with the layout planning and general design considerations for PCBs; and Chapter 4 details design guidelines for specialized circuits such as high frequency circuits and high density interconnects. Special considerations for analog circuits and high power dissipating circuits are also included in this chapter.

Until about a decade ago, artwork generation was carried out manually. The CAD systems available today, with software packages available from a number of vendors, have not only simplified this work, but have also made the artwork design of high density board more convenient. CAD has made it possible to integrate the artwork generation with CAM (Computer-aided manufacturing). The design data transfer mechanism is assuming great importance due to the increasing role of distributed manufacturing facilities. Artwork generation techniques, both manual as well as CAD/CAM based, are covered in Chapter 5.

Chapter 6 deals with the base materials or the laminates, which form the core of the printed circuits. The chapter discusses not only the constructional aspects of PCBs, it also details the usual defects present in them, along with the testing methodology from the point of view of quality assurance.

Image transfer techniques (of the artwork on to the laminate) have undergone tremendous developments, particularly because of the stringent requirements of the fine line printed circuits. The laser direct image transfer method is becoming increasingly popular. Chapter 7 covers the conventional as well as the modern techniques of image transfer.

The next logical step in the manufacture of PCBs is etching. Various techniques of etching, both wet and dry, are explained in Chapter 8, while Chapter 9 details the plating techniques. In addition, Chapter 9 also covers various methods for providing proper surface finish to the conducting pathways, including application of solder mask and conformal coatings.

Precision mechanical operations form an important step in the quality manufacture of PCBs. It is reported that about 85% of all the defects which are discovered in PCBs, are directly or indirectly associated with drilling. In addition, the ever-decreasing size of the holes has made mechanical drilling methods inadequate. Chapter 10 discusses the use of lasers which are now popular in overcoming this limitation.

Multi-layer boards, which have enabled high-density boards, are based on special design and fabrication techniques. These are covered in Chapter 11. Chapter 12 includes special features of flexible PCBs, their design and fabrication techniques, and applications.

Chapter 13 is devoted to soldering and assembly techniques, both manual as well as machine-based automatic systems. It includes rework procedures, especially for boards with surface mount devices and mixed assemblies.

Chapter 14 explains quality and reliability aspects of PCBs and sets out criteria for their acceptability. Various tests on bare boards and assembled boards are detailed along with their limitations and areas of applications. Issues on pollution, associated with the PCB industry, along with their treatment methods, are covered in the last chapter. A brief reference to the end-of-life disposal of PCBs and the concept of *Design for Environment* is also illustrated. The chapter also addresses the vital issue of lead free soldering and the present status of its adoption.

The book provides an exhaustive glossary of commonly used terms. The extensive bibliography will be useful to readers who need specialized information in greater detail. The internet has become an invaluable resource for a wide range of general and technical information, especially from the manufacturers of PCBs and related technologies. References are provided in the text for any material which has been included from the internet.

In conclusion, I would like to thank my wife Mrs. Ramesh Khandpur who has been a source of great inspiration in helping me attain my goals in life. Her wholehearted support never let me slacken in my professional endeavors. Thanks are due to my children and grandchildren who are looking forward to this new publication.

My thanks are also due to Tata McGraw-Hill, New Delhi, for permitting me to use some illustrations from the book *Printed Circuit Boards* by Mr. W C Bosshart and for bringing out this high quality book.

R S Khandpur

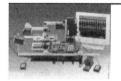

Basics of Printed Circuit Boards

 ## 1.1 Connectivity in Electronic Equipment

Electronic equipment is a combination of electrical and electronic components connected to produce a certain designed function. In the era of vacuum tubes and even later, electronic equipment was constructed by hand wiring and by point-to-point soldering. The wires were stripped of their insulation, tinned and soldered. Each discrete component was installed by hand, electrically and mechanically. The equipment was obviously large, awkward and bulky. It was difficult to meet the demanding requirements for the use of this equipment in aircrafts, the health sector and home emergency uses, thereby necessitating the development of smaller and more compact electronic equipment.

A natural evolution took place in several areas. Smaller components were developed and modular design became popular, basically intended to decrease the time between unit failure and repair due to easy replaceability. The use of miniaturization and sub-miniaturization in electronic equipment design gave birth to a new technique in inter-component wiring and assembly that is popularly known as the ***printed circuit board***. The printed circuit board provides both the physical structure for mounting and holding electronic components as well as the electrical interconnection between components.

Printed circuit board is usually abbreviated as PCB and quite often referred to as *board*. However, in the USA, the term PWB (*Printed Wiring Board*) is more often used instead of PCB.

1.1.1 Advantages of Printed Circuit Boards

There are many good reasons for using printed circuit boards instead of other interconnection wiring methods and component mounting techniques, some of which are as follows:

i. The size of component assembly is reduced with a corresponding decrease in weight.

ii. Quantity production can be achieved at lower unit cost.

iii. Component wiring and assembly can be mechanized.

iv. Circuit characteristics can be maintained without introducing variation in inter-circuit capacitance.

v. They ensure a high level of repeatability and offer uniformity of electrical characteristics from assembly to assembly.

vi. The location of parts is fixed, which simplifies identification and maintenance of electronic equipment and systems.

vii. Inspection time is reduced because printed circuitry eliminates the probability of error.

viii. Printed wiring personnel require minimal technical skills and training. Chances of mis-wiring or short-circuited wiring are minimized.

 ## 1.2 Evolution of Printed Circuit Boards

The history of development of printed circuit boards is not very old. They have been in commercial use only since the early 1950s, even though their concept originated nearly 50 years prior to their commercial use.

Frank Sprague, the founder of Sprague Electric, while still an apprentice, had the idea, in 1904, of eliminating point-to-point wiring. When he conferred with his mentor, Thomas Edison, for implementing his concept, it was suggested that silver reduction, as used in mirror manufacture or the printing of graphite pastes on linen paper, may prove to be suitable to achieve the objective. Subsequently, a number of pioneering efforts were made for the development of printed circuit boards, out of which the following events occupy place of eminence.

The first significant contribution came from Mr. Charles Ducas, who filed a patent application at the US Patent Office on March 2, 1925 for his proposal to mount electrical metal deposits in the shape of conductors directly onto the insulation material to simplify the construction of electrical appliances (Fjelstad, 2001). He used a stencil to form the conductors on the surface of insulation material and applied a conductive paste in the lines desired. After removal of the stencil, the lines were reinforced to the desired thickness by electrolytic metal deposition This development considerably simplified the manufacture of electrical appliances because the electrolytic metal deposition, being a simple process, could be carried out by unskilled operators.

Just 17 days later, Mr. Francis T. Harmann filed a patent for the so-called subtractive method of making PCBs. This development could be considered as the forerunner of etching technology. In April 1926, a patent was granted to Mr. Cesar Pasolini in France for his invention of the additive way of making electrical connections. Mr. Samuel Charles Ryder filed an Australian patent in September 1928, related to the manufacture of inductance coils for use in radio tuning devices or other such applications. He proposed to print or spray the substrate directly with conductive paint during manufacture. Similarly, the patent application filed by Mr. Herbert C. Arlt in July 1935 in America again emphasized the avoidance of wires as the basic purpose of the development.

The major contribution towards the development of modern printed circuit technology was made by Dr. Paul Eisler, who proposed copper clad insulation material in sheet form for use as the base material in circuit board manufacture. According to him, "a resist in the shape of the circuit pattern is printed onto the surface of the copper cladding with the uncovered metal being removed by etching". He also proposed the generation of conductors on both sides of the copper clad base material, with connection between conductors on both sides being made through eyelets. Eisler's work not only gave birth to a method of mass production and an assembly scheme but also offered economy in weight and space, which is especially important in military equipment. Little wonder that Dr. Eisler is often called the Father of printed circuit board technology. However, he gave preference to the use of eyelets rather than the plated-through hole technology, which, with time, became an essential process for the manufacture of double-sided and multi-layer printed circuit boards.

Towards the end of World War II, a technology developed by the US National Bureau of Standards was used in the volume production of US army VT proximity fuses for rockets. Unlike the Print and Etch technique of Eisler, this technology used printed silver paste conductors and graphite resistors that were screen printed onto ceramic substrate. This technique is more commonly associated with today's hybrid circuit technology. It was this technique that ushered in the commercial use of printed circuits.

After World War II, fascinating developments took place in the field of electronics, resulting in a high demand for consumer products like radio and television and simultaneously, the use of electronics for military applications. These developments resulted in the need for reliable circuit boards with increasing complexity. After attaining the level of maximum density, based on contemporary fabrication limitations, single side boards were replaced by double side boards, which allowed wires to cross over each other without the need for additional special jumpers. This was accomplished finally by plated-through holes.

During 1953–55, Motorola introduced the copper metal plating process to provide interconnection between two sides of a board, which was found to be more suitable for the mass manufacturing process. In the 1960s, the electroless method was introduced using catalyst activators, while Photo Circuits, USA, developed the fully additive process in 1964. In this method, the base material does not have copper on it and the copper is plated selectively on the required places for interconnections.

The late 1960s witnessed phenomenal growth in the field of consumer electronics, which necessitated the introduction of automation in fabrication and in testing of bare board/populated circuit boards. Soon thereafter in the 1970s, printed circuit boards were firmly entrenched in the consumer electronic, scientific equipment, medical equipment, air and space, defence, and in nearly all branches of electronics, which later culminated in the personal computer industry. Several new processes were subsequently developed for applications such as photo film lamination, dry film and wet film resist, solder masking, legend printing and CNC drilling, etc. The size of the printed board got considerably reduced with the manufacture of multi-layer and rigid-flexible circuit boards which made use of buried and blind vias plated through hole connection and wet process chemistry.

New developments in component technology, especially in the area of surface mount technology, resulted in several innovations in PCB materials and processes, and today there are constant pressures for improvements in PCB technology in all its aspects. The continuing trend towards high functionality integrated circuit (IC) components with higher input-output (I/O) pin counts of the IC packages has resulted in increased demand for fine-featured PCBs, giving rise to "high density interconnect structures (HDIS)" which are now manufactured by a large number of companies. High frequency electronic systems, with their high speed operations, create a demand for PCBs with lower electrical losses. In addition, higher operating voltages require PCBs with greater resistance to voltage breakdown, high voltage tracking and corona.

PCBs constitute a very important strategic component for electronic products. It is therefore no surprise that the PCB industry worldwide is a booming market, and an annual growth rate of 7 per cent is expected in this industry up to the year 2010. Therefore, printed circuits are likely to continue to be the icon of the electronic industry well through the next decade, which will of course, bring higher functionality/density, improved cost and reliability through more tightly controlled cost-effective processes. The subtractive process-based PCB world is likely to have a transition to an additive process. We can also expect an increase in the use of flexible circuits and a move towards more environment-friendly PCBs without the use of lead. Predictably, the various kinds of printed circuit boards, both those that are currently existing and those to be developed, will retain their function as an essential part of electronics and, in some cases, may even achieve a more significant place in the electronics industry.

1.3 Components of a Printed Circuit Board

The essential components of a printed circuit board are:
- The *base*, which is a thin board of insulating material, rigid or flexible, which supports all conductors and components; and
- The *conductors*, normally of high purity copper in the form of thin strips of appropriate shapes firmly attached to the base material.

The *base* provides mechanical support to all copper areas and all components attached to the copper. The electrical properties of the completed circuit depend upon the dielectric properties of the base material and must therefore, be known and appropriately controlled.

The *conductors* provide not only the electrical connections between components but also solderable attachment points for the same.

When the completed board provides mechanical support and all necessary electrical connections to the components, it is essentially a Printed Wiring Board or Printed Circuit Board. The term *printed* became popular because the conductive areas are usually generated by means of a printing process like screen printing or photo-engraving, which are commonly used to print drawings or inscriptions.

1.4 Classification of Printed Circuit Boards

Printed Circuit Boards may be classified according to their various attributes, often with ambiguous results. They were traditionally divided into three classes according to their use and applications, and were commonly referred to as consumer, professional and high reliability boards.

Consumer PCBs were generally used in consumer products such as radio, television, and cheap test and measuring equipment. They used less expensive base material and allowed greater tolerances for manufacture to keep the cost low. Much importance was not given to good and consistent electrical properties.

Professional boards were made of better quality material to achieve tighter electrical and environmental specifications using controlled fabrication techniques. Higher reliability boards, normally used in strategic applications, were meant to provide the best of electrical properties through the use of high quality base material and tightly controlled manufacturing processes.

The above classification might have been applicable two or three decades ago, but presently, the distinction between consumer and professional markets has disappeared. Many consumer products like compact discs, camcorders or cameras have become more complex, reliable and demanding than what was hitherto considered as professional equipment like personal computers. The advent of surface mount technology and developments in automatic assembly techniques requires that the boards even for the cheapest product must be manufactured to strict mechanical tolerances.

A more simple and understandable classification is now used, which is based on the number of planes or layers of wiring, which constitute the total wiring assembly or structures, and to the presence or absence of plated-through holes. This method of classifying boards has the advantage of being related directly to the board specifications. The important distinguishing constructions of PCBs are detailed below.

1.4.1 Single-sided Printed Circuit Boards

'Single-sided' means that wiring is available only on one side of the insulating substrate. The side which contains the circuit pattern is called the 'solder side' whereas the other side is called the 'component side'. These types of boards are mostly used in case of simple circuitry and where the manufacturing costs are to be kept at a minimum. Nevertheless, they represent a large volume of printed boards currently produced for professional and non-professional grades. Figure 1.1 shows the arrangement of a single-sided board.

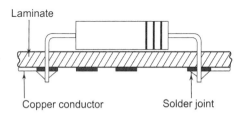

Fig. 1.1 Single-sided PCB

The single-sided boards are manufactured mostly by the 'print and etch' method or by the 'die-cut' technique by using a die that carries an image of the wiring pattern; and the die is either photo-engraved or machine-engraved.

Normally, components are used to jump over conductor tracks, but if this is not possible, jumper wires are used. The number of jumper wires on a board cannot be accepted beyond a small number because of economic reasons, resulting in the requirement for double-sided boards.

1.4.2 Double-sided Printed Circuit Boards

'Double-sided' printed circuit boards have wiring patterns on both sides of the insulating material, i.e. the circuit pattern is available both on the components side and the solder side. Obviously, the component density and the conductor lines are higher than the single-sided boards. Two types of double-sided boards are commonly used, which are:

- Double-sided board with plated through-hole connection (PTH); and
- Double-sided board without plated through-hole connection (non-PTH).

Figure 1.2(a) shows the constructional details of the two types of double-sided boards.

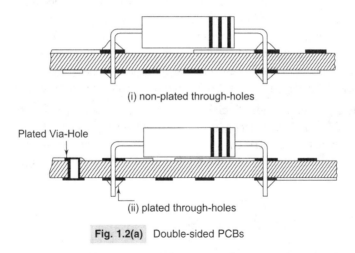

(i) non-plated through-holes

Plated Via-Hole

(ii) plated through-holes

Fig. 1.2(a) Double-sided PCBs

Double-sided PTH board has circuitry on both sides of an insulating substrate, which is connected by metallizing the wall of a hole in the substrate that intersects the circuitry on both sides. This technology, which is the basis for most printed circuits produced, is becoming popular in cases where the circuit complexity and density is high. Figure 1.2(b) shows the configuration of a plated through-hole in a printed circuit board.

Double-sided non-PTH board is only an extension of a single-sided board. Its cost is considerably lower because plating can be avoided. In this case, through contacts are made by soldering the component leads on both sides of the board, wherever required. In the layout design of such boards, the number of solder joints on the component side should be kept to a minimum to facilitate component removal, if required. It is generally recommended that conductors should be realized as much as possible on the non-component side and only the remaining should be placed on the component side.

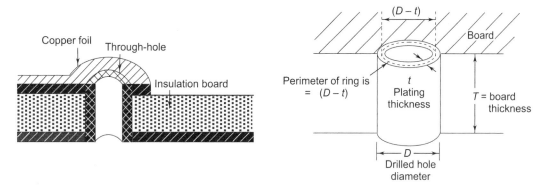

Fig. 1.2(b) Configuration of plated through-hole

The non-plating technique in double-sided boards is shown in Figure 1.3 wherein the interconnection is made by a jumper wire. A formed insulated solid lead wire is placed through the hole, clinched and soldered to the conductor pad on each side of the board. Different types of eyelets are also used for double-sided board interconnection. These are illustrated in Figure 1.4.

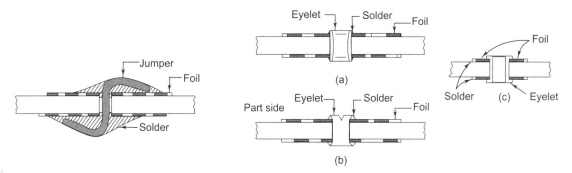

Fig. 1.3 Interconnection with clinched jumper

Fig. 1.4 Interconnections with (a) funnel-flanged eyelet (b) split funnel-flanged eyelet (c) fused-in-place eyelet

1.4.3 Multi-layer Boards

The development of plated through-hole technology has led to a considerable reduction in conductor cross-overs on different planes, resulting in a reduction in space requirements and increased packaging density of electronic components. However, the modern VLSI and other multi-pin configuration devices have tremendously increased the packaging density and consequently the concentration of inter-connecting lines. This has given rise to complex design problems such as noise, cross-talk, stray capacitances and unacceptable voltage drops due to parallel signal lines. These problems could not be satisfactorily solved in single-sided or double-sided boards, thereby necessitating an extension of the two-plane approach to the multi-layer circuit board. A multi-layer board is, therefore, used in situations where the density of connections needed is too high to be handled by two layers or where there are other reasons such as accurate control of line impedances or for earth screening.

The multi-layer board makes use of more than two printed circuit boards with a thin layer of what is known as 'prepreg' material placed between each layer, thus making a sandwich assembly as shown in Figure 1.5. The printed circuit on the top board is similar to a conventional printed circuit

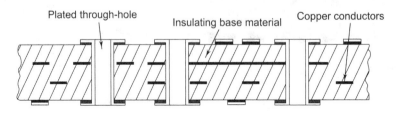

Fig. 1.5 Cross-section of a multi-layer board with four layers

board assembly except that the components are placed much closer to avoid having many terminals, which necessitates the use of additional board layers for the required interconnections. The electrical circuit is completed by interconnecting the different layers with plated through-holes, placed transverse to the board at appropriate places. Multi-layer boards have three or more circuit layers, while some boards have even as many as 50 layers. Figure 1.6 shows the details of the two types of multi-layer boards, one with four-layers and the other with eight-layers.

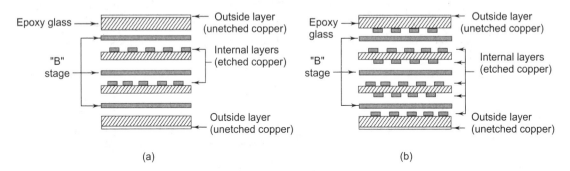

Fig. 1.6 Multi-layer lay-up details (a) four-layer board (b) eight-layer board

By virtue of the multi-layer conductor structure, multi-layer printed wiring has facilitated a reduction in the weight and volume of the interconnections commensurate with the size and weight of the components it interconnects.

The following areas of application necessitate the use of multi-layer printed wiring arrangements:
- Wherever weight and volume savings in interconnections are the overriding considerations, as in military and air-borne missile and space applications;
- When the complexity of interconnection in sub-systems requires complicated and expensive wiring or harnessing;

- When frequency requirements call for careful control and uniformity of conductor wave impedances with minimum distortions and signal propagation, and where the uniformity of these characteristics from board-to-board is important;

- When coupling or shielding of a large number of connections is necessary; the high capacitance distributed between the different layers gives a good de-coupling of power supply which permits satisfactory operation of high speed circuits;

- With multi-layers, all interconnections can be placed on internal layers, and a heat sink of thick solid copper can be placed on the outer surfaces. By mounting the components directly on the metallic surfaces, the problem of heat distribution and heat removal in systems can be minimized. Also, the layout and artwork designs are greatly simplified on account of the absence of the supply and ground lines on the signal planes.

Because of the developments in mass lamination technology, four-layer boards and even six-layer boards can be made with almost the same ease as double-sided boards. With the improvement in reliability and reduction in cost of printed circuit boards, the use of multi-layer boards is no longer limited to only high technology products, but has spread to some of the most common applications like entertainment electronics and the toy industry.

The cost of a printed circuit board depends upon its complexity and the technology used. Figure 1.7 illustrates the relationship between the complexity and cost of printed circuit boards.

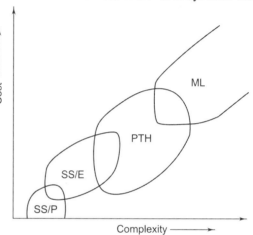

Fig. 1.7 Cost of a printed circuit board depends upon its complexity and on its technology SSIP = single-sided paper base laminate; SSIE = single-sided epoxy glass laminate; PTH = double-sided plated through-hole epoxy glass laminate; ML = multi-layer; (redrawn after Ross and Leonida 1996b)

1.4.4 Rigid and Flexible Printed Circuit Boards

Printed circuit boards can also be classified on the basis of the type of insulating material used, i.e. rigid or flexible. While *rigid boards* are made of a variety of materials, *flexible boards* use flexible substrate material like polyester or polyamide. The base material, which is usually very thin, is in the range of 0.1 mm thickness. Laminates used in flexible boards are available with copper on one or both sides in rolls. *Rigid-flex* boards, which constitute a combination of rigid and flexible boards usually bonded together, are three-dimensional structures that have flexible parts connecting the rigid boards, which usually support components. This arrangement gives volumetrically efficient packaging and is therefore gaining widespread use in electronic equipment. Flexible PCBs may be single-sided, double-sided (PTH or non-PTH) or multi-layer.

1.5 Manufacturing of Basic Printed Circuit Boards

A variety of processes are currently used for manufacturing printed circuit boards. However, most of the processes have identical or similar basic steps. Variations in the basic manufacturing steps are usually made by the manufacturers to improve quality or specific yield.

The most popular process is the '*print and etch*' method, which is a purely subtractive method. In this process, the base material used is copper clad laminate to which all the electronic components are soldered, with one or more layers of etched metal tracks making the connection. The etching process involves achieving a conductive pattern formed on one or both sides of the laminate. The term 'printed wiring' or 'printed circuit' refers only to the conductive pattern that is formed on the laminate to provide point-to-point connection.

Four specific phases of the PCB manufacturing process need to be understood. These are design, fabrication, assembly and test. Historically, these phases have been individual islands of activity relatively isolated from each other (Biancini, 1991). However, with the increasing complexity of the printed circuit boards coupled with the developments in software-based design and testing procedures, the present-day requirements make the circuit designer look beyond the individual element approach and take a holistic approach taking into consideration design for manufacturability, assembly and testability.

1.5.1 Single-sided Boards

The following steps (Figure 1.8) represent, in a simplified manner, the design and fabrication process of a single-sided printed circuit board.

Schematic Diagram

The schematic diagram, also called the circuit or logic diagram, represents the electronic components and connections in the most readable form. The schematic diagram is developed while taking into consideration the specifications of components, interaction between components (especially timing and loading), physical packages and arrangement of connector pin-outs. The circuit diagram will often start on paper and finish in computer-aided design (CAD). The circuit diagram references each part on the printed circuit board with a designator (e.g. IC4) and pin numbers for each connection. A good circuit diagram includes all the essential information required to understand the circuit operation, and has descriptive net and connector labels, including all the parts on the printed circuit board. To this end, the printed circuit board CAD and schematic CAD are tied together through a net-check. In short, the finished circuit diagram, is the main reference document for design.

Artwork Generation

The components and connections in the PCB layout are derived from the circuit diagram, and physically placed and routed by the designer to get the best results in term of board size and its manufacturability. The PCB layout defines the final physical form of the circuit and labelling details are finalized as the layout is completed. When the PCB layout is complete, the track layout information is provided on self-adhesive type crepe material tape stuck on a plastic sheet such as polyester. The layout or artwork is usually enlarged two to four times to improve accuracy. Alternatively, the CAD file is used to generate the artwork on a computer-controlled plotter, or on an electronic transfer medium such as magnetic tape or floppy disc.

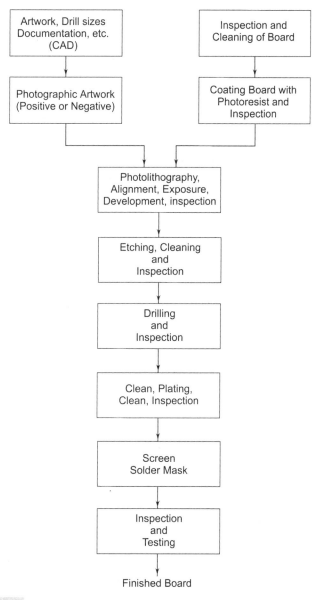

Fig. 1.8 Major steps in the fabrication of a single-sided printed circuit board

The artwork is then reduced to the final size, and a positive or negative print made depending on the requirement of the manufacturer.

Panel Preparation

The raw material for printed circuit boards is a copper clad laminate with copper on one side only. The sheets of the laminate are sheared to provide panels of the required size, keeping it slightly longer than the master pattern of the PCB. The preferred size of panel is 350 × 508 mm. The commonly used laminates for general purpose applications are normally paper base type, whereas epoxy glass laminates are preferred for superior mechanical and electrical properties. The mechanical properties include punching and drilling qualities, flexural strength, flame resistance and water absorption. The important electrical properties include dielectric strength, dielectric constant, dissipation factor, insulation resistance, and surface and volume resistivity. The most commonly used base material is FR-4 epoxy all woven glass laminate, thickness 1.6 mm with copper foil cladding one oz. per sq. ft. (305 g/m^2). This has a foil thickness of 35 microns.

Before any processing can be undertaken on a board, it must be cleaned to get rid of the contaminants, which may be in the form of organic material (oils and greases), particulate (dust and machining particles), and oxides and sulphides on the copper surface. The cleaning is done in cleaning machines as the board is made to pass through de-greasing solvent solution, scrubbing stage, wet brushing and acid wash followed by a series of washes with light quality de-ionized water.

Image Transfer

The next step in manufacturing printed circuit boards is the transfer of original artwork pattern to the copper surface on the card. The artwork may be in the form of a photographic negative or positive. The photographic film consists of a transparent backing of polyester. It is 7 mil (174 microns) thick with a light sensitive silver halide emulsion, 4–8 micron thick. Its maximum sensitivity is at 480–550 nm wavelength. Therefore, processing of the film is usually done in a room with red light. After the image to be printed is available on a photographic film, a screen is prepared and the panel screen printed. All the conductive areas required on the final PCB are covered by the screening ink, which will act as an etch resist during etching. In modern PCB manufacturing facilities, screen printing is confined to only low accuracy image transfer requirements.

A better method is to use a dry film photoresist which is sensitive to ultraviolet light (200–500 nm). The application of the photoresist is carried out in a machine called a laminator. The photoresist is heated to about 110 °C and then pressed to the copper surface of the board. The photoresist may be of positive or negative type. In case of the positive photoresist, the polymerized resist is soluble in the developer and it requires artwork in the form of a positive. The negative type photoresist gets polymerized with ultraviolet light and becomes insoluble in the developer. Here the artwork is in the form of a negative. The coated board is exposed to the ultraviolet light. The resist is then developed, leaving those portions of the copper which are to be retained on the board and is covered by the resist.

Etching

The etching process is the core of the PCB manufacturing process, based on subtractive method which involves removal of copper from undesirable areas in order to achieve the desired circuit patterns. Several chemical processes have been developed and used for etching. The oldest and still used etchant is ferric chloride, which oxidizes copper to cuprous chloride from the areas which are not protected by etch resist. Ferric chloride, however, is not regenerated and is also corrosive. Several other chemicals such as ammonium persulphate, chromic acid, cupric chloride and alkaline ammonia have been used as etchants, with each of them having its own advantages and disadvantages.

Etching is usually done by the immersion, bubble, splash or spray method. The spray etching method is the most common. In this process, the etchant is pumped under pressure from a tank to the nozzles which splash the etchant on the board.

Board Drilling

For small scale production, boards are drilled by using single head manually controlled machines. Jigs are used to ensure that correct drill sizes are used and that no holes are missed. Boards can be stacked so that many of them can be drilled simultaneously. Mass production usually utilizes numerically controlled drilling machines with several heads. The vias and pads have copper etched from the centre to facilitate centering of the drill.

With the increasing miniaturization of electronic components, the need for smaller hole diameters has gone up. Also, a proper drill must be selected for each type of laminate. Tungsten carbide or diamond tipped drills are preferred for fibreglass boards.

Coatings

The base metal conductor used in the fabrication of printed circuit boards is copper. Copper is chosen because of its excellent properties as a conductor of heat and electricity. However, it quickly oxidizes in the presence of air and water. If the copper surface on the printed circuit board is not coated or treated with a protective agent, the exposed area would rapidly become unsolderable. Therefore, all printed circuit boards necessarily use some form of a surface finish on the exposed pads to which electronic components are to be soldered.

The current practice in PCB manufacturing also typically requires circuit traces to be protected with a masking material called soldermask. The soldermask is removed only when electrical access to the circuitry is required for soldering of electrical components. The areas which are not covered with soldermask must be protected with some form of a surface finish. The purpose of the surface finish is normally to protect a copper pad and exposed traces between the time the board is manufactured and when it is subsequently assembled. This would ensure that the board can later be soldered successfully during the assembly process. The most commonly used surface finish processes are detailed below.

Hot Air Solder Level   This process involves the application of tin/lead solder to exposed copper. The solder and exposed copper form an inter-metalic chemical bond that protects the copper from oxidation.

Immersion Precious Metal Plating   This process is based on the plating of the circuit board surface with electroless nickel/immersion gold, silver or tin which provide immunity to corrosion from environmental exposure. Although the solderability of each of the coatings is different, they provide a flat attachment surface which is essential for achieving a reliable solder joint with fine-pitch parts.

Organic Surface Protectant (OSP) Coating   In this process, the circuit board is coated by submersion in a chemical bath containing a nitrogen-bearing organic compound with adhesion to the exposed metal surfaces and not absorbed by the laminate or soldermask. These coatings have a limitation that they break down during a thermal cycle in assembly and are not usually recommended for double-sided circuit boards.

Conformal coatings   Conformal coatings enhance the performance and reliability of printed circuit assemblies that are likely to be subjected to a hostile environment. They are plastic film envelopes which seal out dirt and environmental contaminants. These coatings, which come in the form of acrylics, polyurethanes, epoxies and silicones, are usually applied by spraying, manually or with computer-controlled machines.

Testing

There are two types of PCB tests: bare board test and loaded board tests. The bare board test checks for shorts, opens and net list connectivity, whereas the loaded board tests include analysis of manufacturing defects and in-circuit, functional and combinational tests (Biancini, 1991). With an increase in the track density and the number of through-holes, it has become necessary to test the printed circuit board before assembly. It has been observed that the failure rate in highly populated printed circuits may be as high as twenty per cent. If the boards are not tested at the pre-assembly stage, the failures at a later stage may prove to be extremely expensive in the case of high density and multi-layer boards. Before populating a board with expensive devices such as application-specific ICs and microprocessors, it is cost-effective to first check whether the bare board meets expected quality standards. Bare board testing is thus becoming mandatory for the PCB manufacturers.

It may be noted that at each stage of the manufacturing process, it is necessary to undertake cleaning and it is desirable to carry-out inspection. However, for the sake of simplicity, these stages are not included in the design and description.

1.5.2   Double-sided Plated Through-holes

The processing techniques described for single-sided boards are applicable to most board processing. However, the process for producing double-sided printed through-holes is more complex than the print and etch method. Although there are a number of possible variations, the important steps for their production are shown in Figure 1.9. In the following description, only those steps are explained which differ from similar steps previously described in section 1.5.1.

Panel Preparation: Laminate sheets with copper cladding on both sides are cut to size as per requirement. Although the size of the panel depends upon the capacity of the plating equipment, the preferred size for many manufacturers is 305×406 mm. The laminate commonly used is 1 oz/ft^2 copper foil, epoxy glass type or FR-3.

Hole Drilling: The double-sided board is first drilled, which is followed by the removal of any burs by manual or automatic means. The board is then thoroughly cleaned to remove chips of glass fibre and resin. Cleaning is usually done by using a jet of water under high pressure, of the order of 20–60 atmosphere.

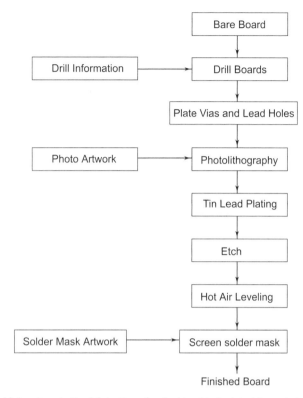

Fig. 1.9 Major steps in the fabrication of a double-sided, plated through-hole board

Electroless Copper Plating: The board is first sensitized by immersing it in a solution of stannous chloride. The stannous ions are absorbed on the board surface, particularly onto the exposed resin of the hole walls. This is followed by immersion of the board in an acidified solution of palladium chloride. The palladium ions are reduced to the colloidal state and form a thin layer which catalyses electroless copper deposition. Electroless copper deposition takes place in a bath with solution containing copper sulphate, sodium hydroxide, formaldehyde, a reducing agent and other special additives. Here, the copper ions are reduced to metallic copper. This results in deposition of copper, whose thickness is determined by the duration of the board in the solution. Usually, a thickness of about 40 microns of copper is built-up on the base copper and on the hole walls.

Image Transfer (*Photolithography*): Both sides of the board are covered with a thin layer of a photoresist, which may be solid or a liquid, and either positive or negative. A solid negative working resist is mostly used. The image transfer process occurs with the resist removed from the area where the tracks are to be kept. This is the reverse of the print and etch process. The copper areas, which will remain on the finished PCB and the hole walls, are unprotected. All other areas are covered by the hardened photoresist. Developing of both sides is usually done in an automatic spray machine.

Tin-Lead Plating: The exposed track areas are electroplated with tin-lead alloy by immersing the board in an electroplating bath. All conductive areas, i.e. all the conductors required on the PCB and within the holes, get plated to a thickness of about 20–25 microns. The minimum thickness should not be less than 10 microns. This metal is used as a resist in the etching process.

Etching: The etching process is similar to the one described in the previous section except that the etchant used must not attack the tin-lead alloy. After etching, the selective areas of the board can be plated with precious metals such as gold or nickel (e.g. tabs) followed by application of surface finish coatings such as: *hot-air levelling*, *soldermasking* and *organic surface protectant.*

The board is then finally inspected and tested as per the user's specifications. It is quite possible that some repairs or re-work may be required on the finished boards. Their acceptance by the users would depend upon the conditions of acceptability initially agreed upon mutually by the manufacturers and users.

1.5.3 Multi-layer Boards

The most widely used method of making multi-layer boards is by laminating or bonding layers of patterned, pre-etched, undrilled copper clad laminates together. After lamination, the subsequent manufacturing processes for multi-layer boards are generally similar to those used for double-sided boards made with the PTH process.

Essentially, the multi-layer boards are produced by bonding together inner layers and outer layers with *prepreg*. Prepreg is a fibreglass fabric impregnated with partially hardened resin. They are formed as if they were a single-sided board. The layers are sandwiched together with unetched copper top and bottom layers. The individual layers, which may be as many as 50, must be arranged in a pressing tool to prevent misalignment of the layers. The stack is laminated to form a single multi-layer board, which can then be processed as double-sided plated through-hole circuit board. The outer layers may consist of either copper foil and prepreg or of single-sided or double-sided copper clad laminates. The inner layers consist of double-sided copper clad, etched and through-plated board material. Bounding is performed in a hydraulic press or in an autoclave (high pressure chamber).

1.5.4 Flexible Boards

Flexible boards are usually made as single-sided boards. They are normally punched and not drilled.

In addition to the print and etch process, there is an alternative technique called 'additive process' which is used for manufacturing printed circuit boards. In this process, there is no copper on the base laminate. The copper is deposited selectively on the base laminate wherever required, as per the design of the circuit.

1.6 Challenges in Modern PCB Design and Manufacture

The electronics market is experiencing phenomenal growth. Even the most conservative estimate indicates that in excess of one trillion US dollars of electronic products are currently being shipped worldwide every year (Maxfield and Wiens, 2000) and are showing an upward trend. This means that electronics is penetrating in newer and newer areas, and that electronic products are getting continuously upgraded.

The main stages (Mentor Graphics, 2001) involved in creating an electronic product at the system level are concept, capture, layout and manufacture as shown in Figure 1.10.

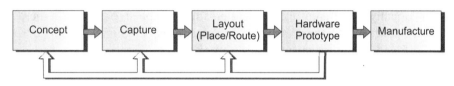

Fig. 1.10   Main stages in the development of an electronic product

The *concept* stage defines the requirement and specifications, and entails deciding on the overall architecture of the design.

The *capture* stage defines the design intent by describing its functionality.

The *layout* step includes determining optimum placements for the components on the circuit boards and routing the tracks that connect them together, besides also accounting for the cables and/ or connectors that tie multiple circuit boards together.

The above steps lead to the development of a hardware prototype. Ultimately, the product is *manufactured* and released into the market.

Not long ago, electronic products were designed and constructed entirely manually. There were no computers and no computer-aided tools to aid design engineers and layout designers. Circuit diagrams were drawn by using pen, paper and stencils. Similarly, placement was performed by using an outline of the board drawn on a piece of paper and cardboard cut-outs to represent the components. The board's copper tracks were then drawn by using different coloured pencils to represent the top and bottom sides of the boards. Similarly, no computer-aided verification tools were available to ensure that the design would function as planned. Thus, the only way to determine if the product would work was to make it and test it, which means a hardware prototype was built and evaluated by hand using the required test equipment.

One can imagine the difficulties experienced by designers at all levels. Simple errors discovered in the prototype could result in changes in the layout which were corrected by cutting tracks with a scalpel and/or adding wires by hand. More serious errors could require changes to the schematic, thereby necessitating an exchange or addition of components. Such changes would require a new prototype to be constructed, resulting in any number of development cycles.

This style of design was extremely time-consuming, expensive and prone to error. As electronic devices and designs grew more complex, automated techniques were developed to aid in the design process. The late 1960s and early 1970s witnessed the introduction of the first design evaluation and verification tools in the form of analog circuit simulators and digital logic simulators. Also, the first computer-aided design (CAD) tools to help digitize, and later layout, circuit boards appeared during this time. These were followed in the late 1970s by computer-aided engineering (CAE) to aid in design capture. During the 1980s, all these tools were gathered together under the umbrella of electronic design automation (EDA).

Today's electronic products are required to be increasingly small, fast, low power, light weight and feature-rich. Furthermore, consumers are demanding evermore sophisticated feature sets which, in turn, require tremendous amount of computing resources. Clock frequencies and signal speeds are rising dramatically. We are currently experiencing an explosive growth in the deployment of wireless-enabled products. All these factors have led to the development of a range of sophisticated CAD/CAM/CAE and design for manufacturability (DFM) tools and systems.

Uptil the mid-1990s, it was common to create circuit boards that were dedicated to a single function: for example, a CPU board or a power supply board. Since each board had a specific function within the overall system, it was correspondingly easy to design and fabricate. In the days of through-hole components, the pin-to-pin spacing was wide and through-holes relatively large, and the task of PCB design was a straight forward exercise. With the advent of surface mount technology, pin pitches began to shrink. The big advantages that surface mount offered at the time were smaller foot-prints and higher pin counts, with as many as 84 per device. While the first surface mount components featured pin pitches of 25 mils, they decreased over time to around 11 mils. Minimum trace widths and clearances decreased accordingly, putting a tremendous strain on PCB design process. With the continuous developments in integration, it is now possible to put very large sub-systems on a single chip in a very small package with hundreds of pins. A number of these sub-systems can then be assembled together to create an extremely complex system on a very small board.

Today, a single board can contain a 3-GHz RF section, analog circuitry, digital devices and power circuit. When all this is integrated, we get an IC in micro-packages like µBGAs with huge pin counts, which are currently as high as 1000 but rising to 2000, 4000 or even higher. It is no surprise then that diminutive ball grid arrays and chip scale packaging have grown in popularity. Fine-pitch attach technology also includes packages such as flip chip (with as many as several thousand I/Os), multi-chip module (MCM), and direct chip attach (DCA). As pin counts routinely exceed 500, chip vendors have adopted the BGA in ever greater numbers and more quickly than was originally anticipated. The dense array of solder balls on the BGA's lower side far outstrips the I/O capacity of the conventional quad flat package (QFP). BGAs are also attractive to board designers and manufacturers, because of their smaller form factor, better electrical performance, and lower power consumption. In 1998, 42 per cent of the PCB designers reported using BGAs and that number is growing. To date, the development of successive BGA generations has also been on the fast track. While early ball grid arrays featured a ball pitch of 1.27 mm, today's fine-pitch BGAs (FPBGA) now feature 0.8 to 0.5 mm ball pitches (Wiens, 2000). Figure 1.11 shows a high density fine-pitch package roadmap. The use of these high density packages leads to super dense, super complex systems, all squeezed together on a 20 layer board using microvia and build-up technologies. All these requirements pose a great challenge to the PCB designers and fabricators. A glimpse of emerging technologies in the PCB field was given by Peace (1991).

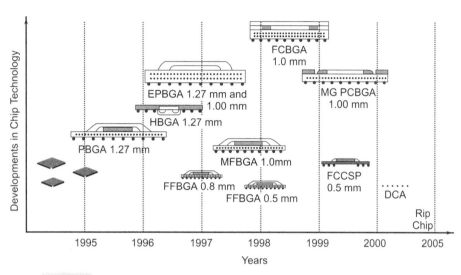

Fig. 1.11 High density fine-pitch package road map (redrawn after Wiens, 2000)

 ## 1.7   Major Market Drivers for the PCB Industry

The last decade witnessed an unparalleled wave of technological innovation and rapid market adoption. Sustaining this blistering pace means decreasing both the time and cost of product design cycles. Figure 1.12 shows the major market drivers, which are driving the modern printed circuit board design process. Development schedules are shrinking and so are the sizes of the products such as the cellphone, laptops and digital cameras. In addition to this, the push for smaller size is accompanied by equally instant demands for more functionality and better performance, so that the products ultimately become smaller, faster, more powerful, feature-rich and reliable.

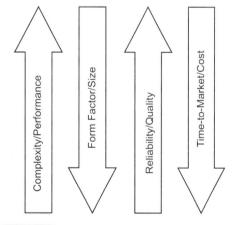

Fig. 1.12 Major market drivers for PCBs indicate that complexity/performance are increasing while size/form factor are falling. Similarly, there is a demand for higher reliability/quality while cost and time to market are expected to decrease

Figure 1.13 shows the developments in the operating frequencies of personal computers (PC) and mobile phones (Okubo and Otsuki, 2003). While the PCs have crossed the 2 GHz mark, the SH-mobile from Hitachi operates at 133 MHz. This is comparable with the first Pentium chips as far as speed is concerned. The operating frequency of mobile phone microcontrollers began to rise sharply from 1999 onwards, when e-mail and Internet access functions spread, and bit width doubled from 16 to 32 bits accompanied by the ever-increasing demand for smaller and smaller cellphones. The complexity of the printed circuit boards in the mobile and digital camera are typical examples of complex printed circuit boards shown in Figure 1.14. No doubt, the challenges for PCB designers are growing. Similarly, the software tools must help designers to deliver better products within the framework of an efficient and cost-effective design process.

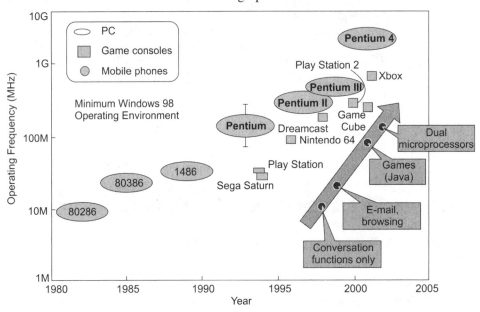

Fig. 1.13 Development in operating frequencies of PCB and mobile phones (redrawn after Okubo and Otsuki, 2003).

It is obvious that traditional electronic interconnect technology is no longer adequate to meet the demands of the new generation of smaller, denser boards. Through-vias are too large and unwieldy to work with BGAs and other miniature, high I/O components. Adding board layers is not an option either. Designers are thus seeking to decrease layer counts for lower production costs and reduced manufacturing times. Advanced, finer-geometry interconnect appears to be the answer for achieving denser routing and at the same time, for lowering the number of layers. The challenge for the PCB designer is thus managing the higher level of complexity that this technological solution brings with it.

In response to the current technological requirements, microvia technology has firmly established itself. Microvias are vias of less than or equal to 6 mils (150 micron) in diameter. Microvias have become the method of choice for routing designs containing BGA and CSP components. The smallest of these vias are 2 mils (50 microns) or 5 mils (125 microns). The proliferation of microvias has also revived the popularity of blind and buried vias to create interconnections through one dielectric layer within a PCB. Microvias are commonly used in blind via constructions where-in the outer layers of a multi-layer PCB are connected to the next adjacent signal layer. Used in all forms of electronic products, they effectively facilitate the cost-effective fabrication of high-density assemblies. The IPC (Institute for Interconnecting and Packaging Electronic Circuits) has selected High-Density Interconnection Structures (HDIS) as a term to refer to all these various microvia technologies. Adopting microvia technology means that products can use the newest, smallest and fastest devices, meet stringent RFI/EMI requirements, and keep pace with downward-spiraling cost targets. All these requirements challenge the system's designers to find better ways to overcome the difficulties encountered in achieving desirable features (Holden and Charbonneau, 2000).

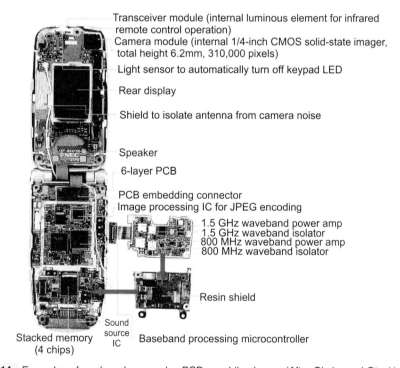

Fig. 1.14 Examples of modern day complex PCBs: mobile phones (After Okubo and Otsuki, 2003).

1.8 PCBs with Embedded Components

The shift towards using PCBs with embedded components, particularly in consumer electronics, is accelerating. This trend has been triggered by Motorola's announcement that the company is using such boards in its GSM mobile phone. Two types of components have been considered for embedding; passive components and ICs. Development is more advanced in the former. Although it is possible to embed ICs with large footprints in smaller boards, test and inspection methods have yet to be established to support this activity. There are, in general, three methods of embedding passive components, which are detailed below.

Embed Existing Passive Components: In this method, there are no restrictions on the type of components that can be embedded, and equipment manufacturers can obtain the resistance or capacitance they require. However, the board may have to be made thicker so as to be able to contain the selected components. In addition, since embedding requires more processes than surface mounting, such as making holes in the board to hold the components, it is difficult to achieve much cost reduction.

Embed Especially Made Thin Passive Components: This method allows much thinner boards, because these components can be as thin as 100μm or less. When the board layers are stacked, the resin flows around these thin components to embed them, so there is no need to open holes in the board.

Make Film Devices Through Printing: This method offers major reductions in the number of packaging processes and also in cost, because the film devices are formed at once. The number of solder bonds drops, providing improved bond reliability, lighter product and lower environmental loading.

The embedded components are likely to result in wearable equipment, such as in the shape of necklaces or bracelets. Figure 1.15 shows the predicted average weight and volume for digital camera, and mobile phone trends for the years 2006 and 2012, along with reduction in weight and volume (Kawai, 2003). PCBs with embedded components are expected to provide benefits such as a reduction in the manufacturing cost and design load.

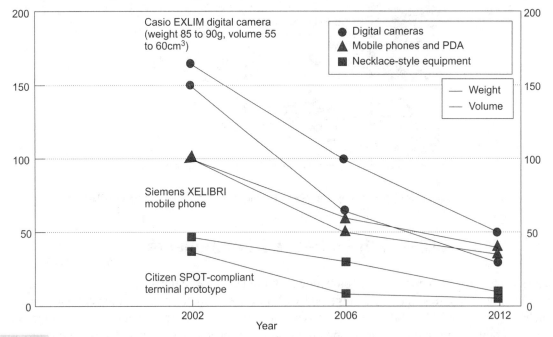

Fig. 1.15 Predictions of average weight and volume for digital camera and mobile phones (redrawn after Kawai, 2003).

1.9   Standards on Printed Circuit Boards

The design, fabrication, assembly and testing of printed circuit boards constitute a complex activity in which several players are involved. For this purpose, it is essential to standardize various aspects of PCB technology, so that there is a universal agreement for producing quality circuit boards. An industry-wide standard, internationally recognized, developed and accepted by consensus among trading partners, serves as the common language of trade.

According to the International Organization for Standardization (ISO), the standards are defined as:

"Documented agreements containing technical specifications or other precise criteria to be used consistently as rules, guidelines or definitions of characteristics, to ensure that materials, products, processes and services are fit for their purpose.

Standards are laid down to define a product so that the quality can be evaluated by using the same parameters. They are essential for any business activity, because without an adequate definition of what is required, no manufacturer is able to ascertain the requirements, especially qualitatively. They help the buyer to monitor the acceptability of the material supplied, i.e. they put the buyer and supplier on common grounds for establishing the criteria of acceptance.

Standards are benchmarks and they completely determine the products, tools and quality requirements. They are designed to serve the public interest by eliminating misunderstandings between manufacturers and purchasers, facilitating interchangeability and improvement of products as well as assisting the purchaser in selecting and obtaining the proper product for his particular need.

Most of the standards are internationally valid and help in reaching a point where-in the components and equipment made in one country will meet the specification mandatory in others, thereby eliminating the need to re-design before selling equipment abroad. They are laid down to achieve repetitive results for satisfying the specified requirements.

Electronics span a global market. Major players throughout the world participate in the market and want to have global standards that meet their desire to produce a product, and use designers anywhere in the world. Thus, international standards, play an important role in bringing a product to the market (Ferrari, 1997).

In the international standardization field, two organizations located in Geneva, Switzerland are: the International Organization for Standardization (ISO) and the International Electrotechnical Commission (IEC).

ISO is primarily concerned with mechanical hardware, quality and numerical standardization. The IEC is concerned with the electronics used in equipment. IEC committees deal with components, connectors, PCBs, surface-mount technology and design automation.

Extensive work has been done at the international level to develop standards and specifications connected with PC boards. The prominent organizations engaged in such activities are:

1. Institute for Interconnecting and Packaging Electronic Circuits (IPC);
2. American National Standards Institute (ANSI);

3. International Electrotechnical Commission (IEC);

4. Department of Defense, USA (DoD); and

5. DIN German Standards.

IPC, a United States-based Trade Association is involved in the creation of relevant standards for the PCB industry, which are referred to the world over. IPC develops standards to facilitate communication between suppliers and customers, issues guidelines with current industry positions on a wide range of subjects, conducts research to solve industry problems, undertakes correlation of industry test methods and encourages new developments in interconnection technology.

The IPCA (Indian Printed Circuit Association/www.ipcaindia.org) also co-ordinates and promotes international standards through education, communication, seminars and workshops. The original IPC Standards can be obtained from IPCA.

 ## 1.10   Useful Standards

The important general standards developed by IPC are:

- *IPC-T-50F*: *Terms and Definitions for Interconnecting and Packaging Electronic Circuits*: Provides description and illustrations to help users and their customers speak the same language, which includes a section on acronyms and an index of terms by technology types.

- *IPC-S-100*: *Standards and Specifications Manual*: A complete compilation of IPC specifications covering all aspects of electronic interconnection technology, from design through assembly to test and includes the latest standards.

- *IPC-M-106*: *Technology Reference for Design Manual*: Covers printed board technology as high density interconnects, flexible printed board design, controlled impedance, usage of photo-tooling for artwork quality and design for reliability (DFR) procedures.

- *IPC-D-859*: *Design Standard for Thick Film Multi-layer Hybrid Circuits*: Covers the requirements and considerations for the design of multi-layer hybrid circuits.

- *IPC-6801*: *IPC/JPCA Terms and Definitions, Test Methods, and Design Examples for Build-Up/High Density Interconnect (HDI) Printed Circuit Boards*: Lists terms specific to HDI and test methods for materials and HDI PCBs, peel strength and thermal shock; also included is a design criteria table and background information for the development of a standard for HDI PCBs.

- *IT-30101*: *High Density PCB Microvia Evaluation*: Details the suitability of microvia technology in high speed/high frequency applications.

- *IPC-MI-660*: *Incoming Inspection of Raw Materials Manual*: Contains background information, applicable specification references and industry-approved test methods for the inspection and evaluation of incoming raw materials, which include laminate, multi-layer board materials, various interconnection substrates, resists and other coatings, processing chemicals, artwork, registration tools, accessories, soldering materials, tooling accessories and other materials.

Electronic Components

 ## 2.1 Basics of Electronic Components

An electronic component is any device that handles electricity. Electronic components come in many different shapes and sizes, and perform different electrical functions depending upon the purpose for which they are used. Accordingly, electronic equipments make use of a variety of components. Some of the basic characteristics associated with electronic components are discussed below.

2.1.1 Active Vs Passive Components

There are broadly two types (Figure 2.1) of components: *passive components and active components.*

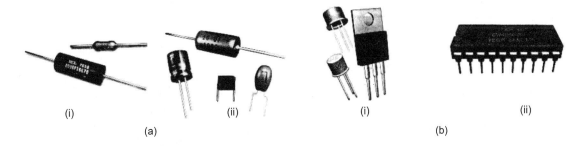

(i)   (ii)   (i)   (ii)

(a)   (b)

Fig. 2.1 Types of components. (a) passive components (i) resistors (ii) capacitors; (b) active components (ii) integrated circuit

Passive Components

A passive device is one that contributes no power gain (amplification) to a circuit or system. It has no control action and does not require any input other than a signal to perform its function. Since passive components always have a gain less than one, they cannot oscillate or amplify a signal. A combination of passive components can multiply a signal by values less than one; they can shift the phase of a signal, reject a signal because it is not made up of the correct frequencies, and control complex circuits, but they cannot multiply by more than one because they basically lack gain. Passive devices include resistors, capacitors and inductors.

Active Components

Active components are devices that are capable of controlling voltages or currents and can create a switching action in the circuit. They can amplify or interpret a signal. They include diodes, transistors and integrated circuits. They are usually semiconductor devices.

2.1.2 Discrete vs Integrated Circuits

When a component is packaged with one or two functional elements, it is known as a *discrete* component. For example, a resistor used to limit the current passing through it functions as a discrete component. On the other hand, an *integrated circuit* is a combination of several interconnected discrete components packaged in a single case to perform multiple functions. A typical example of an integrated circuit is that of a microprocessor which can be used for a variety of applications.

2.1.3 Component Leads

Components can be classified into two types on the basis of the method of their attachment to the circuit board. *Through-hole* components (Figure 2.2a) are those components which have leads that can be inserted through mounting holes in the circuit board. On the other hand, *surface mount* components (Figure 2.2b) are so designed that they can be attached directly on to the surface of the board.

Two types of lead configurations are commonly found in discrete components. The components with *axial leads* (Figure 2.2c) have two leads, each extending from each side of the component like arms. These leads need to be bent for insertion through the holes of a printed circuit board. The other configuration of leads in the components *is radial* (Figure 2.2d) wherein the leads emanate from the bottom of the components like legs.

In the case of integrated circuits, there are a large number of leads which are placed in a row in single line (single in-line package: Figure 2.2e) or in two parallel rows (dual in-line package: Figure 2.2f). These leads can be inserted in the through-holes in the PCB. High density integrated circuits now come in the form of *pin-grid arrays* (Figure 2.2g) that have several rows of round pins extending from the bottom of the component. *Leadless components* (Figure 2.2h) are also available in the surface mount devices in which no metal leads stick out of the component body. They are attached to a circuit board using some type of metallized termination.

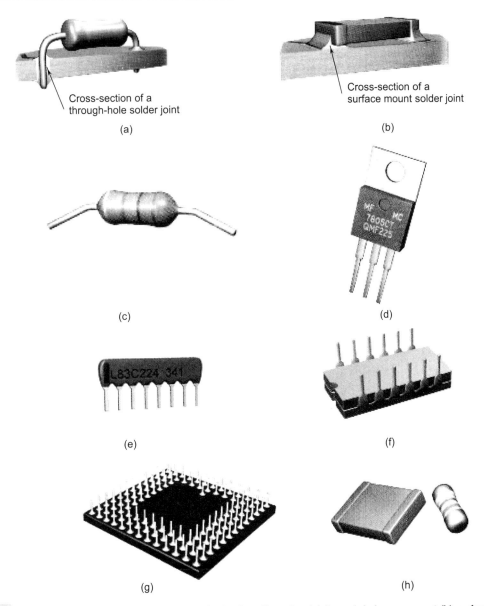

Cross-section of a
through-hole solder joint

(a)

Cross-section of a
surface mount solder joint

(b)

(c)

(d)

(e)

(f)

(g)

(h)

Fig. 2.2 Classification of components based on the lead configuration (a) through-hole component (b) surface mount component (c) component with axial leads (d) components with radial leads (e) single-in-line package (f) dual-in-line package (g) pin grid arrays (h) leadless components.

2.1.4 Polarity in Components

Some components are polarized and therefore have leads which are marked with positive and negative polarity. They must be placed on the board in the correct orientation when connected to the board.

Typical examples are that of electrolytic capacitors and diodes (Figure 2.3). If incorrectly placed with respect to the polarity, the components are likely to be damaged.

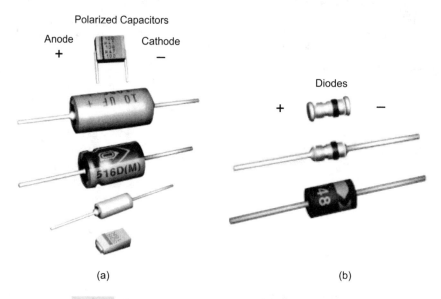

Fig. 2.3 Components with polarity (a) capacitors (b) diodes

2.1.5 Component Symbols

Each discrete component has a specific symbol when represented on a schematic diagram. These symbols have been standardized (Figure 2.4) and specified in the Institute of Electrical and Electronics Engineers (IEEE) standard 315 and 315A (ANSI Y32.2). The integrated circuits are generally represented by a block in the schematic diagram and each one does not have a specific symbol.

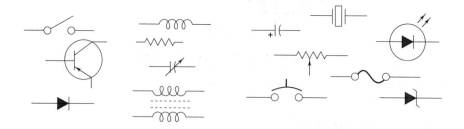

Fig. 2.4 Typical schematic symbols for commonly used components (redrawn after 'Component—identification', IPC-DRM-18F)

The commonly used components are described below in terms of their basic purpose in the circuit, operational aspects, constructional details and their symbols.

2.2 Resistors

The most commonly used component in an electronic assembly is the resistor. It is a passive component which exhibits a controlled value of resistance across its two leads. Resistance, by definition, is the opposition to the flow of current offered by a conductor, device or circuit. It is related to current as follows:

Resistance = voltage/current   (Ohm's Law).

The resistance is expressed in ohms (abbreviated Ω). A 1000 Ohm resistor is typically shown as 1K-Ohm (kilo Ohm), and 1000 K-Ohms is written as 1M-Ohm (megaohm). The resistor is represented by the symbols as shown in Figure 2.5. The symbol is a series of peaks and valleys making a zigzag pattern. The symbol can easily be drawn free-hand. It is also represented as a rectangle with R written inside it.

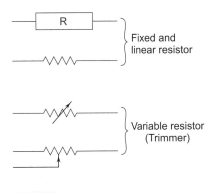

Fig. 2.5   Symbols for resistors

2.2.1 Types of Resistors

There are two classes of resistors; fixed resistors and variable resistors. They are also classified according to the material from which they are made. The most commonly used types of resistors are detailed below.

Carbon Resistors   They are made either by mixing finely ground carbon with a resin binder and an insulating filler or by depositing carbon film onto a ceramic rod.

Most carbon film resistors have low stray capacitance and inductance, so they are usable at higher frequencies. However, their accuracy is limited to 1per cent. In addition, carbon film resistors tend to drift with temperature and vibration (Figure 2.6).

Metal Resistors   They are made of metal film on ceramic rod or metal glaze (a mixture of metals and glass) or metal oxide (a mixture of a metal and an insulating oxide).

Metal film resistors are more stable under temperature and vibration conditions having tolerances approaching 0.5 per cent. Precision metal film resistors with tolerances below 0.1 per cent are also commercially available (Figure 2.7).

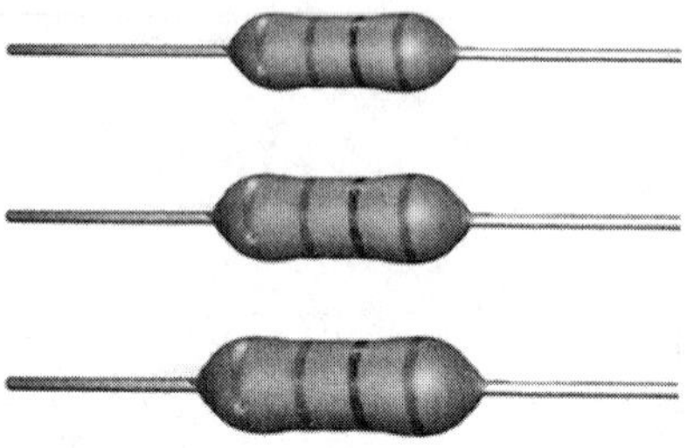

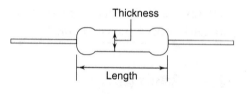

Thickness

Length

Approximate size

Rating power (W)	Thickness (mm)	Length (mm)
1/8	2	3
1/4	2	6
1/2	3	9

From the top of the photograph
1/8W
1/4W
1/2W

Fig. 2.6 Carbon film resistors

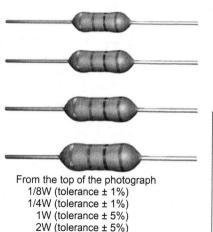

Thickness

Length

Approximate size

Rating power (W)	Thickness (mm)	Length (mm)
1/8	2	3
1/4	2	6
1	3.5	12
2	5	15

From the top of the photograph
1/8W (tolerance ± 1%)
1/4W (tolerance ± 1%)
1W (tolerance ± 5%)
2W (tolerance ± 5%)

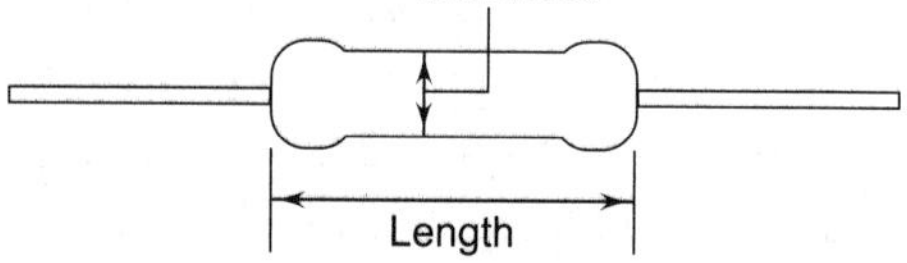

Fig. 2.7 Metal film resistors

Wire-wound Resistors   They are made by winding resistance wire onto an insulting former. They can be made to very close tolerances.

Thick Film Resistor Networks   Thick film resistor networks comprise precious metals in a glass binding system which have been screened on to a ceramic substrate and fired at high temperatures. These networks provide miniaturization, have rugged construction, are inherently reliable and are not subject to catastrophic failures. Networks comprising 1 to 50 resistors, 5 to 20 being typical, are commercially available. Single in-line (SIL) packages, DIP (dual-in-line package) and square packages are commonly available.

Figure 2.8 shows two arrangements for SIL packages. The resistance network is made with many resistors of the same value. One side of each resistor is connected with one side of all the other resistors inside. Eight resistors are housed in the package shown in which each of the leads of the package is one resistor. The ninth lead on the left side is the common lead. A common example of this type of arrangement is to control the current in a circuit powering many light emitting diodes (LEDs). Alternatively, some resistor networks have a "4S" printed on the top of the resistor network. The 4S indicates that the package contains four independent resistors that are not wired together inside. The housing has eight leads instead of nine. To have an idea of the size, for the type with nine leads, the thickness is 1.8 mm, the height 5 mm, and the width 23 mm. For the types with eight component leads, the thickness is 1.8 mm, the height 5 mm, and the width 20 mm.

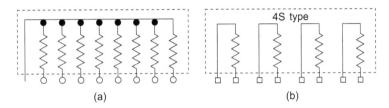

(a)   (b)

Fig. 2.8   Two arrangements for SIL packages

The reason for using such a large range of materials in the construction of resistors is simply a trade-off between cost and a particular performance characteristic, be it low noise, high stability or small size.

2.2.2 Packages

The physical construction of a resistor is often a ceramic or glass cylinder or rectangle on to which the resistor material is deposited. End-caps are used to make the connections, forming a physical pressure contact with the resistor material. Less used configurations are cylindrical with radial leads, arrays or networks of resistors usually having a DIP (dual-in-line) package. The commonly available packages are shown in Figure 2.9.

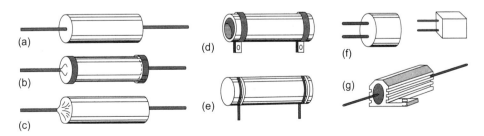

Fig. 2.9   Common packages for fixed resistors: (a), (b), (c) cylindrical package with axial   leads (d) and (e) cylindrical package with radial leads; (f) radial package with radial lead and (g) high-power package, with axial leads and copper body for increased heat dissipation (redrawn after Leonida, 1981)

2.2.3 Characteristics

The main parameters that define a resistor are detailed below.

Resistance: This is the nominal value of resistance between the two leads of the resistor when measured at 25°C.

Tolerance: This is the maximum deviation of value of resistance from the nominal value, usually given as a percentage of the nominal value. For example: for a ± 5% tolerance of a resistor of 500 ohms, the value may vary from 475 to 525 ohms.

Power Rating: This refers to the maximum power that a resistor can dissipate continuously at a temperature of 70°C. This is expressed in watts. Above 70°C, the nominal power rating must be reduced according to the de-rating factor specified by the manufacturer. Most common resistors are normally 0.25 W and 0.5 W. Modern digital circuits have low current requirements and usually use 0.125 W resistors.

Carbon composition and metal resistors are generally available in power ratings of 250 mW, 500 mW, 1 W and 2 W. For dissipating more heat, wire-wound resistors are mostly employed, with the power ratings being up to 25 watts.

Temperature Coefficient: It expresses the extent to which the value of resistance will change with temperature. It is usually expressed in parts per million of the nominal value per degree Celsius (ppm/°C). The temperature coefficient of the most commonly used resistors is in the range of 25 to 500 ppm°C.

Carbon composition resistors have poor stability and relatively poor temperature co-efficient, which is of the order of 1200 ppm/°C. Metal film resistors exhibit comparatively low temperature coefficient (±250 ppm/°C) and good stability, both when stored and under operating conditions.

The critical temperature in a resistor is the *hot spot temperature*, which is the sum of the ambient temperature and the temperature rise caused due to the power being dissipated. Due to uniform construction of the resistor, the maximum temperature is in the middle of the resistor body and it is this temperature, which is known as the *hot spot temperature*.

Stability or Drift: This is a measure of how much the value of a resistor changes with respect to time because of aging. It is normally measured as a percentage change after 1000 hours of operation at 70°C. The stability of a resistor is defined as the percentage change of resistance value with time. It depends upon the power dissipation and ambient temperature.

Noise: Resistors generate white or Johnson noise due to the statistical movement of electrons and depend mainly upon the construction technique. This is usually specified in terms of microvolts/volt. For a given type of resistor, it increases at higher values of resistance and at higher values of frequency. The typical value of noise voltage for a 100-k Ohm resistor at 27°C for a bandwidth of 5 kHz is 8.3 microvolts.

Parasitic Effect: No resistor is ideal. It can be considered as a lumped model represented in Figure 2.10. It consists of a resistor having a shunt capacitor in parallel with an inductance in series.

Consequently, the impedance of a resistor is frequency dependent. However, small fixed film resistors around 20k Ohm in value (< 500 mW) can be considered as ideal to typically 100 MHz and resistors less than 1 K ohm to 300 MHz.

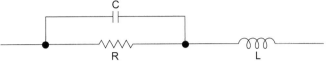

Fig. 2.10 Lumped model of a resistor C = 0.1-2 pf, L = 0.1 μH (for a leaded component)

Maximum Voltage: It represents the maximum dc voltage, which can be safely applied to a resistor on a continuous basis. For most resistors of value around 100 ohms or more, the maximum dc voltage is 1000 volts. Voltage transients above the rated value may induce permanent changes in resistance values.

Identification: The value of resistance is either printed in numbers or is put in the form of colour-coded bands around the body. In the colour code, each number from 0 to 9 has been assigned a colour.

The colour code comes in the form of four-band (Figure 2.11). The first band closest to the end of the resistor represents the first digit of the resistance value. The second band gives the second digit and the third band gives the number of zeros to be added to the first two digits to get the total value of the resistor. The fourth band indicates the tolerance. If the fourth band is absent, the tolerance is ±20%.

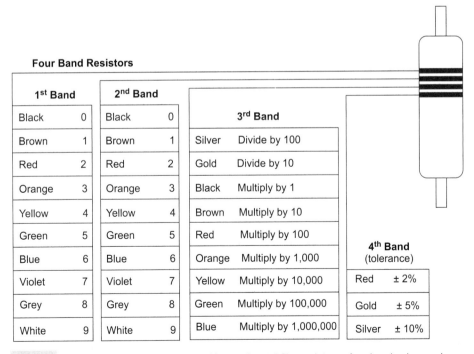

Four Band Resistors

1st Band		2nd Band			
Black	0	Black	0	**3rd Band**	
Brown	1	Brown	1	Silver	Divide by 100
Red	2	Red	2	Gold	Divide by 10
Orange	3	Orange	3	Black	Multiply by 1
Yellow	4	Yellow	4	Brown	Multiply by 10
Green	5	Green	5	Red	Multiply by 100
Blue	6	Blue	6	Orange	Multiply by 1,000
Violet	7	Violet	7	Yellow	Multiply by 10,000
Grey	8	Grey	8	Green	Multiply by 100,000
White	9	White	9	Blue	Multiply by 1,000,000

4th Band (tolerance)

Red	± 2%
Gold	± 5%
Silver	± 10%

Fig. 2.11   Colour code for carbon composition and metal film resistors—four band colour code

In the five-band colour code (Figure 2.12), the first three bands indicate the value, the fourth band indicates the multiplier factor, and the fifth band, the tolerance of the resistor.

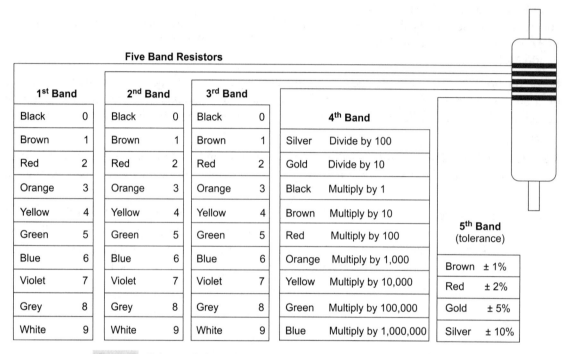

Fig. 2.12 Colour code for carbon and metal film resistors—five band colour code

In the six-band colour code, the sixth band indicates the temperature coefficient of variation of resistance in terms of parts per million per degree centigrade (ppm/°C).

When the value and tolerance of the resistor are printed on the resistor body, the values of the tolerances are coded as follows:

$$F = \pm 1\% \qquad\qquad G = \pm 2\% \qquad\qquad j = \pm 5\%$$
$$K = \pm 10\% \qquad\qquad M = \pm 20\%\pm$$

The following examples illustrate this code :

R 68M is a 0.68 Ω ± 20% resistor

5K 6J is a 5.6 kΩ ± 5% resistor

82KK is 82 kΩ ± 10% resistor

Although it is possible to get resistors of any value, they are generally available in the preferred ranges. The most common series is the E 12 series in which the preferred values are 10, 12, 15, 18, 22, 27, 33, 39, 47, 56, 68, 82. Much closer values are available in E 96 series for ± 1% tolerance values.

2.3 Variable Resistors or Potentiometers

Variable resistors basically consist of a track of some type of resistance material with which a movable wiper makes contact. Variable resistors or potentiometers ('pots' as they are popularly called) can be divided into three categories depending upon the resistive material used (Figure 2.13).

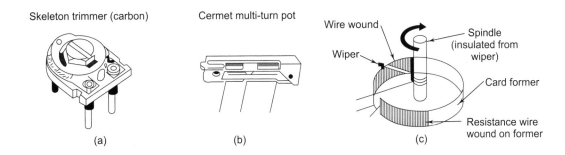

Fig. 2.13   Types of variable resistors: (a) carbon composition (b) multi-turn cermet (c) wire wound

Carbon

Carbon potentiometers are made of either moulded carbon composition giving a solid track or a coating of carbon plus insulating filler onto a substrate.

Cermet

Cermet potentiometers employ a thick film resistance coating on a ceramic substrate.

Wire-wound

Nichrome or other resistance wire is wound on to a suitable insulating former for the construction of wire-wound potentiometers.

Potentiometers can be categorized into the following types depending upon the number of resistors and the control arrangement used:

- *Single Potentiometers:*   Potentiometer control with one resistor;
- *Tandem Potentiometers*:   Two identical resistor units controlled by one spindle;
- *Twin Potentiometers*:   Two resistor units controlled by two independent concentric spindles;
- *Multi-turn Potentiometers*:   Potentiometer with knob or gear wheel for resistance adjustment; they may have up to 40 rotations of spindle; and
- *Potpack*:   Rectangular potentiometers, either single or multi-turn.

Potentiometers are typically used for setting bias values of transistors, setting time constants of RC timers, making gain adjustments of amplifiers, and carrying current or voltage in control circuits. Therefore, they are packaged in such a way that they are compatible with PCB mounting applications.

A variable resistor can be used either as a rheostat or potentiometer. Figure 2.14 shows the difference in the two applications. When used as a voltage divider, the resistor element is connected to a voltage reference source and the slide arm, which is used as the pick-off point, and can be moved to obtain the desired voltage.

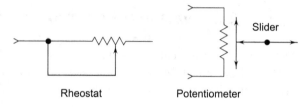

Fig. 2.14 Difference between a potentiometer and rheostat operation

For a variable resistor configuration, the resistor element is connected to the circuit at each end and the slide arm is connected to one of the ends. Alternatively, the entire resistance is in series and the slider is connected to an external circuit. Even in this configuration, it may be used like a potentiometer.

Variable resistors can be constructed to follow one of the following laws:

- *Linear:* The resistance of the pot is distributed evenly over its entire length.
- *Log:* The resistance of the pot varies so as to follow the logarithmic law. In these pots, when the wiper is turned, the resistance increases (from zero) very slowly and gradually until about the half way mark. From then onwards, as the wiper shaft is turned further, the resistance will increase much more rapidly in comparison with the first half of the pot-rotor rotation.

  Figure 2.15 shows the change of resistance value with the angle of rotation for linear and logarithmic type of variable resistors.

- *Sine-Cosine Potentiometers:* As the name implies, the variation of resistance over the track, when the wiper moves, follows the sine-cosine law. The total operative track length over 360 degrees of rotation is divided into four quadrants of 90 degrees each.

Figure 2.16 shows the package shapes of commonly used variable resistors.

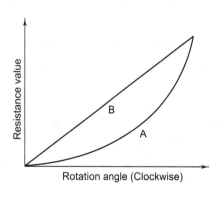

Fig. 2.15 Change of resistance value with rotation angle

Fig. 2.16 Package shapes of variable resistors

2.4 Light-dependent Resistors (LDRs)

Light-dependent Resistors are made of cadmium sulphide. They contain very few free electrons when kept in complete darkness and therefore, exhibit very high resistance. When subjected to light, the electrons are liberated and the material becomes more conducting. When the light is switched off, the electrons are again recaptured and the material becomes less conducting or an insulator. The typical dark resistance of LDRs is 1 MOhms to 10 MOhms. Its light resistance is 75 to 300 ohms. The LDRs take some finite time to change its state and this time is called the *recovery time*. The typical recovery rate is 200 kOhms/sec.

 ## 2.5 Thermistors

Thermistors are resistors with a high temperature co-efficient of resistance. Thermistors with negative temperature co-efficient (fall in resistance value with an increase in temperature) are the most popular. They are oxides of certain metals like manganese, cobalt and nickel. Thermistors are available in a wide variety of shapes and forms suitable for use in different applications. They are available in the form of disks, beads or cylindrical rods. Thermistors have inherently non-linear resistance–temperature characteristics. However, with a proper selection of series and parallel resistors, it is possible to get a nearly linear response of resistance change with temperature over a limited range.

Thermistors with a positive thermo-resistive co-efficient are called *posistors*. They are made from barium titanate ceramic and are characterized by an extremely large resistance change in a small temperature span.

Thermistors have various applications such as excess current limiters, temperature sensors, protection devices against over-heating in all kinds of appliances such as electric motors, washing machines and alarm installations, etc. They are also used as thermostats, time delay devices and compensation resistors. Depending upon the application, the thermistor beads need to be properly protected by sealing them into the tip of a glass tube or placing them inside a stainless steel cover.

 ## 2.6 Capacitors

A capacitor, like a resistor, is also a passive component, which can be used to store electrical charge. Capacitors find widespread applications in the electrical and electronics fields in the form of:

- Ripple filters in power supplies;
- Tuning resonant circuits, oscillator circuits;
- Timing elements in multi-vibrators, delay circuits;
- Coupling in amplifiers;
- De-coupling in power supplies and amplifiers; and
- Spark suppression on contacts on thermostats and relays.

A capacitor (also called a 'condenser') consists of two facing conductive plates called electrodes, which are separated by a dielectric or insulator (Figure 2.17). The dielectric can be made of paper, mica, ceramic, plastic film or foil. To make a practical capacitor, a lead is connected to each plate or electrode. The charge Q which can be stored in a capacitor, when connected to a voltage V across it, is given by:

Q = CV

where C represents the capacitance of the capacitor.

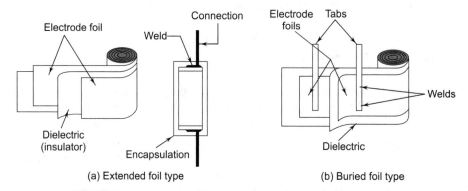

(a) Extended foil type   (b) Buried foil type

Fig. 2.17 Basic capacitor (a) extended foil type (b) buried foil type

Capacitance is measured in farads. A capacitor has a capacitance of one farad when one coulomb charges it to one volt. The farad is too large a unit. The usual sub-units used are microfarad (10^{-6} F) and the picofarad (10^{-12} F).

Capacitors can be fixed or variable. The symbols of various types of capacitors are shown in Figure 2.18.

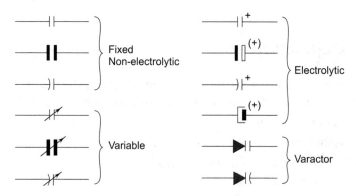

Fig. 2.18 Symbols for capacitors. The polarity mark '+' may be omitted where it is shown in brackets

The value of a capacitor is indicated on the body of the capacitor, either in words or in a colour code. Figure 2.19 shows the capacitor colour code for various types of capacitors.

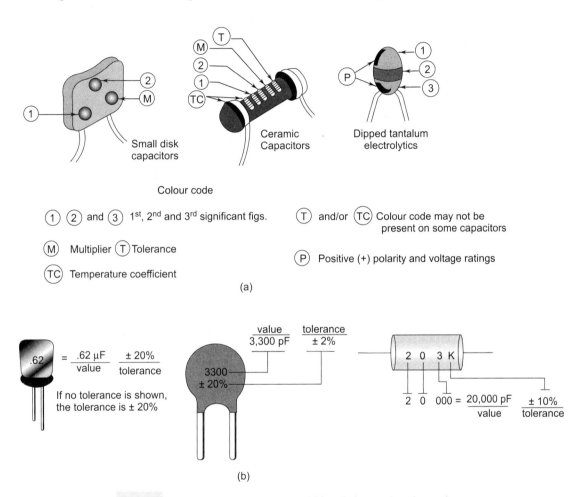

(a)

(b)

Fig. 2.19 (a) capacitor colour code; and (b) code for numbered capacitors

The value of a capacitor is also sometimes written on the body in the form of numbers. Values beginning with decimals are usually measured in microfarads (μF), while all other values are assigned to be in picofarads (pF). Four-digit values are also indicated in picofarads but without a multiplier. Some capacitors are coded with a three-digit number which is similar to the colour band system, with a value and multiplier numbers. For example, 203 means that 2 and 0 are attached to 3 zeros and the value of the capacitor would be 20,000 pF or .02 μF. The tolerance letter codes indicate F = ±1%, G = ±2%, J = ±5%, K = ±10%, M = ±20% and Z = +80 to –20%.

2.6.1 Types of Capacitors

Capacitors are categorized into various types (Figure 2.20) depending upon the dielectric medium used in their construction. The size of the capacitor, its tolerance and the working voltage also depend upon the dielectric used. Some of the common types of capacitors are detailed below.

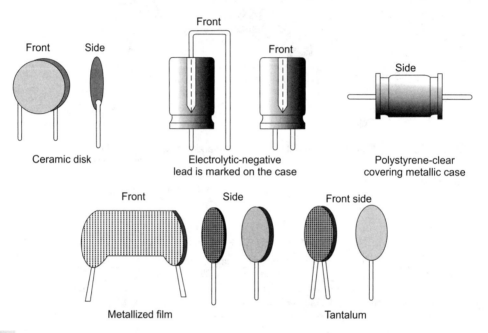

Fig. 2.20 Various types of capacitors. Typical packages for small electrolytic capacitors. The polarity mark '+' may be omitted on non-symmetrical packages. Metal cans may be insulated by a heat-sink sleeve

Paper Capacitors

Paper capacitors make use of thin sheets of paper wound with thin aluminium foils. In order to increase the dielectric strength and to prevent moisture absorption, the paper is impregnated with oils or waxes. The capacitor is normally encapsulated in resin. Paper capacitors tend to be large in size due to the thickness of paper and foil. The thickness is reduced in metallized capacitors (Figure 2.21) by directly depositing the aluminium on the dielectric.

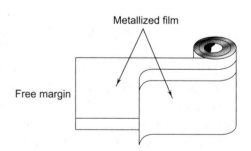

Fig. 2.21 Construction of metallized capacitor

Typical range   :   10 nF to 10 μF

Typical dc voltage :   500 V(max.)

Tolerance   :   ± 10%

Mica Capacitors

A mica capacitor is made by directly metallizing the thin sheets of mica with silver and stacking together several such sheets to make the complete capacitor. The assembly is encapsulated in resin or moulded in plastic.

Typical range : 5 pF to 10 nF

Typical dc voltage : 50 to 500 V

Tolerance : ± 0.5%

Ceramic Capacitors

Ceramic capacitors generally employ barium titanate as the dielectric medium. However, low-loss ceramic capacitors use steatite, which is a natural mineral. A thin plate of ceramic is metallized on both sides and the connecting leads are soldered to it. The body is coated with several layers of lacquer. Modern ceramic capacitors of the monolithic type are made of alternate layers of thin ceramic dielectric and electrodes, which are fired and compressed to form a monolithic block. These capacitors have a comparatively small size.

Typical range : (a) Low loss (steatite)

5 pF to 10 nF

(b) Barium titanate

5 pF to 1 μF

(c) Monolithic

1 nF to 47 μF

Typical voltage range : For a and b

60 V to 10 kV

For c: 60 V to 400 V

Tolerance : ± 10% to ± 20%

Plastic Capacitors

The construction of plastic capacitors is very similar to that of paper capacitors. They are of both foil and metallized types. Polystyrene film or foil capacitors are very popular in applications requiring high stability, low tolerances and low temperature co-efficient. However, they are bulky in size. For less critical applications, metallized polyethylene film capacitors are used. They are commonly referred to as 'polyester capacitors'.

Electrolytic Capacitors

High value capacitors are usually of electrolytic type. They are made of a metal foil (Figure 2.22) with a surface that has an anodic formation of metal oxide film. The anodized foil is in an electrolytic solution. The oxide film is the dielectric between the metal and the solution. The high value of capacity of electrolytic capacitors in a small space is due to the presence of a very thin dielectric layer. Electrolytic capacitors are of the following types:

- *Aluminium:* Plain foil, etched foil and solid; and
- *Tantalum:* Solid, wet-sintered.

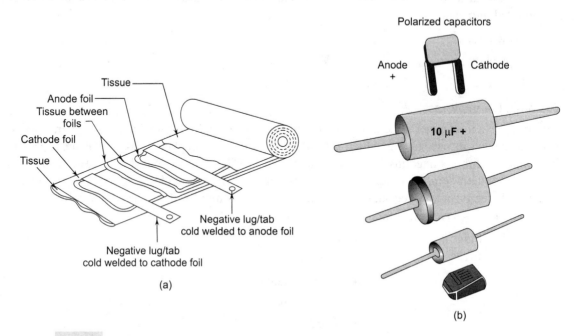

Fig. 2.22 (a) construction of aluminium electrolytic capacitor (b) shapes of electrolytic capacitors

Electrolytic capacitors exhibit a very wide range of tolerances, typically ranging from -20 to $+50\%$. They are usually polarized. Care must be taken not to reverse the voltage applied across it. If a reverse voltage is applied, the dielectric will be removed from the anode and a large current will flow as oxide is formed on the cathode. Sometimes the gases released from the capacitor may build up and cause the capacitor to explode and damage other parts of the circuit.

2.6.2 Packages

Capacitors are available in a large variety of packages, shapes and dimensions. The most common packages are axial, disc, rectangular, tubular, etc., as shown in Figure 2.20.

2.6.3 Performance of Capacitors

The important parameters which characterize a capacitor are delineated below.

Capacitance: This is the nominal value of a capacitor, measured in Farads (or its sub-multiples) at 25 °C.

Tolerance: This refers to the deviation of the actual value of a capacitor from its nominal value. Different types of capacitors have different values of tolerance.

Working Voltage: This is the maximum voltage which can be applied continuously across the capacitor. This is indicated as ac or dc. The maximum voltage that causes permanent damage in the dielectric is referred to as *breakdown voltage*. This is generally twice the working voltage.

Temperature Coefficient: It indicates the change in the value of capacitance with temperature and is expressed as parts per million per degree Celsius (ppm/°C).

DC Leakage: The amount of current which flows through a charged capacitor because of losses due to the conductivity of the dielectric, represents the dc leakage.

Parasitic Effects: The capacitor impedance is a function of frequency: at low frequencies, the capacitor blocks signals and at high frequencies, the capacitor passes signals. Depending on the circuit configuration, the capacitor can pass the signal to the next stage or it can shunt it to ground. The impedance of the capacitor varies with frequency as follows:

$$X_C = \frac{1}{2\pi f C}$$

All capacitors have a self-resonant frequency where-in the parasitic lead and dielectric inductance resonate with the capacitor in a series resonant circuit. Essentially, the capacitor impedance decreases until it reaches self-resonance when its impedance is minimum. Aluminium electrolytic capacitors have a very low self-resonant frequency, so they are not effective in high frequency applications above a few hundred kHz. Tantalum capacitors have a mid-range self-resonant frequency. Thus, they are found in applications up to several MHz and beyond that, ceramic and mica capacitors are preferred because they have self-resonant frequencies ranging into hundreds of MHz. Very low frequency and timing applications require highly stable capacitors. The dielectric of these types are made from paper, polypropylene, polystyrene and polyester. They exhibit low leakage current and low dielectric absorption.

ESR (Equivalent Series Resistance) is an important parameter of any capacitor. It represents the effective resistance resulting from the combination of wiring, internal connections, plates and electrolyte. Figure 2.23 shows capacitor equivalent circuit. ESR is the effective resistance of the capacitance at the operating frequency and therefore affects the performance of tuned circuits. It may result in a totally incorrect or unstable operation of critical circuits such as switch-mode power supplies and

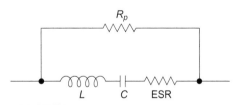

Fig. 2.23 Equivalent circuit of a capacitor

deflection circuits in TVs and monitors. Power supply filter design requires a low ESR capacitor because voltage is dropped across ESR and the current flowing through the capacitor causes power dissipation resulting in self-heating. Electrolytic capacitors tend to have high ESR as compared to other types and it changes, though not for the better, with time.

An ideal capacitor would only have C and no R. Any R in series with C will reduce the capacitor's ability to pass current in response to a variant applied voltage and it will dissipate heat, which is wasteful and could lead to failure of the component. The dissipation factor (DF) is mathematically defined as R/X where R is the resistance in the capacitor and X is the reactance of the capacitor. The higher the R, the higher would be the DF and poorer the capacitor. From the formula, $DF = R/X$, it is clear that DF is an inverse function of X. As X goes down, DF goes up and vice versa. So DF varies proportionately with frequency, which shows that DF is a function of the test frequency. DF is a measure of capacitor quality and the figure is valid only at the frequency of test. R_p models the parallel resistance of a capacitor. Its value is usually in hundreds of Megaohms except for electrolytic capacitors which have comparatively low value.

The quality factor Q serves as a measure of the purity of a reactance, i.e. how close it is to being a pure reactance i.e. having no resistance. This represents as the ratio of the energy stored in a component to the energy dissipated by the component. Q is a dimensionless unit and is expressed as $Q = X/R$. However, Q is commonly applied to inductors; for capacitors the term more often used to express purity is dissipation factor (DF). This quantity is simply the reciprocal of Q.

 ## 2.7 Variable Capacitors

Variable capacitors are constructed by using any one of the dielectrics like ceramic, mica, polystyrene or teflon. Basically, a variable capacitor has a stator and a rotor. The area of the stator is fixed and turning the rotor from 0° to 180° varies the amount of plate surface exposed, thereby varying the value of the capacitor.

In most variable capacitors, the change in capacitance is linear throughout the rotation of the rotor. Figure 2.24 shows a linear increase and decrease in the value of capacitance through 360° rotor rotation.

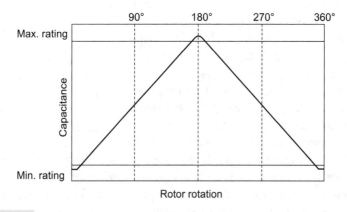

Fig. 2.24 Linear variation of capacitance with rotation in a trimmer capacitor

Variable capacitors are available in the following two configurations:
- *Button type:* This has a variable rotor (Figure 2.25a); and
- *Tubular type:* This has an adjustable core (Figure 2.25b).

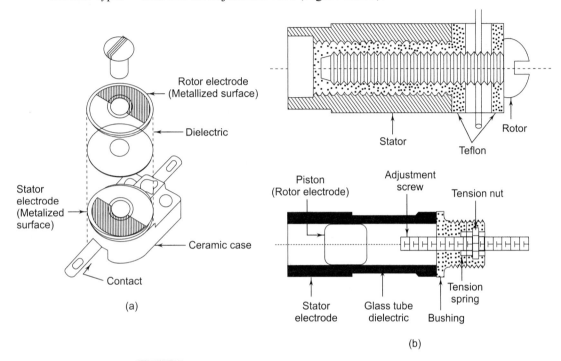

Fig. 2.25 Variable capacitor (a) button type (b) tubular type

It may be noted that adjustments made with a variable capacitor by using a metal screwdriver will alter when the screwdriver is lifted from the turning screw. This is because placing the metal screwdriver on this screw changes the effective area of the metal plated surface of either the stator or, more often, the rotor. In such a case, the use of a non-metallic screwdriver is recommended.

2.8   Inductors

Inductance is the characteristic of a device which resists change in the current through the device. Inductors work on the principle that when a current flows in a coil of wire, a magnetic field is produced, which collapses when the current is stopped. The collapsing magnetic field produces an electromotive force which tries to maintain the current. When the coil current is switched, the induced EMF would be produced in such a direction, so as to oppose the build-up of the current.

Induced emf $e = -L\dfrac{di}{dt}$ where L is the inductance and $\dfrac{di}{dt}$ the rate of change of current

The unit of inductance is Henry. An inductance of one Henry will induce a counter emf (electromotive force) of one volt when the current through it is changing at the rate of one ampere per second. Inductances of several Henries are used in power supplies as smoothing chokes, whereas smaller values (in the milli-or micro-Henry ranges) are used in audio and radio frequency circuits.

The inductors are also sometimes called coils. The symbol of an inductor is shown in Figure 2.26a. Inductors are available in many sizes and shapes (Figure 2.26b).

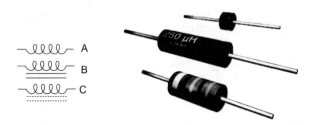

Fig. 2.26   (a) symbol of inductor (b) packages of inductors

The value of an inductor may be printed on the component body or it may be printed with colour bands (Figure 2.27), much in the same way as a resistor. For example, if the first and second bands of an inductor are red (value 2) and the third band is orange (value 3), the value of the inductor is 22,000 μH (micro-Henry). A fourth silver band will indicate its tolerance as ±10%.

Inductor Band Colour Codes

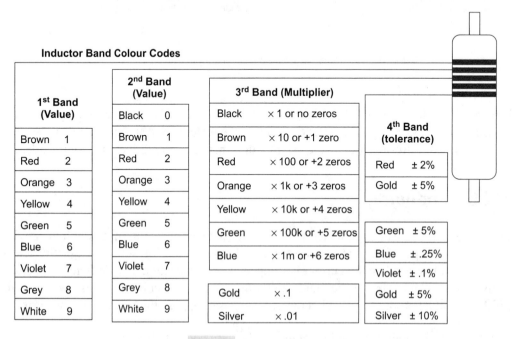

1ˢᵗ Band (Value)		2ⁿᵈ Band (Value)		3ʳᵈ Band (Multiplier)		4ᵗʰ Band (tolerance)	
		Black	0	Black	× 1 or no zeros		
Brown	1	Brown	1	Brown	× 10 or +1 zero		
Red	2	Red	2	Red	× 100 or +2 zeros	Red	± 2%
Orange	3	Orange	3	Orange	× 1k or +3 zeros	Gold	± 5%
Yellow	4	Yellow	4	Yellow	× 10k or +4 zeros		
Green	5	Green	5	Green	× 100k or +5 zeros	Green	± 5%
Blue	6	Blue	6	Blue	× 1m or +6 zeros	Blue	± .25%
Violet	7	Violet	7			Violet	± .1%
Grey	8	Grey	8	Gold	× .1	Gold	± 5%
White	9	White	9	Silver	× .01	Silver	± 10%

Fig. 2.27   Colour code for inductors

The primary use of an inductor is filtering. There are two very different types of filter inductors: the high current inductors wound around a large core are used in power supply filters, and the low current air core inductors are used in signal filters.

The basic components of an inductor are the former (or bobbin), winding wire (with or without separating material) and the core material. *Bobbins* are normally made of moulded plastic and carry the wire and the core. Bobbins usually have termination pins contained within the mould. *Winding* is usually enamelled copper wire whose diameter is calculated to keep the temperature rise under full load to an acceptable level. The core material can be laminated steel, powdered iron or ferrite. The shape of the core is also variable.

The toroidal coil consists of copper wire wrapped around a cylindrical core. It is possible to make it so that the magnetic flux which occurs within the coil doesn't leak out, the coil efficiency is good, and that the magnetic flux has little influence on other components. Toroids are the most efficient cores in ferrites, but they are difficult to wind. Figure 2.28 shows the shapes of some toroidal coils along with the path of magnetic flux in these inductors.

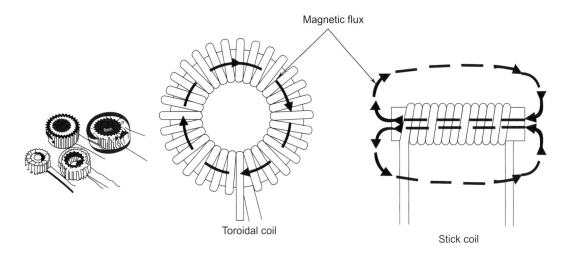

Fig. 2.28   The toroidal coil shapes

Air core coils are frequently used at very high frequencies. In order to provide mechanical protection or reduce high voltage corona effects, inductors may be potted in epoxy. Potting, however, increases the stray capacitance and also makes it difficult and often impossible to repair once it is potted. Figure 2.29 shows some of the common types of inductors used in various applications.

High current inductors require cores to keep the losses within acceptable limits and to achieve high performance. The cores are big and heavy, so they have large weight and size. Switching power supplies require extensive inductors or transformers to control the switching noise and filter the output voltage waveform.

Low current inductors are used for filters in signal processing circuits. An inductive/capacitive filter has sharper slopes than a resistive/capacitive filter, and is thus a more effective filter in some applications. In general, inductors are rarely seen outside power circuits.

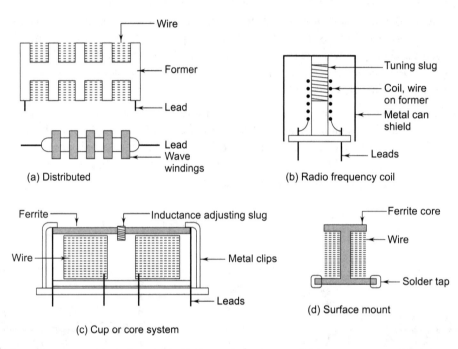

Fig. 2.29 Types of inductors: (a) distributed, (b) shielded radio frequency coil (c) cup or core, and (d) surface mount ferrite inductor (redrawn after Haskard, 1998)

The range of inductor style and shape is considerably larger than for either capacitors or resistors. This is because many organizations need to wind their own inductors to meet their specific demands, which could be for RF coils, audio filters, power supply chokes, etc.

Variable inductors in which the inductance value can be adjusted are also available. The ferrite core of the inductor is made like a screw. The core can be made to move in and out of the inductor by turning it with a screwdriver. A special plastic screwdriver is better to use for adjustment of the coils. By moving the ferrite core in or out of the coil, the value of the coil's inductance can be changed.

2.9 Diodes

A diode is an active component through which the current flows more easily in one direction than in the other. It is made from semiconductor material. As the name implies, diode means a two-electrode device: one electrode is made of n-semiconductor material while the other is p-type. The junction of the two dissimilar materials results in the diode action. The main functions of the diode in a circuit are to act as a switching device, a detector or a rectifier.

Since the diode is a two-element device, its symbol shows the two electrodes (Figure 2.30). The cathode and anode ends of metal encased diodes can be identified on the body. The arrow head of the symbol points in the direction of conventional current flow. In case of glass encased diodes, the cathode end is indicated by a stripe, a series of stripes or a dot. For most silicon or germanium diodes with a series of stripes, the colour code identifies the equipment manufacturer's part number.

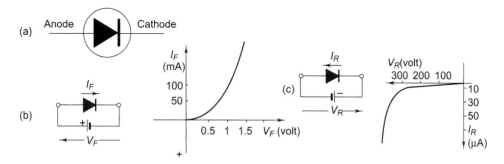

Fig. 2.30 Diode (a) symbol (the circle may be omitted); (b) voltage/current diagram when forward-biased; and (c) voltage/current diagram when reverse-biased (note the different scale).

Conventional diodes normally show a low value of forward resistance and a very high value of reverse resistance. The variation in resistance is due to the non-linear voltage/current characteristics of the diode. Figure 2.31 shows voltage current diagram for a typical semiconductor diode.

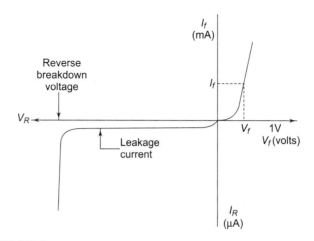

Fig. 2.31 Voltage–Current characteristics of semiconductor diodes

Signal diodes are general purpose diodes, which find applications involving low currents and a wide range of voltages, sometimes extending upto 50 kV. *Switching diodes* change their state from conducting to non-conducting state and vice versa in a very short time when the voltage is reversed. *Rectifiers* are similar to signal diodes, but are more suitable for high currents.

Low and medium power diodes are usually available in axial packages whereas high power diodes are available in a large variety of packages of a vast range of shapes and sizes. Very high power diodes have a thread for mounting on to a PCB or a heat sink. Figure 2.32 shows various shapes of diodes, that are commercially available. The data sheets of the suppliers usually give information about the outline of the diode and include dimensional information. Diode arrays or networks, containing up to 48 devices are also available in packages similar to integrated circuits.

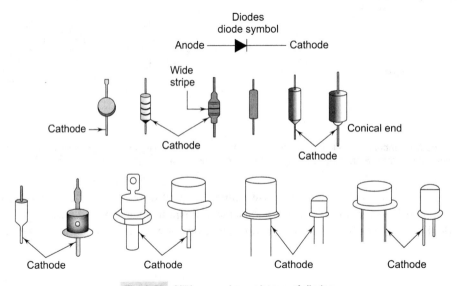

Fig. 2.32   Various package shapes of diodes

A single diode, when used for rectification, gives half wave rectification. When four diodes are combined, full wave rectification can be obtained. Devices containing four diodes in one package are called 'diode bridges'. Diode bridges with large current capacities require a heat sink. Typically, they are screwed to a piece of metal or the chassis of the equipment in which they are used. The heat sink allows the device to radiate excessive heat.

 ## 2.10 Special Types of Diodes

Besides the general purpose semiconductor diodes, there are many other types of diodes which have special characteristics. Following is a description and characteristics of some of the special types of diodes.

2.10.1 Zener Diode

A silicon diode has a very low reverse current, say 1μA at an ambient temperature of 25 °C. However, at some specific value of reverse voltage, a very rapid increase occurs in reverse current. This potential is called breakdown avalanche or the zener voltage and may be as low as 1 volt or as high as several hundred volts, depending upon the construction of the diode.

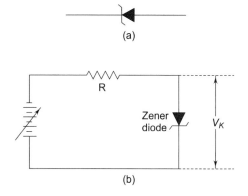

A zener diode has very high resistance at bias potentials below the zener voltage. This resistance could be several Megohms. At zener voltage, the zener diode suddenly shows a very low resistance, say between 5 and 100 Ω.

A zener diode behaves as a constant voltage source in the zener region of operation, as its internal resistance is very low. The current through the zener diode (Figure 2.33) is then limited only by the series resistance R. The value of series resistance is such that the maximum rated power rating of the zener diode is not exceeded.

Fig. 2.33 Zener diode: (a) symbol, (b) use as a constant voltage source

In order to help in distinguishing the zener diode from a general purpose diode, the former is usually labelled with its specified breakdown voltage. Since this voltage is required in the circuit design, the value is generally indicated on the diode. For example, some common values are 6.8 V, 7.2 V, 9.6 V etc.

2.10.2 Varactor Diode

A varactor diode is a silicon diode that works as a variable capacitor in response to a range of reverse voltage values. Varactors are available with nominal capacitance values ranging from 1 to 500 pF, and with maximum rated operating voltages extending from 10 to 100 volts. They mostly find applications in automatic frequency control circuits. In a typical case, a varactor shows 10 pF capacitance at reverse voltage of 5 volts and 5 pF at 30 volts. Figure 2.34 shows different shapes of varactor diodes.

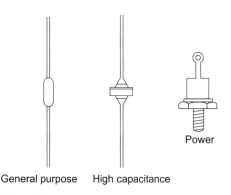

General purpose   High capacitance

Fig. 2.34 Varactor diodes-different types

2.10.3 Varistor

A varistor is a semiconductor device that has a voltage-dependent non-linear resistance which drops as the applied voltage is increased. A forward biased germanium diode shows such types of characteristics and is often used in varistor applications, such as in bias stabilization circuits.

Symmetrical varistor arrangements are used in meter protection circuits (Figure 2.35) wherein the diodes bypass the current around the meter regardless of the direction of current flow. If the meter is accidentally overloaded, varistors do not permit destructive voltages to develop across the meter.

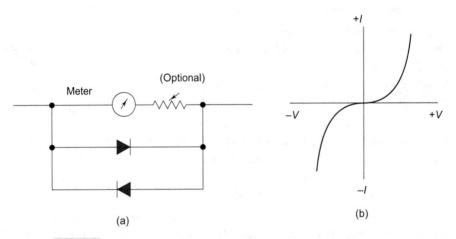

Fig. 2.35   Use of varistor in meter protection (b) varistor characteristics

2.10.4   Light Emitting Diodes (LED)

A light emitting diode is basically a *pn* junction that emits light when forward biased. LEDs are available in various types (Figure 2.36) and mounted with various coloured lenses like red, yellow and green. They are used mostly in displays employing seven segments that are individually energized to form alphanumeric characters.

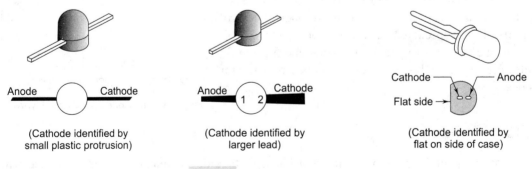

Fig. 2.36   Light emitting diodes

LED displays are encountered in test equipment, calculators and digital thermometers whereas LED arrays are used for specific applications such as light sources, punched tape readers, position readers, etc.

Electrically, LEDs behave like ordinary diodes except that their forward voltage drop is higher. For example, the typical values are; IR (infra-red): 1.2 V, Red: 1.85 V, Yellow: 2 V, Green: 2.15 V. Further, the actual voltages may vary depending upon the actual technology used in the LED.

2.10.5 Photodiode

A photodiode is a solid state device, similar to a conventional diode, except that when light falls on it (*pn* junction), it causes the device to conduct. It is practically an open circuit in darkness, but conducts a substantial amount of current when exposed to light.

2.10.6 Tunnel Diode (TD)

A tunnel diode is a *pn* junction which exhibits a negative resistance interval. The voltage current characteristics of a tunnel diode are shown in Figure 2.37. Negative resistance values range from 1 to 200 ohms for various types of tunnel diodes.

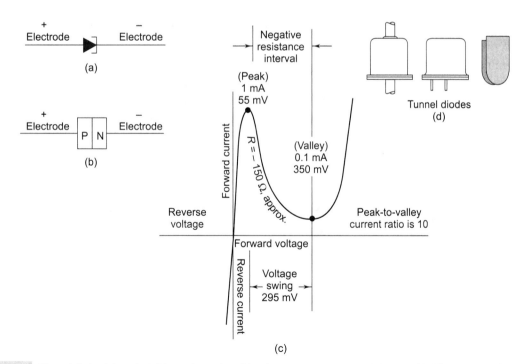

Fig. 2.37 Tunnel diode: (a) symbol, (b) p and n regions (c) voltage–current characteristics, and (d) different types of housings

Tunnel diodes can be utilized in switching circuits. A switching circuit has two quiescent points, i.e. it can be driven from its low current quiescent point to its high current quiescent point by means of pulses.

 ## 2.11 Transistors

2.11.1 Bipolar Transistors

The most commonly used semiconductor device is the transistor having the characteristic to control voltage and current gain in an electronic circuit. These properties enable the transistor to provide amplification, rectification, switching, detection and almost any desired function in the circuit. It is the basic device of all solid state electronics, both as a single component or as an element of integrated circuit.

A transistor is a three-terminal device. The terminals are called base (B), collector (C) and emitter (E). Basically, it is made up of two diodes: a base-emitter diode and a base-collector diode. In normal amplifier operation, the base-emitter diode is forward-biased and the base-collector diode is reverse-biased.

All transistors have leakage current across their reverse-biased base-collector diodes. For silicon transistors, this current is more than several nanoamperes. In germanium transistors, the leakage current may even be several microamperes. Leakage current increases with temperature and doubles about every 10 °C.

More than 500 packages of transistors are listed in the component manufacturers' catalogues. However, only about 100 types are in common use. Metallic packages (TO-3, TO-5 and TO-18) have been in use for a long time. However, they have been mostly replaced in low and medium power applications by cheap plastic packages due to the low cost of the latter. For high power applications, however, metallic packages, both stud or bolt type, are still common, though flat type packages are being replaced by plastic versions, with metallic tabs to improve heat dissipation. Figure 2.38 shows commonly used transistor packages and their terminals.

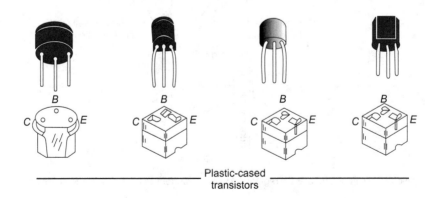

Plastic-cased
transistors

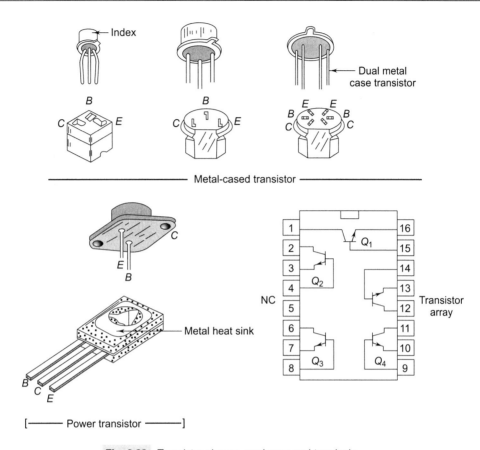

Fig. 2.38 Transistor shapes, packages and terminals

Figure 2.39 indicates the bias polarity required to forward-bias the base-emitter junction. The arrow head distinguishes between the emitter and the collector, and shows the direction of a 'conventional' current flow. Electron flow is opposite from the direction of the arrow points. The figure also compares the biasing required to cause conduction and cut-off in *npn* and *pnp* transistors. If the transistor's base-emitter junction is forward-biased, the transistor conducts. However, if the base-emitter junction is reverse-biased, the transistor is cut off.

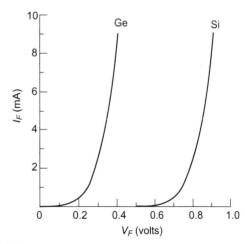

Fig. 2.39 Forward voltage of base-emitter junction in Ge and Si transistors

The voltage drop across a forward-biased emitter-base junction varies with the transistor's collector current. For example, a germanium transistor has a typical forward-bias, base-emitter voltage of 0.2–0.3 V when collector current is 1–10 mA and 0.4–0.5 V when the collector current is 10–100 mA. In contrast, the forward-bias voltage for a silicon transistor is about twice that for germanium types: about 0.5–0.6 V when the collector current is low, and about 0.8–0.9 V when the collector current is high. Figure 2.40 shows the relationship between voltage and current for base-emitter junction in germanium and silicon transistors.

Type	Cutoff	Conduction	
NPN Collector Base ———— Emitter	0 V— +V	+V +V — Control current	Main current
PNP Collector Base ———— Emitter	0 V— −V	−V −V Control current	Main current

Fig. 2.40 Three basic biasing arrangements in transistors

The three basic transistor circuits along with their characteristics are shown in Figure 2.41. When examining a transistor stage, just determine if the emitter-base junction is biased for conduction (forward-biased) by measuring the voltage difference between the emitter and the base.

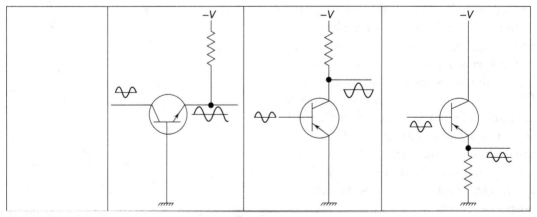

(Contd.)

Fig. 2.41   *(Contd.)*

Characteristic	Common base	Common emitter	Common collector
Input impedance	30 Ω – 50 Ω	300 Ω – 1500 Ω	20 kΩ – 500 kΩ
Output impedance	300 kΩ – 500 kΩ	30 kΩ – 50 kΩ	50 Ω – 1000 Ω
Voltage gain	500 – 1500	300 – 1000	< 1
Current gain	< 1	25 – 50	25 – 50
Power gain	20 dB – 30 dB	25 dB – 40 dB	10 dB – 20 dB
			(Emitter follower)

Fig. 2.41   Transistor amplifier characteristics

A common problem in transistors is the leakage current, which can shunt signals or change bias voltages, thereby upsetting circuit operation. This problem is particularly serious in direct-coupled or high frequency stages. Leakage current is the reverse current that flows in the junction of a transistor when specified voltage is applied across it, with the third terminal being left open. For example, I_{CEO} is the dc collector current that flows when a specified voltage is applied from collector to emitter, with the base being left open (unconnected). The polarity of the applied voltage is such that the collector-base junction is reverse-biased. Obviously, in a transistor, six leakage paths are present (with the third electrode open), as shown in Figure 2.42.

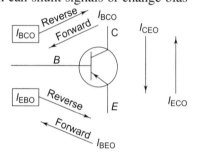

Fig. 2.42   Leakage paths in a transistor

Figure 2.43 shows typical bi-polar transistor junction resistance readings. The polarity of the ohmmeter to be applied on the various transistor leads is also indicated in the figure.

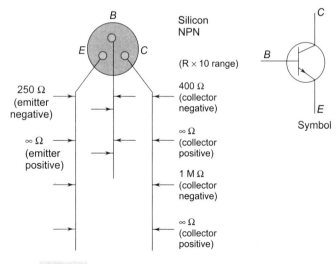

Fig. 2.43   Bi-polar transistor junction resistance values

2.11.2   Power Transistors

The junctions of the power transistors have comparatively larger areas than small signal transistors and have the following characteristics:

- Forward resistance values are generally lower than those for small signal silicon transistors.
- Similarly, they have lower reverse resistance values. The test results with an ohmmeter on a silicon power transistor are shown in Figure 2.44

Power transistors are usually mounted on the heat sinks or heat radiators (Figure 2.45). They are sometimes mounted on the chassis using silicone grease to increase heat transfer.

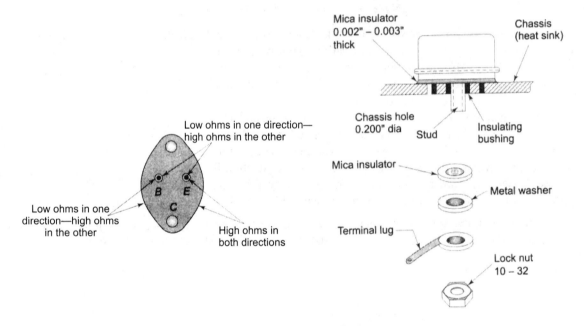

Fig. 2.44   Power Transistor: junction resistance values

Fig. 2.45   Power Transistor: mounting arrangement

2.11.3   Darlington Transistors

A Darlington is a special type of configuration usually consisting of two transistors fabricated on the same chip or at least mounted in the same package. Darlington pairs are often used as amplifiers in input circuits to provide a high input impedance. Darlingtons are used where drive is limited and a high gain, typically over 1000, is needed. In this configuration, (Figure 2.46) the

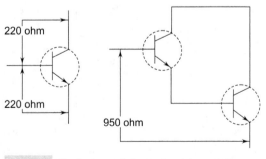

Fig. 2.46   Darlington pair forward resistance values

emitter base junctions are connected in series and the collector terminals are connected in parallel. A Darlington configuration behaves like a single transistor where-in the current gains (h_{fe}) of the individual transistors it is composed of are multiplied together and the base-emitter voltage drops of the individual transistors are added together.

2.11.4 Field-effect Transistors

Field-effect transistors, like bipolar transistors, have three terminals. They are designated as: source, drain and gate (Figure 2.47), which correspond in function to the emitter, collector and base of junction transistors. Source and drain leads are attached to the same block (channel of n or p semiconductor material). A band of oppositely doped material around the channel (between the source and drain leads) is connected to the gate lead.

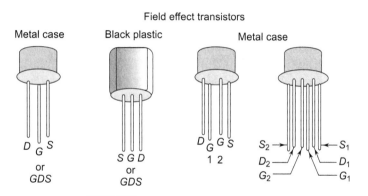

Fig. 2.47 FET packages and terminals

In normal junction FET operation, the gate source voltage reverse-biases the *pn* junction, causing an electric field that creates a depletion region in the source-drain channel. In the depletion region, the number of available current carriers is reduced as the reverse-biasing voltage increases, making source drain current a function of gate-source voltage. With the input (gate-source) circuit reverse-biased, the FET presents a high impedance to its signal source. This is in contrast to the low impedance of the forward-biased junction bipolar transistor base-emitter circuit. Since there is no input current, FETs emit less noise than junction transistors. Figure 2.48 shows the schematic symbol and biasing for n-channel and p-channel depletion mode field-effect transistors. Figure 2.49 shows FET amplifier characteristics.

Conversely, most MOSFET transistors, including those in the CMOS integrated circuits, are 'Enhancement Mode' type devices. With zero gate-to-source bias, these devices are off, and are increasingly turned on by the application of increasing gate-to-source bias (positive for n-channel, negative for p-channel).

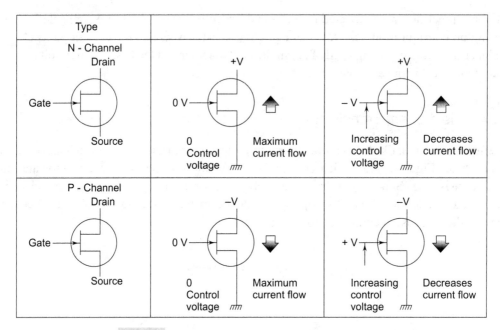

Fig. 2.48 Biasing arrangement in field-effect transistors

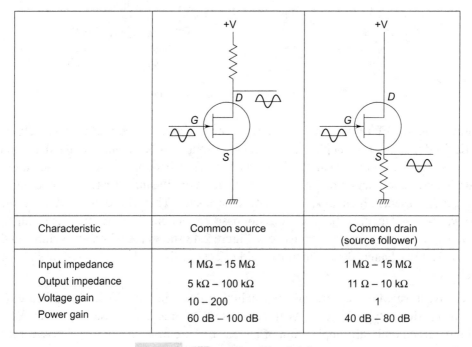

Characteristic	Common source	Common drain (source follower)
Input impedance	1 MΩ – 15 MΩ	1 MΩ – 15 MΩ
Output impedance	5 kΩ – 100 kΩ	11 Ω – 10 kΩ
Voltage gain	10 – 200	1
Power gain	60 dB – 100 dB	40 dB – 80 dB

Fig. 2.49 FET amplifier characteristics

There are three different types of field-effect transistors including;

(a) Junction gate;

(b) Insulated gate (non-enhanced type); and

(c) Insulated gate (enhanced type).

Each type comes with either n-channel or p-channel.

The junction gate and non-enhanced type insulated gate FETs are basically 'ON' devices like vacuum tubes. These two devices must be biased off. On the contrary, the enhanced type insulated gate FET is basically an 'OFF' device and must be biased on.

Figure 2.50 shows the junction resistance readings of a junction FET (JFET). The forward and reverse readings occur between the gate and the source or between the gate and drain only. The resistance between the source and the drain is the same irrespective of the ohmmeter polarity. It may be remembered that the gate source and gate drain junctions are non-linear and that the resistance values will change depending on the range used.

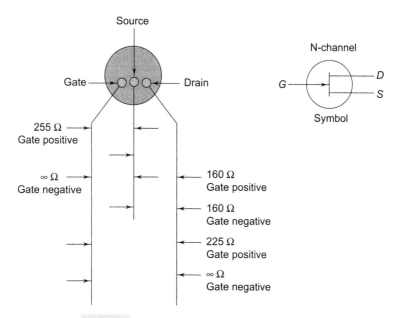

Fig. 2.50 Junction resistance readings of a JFET

A JFET can be removed from or inserted into a circuit without any special precautions other than the need to guard against over-heating during soldering and de-soldering operations. However, in the case of insulated gate FET, it may be remembered that the insulation is between the gate and channel, which is, in fact, a delicate capacitor. The insulation is so thin and the gate so small that it can be easily ruined (Figure 2.51). It is necessary to watch out for the static charge from fluorescent lights if the gate lead is left open. This can happen before the FET is installed, or if after removing from its insulated case, it is left on a table with its shorting wire removed. A shorting wire is usually a small piece of wire wrapped around all the leads.

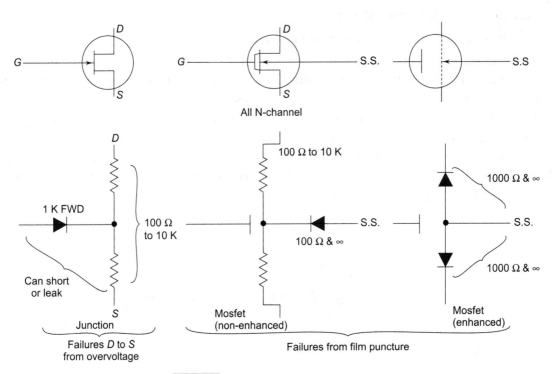

Fig. 2.51 Failures in field-effect transistors

2.11.5 Insulated Gate Bipolar Transistor (IGBT)

Prior to the development of IGBTs, power MOSFETs were used in medium or low voltage applications which require fast switching, whereas bipolar power transistors and thyristors were used in medium to high voltage applications which require high current conduction. A power MOSFET allows for simple gate control circuit design and has excellent fast switching capability. On the other hand, at 200 V or higher, it has the disadvantage of rapidly increasing on-resistance as the break-down voltage increases. The bipolar power transistor has excellent on-state characteristics due to the low forward voltage drop, but its base control circuit is complex and fast switching operation is difficult as compared with the MOSFET. The IGBT has the combined advantages of the above two devices.

The IGBT structure is a combination of the power MOSFET and a bipolar power transistor as shown in Figure 2.52 The input has a MOS gate structure, and the output is a wide base PNP transistor. The base drive current for the PNP transistor is fed through the input channel. Besides the

PNP transistor, there is an NPN transistor, which is designed to be inactivated by shorting the base and the emitter to the MOSFET source metal. The four layers of PNPN, which comprise the PNP transistor and the NPN transistor, form a thyristor structure, which causes the possibility of a latch-up. Unlike the power MOSFET, it does not have an integral reverse diode that exists parasitically, and because of this, it needs to be connected with the appropriate fast recovery diode when needed.

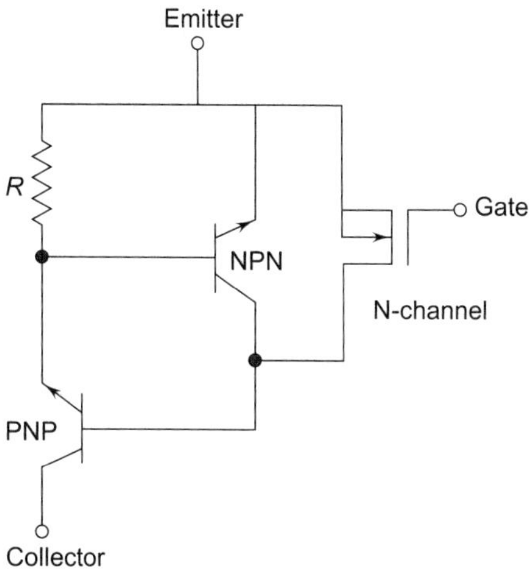

Fig. 2.52 Structure of IGBT (Insulated Gate Bipolar Transistor)

2.11.6   Transistor Type Numbers

Over the years, manufacturers have issued about 50,000 to 100,000 separate transistor type numbers.

A vast majority of these, are, however, no longer in use. Some of the commonly used type numbers are detailed below.

Joint Electron Device Engineering Council (JEDEC)

The transistor markings in this case take the following form:

Digit, letter, serial number, (suffix),

where the letter is always 'N', the serial number runs from 100 to 9999 and tells nothing about the transistor except its approximate time of introduction. The (optional) suffix indicates the gain (h_{fe}) group of the device. For example:

A = low gain   B   = medium gain
C = high gain   No suffix   = ungrouped (any gain)

Examples: 2N904, 2N3819, 2N2221A.

The data sheets give information on the actual gain spread and groupings. The reason for gain grouping is that the low gain devices are usually cheaper than the high gain devices, resulting in savings for high volume users.

Japanese Industrial Standard (JIS)

These take the following form:

Digit, two letters, serial number, (suffix).

The letters indicate the application area according to the following code:

SA = PNP HF transistor   SB = PNP AF transistor
SC = NPN HF transistor   SD = NPN AF transistor
SJ   = P-channel FET/MOSFET   SK = N-channel FET/MOSFET

The serial number runs from 10 to 9999.

The (optional) suffix indicates that the type is approved for use by various Japanese organizations. Since the code for a transistor always begins with 2S, it is sometimes omitted. For example, a

2SC733 would be marked C733. The typical examples of JIS based transistor markings are: 2SA1187, 2SB646, 2SC733.

Pro-electron System

This European system adopts the following form:

Two letters, (letter), serial number, (suffix).

The first letter indicates the material as follows:

A = Germanium (Ge)   B = Silicon (Si)

C = Gallium Arsenide (GaAs) R = Compound materials

The majority of transistors are of silicon, and therefore, begin with a B. The second letter indicates the device application:

C = Transistor, AF, small signal   D = Transistor, AF, power

F = Transistor, HF, small signal   L = Transistor, HF, power

U = Transistor, power, switching

The third letter indicates that the device is intended for industrial or professional rather than commercial applications. It is usually a W, X, Y or Z. The serial number runs from 100 to 9999. The suffix indicates the gain grouping, as for JEDEC.

Examples: BC108A, BAW68, BF239, BFY51.

Old Standards

Some of the old numbers use OC or OD followed by two or three numerals, (e.g. 0C28) or CV numbers (UK) like CV 7. They are no longer used with modern transistors.

Manufacturer's Codes

Apart from the above, manufacturers often introduce their own types, for commercial reasons, or to emphasize that the range belongs to a special application. Some common brand specific prefixes are:

TIS = Texas Instruments, small signal transistor (plastic case)

TIP = Texas Instruments, power transistor (plastic case)

MPS = Motorola, low power transistor (plastic case)

MRF = Motorola, HF, VHF and microwave transistor

RCA = RCA.

2.12   Thyristors

Thyristor is the generic name for the solid state devices that have electrical characteristics similar to that of the thyratron. This family of components is mostly used in all solid state power control and switching circuits, thereby replacing the old relay circuits. The following three types of thyristors are widely used:

- Silicon-controlled rectifier (reverse blocking triode thyristor);

- Triac (bidirectional triode thyristor); and
- Four-terminal thyristor (bilateral switch).

Thyristors are used extensively in power control circuits. They are particularly suited for ac power control applications such as lamp dimmers, motor speed control, temperature control and invertors. They are also employed for over-voltage protection in dc power supplies.

The thyristor is basically a four-layer *pnpn* device (Figure 2.53) and can be represented as a

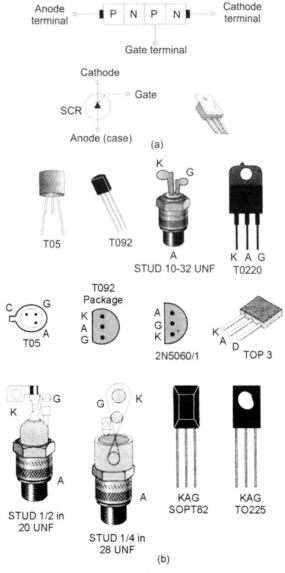

Fig. 2.53 (a) Silicon-controlled rectifier, device structure and symbol; and (b) different forms of SCR packages

two-transistor combination structure. The two transistors are cross-connected: one is NPN and the other is PNP. The base of the NPN transistor is connected to the collector of the PNP transistor while the base of the PNP transistor is connected to the collector of the NPN transistor. The device is normally off, but a trigger pulse at the gate switches the thyristor from a non-conducting state into a low resistance forward conducting state. Once triggered in conduction, the thyristor remains on unless the current flowing through it is reduced below the holding current value or it is reverse-biased. This means that the thyristor has extremely non-linear voltage-current characteristics (Figure 2.54).

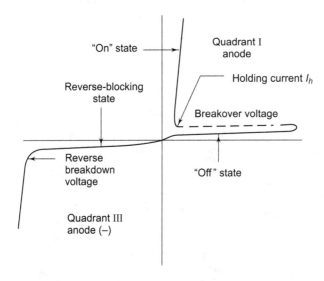

Fig. 2.54 Voltage-current characteristics of SCR

In case of a motor speed control or light dimmer, the exact time when the thyristor is triggered relative to the zero crossings of the ac power is used to determine the power level. Triggering the thyristor early in the cycle results in the delivery of high power to the load. Similarly, if the thyristor is triggered late in the cycle, only a small amount of power is delivered to the load. The advantage of thyristors over simple variable resistors is that they dissipate very little power as they are either fully 'on' or fully 'off'.

The *triac* is similar in operation to two thyristors connected in reverse-parallel, but with a common gate connection. This means that the device can pass or block current in both directions. Obviously, it can be triggered into conduction in either direction by applying either positive or negative gate signals.

Many transistor packages are used for packaging SCRs and TRIACs. Special packages have however, been also developed. The most common packages are TO-49, TO-118, TO-65 and TO-200. These are shown in Figure 2.55.

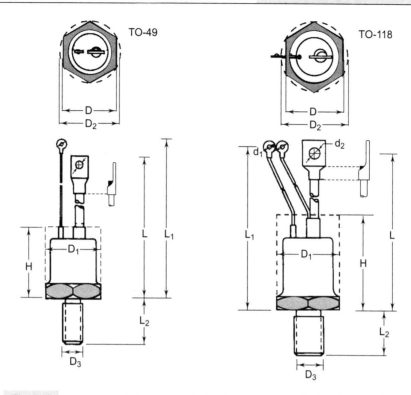

Fig. 2.55 Typical threaded packages for SCR and TRIACs, with flexible terminals

 ## 2.13 Integrated Circuits (ICs)

An integrated circuit contains transistors, capacitors, resistors and other parts packed on a single chip. Its function is similar to that of a circuit made with separate components. However, the components are formed in a miniature form, so that they can be packed in high density on a base of silicon. There are many kinds of integrated circuits and they come in several forms of packages. The number of pins on the integrated circuits differ from package to package depending upon the function of the IC. The IC can be attached directly to a printed circuit board with a solder. However, some designers prefer to use an IC socket as it is easy to exchange it should the IC fail. Broadly, there are two types of integrated circuits: linear integrated circuits and digital integrated circuits.

 ## 2.14 Linear Integrated Circuits

Linear integrated circuits are characterized by an output that is proportional to its input. There are many families of linear integrated circuits such as: (I) operational amplifiers; (II) differential amplifiers; (III) instrumentation amplifiers; (IV) audio and video amplifiers; (V) wide band amplifiers; (VI) radio frequency amplifiers; (VII) voltage/current regulators; and (VIII) analog-to-digital (A/D)

and digital-to-analog (D/A) converters. Most of these provide, in a single package, a circuit equivalent to many discrete components.

However, an important class of linear integrated circuits is operational amplifiers (op-amps). These amplifiers were originally utilized in analog computers to perform various mathematical operations such as addition, subtraction, integration and differentiation. Op-amps are now used to perform a variety of precise circuit functions. There are over 2500 types of commercially available op-amps. Most are low power devices with a power dissipation of upto 1W. However, they differ according to their voltage gain, temperature range, noise level and other characteristics.

2.14.1 Operational Amplifiers (Op-amp)

An operational amplifier is a complete amplifier circuit constructed as an integrated circuit on a single silicon chip. Inside, it contains a number of transistors and other components packaged into a single functional unit. It has a balanced arrangement in the input and is characterized by extremely high dc (static) and low frequency gain, a very high input impedance, a low closed loop output impedance and a fairly uniform roll-off in gain with frequency over many decades. The linear roll-off characteristic of an operational amplifier gives it the universality, and ability to accept feedback from a wide variety of feedback networks with excellent dynamic stability. The particular application of an op-amp is obviously determined by the device and its external circuit connections.

Symbolic Representation

The op-amp is symbolically represented as a triangle (Figure 2.56) on its side. In digital circuit symbols, the inverter is represented as a triangle, but the op-amp symbol is much larger. The triangle indicates the direction of signal flow. It is associated with three horizontal lines, two of which (*A* and *B*) indicate signal input and the third (*C*), the output signal connections.

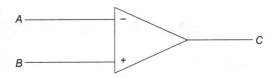

Fig. 2.56 Symbolic representation of an op-amp

The signal input terminals are described by minus (–) and plus (+) signs inside the triangle. The (–) input is called the inverting input, because the output voltage is 180 degrees out of phase with the voltage to this input. On the other hand, the (+) input is called the non-inverting input because the output voltage is in phase with this voltage applied to this terminal. The names 'inverting' and 'non-inverting terminals' have been given to indicate the phase of output signal in relation to the voltage applied at the inputs. Figure 2.57 shows the operation of the op-amp as an inverting and non-inverting amplifier.

Operational amplifiers are available both in metal as well as dual-in-line epoxy packages. Sockets for both types of packages are available and are mostly used in the printed circuit boards, though the

IC can be directly soldered on the board. The numbering convention for the pin numbers on the IC and the socket are shown in Figure 2.58.

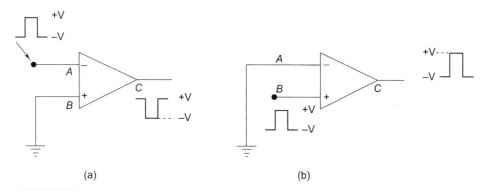

(a)   (b)

Fig. 2.57 (a) Inverting operation of an op-amp; and (b) Non-inverting operation of an op-amp

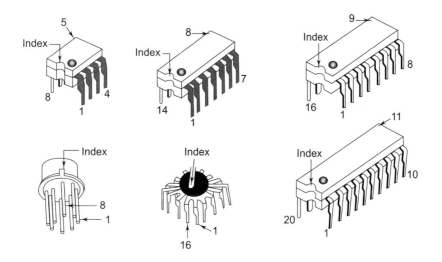

Fig. 2.58 Commonly available linear IC packages

Op-amp Identification

In general, the op-amps carry the following three types (Figure 2.59) of information on the pack:

uA 741	T	C
Device type	*Package type*	*Temperature*

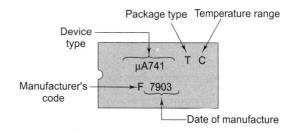

Fig. 2.59 Op-amp identification

Device Type

This group of alphanumeric characters defines the data sheet which specifies the functional and electrical characteristics of the device.

Package Type

One letter represents the basic package style. The various letters are:

D = Dual in-line package (hermetic, ceramic);
F = Flat pack;
H = Metal can package;
J = Metal power package (TO-66 outline);
K = Metal power package (TO-3 outline);
P = Dual-in line package (moulded);
R = Mini DIP (hermetic, ceramic);
T = Mini DIP (moulded); and
U = Power package (moulded, TO-220 outline).

Temperature Range

The three basic temperature grades in common use are:

C = Commercial $0\,°C$ to $+ 70\,°C$
M = Military $-55\,°C$ to $+ 125\,°C$
                 $-55\,°C$ to $+ 85\,°C$
V = Industrial $-20\,°C$ to $+ 85\,°C$
                 $-40\,°C$ to $+ 85\,°C$

Examples:

- uA 710 FM

  This code indicates a µA 710 voltage comparator in a flat pack with military temperature rating.

- uA 725 HC

  This number code indicates a µA 725 instrumentation operational amplifier, in a metal can with a commercial temperature rating capability.

In addition, the year of manufacture, batch number and manufacturer's identification are also given on the device.

Power Supply Requirements for Op-amps

Op-amps need to be powered with dc power supply, like any other transistor amplifier. The power supply should be of proper voltage regulation and filtering for correct operation of the op-amp.

The power supply leads on the op-amp are marked +V and –V to which positive and negative supply voltages should be connected respectively, with reference to ground. The positive and negative supply voltages are usually symmetrical, i.e. the two voltages are equal but opposite in sign. The most commonly used voltage to power op-amps are +15V and −15V. However, this is not always the case. Therefore, it is advisable to consult the manufacturer's data manuals on the op-amp of interest to determine the power supply requirements.

It may be noted that usually on the circuit schematics, the power supply leads are not shown on the op-amps. It is assumed that the reader is aware that dc voltage is necessary for operation of the op-amp.

2.14.2 Three-terminal Voltage Regulator

These are integrated circuits with three pins in which the input is applied between the centre leg and the input terminal, and the output is taken between the output terminal and the centre leg. They are available for fixed voltages like 5 V, 9 V, 12 V, etc. Most of the modern equipment makes use of these regulators.

A typical three-pin regulator for +5 volts is 78L05. The size and form of this regulator are similar to that of a 2SC1815 transistor. This is shown in Figure 2.60. 7805 is again a +5 volt three-pin regulator but with the current capacity of 500 mA to 1 A depending upon the heat sink used. The input voltage in both the types of 7805 is +35 volts. There are many types of three-pin regulators with different output voltages such as: 5 V, 6 V, 8 V, 9 V, 10 V, 12 V, 15 V, and 18 V.

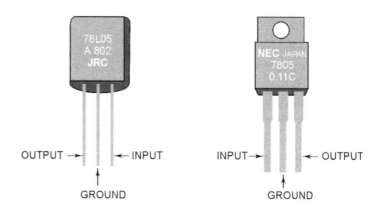

Fig. 2.60 (a) Three terminal voltage regulator 78L05 (b) Three terminal voltage regulator 7805

The component leads of different kinds of regulators must be confirmed with the data sheets.

 ## 2.15 Digital Integrated Circuits

Digital integrated circuits are used extensively in all branches of electronics from computing to industrial control, electronic instruments, communication systems and medical equipment. In fact, there does not seem to be any area in electronics where-in digital circuits are not or will not be used in some form or the other. The basic reason for this is that digital circuits operate from defined voltage levels, which reduces any uncertainty about the resulting output and the behaviour of a circuit. Many circuits operate with voltages that can only be 'on' or 'off', e.g. a light can be 'on' or 'off', a motor can be running or stopped, or a valve can be open or shut. All these are digital operations and would need digital circuit elements for their operation and control.

Digital circuits cover a wide range of applications, from high current industrial motors to microprocessors. However, the basic elements of all digital circuits are logic gates that perform logical operations on their inputs.

2.15.1   Logic Circuits

Logic circuits are decision-making elements in electronic circuits. They are the basic building blocks of the circuits that control data flow and processing of standard signals. In most systems, which use logic, the output function represents a voltage level, which is high or low.

There are several ways to represent two state 'yes' and 'no' decisions. Some of these are given below:

Yes	No
Open	Closed
1	0
Positive	Negative
True	False
High	Low
ON	OFF

2.15.1.1   Logic Convention

In digital circuits, 0 and 1 are represented by two different voltage levels, often called HIGH and LOW. The logic convention usually employed to relate these two entities is given below. (Figure 2.61).

In the positive logic convention, logic 1 is assigned to the most positive (HIGH) level of the voltage and logic '0' to the least positive (LOW) level.

In the negative logic convention, logic 1 is assigned to the most negative (LOW) level and logic 0 to the least negative (HIGH) level.

It is important to understand this convention for the interpretation of digital data. For example, suppose 1001 (binary) data is presented on a set of binary coded decimal output lines. In positive logic, this would mean 1001 (binary) = 9 (decimal) while in negative logic, the same would mean 0110 (binary) = 6 (decimal).

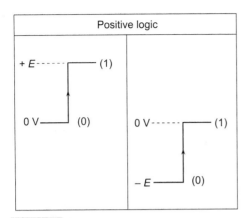 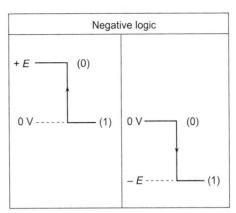

Fig. 2.61 Logic convention usually employed to represent two levels, high and low, in digital circuits

A gate is the elementary logic function of the digital circuit, which is designed to perform an operation of Boolean algebra. The most common gates are AND, NAND, OR, NOR, NOT, etc. The number of gates within a single IC depends mainly upon the number of pins of its package. For example, a 14 lead package can contain four-input-NAND or three-input-NAND or four two-input-NAND gates.

2.15.1.2 The Logic Families

The nature of the basic logic elements depends upon the properties of the electrical components used to realize them. In the early days of digital techniques, when diodes were largely used in the circuitry, it was natural to take the AND or OR gates as the basic elements. Later, when the transistor came to the fore, it became natural to basic logic circuits on the NAND and NOR gates. This is because the output signal of the transistor is opposite to the sign of its input. The most popular and most widely used circuits in modern digital equipment are the transistor-transistor logic, and complementary metal-oxide semiconductor logic families. The logic circuits have become increasingly complex. The developments in integrated circuit technology have solved the problem of bulk besides creating possibilities of obtaining several functions on one chip. It may be noted that the logic function of any IC gate is the same irrespective of the technology employed in fabricating the gate.

Transistor-Transistor Logic (TTL):   The most popular and most widely employed logic family is the transistor-transistor family. The various logic gates are manufactured in the integrated circuit form by most manufacturers of semiconductors. The basic element in TTL circuits is the bipolar transistor. TTL technology makes use of multiple-emitter transistors for the input devices. TTL gates use a totem-pole output circuit. Another type of output circuit is the open collector output in which an external pull-up resistor is required to get the proper HIGH and LOW level logic outputs.

The popularity of the TTL family rests on its good fan-in and fan-out capabilities, high speed (particularly Schottky TTL version), easy interconnection to other digital circuits and relatively low cost. The main characteristics of TTL logic are: propagation delay 10ns, flip-flop rate 20 MHz, fan-out 10, noise margin 0.4 V, dissipation per gate 10 mW. The standard TTL gates are marketed as 74 series which can operate up to 70 °C. However, 54 series are operatable up to a temperature of 125 °C. Most IC packages contain more than one gate. For example, IC 7400 is a quad two-input NAND gates whereas 7420 is a dual four-input NAND gates. There are various types of TTL families, which mostly differ only in speed and power dissipation. They are highlighted below.

Schottky TTL: The gates in the family are faster than those in standard TTL and consume much less power. Schottky TTL logic gates are available in the integrated form as 74S/54S series. A low power Schottky TTL series is also available as 54LS/74LS.

Emitter-coupled Logic (ECL Family): The ECL family provides another means of achieving higher speed of the gate. This differs completely from the other types of logic families in that the transistors, when conducting, are not saturated with the result that logic swings are reduced. For example, if the ECL gate is operated from 5 volts, the logic is represented by 0.9 V and logic 1 by 1.75 V.

CMOS Logic Families: The complementary metal-oxide semiconductor (CMOS) logic families offer significant advantages over bipolar transistor-based logic circuits, particularly as they feature very low power dissipation and good noise immunity.

The great advantage of CMOS technology is the possibility of high density packing of a large number of devices. The technique is most suitable for the construction of large scale integrated circuits rather than simple gates and flip-flops. Commercially available CMOS gates are available as 4000 series. For example, quad two-input AND gate in CMOS comes as 4081 (7408 TTL) while quad two-input NOR gate as 4001 (7402 TTL).

CMOS Digital Integrated Circuits: Modern electronic equipment, which makes use of logic circuits, generally employs CMOS circuits. CMOS stands for complementary symmetry metal-oxide semiconductor and is alternatively termed COS/MOS.

CMOS, from a black box point of view, operates fundamentally the same way as the conventional, bipolar TTL family of logic inverters, flip-flops, NAND, AND, OR, NOR circuits, etc. Figure 2.62 shows the configurations of TTL and CMOS inverters. A typical TTL is made up of transistors and registers whereas CMOS is totally semiconductor material, resulting in greater simplicity and much less power consumption. The CMOS inverter is made from an n-channel and a p-channel MOSFET connected as shown, with their working as follows:

- *n*-channel MOSFETs, which are turned on by a positive gate voltage; and
- *p*-channel MOSFETs, which are turned on by a negative gate voltage

'ON' source-to-drain resistance equals 1000 ohms typically whereas 'OFF' source-to-drain resistance equals 10,000 megaohms.

Placing +12 V (logic 1) on the input turns the *n*-channel 'ON' and *p*-channel MOSFET 'OFF'. The output is then essentially at ground or a logic '0' level, which completes the inversion. On the other hand, placing a ground level (logic 0) on the input turns the *p*-channel MOSFET 'ON' and the *n*-channel MOSFET 'OFF'. The output is then essentially at V_{DD} or 12 V or logic 1 level, which completes the inversion.

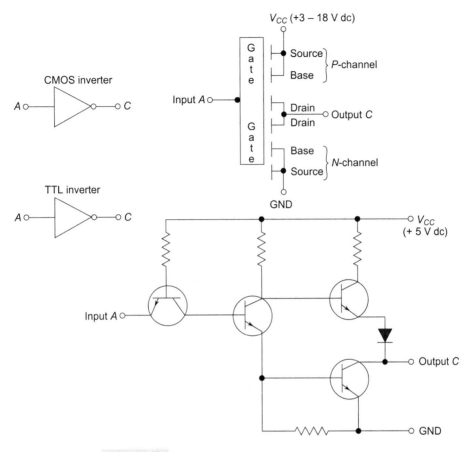

Fig. 2.62 Configuration of TTL and CMOS invertors

2.15.1.3 Categories of Integrated Circuits Based on Packing Density

SSI (Small scale integration) refers to integration levels typically having about 12 equivalent gates on chip. They are available in 14 or 16 pin DIP or Flat packs.

MSI (Medium scale integration) means integration typically between 12 and 100 equivalent gates per IC package. It is available in 24-pin DIP or Flat pack or 28-pin ceramic chip carrier package.

LSI (Large scale integration) implies integration typically up to 1000 equivalent gates per IC package. It is includes memories and some microprocessor circuits.

VLSI (Very large scale integration) means integration levels with extra high number of gates, say up to 1,00,000 gates per chip. For example, a RAM may have more than 4000 gates in a single chip, which is why it comes under the category of VLSI device.

2.15.1.4 Logic IC Series

The commonly used logic IC families are:

(a) Standard TTL (Type 74/54)

(b) CMOS (Type 4000 B)

(c) Low power TTL (74L/54L)

(d) Schottky TTL (Type 74S/54S)

(e) Low power Schottky TTL (Type 74LS/54LS)

(f) ECL (Type 10,000)

2.15.1.5 Packages In Digital ICs

Digital ICs come in several different packaged forms. Some of these forms are shown in Figure 2.63. The various packages are listed below.

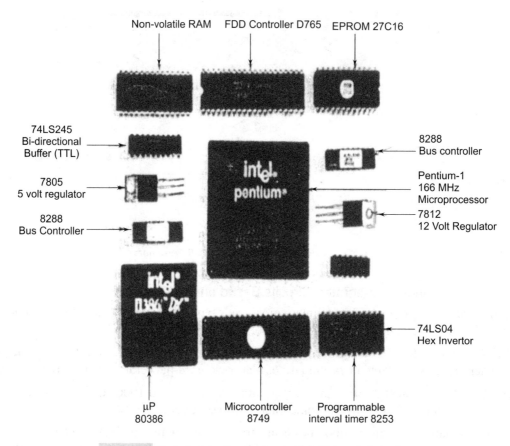

Fig. 2.63 Typical packaging systems in digital integrated circuits

Dual-in-Line Package (DIP): Most TTL and MOS devices in SSI, MSI and LSI are packaged in 14-,16-,24- or 40- pin DIPs.

Mini Dual-in-Line Package (Mini DIP): Mini DIPs are usually 8-pin packages.

Flat Pack: Flat packages are commonly used in applications where light weight is an essential requirement. Many military and space applications use flat packs. The number of pins on a flat pack varies from device to device.

TO-5, TO-8 Metal Can: The number of pins on a TO-5 or TO-8 can vary from 2 to 12.

All the above styles of packaging have different systems of numbering pins. For learning about how the pins of a particular package are numbered, the manufacturer's data sheet on package type and pin numbers must be consulted.

2.15.1.6 Identification of Integrated Circuits

Usually the digital integrated circuits come in a dual-in-line (DIP) package. Sometimes, the device in a DIP package may be an analog component—an operational amplifier or tapped resistors and therefore, it is essential to understand as to how to identify a particular IC.

In a schematic diagram, the ICs are represented in one of the following two methods:

- The IC is represented by a rectangle (Figure 2.64) with pin numbers shown along with each pin. The identification number of the IC is given on the schematic.
- The IC is represented in terms of its simple logic elements. For example, IC 74 LS 08 is quad two-input AND Gate and when it is represented in a schematic, it is listed as ¼ 74 LS 08 (Figure 2.65).

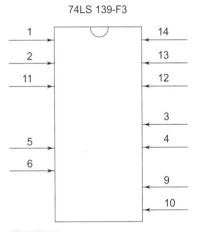

Fig. 2.64 Representation scheme for digital ICs

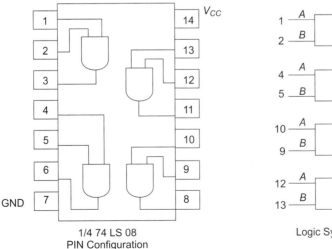

1/4 74 LS 08
PIN Configuration

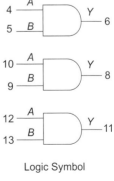

Logic Symbol

Fig. 2.65 Representation of IC in terms of its logic elements

An IC can be identified from the information given on the IC itself. The numbering system, though standardized, has some variations from manufacturer to manufacturer. Usually, an IC has the following markings on its surface (Figure 2.66).

Core Number: This identifies the logic family and its functions. In 74 LS 51, the first two numbers indicate that the IC is a member of the 7400 series IC family. The last letters give the function of the IC. Letters inserted in the centre of the core number show the logic sub-family. Since TTL is the most common series, no letter is inserted in the centre of the core number. In case of other families, the following letters are used:

C = CMOS
L = Low Power
LS = Low Power Schottky
S = Schottky
H = High Speed

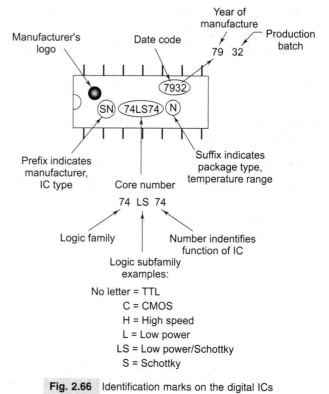

Fig. 2.66 Identification marks on the digital ICs

The same numbered ICs in each family perform the same function and have the same pin numbers. They are however, not interchangeable because of differences in timing and power requirements.

The *Prefix* to the core number identifies the manufacturer. For example, SN shows a device from Texas instruments. *Suffix* to the core number indicates package type, temperature range, etc.

In some ICs, marking is also provided for the year of manufacture and production batch. For example, 9834 indicates that the device was produced in 1998 in the 34th batch.

The manufacturer's logo (trade mark of the manufacturer) is also printed along with other information about the IC. Further detailed information about the ICs can be obtained from several sources listed below.

- Most of the IC manufacturers publish data books and product information data sheets. For most common series like 7400, you can get information from several sources. The information is provided on pin-outs, truth tables, etc.
- The IC Master, published regularly, serves as a reference book. This publication is usually available in a good technical library.

2.15.1.7 IC Pin-Outs

Packages for digital ICs generally fall into three categories, which are delineated below.

Metal Case TO Type Packages: In these, all the pins are in a circle and are numbered in counter clock-wise direction when viewed from the bottom. The pin closest to the orientation tab has the highest number. The most commonly used metal packages are TO-73 (12 leads), TO-99 (8 leads) and TO-101 (12 leads).

Dial-In-line Packages (DIL or DIP):   They have all leads arranged in two parallel lines. The most common package is T-116, which has 14 pins. Another most common package is a 16-pins DIP, which has the same overall dimensions as that of 14- pin- DIP.

Flat Packages: They are smaller than the TO-16 packages and therefore allow higher packing density. They can be assembled on both sides of a PCB by reflow soldering with their strip-like leads on pads without horse. They are preferred for use in military equipment. Common flat packages for integrated circuits are shown in Figure 2.67.

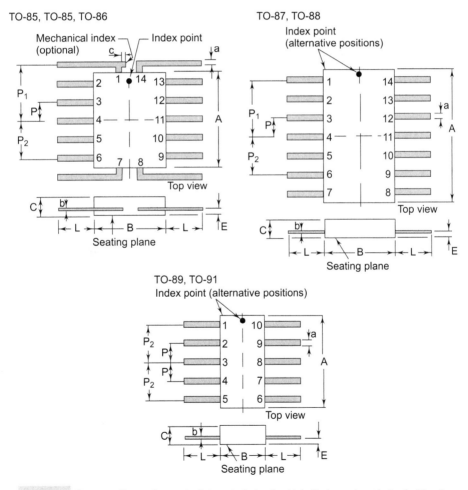

Fig. 2.67 Common flat packages for integrated circuits. Note that spacing of pins is 50 mils

The technical information on the IC includes a pin connection diagram which shows the signals that are connected to each pin on the IC (Figure 2.68). One end of the IC is marked with a white dot or a notch on the plastic. Pin number 1 is always the upper left hand pin on the end of the IC that includes the notch. The numbers run down the left side of the IC and up the right side.

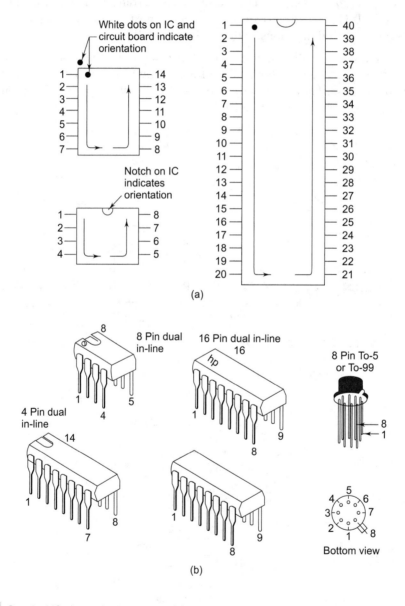

Fig. 2.68 Standard IC pin numbering system: (a) numbering system scheme; and (b) actual representation

In the pin connection diagram, the pin for supply voltage is indicated by V_{CC}. In most ICs, this voltage is +5 V_{dc}. The pin connection for ground is indicated by GND on the diagram. In general, in digital ICs, the pin with the highest number is V_{CC} and the pin with half that number is GND. This is, however, not always true.

 ## 2.16 Microprocessors

A microprocessor, also known as a CPU or Central Processing Unit, is a complete computation device that is fabricated on a single chip. A chip is basically an integrated circuit. Generally it is a small, thin piece of silicon onto which the transistors making up the microprocessor have been etched. A chip might be as large as an inch on a side and can contain as many as 10 million transistors. Simpler processors may contain only a few thousand transistors etched onto a chip.

The first microprocessor to make it into a home computer was the Intel 8080, a complete eight-bit computer on one chip introduced in 1974. The first microprocessor to make a real revolution in the market was the Intel 8088, introduced in 1979 and incorporated into the IBM PC (International Business Machines—Personal Computer), which appeared in the market around 1982. If you are familiar with the developments and history of the PC, then you must be knowing about the PC market moving from 8088 to the 80286, to the 80386, to the 80486, to the Pentium, to Pentium II, Pentium III and the now Pentium IV. All these microprocessors are made by Intel and are improvements on the basic design of the 8088. The new Pentiums can execute any piece of code that ran on the original 8088, but the Pentium II runs about 3000 times faster than 8088.

The microprocessor is the most important component of a microcomputer. It executes a collection of machine instructions that tell the processor what to do. Based on the basis of the instructions, a microprocessor does the following three basic things:

- Using its Arithmetic/Logic unit (ALU), a microprocessor can perform mathematical operations like addition, subtraction, multiplication and division.
- It can move data from one memory location to another.
- It can make decisions and jump to a new set of instructions based on these decisions.

Figure 2.69 shows a simplified diagram of a microprocessor capable of carrying out the above functions. This microprocessor has:

- An address bus (that may be 8, 16 or 32 bits wide), which sends an address to memory;
- A data bus (that may be 8,16 or 32 bits wide) which can send data to memory or receive data from memory;
- An RD (read) and WR (write) line to tell the memory whether it wants to set or get the addressed location;
- A clock line that lets a clock pulse sequence the processor; and
- A reset line that resets the programme counter to zero and restarts execution.

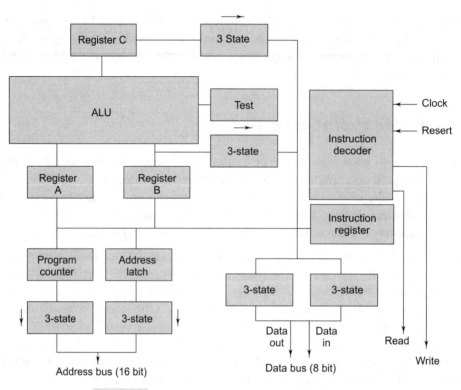

Fig. 2.69 Simplified block diagram of a microprocessor

Most practical microprocessors use 40 or more pins, and even keeping to that number requires that the eight data pins be used for both reading and writing. Figure 2.70 shows the representation of a typical CPU and the different types of packages in which they are available.

The function of the microprocessor is to manipulate data in accordance with the instructions stored in the memory. For this, the microprocessor transfers data and internal state information via an 8-bit bi-directional three-state data bus. Memory and peripheral device addresses are transmitted over a separate 16-bit three-state address bus. Timing and control outputs are given out for synchronization and control inputs like reset, hold and, ready and interrupt are used to perform specific functions.

In the context of microprocessors, the following terms must be understood:

- *Microns:* This refers to the width, in microns, of the smallest wire on the chip. For comparison, a human hair is 100 microns thick. As the feature size on the chip goes down, the number of transistors rises. Modern microprocessors are based on 0.25 micron or less technology.
- *Clock speed:* This is the maximum rate that the chip can be clocked. The clock speed of modern microprocessors is over 2 GHz.

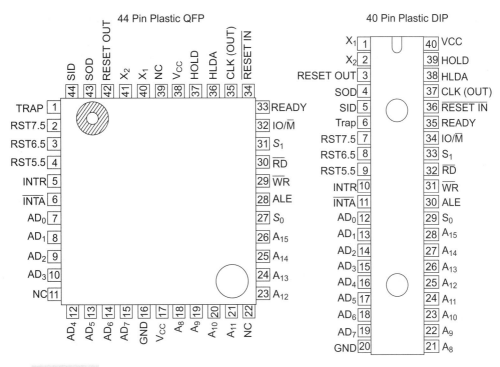

Fig. 2.70   Typical CPU pin-configuration (Intel 80C85) and different packages of 80C85

- *Data Width:*   This is the width of the ALU. An 8-bit ALU can add/subtract/multiply/etc. two 8-bit numbers, while a 32-bit ALU can manipulate 32-bit numbers. An 8-bit ALU would have to execute four instructions to add two 32-bit numbers, while a 32-bit ALU can do it in one instruction. In many cases, the external data bus is of the same width as the ALU, but not always. The 8088 had a 16-bit ALU and an 8-bit bus, while the modern Pentiums fetch data 64 bits at a time for their 32-bit ALUs.

- *MIPS:*   This stands for Millions of Instructions Per Second, and is a measure of the performance of a CPU. The MIPS value of modern processors is around 1000, as compared to 1 MIPS for an 80286 processor.

 ## 2.17 Semiconductor Memories

Memory technically implies any form of electronics storage. A digital memory is an array of binary storage elements arranged in a manner that it can be externally accessed. The memory array is organized as a set of memory words. Each word consists of a number of single bit storage elements called *memory cells*. The word length of a memory word is typically one, four or eight memory cells. Therefore, 1 bit, 4 bits or 8 bits (byte) of information can be stored by the memory word respectively. The memory capacity is the product of the number of memory words and the number of memory cells in each word. It is measured in bits and frequently expressed in kilo bits where *1 kilo bit = 2^{10} = 1024*.

All microprocessor-based systems make use of two types of memory—ROM and RAM. ROM stands for Read-Only Memory whereas RAM stands for Random Access Memory.

2.17.1 Random Access Memory

Random Access Memory (RAM) is used in a microprocessing system to store variable information. The CPU (central processing unit) under programme control can read or change the contents of a RAM location as desired. RAMs constitute a generic category that encompasses all memory devices in which the contents of any address can be accessed at random in essentially the same time as any other address.

There are two types of RAMs: static and dynamic. Both dynamic and static MOS random access memories are popular—the dynamic ones for their high circuit density per chip and low fabrication costs, and the static RAMs for single power supply operation, lack of refresh requirements and low power dissipation.

Dynamic RAM

In the dynamic RAM, information is stored as electrical charge on the gate capacitance of MOS transistors. Since these capacitors are not perfect, the charge will leak away and the information is likely to be lost with time if the charge is not periodically refreshed. This can be done in several ways and depends upon the type of device in use.

Static RAM

Static RAM does not need to be refreshed, as the memory cells are bi-stable and similar in design to conventional flip-flops. In general, a static RAM consumes more power than its dynamic counterpart. However, it requires less support circuitry. Also, there are no problems of synchronizing the memory refresh cycles with normal CPU read and write operations.

When the information is stored in the memory, it is written into the memory. When information is retrieved from a semiconductor memory, it is read from the memory. These are the only two functions that are done to static memories. Writing information into a memory is done in a write 'cycle'. Reading information from a memory is done in a read cycle. The term 'cycle' means a fixed period of time required to perform the function of writing into or reading from a memory. In fact, the electrical data or information is stored as a level of dc voltage. One dc voltage level corresponds to a '1' stored in the memory. A different dc voltage level corresponds to a '0' being stored in the memory.

In a semiconductor memory (Figure 2.71), data is entered on an input pin on the physical device labelled 'data in'. Data being read from a memory, are read on a device output pin labelled 'data out'. Therefore, a one-bit memory device will have four major physical connections: power input (V_{CC}), Data input (D_1), Data output (D_0) and read cycle or write cycle (R/W). If we want 16 bits of information storage, four address pins will be required ($2^4 = 16$) in the memory device (Figure 2.72). Under those conditions, no particular sequence will be needed to read or write information in the memory.

Fig. 2.71 Block diagram of one-bit memory   **Fig. 2.72** Block diagram of 16-bit memory

Figure 2.73 shows the pin configuration, logic symbol and block diagram of a typical $1 \text{ K} \times 1$ bit static random access memory, 2125A from Intel. It is packaged in 16-pin dual in-line package and operates on single $+ 5\text{V}$ supply. This is directly TTL compatible in all respects: inputs and outputs. It has a three-state data output and can be used for memory expansion through chip select (CS) enable input. Besides ten address input lines for addressing all the 1024 words, it also has control to choose either READ or WRITE operation. When chip enable (CE) is high, the D_{out} is a high impedance state.

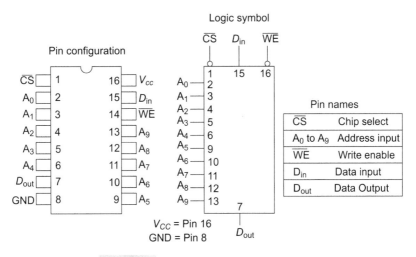

Fig. 2.73 Pin configuration of Intel 2125A

2.17.2   Read Only Memory

In a microprocessor-based system, ROMs are normally used to hold the program of instructions and data constants such as look-up tables. Unlike the RAM, the ROM is non-volatile, i.e. the contents of the memory are not lost when the power supply is removed. Data stored in these chips is either unchangeable or requires a special operation to change. This means that removing the power source from the chip will not cause it to lose any data.

The following are the five basic ROM types:

- ROM;
- PROM;
- EPROM;
- EEPROM; and
- Flash Memory.

Mask-programmed ROMs: They are programmed by the manufacturer to the user's requirements. This type of ROM is only used if a fairly large number of units are required, because the cost of preparation of creating the bit pattern on the chip is quite high. The contents of these ROMs cannot be altered after manufacture. Once the chip is made, the actual chip can cost very little money. They use very little power, are extremely reliable and, in the case of most small electronic devices, contain all the necessary programming to control the device.

Programmable Read Only Memory (PROM):   This is programmed by the user. Selectively fusing (open-circuiting) the metal or polysilicon links in each memory cell sets that cell to a fixed state. The process is irreversible. In one form of PROM, the information is stored as a charge in a MOSFET cell. Blank PROM chips can be coded by anyone with a programmer. The process is known as **'burning the PROM'**.

The contents of a PROM can be erased by flooding the chip with ultraviolet radiation. Following this process, a fresh pattern can be entered. PROMs are used in the microprocessor-based systems during the system development phase and on the production system when the total production run is not high enough to justify the use of mask-programmed ROMs.

PROMs can only be programmed once. They are more fragile than ROMs. A jolt of static electricity can easily cause damage to the chip. But blank PROMs are inexpensive and are great for prototyping the data for a ROM before committing to the costly ROM fabrication process.

Erasable Programmable Read Only Memories (EPROM):   These devices provide the facility of re-writing the chips several times. EPROMs are configured using an EPROM programmer that provides voltage at specified levels, depending upon the type of EPROM used.

For erasing the chips of its previous contents, an EPROM requires a special tool that emits a certain frequency of ultraviolet (UV) light. Because the UV light will not penetrate most plastics or glasses, each EPROM chip has a quartz window on top of the chip. The EPROM is kept very close to the eraser's light source, within an inch or two, to work properly. An EPROM eraser is not selective, it will erase the entire EPROM. The EPROM must be removed from the device it is in and placed under the UV light of the EPROM eraser for several minutes. An EPROM that is left under UV light too long can become over-erased. In such a case, the chip cannot be programmed.

Electrically Erasable Programmable Read Only Memories (EEPROM) or Read-Mostly Memories (RMM): They are designed such that the contents of these memories can be altered electrically. However, this is a fairly slow process. It often requires voltages and circuit techniques that are not commonly found in normal logic circuitry.

Flash Memory: This is a type of EEPROM that uses in-circuit wiring to erase by applying an electrical field to the entire chip or pre-determined sections of it called *blocks*. Flash memory works much faster than traditional EEPROM because it writes data in chunks, usually 512 bytes in size, instead of a byte at a time.

Figure 2.74 shows a typical symbol of ROM, for storing 1024 8-bit words. This is also called a 1K × 8 ROM where 1K represents 1024. Similarly, a 2048 × 8 can be written as a 2 K × 8 and so on. Since 1 K ROM stores 1024 different words, it needs 10 address inputs (2^{10} = 1024). The word size is 8-bits, so there are eight output lines. The memory chip is enabled or disabled through the chip select (CS) input. ROMs do not provide for data input or read/write control because they do not normally have the write operation. Some ROMs do provide for special input facilities for initially writing the data into the ROM which is generally shown on the symbol.

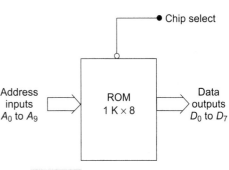

Fig. 2.74 Typical symbol of ROM

The Intel 2716 is a 16,384 (2K × 8) bit ultraviolet erasable and electrically programmable read only memory. Figure 2.75 shows the pin configuration and block diagram of the 2716 EPROM.

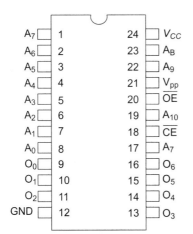

A₇	1	24	V_{CC}
A₆	2	23	A₈
A₅	3	22	A₉
A₄	4	21	V_{pp}
A₃	5	20	$\overline{OE}$
A₂	6	19	A₁₀
A₁	7	18	$\overline{CE}$
A₀	8	17	A₇
O₀	9	16	O₆
O₁	10	15	O₅
O₂	11	14	O₄
GND	12	13	O₃

Pin names

$A_0 - A_{10}$	Address
$\overline{CE}$	Chip enable
$\overline{OE}$	Output enable
$O_0 - O_7$	Outputs

Fig. 2.75 Pin configuration of Intel 2716 EPROM

2.18 Microcontrollers

A microcontroller is basically a single-chip microcomputer provided in a single integrated circuit package, which contains a CPU, clock circuitry, ROM (flash), RAM, serial port, timer/counter and I/O circuitry. As such, unlike conventional microprocessors it does not require a host of associated chips for its operation.

Most of the microcontrollers come in 40-pin DIP packages; the pin-out consists essentially of upto 32 I/O lines with the remainder being used for power, reset interrupt and timing. The instruction set of a single chip microcomputer generally bears a close resemblance to that of the microprocessor family to which the microcontroller belongs.

Microcontrollers have been available for a long time. Intel introduced 8048 (MCS-48, an 8-bit microcontroller) in 1976. It has developed upwards and its new version is the 8051 family, which is a high performance 40-pin DIP package. Microcontroller's functionality has increased tremendously in recent years. Today, one gets microcontrollers which are stand-alone for applications in data acquisition systems and control. They have A/D converters on the chip which enable their direct use in instrumentation.

One of the popular microcontrollers available in the CMOS technology is 89C51, which has 4K bytes of Flash programmable and erasable read-only memory (EEPROM). Its instruction set and pin-out are compatible with industry standard MCS-51. Flash allows the programme memory to be re-programmed in-system or by a conventional non-volatile memory programmer. This microcontroller is available in both DIP or QUAD package.

 ## 2.19 Surface Mount Devices

The conventional through-hole mounting technology used for printed circuit assemblies is being increasingly superseded by surface mount technology. Instead of inserting leaded components through the holes, special miniaturized components are directly attached and soldered to the printed circuit board. The surface mounted components and their packing are particularly suitable for automatic assembly. The advantages of surface mounting are rationalized production, reduced board size and increased reliability. The prime motivation for introducing surface mount technology (SMT) is, thus, density increase and board area reduction due to continuous demand and market trend for miniaturization in electronic assembly, particularly in portable products.

Figure 2.76 shows the conventional through-hole technology in which the components are placed on one PCB side (component side) and soldered on the other side. In surface mount technology, the components can be assembled on both sides of the board (Figure 2.77). The components are attached to the PCB by solder paste or non-conductive glue and then soldered.

In addition, there are hybrid circuits consisting of thick and thin film circuits, which are basically leadless components. They are reflow soldered on to the ceramic or glass substrate in addition to the components already integrated on the substrate. The mounting of the hybrid circuits is shown in Figure 2.78.

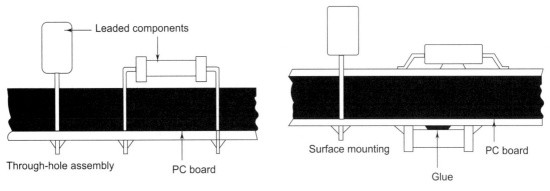

Fig. 2.76 Conventional through-hole technology-components are placed on one side and soldered on the other

Fig. 2.77 Surface-mount technology-components are mounted on both the sides

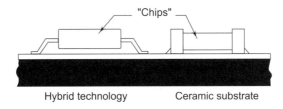

Fig. 2.78 Hybrid technology-consisting of components integrated on the substrate and soldered chips

Since not all component types are available in the surface mount version, a combination of leaded and surface mounted components is generally seen in manufactured electronic equipment.

The 100 mil lead spacing used in DIL packages has a severe limitation in terms of the number of pins that could be accommodated in high density VLSI circuits, with 64 being the limit. The development of miniature systems has forced new package types to appear specially in the surface mount type. The small outline (SO) package with gull leads is available in which pins are on a 50 mil grid. For higher pin counts, the quad package with J leads has become a popular package in the industry. With the pin numbers going up to 300+, the pin spacing has been decreasing from 50 to 30 and then to 25 mil.

2.19.1 Surface Mount Devices

The abbreviation SMD for Surface Mount Device is the most common designation for the components used in surface mount technology. SMDs are designed with soldering pads or short leads, and are much smaller than comparable leaded components. In contrast to conventional components, the leads of which must be inserted into holes, SMDs are directly attached to the surface of the PCB and then soldered.

Resistors, ceramic capacitors and discrete semiconductors represent 80 per cent of the total available SMDs. Normally, in the SMDs, the cubic shape prevails over cylindrical versions, as the latter can only have two pins thus being exclusively suitable for resistors, capacitors and diodes. If development of a special SMD package is not possible for electric or economic reasons, the DIP package can be converted into a surface mountable version by bending the leads.

Lead Styles   SMDs are constructed with different types of lead styles. The commonly used lead styles are shown in Figure 2.79. The shapes of the different SMD components with different lead styles are shown in Figure 2.80.

Type	Drawing	Components
Gull-wing		SOIC QEP TSOP
J-lead		PLCC SOJ
Ball		BGA Chip Scale Flip Chip (Bump)
Metallized Terminations		Capacitors Resistors Ferrites

Fig. 2.79 SMD lead styles

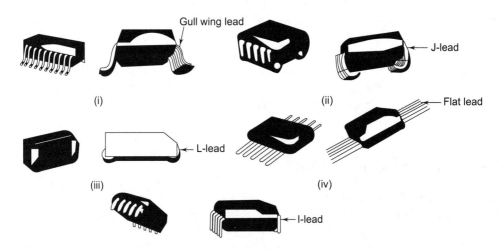

Fig. 2.80 Shapes of components with different lead styles (i) gull wing lead: metal lead that bends down and away (ii) J-led: metal lead that bends down and underneath a component in the shape of letter J (iii) L-lead: inward formed underneath a component (iv) flat lead: protrudes directly out from the body of the component (v) I-lead: a through-hole lead cut short for surface mounting

Lead Pitch: The lead pitch in a SMD is measured from the centre-to-centre of leads, and is not the air gap between the leads as shown in Figure 2.81.

Component Packaging: Automated assembling of printed circuit boards is carried out by pick-and-place machines. For this purpose, proper packaging is required to protect the components, particularly the SMDs, from damage during transport. The various packaging methods to provide proper feeders to receive the components are trays, tubes, tape and reel and bulk feed cassettes. They are shown in Figure 2.82.

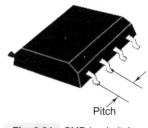

Pitch

Fig. 2.81 SMD lead pitch

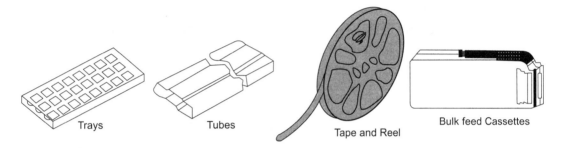

Trays   Tubes   Tape and Reel   Bulk feed Cassettes

Fig. 2.82 SMD packaging techniques

Chip Size Codes: The size of chip components is defined by a four-digit code. The thickness of the component is not defined in the size code. Figure 2.83 illustrates the examples of specifying the size of the components. The size code may be stated in inches or in the metric system. For example, the size code of ceramic capacitors and resistors is usually stated in inches. Tantalum capacitors are stated in metrics.

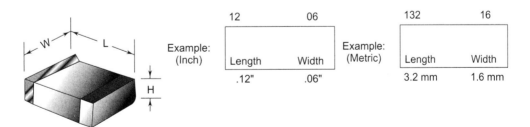

12	06
Length	Width
.12"	.06"

Example: (Inch)

132	16
Length	Width
3.2 mm	1.6 mm

Example: (Metric)

Fig. 2.83 Chip size codes

2.19.2 Surface Mounting Semiconductor Packages

With developments in the technology of surface mounting, a range of packaging types has emerged. The commonly used packages in semiconductor SMDs are detailed below.

SOIC (Small Outline Integrated Circuit) This is a plastic package, available in 6, 8, 10, 14, and 16 versions with a body width of 4 mm, and in 16, 20, 24, and 28 pin versions with a wider body of 7.6 mm. The leads are on standard 1.27 mm centres and are formed outwards so that the tips of the leads lie in contact with the PCB (Figure 2.84). In addition, there are also two Very Small Outline (VSO) packages with 40 and 56 leads at a pitch of 0.762 mm.

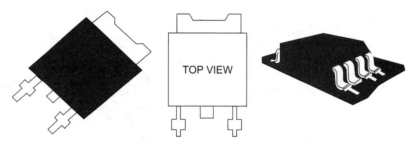

Fig. 2.84 Small outline integrated circuits (SOIC)

SOT (Small Outline Transistor) Packages They are used for discrete transistors and diodes. The most common packages are the SOT-23 and the SOT-89 (now renamed as TO-236 and TO-243 respectively). The construction of a typical SOT-23 is shown in Figure 2.85 with the standard dimension of the package. The package has three leads, two along one edge and a third in the centre of the opposite edge. Semiconductors on larger chips (up to about 1.5 mm square) are packaged in the SOT-89 format (Figure 2.86). Its three leads are all along the same edge of the package but the centre one extends across the bottom to improve the thermal conductivity.

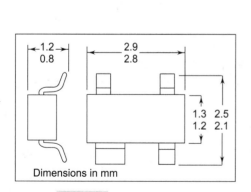

Fig. 2.85 SOT-23 package

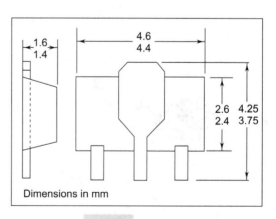

Fig. 2.86 SOT-89 package

For high power applications, SOT-194, as shown in Figure 2.87, has been developed which can allow power dissipation of up to 4 watts when used with a suitable heat dissipating interconnection substrate. An outline of a four - lead package, SOT-143, is shown in Figure 2.88. This is used for dual gate devices.

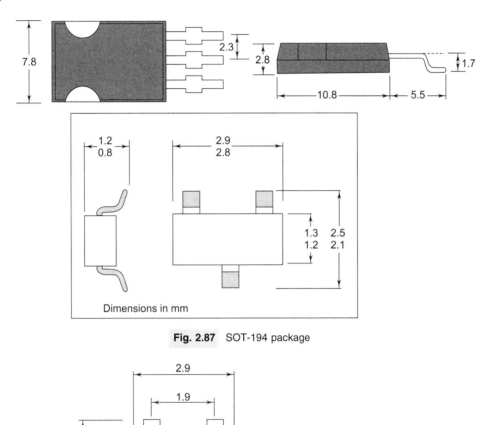

Fig. 2.87 SOT-194 package

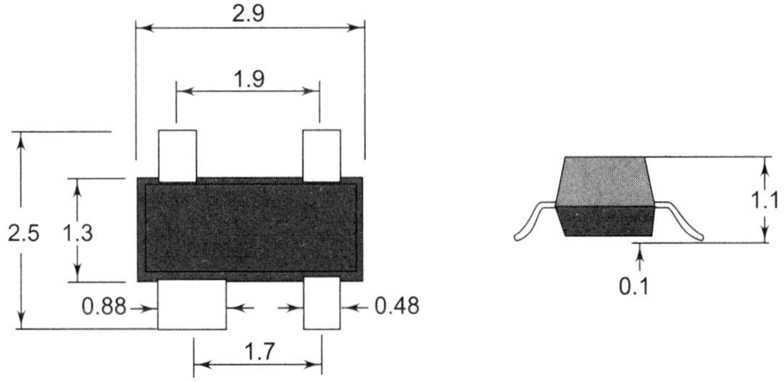

Fig. 2.88 SOT-143 package

Cylindrical Diode Packages: The two most popular packages developed specially for diodes in the cylindrical shape are:

- SOD (Small Outline Diode) package is specifically designed for small diode chips, limited to a power dissipation of 250 mW. A typical example of this type of device is SOD-80 whose dimensions are shown in Figure 2.89.

- MELF (Metal Electrode Face Bonded) package in larger cylindrical encapsulation is used when more power handling capability is required. The typical dimensions of MELF diode are shown in Figure 2.90. The SOD-80 package is also sometimes referred to as Min MELF.

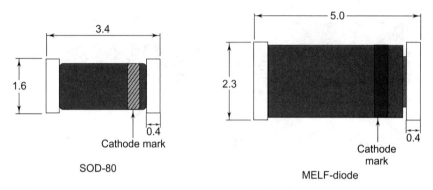

Fig. 2.89 Small outline diode package

Fig. 2.90 High power diode package

LCCC (Leadless Ceramic Chip Carriers)   The term chip carriers refers to those IC packages that are square or nearly square with their terminations brought out on all four sides. LCCC are those devices which do not carry any leads and excess packaging material. These packages are suitable for direct soldering or for attachment by sockets with added leads. In these devices, the IC chip is bonded to a ceramic base and connections are brought out with wires to solderable contact pads as shown in Figure 2.91.

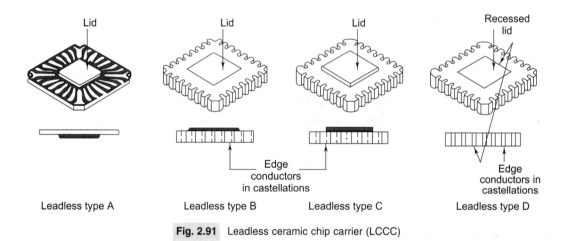

Fig. 2.91 Leadless ceramic chip carrier (LCCC)

LCCC are commonly available in 18, 20, 28, 32, 44, 52, 68, 84, 100, 124 and 156 termination versions. The component height is typically 1.5–2.0 mm. The pitch between the terminations is always 1.27 mm.

PLCC (Plastic Leaded Chip Carriers) They are available in a wide range in the same sizes and formats as the LCCCs. The leads of these devices also have a pitch of 1.27 mm. The majority of PLCCs are available with 'J' leads that are folded underneath the package. This is shown in Figure 2.92. Since the 'J' leads are tucked under the device, and are not protruding, they present difficulties for the inspection and testing of circuits.

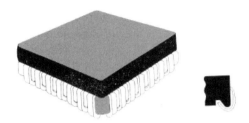

Fig. 2.92 J-leaded PLCC package

Flat Packs and Quad Packs The flat pack package has its lead frame coplanar to the body of the package. The original flat pack had leads on two sides of the body, but presently they are available with leads on all four sides, and are accordingly called quad pack. The quad packs are usually high lead-count plastic packages in the range of 40 to 200. The size of the package remains the same, but the pitch of the leads varies with the lead count. For example; the pitch is 1.0 mm on packages of up to 64 leads, 0.8 mm on the 80 lead and 0.65 mm on the 100-pin package. The shape of the package is shown in Figure 2.93.

LGA (Land Grid Arrays) In these devices, the pins emanate from an array on the under side of the package (Figure 2.94) rather than its periphery. The surface mounting version of the leaded grid array is the land grid array, whereby the pins are substituted by an array of solderable pads on the base. They are available in various sizes, pad sizes and pad densities to meet the requirements of different lead-out arrangements.

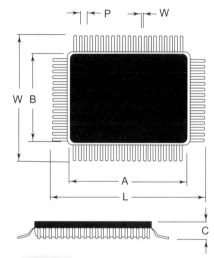

Fig. 2.93 Quad flat-pack package

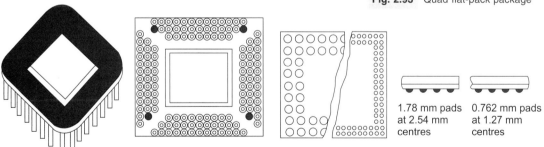

Fig. 2.94 Land Grid Array (LGA)

BGA (Ball Grid Arrays) These refer to actually any type of IC package that routes from the die and connects to the PCB via solder bumps. The package provides solutions for high performance, high pin count applications, with pin counts now nearing 1000, with 300 to 600 being standard. A typical BGA package consists of a substrate with two metal layers and through-hole vias. The IC die is mounted on top of the substrate, and is enclosed in a plastic mould (Figure 2.95). Ball pitches are decreasing from 1.27 mm to 1.0 mm and as low as 0.5 mm for new chip scale packaging. For high reliability applications, several vendors offer a ball grid array package

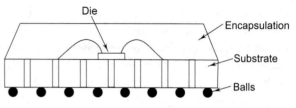

Fig. 2.95 Plastic base grid array package

with a ceramic substrate. Multi-chip packages are usually BGAs or QFPs containing two to four die.

Different ball patterns available on BGA are full grid, peripheral, stagger and thermal via (Figure 2.96). Acronyms for other "Grid Array" are:

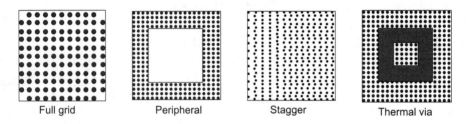

| Full grid | Peripheral | Stagger | Thermal via |

Fig. 2.96 Different pattern types of ball grid arrays

CBGA	-	Ceramic Ball Grid Arrays; for high temperature requirements.
fBGA	-	Flex BGA; uses a flex polyimide substrate.
SBGA	-	Super BGA; has metal heat spreader on top.
PBGA	-	Plastic BGA; Industry standard BGA.
LGA	-	Land Grid Array; pads without the balls.
CGA	-	Column Grid Array; solder columns instead of balls.
CSP	-	Chip Scale Package ; Fine-pitch BGA. Package is max. 120%>chip size.
µBGA	-	Chip Scale Package; trademark of Tessera
Flip Chip	-	Die with solder bumps; very small

COB (Chip-on-board) With chip-on board, a bare (unpackaged) semiconductor is attached with epoxy directly to a PCB, wire bonded and then encapsulated with polymeric materials. It offers high packaging density and fast signal speed by means of wire bonding the chip directly onto the board.

2.19.3 Packaging of Passive Components as SMDs

In order to utilize the full potential of surface mount technology, it is desirable that components other than semiconductors must also be surface mounted. These components include resistors, capacitors, inductors etc. The capacitors and resistors are available in cubic dimensions and are often referred to as 'chips'. The most common chip components are resistors, capacitors and diodes.

However, every kind of two-terminal devices can be had in the chip form. The most commonly available package is in the rectangular form and these devices have their solderable terminations only on the end face, or on the top and bottom faces as well as the end or on the sides in addition. Figure 2.97 shows typical examples of different types of metallizations on chip component.

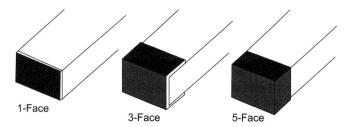

1-Face   3-Face   5-Face

Fig. 2.97 Difference between one-face, three-face and five-face metallization on chip components

Typical packages of different types of capacitors, resistors and inductors are shown in Figure 2.98.

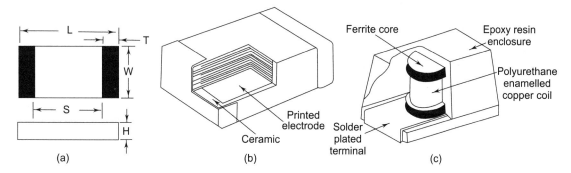

(a)   (b)   (c)

Fig. 2.98 Surface mount: (a) resistors; (b) capacitor; and (c) inductors

2.20 Heat Sinks

Electronic components, particularly semiconductor devices, show temperature-dependent characteristics. They can be permanently damaged by over-heating if the temperature of the device is not maintained within specified limits. It is, therefore, necessary to determine how much heat an active component will generate and also to make arrangements for its dissipation. This is usually done by providing a heat sink.

Discrete power devices are usually supplied in packages which can be easily assembled on to a heat sink. In such cases, it is necessary to ensure good contact between the package and its heat sinks, specially by interposing a thin layer of silicone grease to improve heat transfer where the metal surfaces may not be meeting, and by applying sufficient pressure through the screw connection.

Power DIP packages are usually provided with a metal slug in the body to which good thermal contact can be made with an external heat sink. Such an arrangement is shown in Figure 2.99.

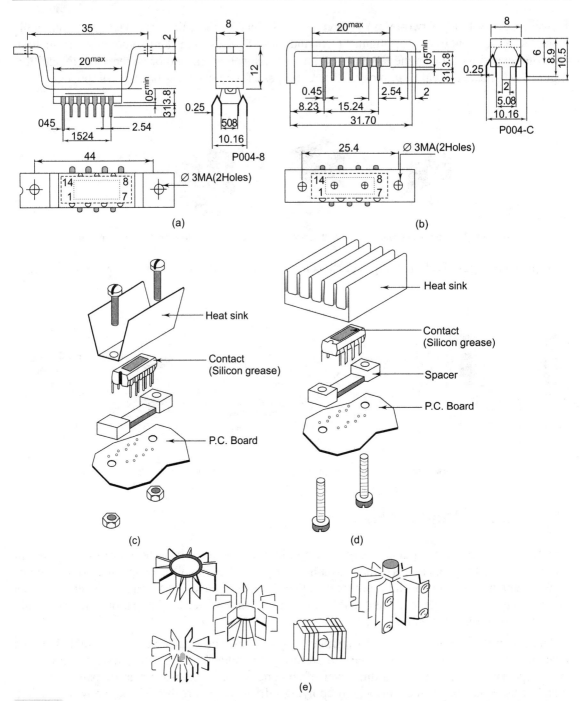

Fig. 2.99 Heatsinks: (a) and (b) show two DIPs with an inserted metallic heatsink, to which an external heatsink has been soldered by the supplier; and (c) and (d) show typical assemblies suggested by the manufacturer (SGS-ATES) for greater dissipation; (e) radiating fin heat sinks. Dimensions are in millimeters.

Some DIP packages have two side tabs to which the heat sink can be attached. It is sometimes possible to solder these tabs to a large copper area on the component side of PCB and to use that as a heat sink.

2.21 Transformer

A transformer is an electrical device which, by electromagnetic induction, transforms electric energy from one circuit to another at the same frequency, but usually at a different voltage and current value. In electronic equipment, transformers are generally used to provide the required ac voltage to the circuit by appropriately transforming the mains voltage. When a transformer transfers electric energy at a different level, it is called either a step-up (increase in the voltage ratio) or a step-down (decrease in the voltage) transformer. If there is no level change, then it is said to have a 1:1 voltage ratio.

A transformer is so constructed that one winding (primary) induces voltage into a second winding or windings (secondary). Accordingly, the transformer symbol (Figure 2.100) shows its construction with two or more windings wound on a common core. The windings are adjacent to each other to obtain magnetic or inductive coupling. The input voltage is applied to the primary winding, and the output is taken from the secondary winding. The primary is normally shown on the left, so that the signal flows from left to right, with the output at the right.

The transformer may have either an air or an iron core. The air core transformer is used for coupling signals between stages at higher frequencies. The power transformer is usually of iron core and has two or more secondary windings. These windings are intended for generation of different voltage levels required in some electronic circuits.

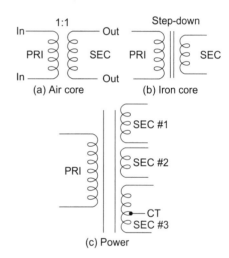

Fig. 2.100   Transformer (a) air core (b) iron core (c) power transformer

2.22 Relays

A relay is an electromechanical device. It depends upon the energizing of a magnetic coil (solenoid) in one circuit to control the opening or closing of contacts in a completely different circuit. The relay solenoid and the contacts of that relay may appear in different locations on the diagram.

The symbols for a relay resemble its actual construction and circuit operation. The relay in Figure 2.101a contains two normally open contacts and one normally closed contact. The contacts of relay

Figure 2.101b are normally closed. In a schematic presentation, a relay is usually shown de-energized. When the relay energizes, the normally open contacts close, and the normally closed contact opens. Another symbol for the relay (solenoid) is like the symbol for an inductor. The switch symbol Figure 2.101c is the most commonly used symbol for a relay.

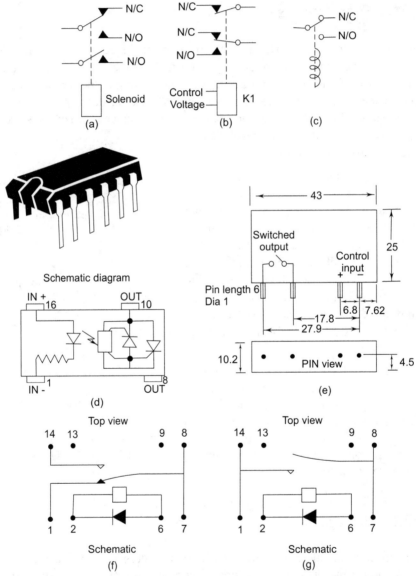

Fig. 2.101 Symbol and types of relays (a) relay with two normally open contacts and one normally closed contact (b) relay with normally closed contacts (c) another symbol for relay (d) solid state relay with optical isolation of inputs/output terminals, dual-in-line package (e) single-in-line package (f) reed relay with normal connection between 1–8 changes over to 14–8 when operated (g) reed relay with normally open between 14–8, closes after relay operation.

Solid State Relays

With a sealed construction and no moving parts, the solid state relays are particularly suited to ac switching applications requiring long life and high reliability. The switching is silent, causes no arcing and is unaffected by vibration and corrosive atmospheres. The control input is optically isolated (Figure 2.101d) from the zero voltage switching circuit which produces virtually no RF interference. They are operated by a TTL open collector. The output circuit is 'normally open'. These relays are available as SIL (single-in-line: Figure 2.101e) or DIL (dual-in-line) PCB mounting type packages.

Reed Relays

Encapsulated reed relays which incorporate a coil (Figure 2.101 f and g) to operate the contracts, when energized, are available in DIL (dual-in-line) package. These relays internally incorporate a diode across the coil to protect the driver circuitry from back EMF. The relays normally have open contacts with ac mains (230 V) switching capacity. They are normally used with mains Triac Trigger devices.

 ## 2.23 Connectors

The connector on a printed circuit board provides a route for all input and output signals, voltages and grounds. The connectors on the board are usually of male contact type which can be 'plugged-into' a female receptacle. The male or plug connector symbol is shown in Figure 2.102a. The female receptacle symbol is shown in Figure 2.102b. The plug or male contacts always constitute the removable part of a connector assembly and the receptacle is the stationary or fixed part (Figure 2.102c). The male or pin contacts are not wired to voltage or power sources to protect the personnel from touching the pins and a possible shock hazard, when the plug is separated from the female part.

The symbol for a male contact looks like the head of an arrow, and for a female contact, like the tail of an arrow. The type of contact can be shown graphically as either a plug or a jack. The plug must be shown at the top of the diagram with a dotted line joining each contact.

The connector contacts are numbered on the diagram for convenience. They show the location of circuits, not necessarily in alphabetical or numerical order. Various types of connectors are shown in Figure 2.102(d to j)

 ## 2.24 Useful Standards

IPC-M-109: Component Handling Manual: Includes the latest editions of IPC standards and guidelines related to the classification and use of moisture-sensitive components, including how to package, handle, store and test them, so that all components will be compatible with the assembly process.

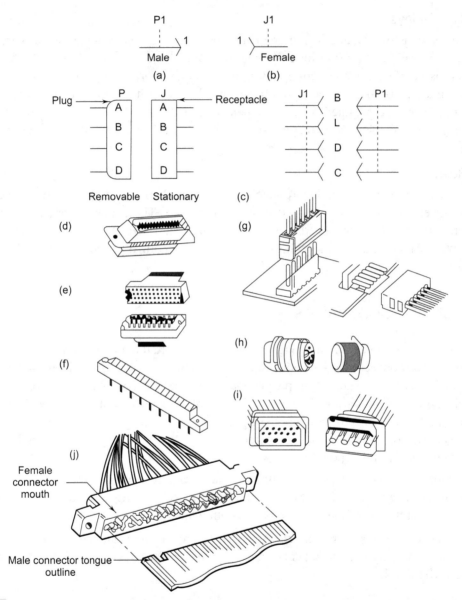

Fig. 2.102 Connectors (a) connector with male contacts (b) female socket of the connector (c) plug and receptacle part of the connector (d) chassis plug-in-unit (e) chassis connector (f) printed circuit connector (g) crimp housing and PCB headers (h) cabling and harness (i) rack and panel (j) female connector for PCB

IPC/JEDEC J-STD-020B: Moisture/Reflow Sensitivity Classification for Non-hermetic Solid State Surface Mount Devices: Identifies the classification level of non-hermetic solid state surface mount devices that are sensitive to moisture-induced stress; also covers components to be processed at higher temperatures for lead free assembly.

IPC/JEDEC J-STD-033A: Handling, Packaging, Shipping and Use of Moisture/Re-flow Sensitive Surface Mount Devices: Provides surface mount device manufacturers and users with standardized methods for handling, packing, shipping and use of moisture/re-flow sensitive SMDS.

IPC-DRM-18F: Component Identification Desk Reference Manual: Contains colour 3-D graphics and full descriptions of more than 40 of the most commonly used through-hole and surface mount components, connectors and other hardware used in electronic assembly today; also includes a terminology section with quick facts on polarity, orientation, packaging, lead styles, class ID letters and component reference designators (CRDs).

IPC-TA-723: Technology Assessment Handbook on Surface Mounting: Contains 71 articles on all aspects of surface mount technology and includes topics such as general SMT overflow, packaging concepts, design considerations, packaging components' process material, process considerations, quality assessments and reliability.

Layout Planning and Design

 ## 3.1 Reading Drawings and Diagrams

3.1.1 Block Diagram

All electronic equipments can be considered as systems comprising a set of interacting elements responding to inputs to produce outputs. It is quite possible that a system may be too complex to be analysed in detail. It is therefore, necessary to divide it into sub-systems and then integrate them. Each sub-system would then represent a functional block, and the combination of all the blocks would constitute the functional 'Block Diagram' of the equipment. A block is only a 'black box' with certain inputs and outputs, but performing a definite function. The lines interconnecting these blocks indicate the signal flow from block to block or circuit to circuit. Understanding of the circuit function becomes easy with a block diagram. Figure 3.1 shows a typical block diagram representation of a simple recorder.

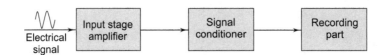

Fig. 3.1 Concept of a block diagram. It shows various sub-systems in an equipment

The integrated circuits such as microprocessors, counters, etc. are represented as individual blocks. These blocks are labelled with pin numbers, signals and associated interconnecting wires.

3.1.2 Schematic Diagram

A schematic diagram is a graphical representation of interconnections of various electronic, electrical and electromechanical components of an equipment. The schematic is the first step in an electronic circuit design because it displays and identifies the components that make up the equipment. Further, the first step in designing a printed circuit is to convert the schematic diagram in to an art master. Therefore, for any printed circuit designer, it is important to learn to read and interpret the schematic diagram. However, the schematic diagram does not show any of the mechanical details of the printed circuit board.

The schematic provides the most broadly used view of the design and includes all components. In addition (Mentor Graphics, 2002):

- It gives visibility into the status of all parts of the design process;
- Schematics are the primary source for developing deliverables to product design and manufacturing groups;
- Design variants are built around slightly differing schematics;
- Test departments rely on schematics;
- Field service relies on schematics; and
- Bills-of-materials are generated from schematics.

In short, a schematic is the focal point for a product's electronic data and can be viewed as a set of crucial business documents that capture the decisions affecting all aspects of the product.

Usually, every assembly in an equipment is assigned an assembly number which appears on the schematic diagram. The schematic diagram shows various components by means of symbols which are so arranged that they show the working of the circuit in a clear way. The component symbols are usually governed by various standards, which do vary widely. Therefore, it is advisable to first find out which standard has been followed before attempting to read a schematic diagram. The schematic diagram is also called the 'circuit diagram'.

In a schematic diagram, the symbol represents either what the component does in the circuit or how it is physically constructed. For example, a capacitor can be charged to store electricity similar to a battery. A picture of two parallel plates of equal length and separated by a given space has thus been adopted as the symbol of capacitor. In most cases, electronic symbols have been evolved logically from their circuit application, their construction, or from a combination of both.

All electronic components have been designated when represented on a schematic diagram. The common classification from ANSI (American National Standards Institute), IEEE (Institute of Electrical and Electronic Engineers) and IEC (International Electrotechnical Commission) is given in Table 3.1.

Table 3.1   Reference Designators (Reference: Component Identification, IPC-DRM-18 F, Desk Reference Manual

Component	ANSI/IEEE	IE C	Component	ANSI/IEEE	IEC
Amplifier	AR	A	Jumper	W,P or R	
Capacitor	C		Microprocessor	U	
Pack	C				G
network	C				
Polarized	C		Oscillator	Y (crystal or G (other)	
Variable	C				
			Relay	K	
Connector	J or P		Resistor	R	
Crystal	Y	B	Pack	R	
			network	R	
Delay Line	DL		Potentiometer	R	
			Variable	R	
Diode	D or CR	V			
Light Emitting Diode	DS (display)	E	Thermistor	RT	
Voltage Rectifier	D or CR	V	Varistor		
Zener Diode	D or VR	V	Asymmetrical	D or CR	
			Symmetrical	RV	
Filter	FL	Z	Socket	X, XAR, XU, XQ, etc.	
Fuse	F				
Header	J or P		Switch	S	
Inductor, Choke	L		Test Point or Pin	TP	V
Integrated Circuit	U, IC		Transistor	Q	
Insulated Jumper	W or P				
			Transformer	T	
Battery	BT		Voltage Regulator	VR	
Meter, Instrument	M		Antenna, Spark Gap, Shield	E	
Plug, Connector Male	P				
			Attenuator	AT	
Power Supply	PS		Motor, Fan, Synchro	B	
Test Point	TP				

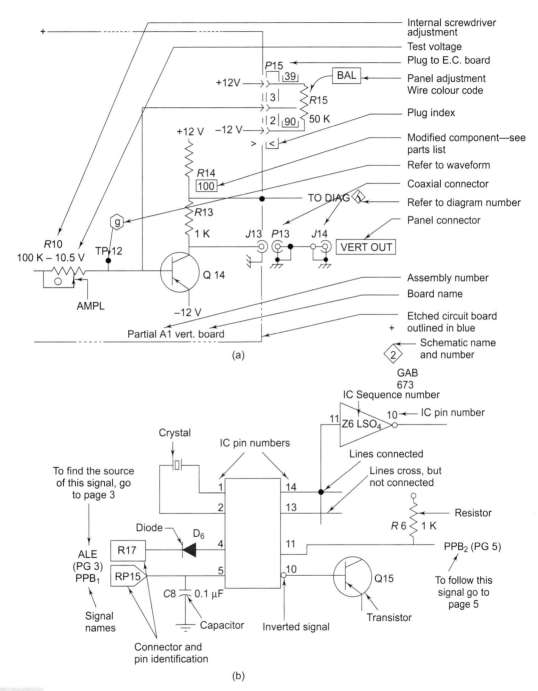

Fig. 3.2   Schematic circuit with symbols as per the American National Standard Institute (a) Circuit with discrete components (b) Circuit with integrated circuit

Figure 3.2 shows a typical schematic diagram and demonstrates its basic parts. Each component is represented as a symbol in the diagram along with its reference designation. Electronic components shown on the circuit diagram are generally in the following units unless mentioned otherwise:

Capacitors = Values one or greater are in the picofarads (pF)
         = Values less than one are in the microfarads (µF)
Resistors = Ohms (Ω)

Guidelines have been developed over the years for drawing schematic diagrams. The main features of these guidelines are:

- Signal flow moves from left to right across the page with inputs on the left and outputs on the right.
- Electronic potentials (voltages) should increase as you move from the bottom to the top of a page. For example, in the figure, +12V supply is shown upwards while the –12V is indicated downwards.
- Use the 'unit number'convention for assigning a unique IC package identification. For example, U1 with its internal gates identified by letter suffixes; U1A, U1B, etc. Only one of the common gates need show the power connections. Power connections are often omitted, but it is better to include as a reminder as well as to make your schematic complete.

 ## 3.2 General PCB Design Considerations

The basic function of a printed circuit is to provide support for circuit components and to interconnect the components electrically. In order to achieve these objectives, various printed wiring types have been developed. They vary in base material (laminate), conductor type, number of conductor planes, rigidity, etc. It is therefore expected that the printed circuit designers are adequately familiar with the variations and their effect on cost, component placement, wiring density, delivery cycles and functional performance. No finished product is ever better than its original design or the material from which it is made. The manufacturing process, at best, can reproduce the design. The same is true with printed circuit boards. The need for formalizing design and layout methods and procedures thus assumes critical importance.

Design and layout broadly includes the perspective of total system hardware, which includes not only the printed circuit but each and every component in its final form. Design and layout considerations must also address the relations between and interactions of the components and assemblies throughout the system.

Board design is an extremely important aspect of printed circuit board technology. Quite often, designers underestimate the time and effort required to do a good job. This can cause delay in production start-up and much hidden cost during the life of the product (Ross and Leonida, 1996d).

The technical requirements that are likely to affect the design of an electrical equipment are mechanical, electrical, functional and environmental.

Mechanical design requirements include size, shape and weight; location of components and their mounting, dimensional tolerances, shielding and equipment marking.

Electrical design requirements have such parameters as circuit function and wiring distribution, component selection with respect to electrical ratings, size and tolerance, internal and external interconnections.

Functional design parameters include reliability, maintainability, accessibility, and human engineering (displays, controls).

Environmental design takes into account factors such as mechanical shock and vibration, temperature extremes, salt spray and fungus proofing and operations in space or underwater.

All the above factors are not necessarily inherent in printed circuit boards, but by careful design, proper selection of materials and manufacturing techniques, it is possible to optimize most of the above parameters. The following factors should be taken into consideration while designing printed circuit boards including multi-layer boards. Many factors influence the design of PCBs.

3.2.1 Important Design Elements

The design inputs which should be provided by the equipment designer to the PCB designer are called design elements. They are:

- Type of circuit (analog or digital, etc.);
- Board size
- Number of layers
- Pad stack sizes
- Hole sizes
- Layer thickness
- Board thickness
- External connections
- Mounting holes
- Supply and ground layer thickness and
- Component details with specifications.

3.2.2 Important Performance Parameters

These are:

- Tensile strength;
- Flexural strength;
- Shock and vibration;

- Thermal shock and temperature cycling;
- Moisture resistance;
- Fungus resistance;
- Salt spray;
- Warp or twist;
- Dielectric breakdown voltage;
- Solderability and re-solderability;
- Insulation resistance (surface and bulk);
- Flame resistance;
- Conductor temperature rise;
- Machinability; and
- High attitude considerations.

 ## 3.3 Mechanical Design Considerations

3.3.1 Types of Boards

The layout design considerations in most commonly used board types are detailed below.

- *Single-sided Boards:* These are mostly used in applications where costs have to be kept at a minimum. When designing layout, to jump over conductor tracks, either components or jumper wires are used. If their number is too large, double-sided PCBs should be considered.
- *Double-sided Boards:* Double-sided boards can be made with or without PTH. Since PTH boards are expensive, they are chosen where the circuit complexity and density necessitate their use.

In the layout design, the conductors on the component side must be kept minimum in number to ensure easy sourceability.

In PTH boards, via holes should be utilized only for through contacts and not for component mounting. The number of via holes should be kept minimum for reasons of economy and reliability.

In order to take a decision on the number of sides, single-sided or double-sided, it is important to take into consideration the component surface area (C), which is a fairly constant percentage of the total PCB area (S), useful for mounting components. It may be noted that 'S' is normally computed on one side of the board. Table 3.2 shows the usual range of the ratio S:C for the most common types of PCB.

Table 3.2 PCB-to-component Area Ratio (Leonida, 1989)

Board Type	Single-sided	Double-sided PTH
Discrete components (ICs no more than 5% of the area)	2–3	1.5–2
Mixed (ICs from 35 to 50%)	2.5–4	2–3
IC board (discrete components no more than 20%)	4–6	2–3

The values shown in the table are to be taken as guidelines only because they depend upon the standards selected for conductor width, land diameter, minimum clearance, etc.

In general, the decision about the number of sides must be fully cost-effective. As a thumb rule, the double-sided PCB with plated through-hole costs 5 to 10 times more than the single-sided PCB. Also, the cost for component assembling is another important consideration. The approximate cost of assembling (manual) the components on a PCB is 25–50 per cent of the cost of a single-sided PCB and 15–30 per cent of the cost of a double-sided PCB with PTH.

The PCB provides mechanical support and connectivity to the components mounted on it. It is therefore necessary for the PCB designer to know the overall physical size of the board (outline dimensions), position of mounting holes, height restrictions and related details. The following are the main mechanical design considerations for the PCB:

- Optimal board size compatible with the PCB manufacturing process;
- Position of board mounting holes, brackets, clamps, clips, shielding boxes and heat sinks;
- Proper fixation arrangement for heavy components;
- Proper hole diameter for component mounting;
- Assembled board to withstand the mechanical stress and vibrations occurring in transportation;
- Type of installation of the board (vertical/horizontal);
- Method of cooling; and
- Specific locational requirements of components like front panel operated components such as push buttons, variable resistors, etc.

3.3.1.1 Determining the Component Area

The component area on the board is calculated by adding the contribution of each single component. Each component is considered in its orthographic projection on the board. The dimensions of the component are obtained from the manufacturer's catalogue or by actually measuring the same.

The components are considered as simple geometrical figures, for example, an integrated circuit in a DIP package is a rectangle and so is an axial component. However, for an axial component mounted horizontally, the width of the rectangle will be its diameter whereas the length will be its body plus a portion of the leads. Similarly, a TO-18 packaged transistor will be represented as a circle.

The parts to be mounted on the PCB should be detailed on the parts list. Each part should be identified by a unique reference designator and a part description. For example, a resistor might be shown as reference designator R1 with a description ¼ watt carbon film resistor. Any additional information useful to the assembly process can be included on this list, such as mounting hardware, part spacers, connector shrouds, or any other material not shown in the schematic diagram. Part manufacturers provide data sheets to be used by the circuit designer to select parts for the circuit. For designing the PCB, these sheets should also have the physical dimensions of the part included.

3.3.1.2 Volume Computing

Some equipment may have to fit into an existing enclosure, which can limit the board size or leave the designer with a choice from only a small number of preferred board sizes. In many cases, the PCB size and the number are dictated by the exterior design concept of the equipment. For example, for a paging system receiver, the PCB has to be very small because the receiver is carried in the pocket. Similarly, the PCB of an implantable pacemaker must also be very small as the pacemaker is implanted in the body. Unless there is a mandatory requirement to use a standard enclosure, the enclosure or case should be designed to fit the system to avoid forcing the system into an enclosure size that may dictate the use of a non-standard or non-optimum board size. In such cases, the volume available for an electronic assembly is calculated with care. Rather than the actual volume, it is important to know the maximum volume that the board can occupy in the worst condition, including the safety clearances.

3.3.1.3 Accessibility for Adjustable Components

Adjustable components, usually variable resistors, are common in many printed circuit board assemblies. These components should be mounted on a PCB in a such a manner that there is an easy access to such components.

3.3.1.4 Horizontal or Vertical Mounting of Components

Axial-leaded components can be mounted either horizontally or vertically. Vertically assembled axial components require less surface area, resulting in a smaller PCB. However, this technique has certain disadvantages such as lower reliability, increased difficulty in component forming and manual assembly. In addition, the increased density of conductors can limit the packing density. Vertical mounting should, therefore, be adopted only when the area is limited and there are limits to volume, and of course, in cases when some of the components of the board have a height which is greater than that of the axial components.

3.3.1.5 Board Size

A functional printed circuit board is not a product in itself. It always requires connections to the outside world to get power, exchange information, or display results. There may be a need to fit it into a case or slide it into a rack to perform its function. There may be areas that may require height restrictions on the board. Tooling holes and keep-out areas may be required in the board for assembly or manufacturing purposes. All these factors need to be defined before the board can be designed, including the maximum dimensions of the board and the locations of connectors, displays, mounting brackets or any other external features.

In order to avoid interconnections between different PCB boards through terminations, cables or connectors, it is preferable to accommodate all the circuitry on a single board. This approach, in many cases, can result in disproportionately large-sized boards, requiring more and more space for interconnections, thus leading to more functional disturbances. On the other hand, too many small boards forming one complete circuit can lead to higher cost but the resultant unreliability caused by several external interconnections and connectors. In general, an equipment or system consisting of smaller boards is easier to repair and service because of its modular structure and convenience of isolating signal flow paths.

When designing a layout, it is a good practice to provide for a 5 to 10 per cent area for any modification which may be required on the board. But it is normally done at the prototyping or batch production stage. The layout design is best worked out when the physical design of the final product shape is available to match the board profile and size according to the same requirements.

While working on the board size, locational constraints in respect of the following components are encountered:

- Connectors or connecting tabs;
- Fixtures or anchoring areas; and
- Control or adjusting devices such as switches and potentiometers.

Besides these, mounting holes along with the safety areas around them are fixed. In this way, PCB net area is worked out from the gross area, taking into consideration the following geometrical constraints:

- *Locating holes (for assembly)*: The preferred diameter is 3.175 mm, their position is usually close to the longest edge of the board, with the widest possible span. Their centre should be at least 1.5 times the hole diameter from the edge of the board, but not less than 2 mm. The safety area around them should be of circular form.
- *Mounting holes:* Safety areas need to be provided around the fixing holes, keeping in view the size of the washers and screws, unless insulating washers are used.
- *Edges:* Usually a 2–5 mm safety strip is provided along all edges. This is necessary because trimming of the board could cause de-lamination of copper areas too close to the edge.

It is generally not possible to prescribe a universally applicable standard for the board size, though the ideal board size should be neither very large nor very small. However, in industrial applications, the generally adopted rack size is 19", and obviously the printed circuit board sizes should follow this standard. Standard board sizes allow the interchange of cards and the production of a standard range of blank prototyping boards, frames, fittings and modules for packaging systems. Standard PCB sizes have been suggested by different organizations. For example, the DIN standards (German Standards Organization) shown in Table 3.3 are widely accepted, especially in European countries.

Table 3.3 Standard Sizes of PCBs

Front	Printed Circuit Board			
Height	**Height**	**Length**	**Length**	**Length**
132.5	100	100	160	220
177.0	144.5	100	160	220
221.4	188.9	100	160	220
265.9	233.4	100	160	220

The IBM personal computer and its clones have meant that the IBM card and its half card sizes have also become standards in industry.

The use of a panel size smaller than the largest sub-multiple of the full size sheet is recommended. One common panel size is 460 mm × 610 mm. There is also a size standardization: for example: 1 = 60 mm. 2 = 120 mm, 3 = 180 mm and 4 = 240 mm.

Lengthwise, these are A = 80 mm, B = 170 mm, C = 260 mm and D = 350 mm.

Several factors affect the selection of board size and shape and therefore, the final choice is probably a compromise amongst conflicting factors.

3.3.1.6 Dimensions and Tolerance

All board outlines must be expressed to include some tolerance.The most commonly adopted tolerances are +0.25 and +0.50 mm. The tolerance is important because calculating the actual working board area to its smallest size is the objective.

The tolerance is particularly important in conductors/traces on very dense boards, where a trace runs within 1 mm of the board edge. The tolerance could be significant because in this case, the trace could come with 0.50 mm of the board edge which is a poor design practice. The traces could be damaged when cutting the board outline.

Dimensional tolerances are needed to reduce manufacturing cost. It is a common practice to process a number of smaller boards on a larger panel and then separate them in the last steps of the manufacturing cycle. This practice can also be used in the manufacture of multi-layer boards. If the tolerances are wide, larger panels with more small individual circuits can be the manufactured without any difficulty. This results in a considerable reduction in the manufacturing cost.

The final finished board dimensions affect the board cost. A 25 mm margin must be allowed around the final size of the board to accommodate tooling holes. The final size of the board must be limited so that a good yield is obtained from each standard size copper clad base material sheets which are normally available in 24 × 36 inch sheets in multiple of 1ft. increments. Such a design would result in lesser scrap and consequently a lower cost.

3.3.1.7 Partitioning

Minimizing the total number of interconnections can help in deciding board size. This requires a judicious action as partitioning equipment in different ways can result in very different numbers of interconnections. Generally the block diagram of the circuit is examined to determine the points at which it can be divided so as to break the smallest number of connections which may decide the size of the PCB.

The decision regarding locating components on one or more than one boards is a function of the shape of the available volume, modularity of some parts and maintenance requirements. Although single board solution is usually more economical as one can save in terms of the total required PCB, yet there are some advantages in splitting the board into two or more boards. These are: easier layout design, manufacture testing and maintenance.

Notwithstanding these advantages, the complexities of modern digital circuit are resulting in increasingly larger boards, particularly for professional boards, including multi-layer boards. A few years ago, professional PTH boards had maximum dimensions of about 200×250 mm whereas now boards of the sizes 400×500 mm are quite common. This has become possible partly with the introduction of CNC (computerized numerically controlled) machines (mainly for PCB drilling and component insertion), which provide very high accuracy positioning (25 microns) over an area as large as 610×610 mm.

From the above points, it is clear that board size must inevitably be a compromise, unless the board is to be fitted into a pre-decided enclosure or pre-designed frame. In that case the size is not really important, and only the question of packaging density and the type of connectors to be used needs to be settled.

3.3.2 Board Mounting Techniques

Various techniques are available to mount the printed circuit board to the chassis or to the next assembly. For providing good mechanical stability, the board should be supported within 25 mm of the board edge on at least three sides. As a general practice, boards between 0.031 and 0.062 inch (0.785 and 1.57 mm) thick should be supported at intervals of at least 10 cm. The choice of board mounting technique would depend upon the following factors:

- Board size and shape
- Input/output terminations
- Board removal requirements
- Heat dissipation requirements
- Shielding required
- Type of mounting hardware
- Available equipment space and
- Type of circuit and its relation with other circuits.

3.3.3 Board Guiding and Retaining

The most convenient method for use with plug-in printed circuit assemblies is that of card guides. They provide a quick connect/disconnect capability with the convenience of testing the board out of the units by means of extender cards. The type of card guide depends upon the shape of the board and the degree of accuracy needed to ensure proper mating alignment. Some of the commonly used card guides are shown in Figure 3.3. There must be sufficient area to allow room for the card guide along the edge of the board.

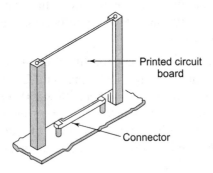

(a) Grooved posts and angle clamps
   The posts serve as guides as the board is lowered to the connector, then the clamps provide positive retention.

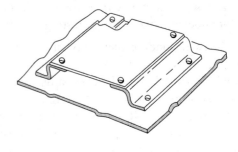

(c) Z angle brackets
   Z angle strips can be cut to receive any size board.

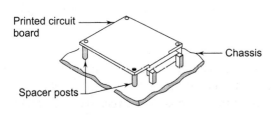

(b) Mount on tapped spacers
   Any size board can be accommodated by this simple mounting. It can be a space-saver on walls and doors of enclosures.

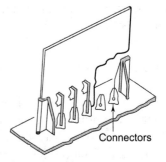

(d) Sheet-metal support guides
   Although not positively retained in place, boards are quickly removed and replaced.

Fig. 3.3 Methods of mounting boards using card guides (a) grooved posts and angle clamps, (b) mount on tapped spacers, (c) z-angle brackets (d) sheet metal support guides (Lindsey, 1985)

If the electrical interface does not require a connector or a card guide is not practical, then mounting holes may be provided on the board so that it can be installed with screws, stand-offs or other mechanical fasteners. However, sufficient clearance should be provided so that components or conductors will not interfere or short-out to the mounting hardware. Adequate clearance

(1.5 to 2.5 mm) must be provided for all leads, wires, hardware, etc., to ensure that protruding leads through the non-component side of the printed circuit board do not short-out to the adjacent board. Additional clearances are required in case of exceptionally large or thick boards to ensure that flexing of the board under vibration and shock would not cause the leads to short-out.

Stiffners are used on PC boards to ensure that the board remains flat under mechanical stresses. Mounting the boards on a frame or the use of stiffeners can obviate this problem. These devices also avoid the warpage that might develop in the board from wave or flow soldering. Figure 3.4 shows the types of commonly used stiffeners.

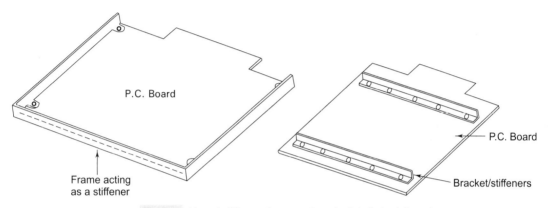

Fig. 3.4   Use of stiffeners for mounting of printed circuit boards

3.3.4   Input/Output Terminations

The most common method of providing electrical interface between the PCB and the associated equipment is by the use of connectors, terminals and cables. The type of interface to be used for any particular board is generally decided during the mechanical design of the equipment. The most commonly used interface comprises the *connectors* which are available in numerous types. Single part printed circuit connectors are the most common type in which one-half uses one edge of the printed circuit board as the plug which has printed and plated conductors as the connector contacts. The other half of the single part printed circuit board connector is usually an assembly of mating contacts in plastic moulded receptacle.

Double part contactors are made of self-contained multiple contact male (plug) and female (receptacle) assemblies. Usually the male part is mounted on the printed circuit board and the female part to an interconnecting board such as a motherboard or back plate.

Single or double part connectors usually have a limited number of contacts and can be mounted generally at the edges in particular directions. These limitations can be overcome by the use of discrete connectors.

3.3.5 Board Extraction

A number of techniques have been used to achieve printed board extraction, resulting in the development of many types of extracting tools. The basic requirement of such tools is that they must use a minimum of board space, thereby maximizing the amount of board available for components and also should not cause any damage to the board, components and connectors. The most commonly used types of extractors are hook type, finger hole and pull tab.

3.3.6 Testing and Servicing

Depending upon the product complexity, the quality of incoming components and the manufacturing process, there is a likelihood that a certain number of finally assembled boards may not work. Therefore, the board design must take into consideration the level at which the test must be performed and to make such testing as simple as possible.

Smaller boards are preferred to achieve efficient testing and repair of PCBs. In large boards, the isolation of defective parts becomes difficult because it is usually not possible to interrupt or influence the signal flow. If the complete circuit is realized on different smaller cards, it is easier to isolate the defective card, repair it or exchange the same with a working card. Another advantage in having an electronic circuit split into sub-units on different boards is the possibility to use the common subunits like power supply for other equipment, if it is so designed.

3.3.7 Mechanical Stress

Comparatively small PCBs with a size of less than 100×150 mm and the standard thickness of 1.6 mm will hardly pose any problem in mechanical strength if assembled with the usual electronic components. Care is needed with bigger board sizes or thinner laminates or if heavy parts like transformers have to be mounted on the board. As a general guideline, heavy parts should be mounted near a supporting device like a card guide, connector or stiffener.

3.3.8 Board Thickness

There is no standard rule for the optimum thickness neither for the printed wiring nor for the number of multilayer conductive layers. Occasionally, the limiting factor for printed wiring thickness is the diameter of the smallest hole, especially when the holes are plated though.

The final board thickness will depend upon the number of conductor layers and on the electrical layer-to-layer spacing requirements of the design. In multi-layer boards, the increase in cost is not directly proportional to the increase in the number of conductor layers. For example, doubling the number of layers from four to eight will probably increase cost by only 30 per cent. However, if the number of conductor layers exceeds 10, the extra layer costs increase at a rapid rate.

Printed board thicknesses can vary from 0.50 to 6.5 mm, but most rigid boards have thicknesses of 1.5 mm.

3.3.9 Important Specifications and Standards

Design principles and recommendations must be used in conjunction with the detailed requirements of a specific interconnecting standard. The standard "*ANSI/IPC 2221 Generic design*" identifies the generic physical design principles as well as the material selection. Besides this, the performance requirements of the finished rigid printed boards are specified in *IPC-6012*.

An understanding of appropriate design principles that address pertinent manufacturability, testing and quality issues must be applied during the design process. The modern PCB designer has to therefore know/understand the manufacturing and assembly process of printed circuit boards.

 ## 3.4 Electrical Design Considerations

3.4.1 Conductor Dimensions

In general, conductor width is determined by:
- Component packing density;
- Minimum spacing between conductors and components; and
- Geometrical constraints due to component outlines.

In former times, the current carrying capacity of PCB conductors was often disregarded because conductor dimensions were usually much larger than needed for carrying the currents involved. However, with higher packaging density and thermal considerations, the conductor width has to be determined or at least checked according to the required current carrying capacity.

In most electronic circuits, comparatively small currents are normally flowing for which the conductor resistance can practically be neglected. However, when we deal with supply and ground lines, especially in case of high speed signals and in some cases, digital circuitry, much broader conductors than ohmically necessary have to be provided between the supply and ground lines. Therefore, it is necessary to understand the factors which govern the choice of appropriate conductor width which determines its current carrying capacity. Obviously, ohmic resistance of the conductor may be problem when conductive paths are unusually long or when voltage regulation is critical.

3.4.2 Resistance

The copper printed tracks on a PCB have a finite resistance which introduces a voltage drop proportional to the current flowing in that particular conductor. The resistance of a conductor considered as a metal section having a rectangular cross-section depends upon the specific resistivity of copper, which is 1.724×10^{-6} ohm cm at 20°C. It will be useful to know the resistance of a 1mm wide copper conductor per cm of length. A standard copper foil of 35 μm thickness (without any plating) may be assumed.

$$R = \rho L/A \text{ ohms}$$
$$\rho = \text{resistivity (ohms cm} \times 10^{-6})$$
$$L = \text{conductor length (cm)}$$
$$A = \text{area of cross-section of the conductor (cm}^2)$$
$$\rho \text{ (copper)} = 1.724 \times 10^{-6} \text{ (at 20° C)}$$
$$L = 1 \text{ cm}$$
$$A = 35 \times 10^{-4} \times 0.1 \text{ cm}^2$$
$$R = 1.724 \times 10^{-6} \times 1$$
$$= 35 \times 10^{-4} \times 0.1$$
$$= 0.0049 \text{ ohms}$$
$$\approx 0.005 \text{ ohms} \quad \approx 5 \text{ m ohms}$$

This shows that a 1 mm wide conductor of 1 cm length for standard copper foil of 35 μm thickness is 5 m ohms. Similarly, a 0.5 mm wide conductor of 10 cm length will have a resistance of $5 \times 10 \times 2 = 100$ m ohms, since half the conductor width gives double as high resistance. It may be noted that the resistivity factor of the material is valid for 20° C and it will rise with temperature.

Table 3.4 shows the specific resistivity of some materials often used in electronics.

Table 3.4 Specific Resistivity of Some Common Materials

Material	$p/\Omega mm^2 cm^{-1}$	Material	$p/\Omega mm^2 cm^{-1}$
Silver	1.6	Nickel	43
Copper	1.8	Glass	5×10^{19}
Aluminium	3.0	Quartz	10^{23}

The specific resistivity p of copper can be assumed to be:

$$p = 1.78 \times 10^{-6} \text{ ohm-cm for a temperature at 25 °C}$$

The resistance (at 25 °C) of a copper conductor with 0.3 mm conductor width and 35 μm copper thickness and 500 mm length will be:

$$R = 17.8 \times 10^{-3} \times \text{Length/A with} \qquad R \text{ in Ohms}$$
$$\text{Length and width in mm}$$
$$\text{Thickness in microns}$$

$$R = 17.8 \times 500/0.3 \times 35 \times 1000$$
$$= 17.8/0.3 \times 35 \times 2$$
$$= 0.85\,\Omega$$

It is known that when current flows through a conductor, its temperature increases due to the joule effect. The copper resistance increases with temperature. Thus, all values of resistance at 20 °C have to be revised in relation to the possible increase in temperature within the conductors due to the joule effect or to any heating caused by components. The temperature difference within electronic equipment and operating temperature may be as high as 60 °C or more. This would significantly increase the conductor resistance and must be accounted for in the design. It is known that

$$R_t = R_o\,[1 + \alpha\,(T_1 - T_o)]$$
R_t = resistance at temperature T_1
R_o = resistance at temperature T_o
α = temperature coefficient of conductivity

To illustrate this, let us assume that inside temperature in an electric equipment is 80 °C while the outside temperature is 20 °C. Taking temperature coefficient for copper conductivity as +0.0039, the resistance of 0.5 mm conductor, 10 cm long (with 100 ohm resistance), would be

$$R_t = 100\,[1 + 0.0039\,(80 - 20)]$$
$$= 100\,[1 + 0.0039 \times 60] = 123.4 \text{ m ohm}$$

This shows that the resistance has increased by 23.4 per cent.

All conductors which carry current will be at a temperature higher than temperature rise which depends upon the current carried, conductor width and copper thickness.

Taking the thickness of the copper conductor, temperature of the operation and the current carrying requirement, the figure offers some guidance for choosing the appropriate conductor width. It can be used to find the current carrying capacity of a given conductor or to determine which conductor width is needed for carrying a given current. In both cases, the maximum temperature rise must be established. The values usually taken are 40 °C (72 °F) for forced air cooled equipment and 10 or 15 °C (18 or 27 °F) for equipment in which no forced air circulation is used (Figure 3.5).

To illustrate, if a minimum conductor width is to be determined, to carry an 8 Amp current on a board with 35 μm copper-clad at 30° rise above ambient (room) temperature, proceed as follows:

i) Find 8 Amp in the left top scale.
ii) Follow the 8 Amp line to intersect the 30 °C graph line.
iii) Follow a vertical line from that point down to intersect the 35 μm copper graph line.
iv) Draw a horizontal line to the left from that intersection to find the minimum conductor width which, in this case, is 2.6 mm.

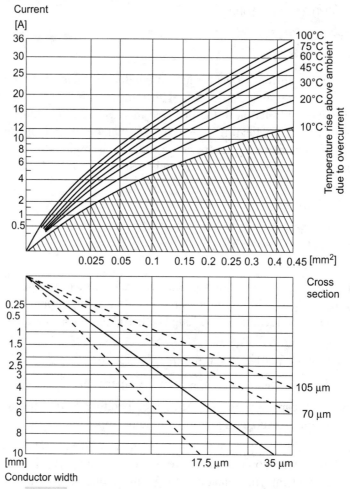

Fig. 3.5   Conductor widths for safe operating temperatures

In all critical cases, the cross-sectional area of the track should be calculated by using the worst case values. However, the manufacturing tolerances for the PCB conductor width can be as high as 30 per cent if the width is less than 0.5 mm, 20 per cent if the width is between 0.5 and 1.0 mm and 10 per cent if the width is more than 1.0 mm.

For determining conductor resistance, a nomograph shown in Figure 3.6 can be used. By aligning the width of the conductor (left line) and the copper foil thickeners (centre line), the resistance of a unit length of the resistance can be determined on the right line. Conversely, if the resistance of the conductor per unit length is known, the nomograph can be used to obtain the width of the conductor for a given foil thickness. Similarly, the required copper foil thickness can be known for a given width of the conductor.

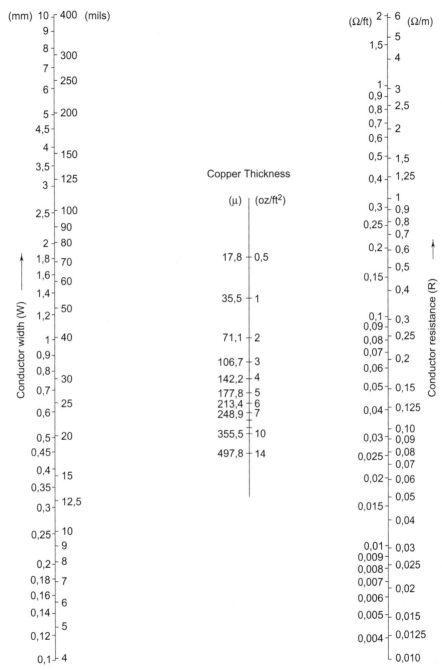

Fig.3.6 Nomograph for determining the resistance of a printed board conductor as a function of conductor width and copper thickness. Scales are logarithmic (redrawn after Leonida,1989).

The maximum recommended current carrying capacity of traces is given in Table 3.5.

Table 3.5 Recommended Current Carrying Capacity of Traces

Conductor width (in)	Current (Amps)			
	½ oz	1 oz	2 oz	3oz
0.005	0.13	0.50	0.70	1.00
0.010	0.50	0.80	1.40	1.90
0.020	0.70	1.40	2.20	3.00
0.030	1.00	1.90	3.00	4.00
0.050	1.50	2.50	4.00	5.50
0.070	2.00	3.50	5.00	7.00
0.100	2.50	4.00	7.00	9.00
0.150	3.50	5.50	9.00	13.00
0.200	4.00	6.00	11.00	14.00

For conversion from inches to millimeters: 1 inch = 25.3994 mm.

3.4.3 Capacitance Considerations

Capacitance is a parameter of considerable importance, particularly in the design of PCBs at high frequency. The capacitance comes into play in the following two situations:

- Capacitance between conductors on opposite sides of the PCB; and
- Capacitance between adjacent conductors.

Capacitance between Conductors on Opposite Sides of the PCB: Two PCB conductors lying one above another (Figure 3.7) and separated by a dielectric (laminate) form a capacitor whose approximate capacitance can be calculated from the basic capacitor formula:

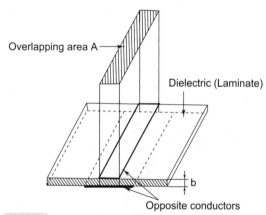

Fig. 3.7   Capacitance between two conductors separated by a dielectric (conductors on opposite sides of the PCB)

$$C = 0.886 \times \varepsilon \times A/b \text{ (pF)}$$

A = total overlapping area (cm^2)

b = thickness of dielectric (mm)

ε = relative dielectric constant, whose value is normally available from the manufacturer's of the PCB laminates.

It may be noted that this formula gives only an approximate value. However, when the conductor width is at least 10 times greater than the dielectric separation, the equation is generally in close agreement with empirical values. The capacitance coupling between conductors can be minimized by limiting the length of conductors running in the same vertical plane.

In order to reduce the magnitude of spikes on supply and the ground lines of fast switching circuits, it is advisable to run these lines exactly on opposite sides of the PCB. The supply and ground lines form a distributed de-coupling capacitor across the PCB. The TTL circuit realized on the PCB will give narrow current spikes on the supply line which will be drawn from the de-coupling capacitor without disturbing the supply voltage.

Capacitance between Adjacent Conductors: It is a function of conductor width, thickness and spacing as well as the dielectric constant of the board material. For all practical purposes, the value of coupling capacitance (pF/cm) for a G-10 laminate with a dielectric constant of 5.4 and conductor thickness of 35 μm is given in Figure 3.8.

To illustrate a practical case, let us assume two conductors of 1mm width which are running parallel with 1mm spacing for a length of 100 mm. From the figure, it is found that the coupling capacitance comes out to be 0.4pF/cm. This gives a total effective capacitance as $0.4 \times 10 = 4$ pF.

The capacitance coupling between adjacent tracks is normally undesirable. Therefore, ways and means need to be found out to minimize the same. The following precautions help to reduce the coupling by a factor of 3 to 10:

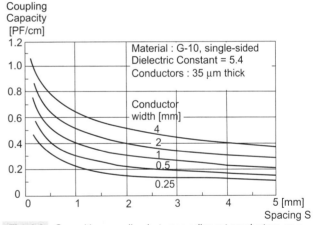

Fig. 3.8   Capacitive coupling between adjacent conductors as a function of spacing

- Keep critical conductors narrow and provide sufficient spacing between them.
- Run a ground line between the critical conductors, if possible. The broader the ground line, the better would be the result.
- Where such a ground line has been provided, the two signal conductors should run as close to the ground line as possible. This would keep the capacitance coupling to ground high, while coupling between the signal lines at the same time becomes less.

For critical high frequency circuitry, the electrical characteristics of single-sided circuits made of epoxy glass or epoxy paper are not adequate and associated ground plane micro-strip line construction is a must. The method of achieving transmission line capability with double-sided printed wiring is called 'microstrip'.

3.4.4   Inductance of PCB Conductors

In designing the conductor patterns for fast signal or high speed logic circuits, the inductive couplings are also of major concern. In logic circuits operating at a clock rate of only 10 kHz, high frequency components of the rectangular shaped signals can often cause problems. Therefore, in such situations, it is important to know the inductance of a conductor arrangement.

The calculation of the inductance of conductors is a rather complex procedure. However, for a given type of copper clad board having 35μm copper conductor thickness, the inductance of

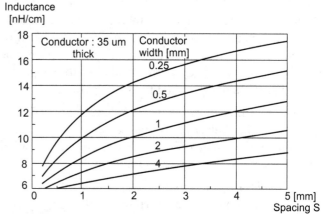

Fig. 3.9   Inductance of parallel running conductors (redrawn after Bosshart, 1983)

parallel running conductors for different conductor widths can be calculated using Figure 3.9.

3.4.5   High Electrical Stresses

The increasing density of interconnection in printed circuit boards demands that the designer progressively decrease spacing and the size of conductive parts such as line sections, PTH diameters, pad areas, etc. Therefore, the increasing level of integration is naturally accompanied by an increase in the electrical stresses in the PCB. Obviously, the high electrical stresses (a few kV/mm) can start the degradation mechanisms depending upon the electrical stress level, environmental conditions and the presence of thermal and mechanical stresses (Travi, et al., 1996). Therefore, in order to keep the electrical stresses at a level appropriate to achieve the desired level in insulation reliability, the designers have to provide for insulation adequate distances which may be higher than what is economically desirable.

 ## 3.5 Conductor Patterns

The manufacturability and reliability of a PC board depends, to a large extent, upon the basic design of the PC board in terms of conductor width, thickness, spacing, shapes and routing, etc. The design can be done manually or with a computer, but the basic rules in both the cases are fairly constant. The three basic rules for layout design are:

- No interference between the components;
- Conductors not to cross each other; and
- Sufficient spacing between any two close conductors.

As a general rule, in each hole, only one lead will be present and each lead has to pass through a hole. It cannot be soldered to another lead, regardless of how well this joint can be made. Axial components will generally have both leads parallel to the body axis, unless specified otherwise.

 ## 3.6 Component Placement Rules

Following are the rules for component placement:

- In a highly sensitive circuit, the critical components are placed first and in such a manner as to require minimum length for the critical conductors.
- In a less critical circuit, the components are arranged exactly in the order of signal flow. This will result in a minimum overall conductor length.
- In a circuit where a few components have considerably more connecting points than the others, these key components have to be placed first and the remaining ones are grouped around them.
- The general rule is to place first components, whose position is fixed for the final fitting and interconnections, e.g. connectors, heat sinks, etc. Then place the components which are connected to these fixed components.
- Components should be placed on the grid of 2.5 mm.
- Among the components, larger components are placed first and the space in between is filled with smaller ones.
- All the components should be placed in such a manner that disordering of other components is not necessary if they have to be replaced.
- Components should be placed in a row or a column, so that it gives a good overview.

3.6.1 Conductor Width and Thickness

The conductor width is determined by (Ross and Leonida, 1997):

- The component packing density;
- The minimum spacing between conductors and/or components; and
- Geometrical constraints due to component outlines or fan-out.

The width of a conductor is basically a function of the current carried and the maximum allowable heat rise due to resistance. Narrow conductors result in high resistance. Wide conductors are, therefore, desirable for low impedance signals where series resistance and inductance are to be minimized and ray capacitance is unimportant. On the other hand, in case of high impedance signal lines, where stray capacitance is required to be limited to a low value, narrow conductors are used. Normally, the conductor width should be as generous as possible to take care of any variation in the etching process as well as any scratches in the artwork caused unintentionally. Figure 3.10 shows the current carrying capacity of etched copper conductors for rigid boards. Generally, for 35 μm and 70 μm thickness conductors, a nominal 10 per cent de-rating is allowed to provide for normal variations in etching methods.

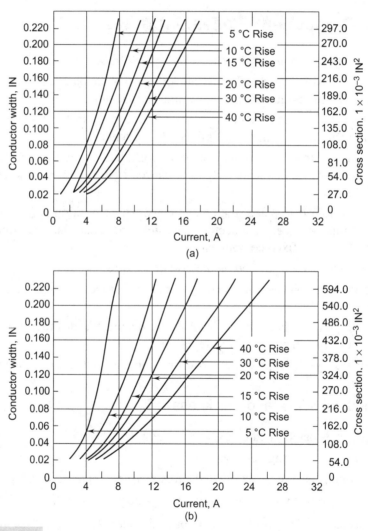

Fig. 3.10 Temperature rise versus current for (a) 1 oz copper (b) 2 oz copper

The chemical and photographic processes used to produce a PCB put requirements on the minimum width of trace (conductor) and the minimum spacing between traces. If a trace is made smaller than this minimum width, there is a chance that it will open (no connection) when manufactured. If two traces are closer together than the minimum spacing, there is some chance that they will short when manufactured. These parameters are usually specified as X/Y rules, where X is the minimum trace width and Y is the minimum trace spacing. For example, 8/10 rules would indicate 8 mil minimum trace width and 10 mil minimum trace spacing. Typical modern process rules are 8/8 rules with values as small as 2/2 rules being available.

3.6.2 Conductor Spacing

The conductor spacing considerations are generally based on voltage breakdown or flashover between adjacent conductors. The conductor spacing is determined by the peak voltage difference between adjacent conductors, capacitive coupling parameters and the use of a coating. The recommended conductor spacing designs are:

General designs: 0.025 mm voltage limitation 400 V dc high density:

Preferred ac or peak minimum is 0.015 mm, voltage limitation 50V dc or ac peak.

The minimum by exception is 0.010 mm.

In case of preferred minimum or minimum by exception designs, the printed board assembly must be conformal coated or a suitable solder mask must be applied.

The minimum permissible spacing should be applied only where there is no way to avoid it, otherwise higher spacing should be given. This is to minimize the reject rate during production.

Additional considerations in deciding conductor spacing are:

i) Critical-impedance or high frequency components should be placed very close together to reduce critical stage delay.

ii) Transformers and inductive elements should be isolated to prevent coupling.

iii) Inductive signal paths should cross at right angles.

iv) Components which produce any electrical noise from movements of magnetic fields should be isolated.

There are no standards for minimum spacing between any two leads that are not electrically connected. However, an average 2 mm is generally taken where voltages are in excess of 30 volts or where a short will cause failure of critical nature or expensive components, when an increase in spacing becomes necessary.

Most modern circuits make use of semiconductor devices operating at voltages normally upto 24 volts. Therefore, the problem of breakdown due to narrow spacing between conductors almost does not exist. However, for a normal uncoated epoxy card, operating at an altitude of less than 1000 meters, the spacing should be 0.002 mm/volt. This figure should be increased further for harsher conditions to ensure adequate safety. The recommended minimum conductor spacing units for different voltages are given in Table 3.6.

Table 3.6   Recommended Minimum Conductor Spacing

Voltage between conductors dc or ac Peak (volts)	Uncoated traces 0-10,000 ft IPC ML 910A	Uncoated above 10,000 ft Alt.	Coated and Internal Layers MIL STD 275 and IPC ML 910A
0–15	0.015 (0.38 mm)	0.025" (0.64 mm)	0.005" (0.13 mm)
16–30			0.010" (0.25 mm)
31–50			0.015" (0.38 mm)
51–100		0.060" (1.52 mm)	0.020" (0.51 mm)
101–150	0.025" (0.64 mm)		
151–170		0.125" (3.18 mm)	
171–250		0.250" (6.35 mm)	
250–251	0.050" (1.27 mm)		0.030" (0.76 mm)
251–301			
500	0.100" (2.54 mm)	0.500" (12.70 mm)	0.060 (1.52 mm)
500 +	0.0002 in/volt (0.0051 mm/volt)	0.0010 in/volt (0.0030 mm/volt)	0.00012 in/volt (0.00305 mm/volt)

3.6.3 Conductor Shapes

While deciding the layout, sharp covers and acute angle bends in conductors should be avoided as far as possible. The rounded contours will not only minimize conductor cracking, foil lifting and electrical breakdown, but also greatly facilitate solder distribution. The process may be more expensive from the drafting standpoint. Rounded corners at conductor bends and smooth fillets at the junction of conductors and terminal areas are desirable.

A trace that extends in a straight line is relatively clean, one that extends straight and then turns 180 degrees back on itself looks just like an antenna. A line that makes a right angle turn also begins to look like, and have the characteristics of, an antenna. It is admittedly not a really good antenna.

Therefore, boards should never have signal lines that turn more than 45 degrees and all trace corne. should be mitred as shown in Figure 3.11.

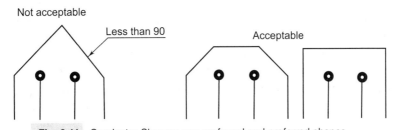

Fig. 3.11   Conductor Shapes: non-preferred and preferred shapes

3.6.4   Conductor Routing and Locations

For error-free operation of the circuit, it is necessary to provide proper routing of the printed conductors and to locate various components in a manner that they do not cause electrical or mechanical disturbance.

The conductor routing rules as per ANSI/IPC-2221 are as follows:

- Conductor length should be the shortest possible. In simple circuits where lots of space is available, conductors can be run in any direction so as to give the shortest interconnection length.
- Conductors forming sharp angles should be avoided as this creates problems in etching.
- Where one or several conductors have to pass between pads or other conductive areas, the spacing has to be equally distributed.
- Minimum spacing is applied only where it cannot be avoided, otherwise higher spacing should be given.
- In a double-sided PCB, it is a normal practice to draw the tracks on the component side in the direction of the Y-axis and tracks on the solder side in the direction of the X-axis.
- Distribute Maximum number of tracks should be distributed on the solder side and drawn in the direction of X-axis.

Some other practical suggestions are:

- Use minimum number of layers of wiring.
- Provide maximum line width and terminal area commensurate with the density of packaging.
- Avoid sharp angles and bends in conductors to minimize electrical and mechanical problems.
- By careful routing of individual sections of each circuit, provide an exclusive conductor to a single ground at essentially a single point. Grounding circuitry is of major concern in determining the internal circuitry layout.

3.6.5 Supply and Ground Conductors

The primary consideration in a power or ground conductor is to provide a direct connection from the device to the power supply. As this is not always possible, the next best step is to increase the width of the conductor to provide enough mass to accomplish essentially the same results. It may be remembered that supply and ground lines on a printed circuit board are not just conductive links. The width of these conductors and their layout play an important role in imparting stability to the circuit voltages. In some cases, resistive losses in these conductors may result in unstable supply voltage or ground system. Therefore, it is advisable to be fully aware of the possible circuits on a PCB for designing the conductors with an adequate width. The problems are more serious in case of digital and high frequency signal circuits.

When several supply voltages are used in a circuit, it must be ensured that the ground conductor has the capacity to carry the combined load under the worst case condition. A rule of thumb for deciding the width of conductors for various purposes is:

$$W_{ground} > W_{supply} > W_{signal}$$

Where

W_{ground} = conductor width of ground line

W_{supply} = conductor width of supply line

and W_{signal} = conductor width of signal line.

The fundamental rule for TTL circuits is

$$W_{ground} \geq 2\, W_{supply}$$
$$W_{supply} \geq 2\, W_{signal}$$

Also, for TTL circuits, it is advisable to utilize all the unused board area for ground conductors of a ground plane.

The distribution of voltage or power and ground planes is therefore a critical design element in the layout of a complex or high I/O semiconductor package. Voltage and ground conductors are commonly designed as full metal planes. The main problems associated in the design of voltage and ground conductors are (Braun, 2002):

- Power losses;
- Power and voltage level variation;
- Efficiency; and
- Interconnection cross-talk.

These problems are addressed in the design rules as per ANSI/IPC-2221. These are:

- The analog and digital circuits on the same PCB should strictly have an independent ground network to avoid power losses.
- All digital signals and components should be located away from analog circuitry. All high speed digital traces should take the most direct route over the digital ground or power plane.

- Voltage and ground signal conductors have to be provided with sufficient width to keep resistance and inductance low and to carry the required current, otherwise the conductor will act like a fuse.

- While connecting voltage and ground conductors, priority should be given to the component with the highest power consumption, so that the power consumption along the supply line should continuously decrease.

- The spacing between voltage and ground conductors should be as large as possible to avoid cross-talk problems, i.e. interfering of currents.

- All I/O voltage and ground conductors should have minimum conductor length to achieve more efficiency in circuit.

- To increase the performance speed of semiconductor die and packages, many power and ground I/Os are required. Normally, a PCB with 40-60 per cent voltage and ground conductors has a good performance speed.

- Avoid switch mode power supplies near ADCs, DACs and analog circuits. Sometimes, it is easier to use a separate 5V three-terminal regulator, near the chip, for the analog supply. A 22 μF tantalum or aluminum capacitor at the board edge helps to reduce power supply noise.

- Watch out for the external magnetic field of inductors and transformers. Use electrostatic and magnetically shielded components, if necessary. RF de-coupling chokes can be mounted at right angles to minimize mutual inductance. Power transformers should be mounted off the board and oriented, with the most intense area of their external field, away from critical analog circuits. Use toroidal power transformers to reduce magnetic external fields.

- Consider carefully the presence of Programmable Logic Devices (PLDs) and Very Large Scale Integrated (VLSI) logic chips on the same printed circuit boards. These chips frequently include lots of synchronous logic and generate large switching currents that can infiltrate the rest of the board. Make sure they are well bypassed at the chip pins. This will not only ensure their reliable operation but reduce noise on the supply line.

In a complex circuit, it is desirable to split the ground connection to achieve optimal efficiency of the circuit. A common ground may give rise to detrimental voltage drops along the conductor, introducing noise and false signals and resulting in malfunctioning of the circuit. Therefore, the designer must either decrease the length or increase the width of the ground conductor as far as possible.

In order to provide adequate shielding, particularly in the case of high frequency shielding, it is desirable to provide a ground plane. In the PCB world, a plane is a solid sheet of copper. It is a ground plane if it is connected to ground and it is a power plane if it is connected to a power supply voltage. But since there are usually many bypass capacitors between power and ground, the distinction between power and ground has no significance for ac signals, ac which can and do travel on either type of plane.

The provision of a ground plane ensures a high frequency return line of low inductance. Mutual capacitances can be minimized if the signal line is placed close to the ground plane or is laid in the latter. Therefore, wide ground planes and small conductor spacing gives low impedance and less interference, and in turn, greater stability. Figure 3.12 shows how a ground plane is provided in the PCB layout.

Continuous conductive area

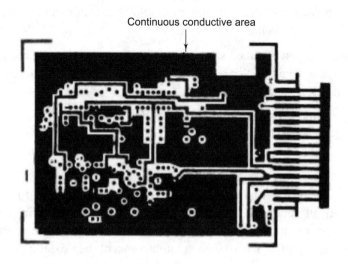

Fig. 3.12 Ground plane in a PCB layout. Usually large left out areas are converted into ground plane

 ## 3.7 Fabrication and Assembly Considerations

Certain limits should be taken into account in order to maximize manufacturability and thereby minimize cost. Also, the human factors should be considered before designing is undertaken. These factors are delineated below.

- Conductor spacing less than 0.1mm will not work with the etching process, because the etchant fluid does not circulate efficiently in narrower spaces resulting in incomplete metal removal.
- Features with a conductor width smaller than 0.1 mm will lead to breakage and damage during etching.
- The land size should be at least 0.6mm greater than the hole size.

The following limitations determine the layout techniques:

- Size capability of reprographic camera for film master production;
- Artwork table size;
- Minimum or maximum board processing size;
- Drilling accuracy; and
- Fine line etching facilities.

The following parameters are taken into considerations for design from the point of view of assembly of printed circuit boards:

- Hole diameter shall be expressed in terms of maximum material conditions (MMC) and least material conditions (LMC) limits. The diameter of an unsupported component hole shall be such that the MMC of the lead subtracted from the MMC of the hole provides a clearance between a minimum of 0.15 mm and a maximum of 0.5 mm. Also, for flat ribbon leads, the difference between the nominal diagonal of the lead and the inside diameter of the unsupported hole shall not exceed 0.5 mm and shall be not less than 0.15 mm.
- Properly locate smaller components so that they are not shadowed by large components.
- Solder mask thickness should not be greater than 0.05 mm.
- Screen print legend must not interface with any solder pad.
- The top half of the board should be a mirror of the bottom half of the board to achieve a balanced construction, because asymmetrical boards tend to warp.

One of the important considerations from the point of view of PCB assembly is that adequate attention must be given to the possibility of shorts being generated by an inserted component deviating from its theoretical position before soldering. As a rule of thumb, the maximum allowed inclination for a component lead is that it should remain within 15° of its theoretical position. It can go upto 20° provided the difference between the hole and the lead diameter is high. In vertical mounted components, the inclination can go up to 25° or 30°, resulting in significant reduction in packing density. Figure 3.13 shows TO-18 transistor package with leads at different angles.

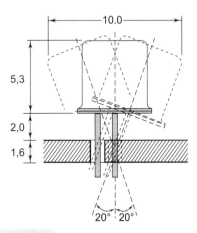

Fig. 3.13 Deviation from the vertical of a TO-18 package, mounted at a distance of 2 mm from the PCB. If the hole diameter is 1mm, the deviation can be upto 20° without any deviation of the leads

Multiple board assembly normally makes servicing at the field level easier as board level replacement can be easily carried out. However, this is possible only if each individual board performs a specific function. In such cases, the board replacement involves no major dismantling with minimal soldering/de-soldering. The design of the printed circuit boards must therefore take into consideration the *maintainability* aspects.

Soldering techniques and equipment for assembly also impose many restrictions on the board design and layout. For example, in wave soldering, the maximum sizes of the slots, edge clearances and handling clearances are important parameters. Also, the designers must be aware of what the final product will be and try to protect its most sensitive parts, as far as possible. For example, any high voltage circuit must be protected to prevent contact from outside. Careful location of components on the boards and of the boards in the product can help to minimize the likelihood of damage by external agents.

 ## 3.8 Environmental Factors

The reliability of an equipment, to a large extent, would depend upon the reliability of the basic printed circuit board. It is therefore, expected that the PCB should withstand exposure to the environmental requirements without either physical damage or change in operating characteristics. Also, besides serving as an electrical connection medium, printed wiring planes provide mechanical support for the active and passive components they are interconnecting. Thus, they become an integral part of the package or assembly and must therefore be able to withstand the environmental stresses associated with the entire structure. The important environmental factors in the design of printed circuit boards are detailed below.

3.8.1   Thermal Considerations

The PCB designer should keep in view the following points to ensure proper cooling of the electronic packages:
- Use of high temperature components, where possible;
- Thermal isolation of temperature-sensitive components from high heat-emitting sources; and
- Ensuring proper conductive cooling; the heat removal can be achieved by all the three modes of heat transfer, i.e. conduction, convection and radiation.

Removal of heat by conduction is achieved by:
- Use of materials with high thermal conductivity;
- Adopting the shortest/direct path to the heat sinks;
- Ensuring good thermal coupling between parts involved in the conduction path; and
- Designing the printed conductor in the thermal path as large as possible.

Cooling through convection can be increased by:
- Enhancement of the surface area available for heat transfer; and
- Replacement of laminar flow with turbulent flow, thereby increasing the heat transfer coefficient and ensuring a good scrubbing action around the parts to be cooled.

Heat transfer through *radiation* can be enhanced by:

- Use of materials with high emissivity and absorptivity;
- Raising the temperature of the radiating body;
- Lowering the temperature of the absorbing body; and
- Arranging the geometry to minimize back reflection to the radiating body itself.

In order to eliminate local hot spots that can damage the board or adjoining components, special consideration should be given to the placement of power transistors or high wattage resistors. In general, such components should be mounted close to the frame which serves as the heat sink.

In order to maintain the components below their maximum operating temperature, proceed as follows:

- i) Analyse the circuit and obtain the maximum power dissipation for each component.
- ii) Determine the maximum operating surface temperature to be expected. The maximum allowable temperature is dependent on the insulation present as well as the components themselves. The design is worked out keeping these parameters in mind.

3.8.2 Contamination

Printed circuit boards must be protected against dust, dirt, contamination, humidity, salt spray and mechanical abuse. There are many insulating compounds that can be applied as protective coatings. The commonly used compounds are polyurethanes, silicones, acrylics, polystyrenes and varnishes. The following are the broad technical considerations involved in the selection of protective coating:

- i) Ability to prevent corrosion and provide protection to the board;
- ii) Flexibility — resistance to cracking during shock;
- iii) Easy application and processing;
- iv) Transparency — to enable viewing of the board's component marking; and
- v) Easily removable for repairing the printed wiring assembly — minimum effect due to its thickness on important electrical properties such as dissipation factor, dielectric constant.

The thickness of the coatings on PCBs for military applications is governed in accordance with MIL-I- 46058. For general applications, the thickness is typically 0.075 mm minimum and 0.25 mm maximum.

3.8.3 Shock and Vibration

Vibration, flexing and bowing are the problems usually encountered on larger boards. The effects of vibration and warping can be minimized in exactly the same way as those met in any other form of engineering and similar solutions can be used. Parts that might be susceptible to failure because of shock or vibration should be located as near to the supported areas of the board as possible. Clamping or strapping may be required for properly holding the components in place.

One of the commonly encountered problems due to vibration or bowing is the possibility of components with electrically live cases coming in contact with the soldered joints on the back of the adjoining board. Such a danger is avoided by fitting the board with spacers higher than any of the electronic component/packages on it at suitable places.

The design of PCBs that will be subjected to vibration while in service requires a special consideration (IPC 2221-p.27/28) for board layout. Bulky or heavy components need particular attention. Unless they have many leads, it is necessary to anchor them to the board with some device. Such arrangements are shown in Figure 3.14.

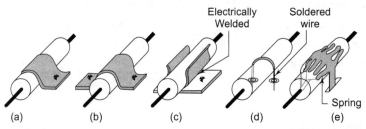

Fig. 3.14   Common methods of fastening heavy components to the board (a) single-sided clamp (b) double-sided clamp (c) electrically welded holder (d) fixing with soldered wire (e) holding with a spring loaded tape

The following guidelines should be observed during the design of printed boards to eliminate vibration induced failures of the PCB assemblies:

- The mounting height of free-standing components should be kept to a minimum.
- Positive support of all components with a weight of more than 5 g per lead has to be considered when the board will be subjected to vibration.
- Vibration isolators should be considered for mounting of units, whenever practical.
- Board stiffeners and/or metal cores should be considered to reduce the board deflection.

If the equipment in which the PCB is mounted is subject to shocks, vibration, etc. the leads generated by it can easily lift the pads of single-sided PCBs. One method to prevent such a failure is to enlarge the pad area to cater for large components. Alternatively, it is recommended to provide two dummy pads as shown in Figure 3.15. If one pad de-laminates, the board can be easily repaired by soldering the component to the dummy lead. An even better practice is to use a funnel eyelet into the holes used for mounting heavy components. This provides much larger mechanical resistance to loads applied to the PCB through a component lead.

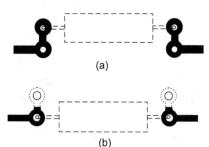

Fig. 3.15   Use of dummy pads to help in repairing boards. The heavy component is assembled as in (a); if a pad delaminates, it can be mounted again as in (b)

3.9   Cooling Requirements and Packaging Density

3.9.1   Heat Sinks

Thermal management is an important aspect of the design of printed circuit boards. The design should accommodate the problems of heat distribution and heat removal in systems utilizing integrated circuits. For example, component density is often higher with SMT, resulting in greater power dissipation per square inch of PCB. In addition, closely spaced components make forced air cooling less efficient. Therefore, the air flow arrangement must be so designed that it can deliver the volume of air required to restrict the temperature rise of the board within the permissible limit. Reliability can thus get degraded unless special attention is paid to thermal management. For example, in multi-layer boards, all interconnections can be placed on internal layers and a heat sink of thick, solid copper or another material can be placed on the outer surfaces. Components can then be mounted directly on the metallic surface.

Sufficient free space should be provided around the heat sinks to improve efficiency. No bulky component should be mounted near the heat sink which may obstruct the free air flow. Generally, heat-generating components are raised to a higher level above the board. This prevents damage to the component and the board itself. In a vertically mounted PCB, two heat sinks should not be designed and mounted one above the other. In order to ensure maximum exchange of heat in heat sinks with unidirectional slots, the air flow must always pass the heat sink in the same direction as the slots are made.

3.9.2   Packaging Density

There is no simple formula to suggest the optimum packaging density on a PCB. For example, if the density is very low, a larger PCB area or a higher number of PCBs will be required to realize the same circuit. This will result in more volume of the equipment, more connectors and wiring with more parasitic influences on the working of the circuit, thereby degrading reliability while pushing up the cost. On the other hand, a very high packaging density will give higher circuit temperatures, more cross-talk, difficult servicing and maintenance, and probably a higher reject rate in PCB production. This again brings down the reliability and makes the cost higher.

The packaging density is usually dictated by:
- Purpose, use and application of equipment — whether fixed installation, portable or airborne;
- Heat generated and cooling arrangement — natural air flow, forced cooling or hermetically sealed unit;
- Type of components on board;
- Component technology-whether discrete, SSI, LSI, VLSI, or SMT; and
- Type of PCB used (interconnection density) — whether single-sided, double-sided, or multi-layered.

Multi-layer boards are preferred when the component densities possible with double-sided boards are not adequate. For example, in a double-sided board, a usual maximum is 2.0 TO cans per square inch which can be increased to more than 3. In some designs, it is possible to double the component density in the multi-layer boards as compared to double-sided boards, without appreciably increasing the volume of interconnections.

As general design guidelines, the packaging density of a PCB can be estimated from the number of component mounting holes per square inch of usable surface. This figure is 3–10 holes/in^2 for single-sided boards, 10-20 holes/in^2 for double-sided boards and more than 20 holes/in^2 for multi-layer boards.

The packaging density is basically governed by:
- Board outline, size and form;
- Type of housing or enclosure in which the PC board will be finally mounted;
- Methods for mechanical attachment, i.e. card guides, stand-offs, etc.;
- Input/output termination, i.e. connector type, cable wire, etc.;
- Degree of support, i.e. retention and fastening;
- Card removal requirements, i.e. card extractors or special extract tools to aid in the removal of the board from enclosure;
- Desired accessibility for adjustable components;
- Heat dissipation requirements;
- Shielding requirements, i.e. circuit compatibility with other circuits and the environment;
- Type of circuit and its relationship to other circuits, i.e. the placement and area required;
- Environmental considerations such as shock and vibration humidity, salt spray, dust and radiation etc.; and
- Manufacturability, i.e. cost and case of manufacturing.

3.9.3   Package Style and Physical Attributes

Every electronic system consists of various parts including electronic components, interfaces, electronic storage media and the printed board assembly. The complexity of these systems is reflected in both the type of components used and their interconnecting structure.

Components are generally grouped into the following categories:
- Axial lead components;
- Radial lead components;
- Surface mounted devices; and
- Electromechanical components.

Axial Lead Components
They are the most common type. They include resistors, some types of capacitors and diodes. For fixing the components, the leads are bent approximately 90° and inserted into the holes on the PCB

and soldered. The span between the two leads of these depends on the length of the body, lead diameter and the length of the lead upto the bend. If the axial component is required to be mounted vertically, one lead is bent vertically like a hairpin and is generally insulated with a sleeve.

Radial Components

They have leads perpendicular to their body and include certain capacitor types, variable resistors, active devices like ICs, transistors and some electromechanical components like switches, relays, connectors, etc. Transistors in TO packages are usually soldered to a cluster of pads (Figure 3.16). ICs are often mounted on a base, which have leads spread so that they fall on a grid. Dual-in-line package (DIP) is the most commonly used IC package, standard DIP packages have 8, 14, 16, 20, 24, 28, 40, and 64 pins. They have two rows of leads with a lead separation of 2.5 mm within the row and a spacing of 7.5/10/15 mm between the rows.

| TO-5 | TO-18 | TO-100 |

Fig. 3.16   Footprints TO-100/TO-5/TO-18

Axial and radial components belong to the through-hole components type as they rely on a lead being inserted into drilled hole for the mechanical holding, and the soldering of the lead to the solder pad for the electrical connectivity.

The more complex components, as judged by the amount of input/output terminals they possess, the more complex is the interconnecting substrate.

Many peripheral leaded, lower I/O count devices such as memory and logic devices are being converted into area array packaging formats as either BGAs or fine pitch BGAs.

Surface Mount Technology (SMT)

Surface mount devices have leads with flat surfaces. These are soldered to solder pads which are called lands. The component is placed on the surface of the board instead of being inserted into them. Surface Mount Technology (SMT) has advanced to a stage wherein the majority of electronic components manufactured today are only available in SMT form.

BGA Packaging

Array packages such as BGA and fine pitch BGAs are now the latest technology component packages for I/O devices like memories, processors, and FPGAs. Ball and column grid arrays were standardized in 1992 with 1.5, 1.27 and 1.0 mm pitch. Fine pitch BGA array packages standards have established pitches of 0.8, 0.75, 0.65, and 0.5 mm. Area array packaging has the intrinsic value of making a coherent design. The signal I/O count for high performance BGAs is about 2.5 times of what is commonly required for BGAs used in hand-held products. BGA packaging is more useful for high frequency PCB design.

Electro Mechanical Components

This category of components includes relays, transformers, connectors, etc. In general, they do not follow any specific pattern of pin configuration, but have standard grided footprint, except transformers.

 ## 3.10   Layout Design

The printed circuit layout is basically a sketch that represents the printed wiring on the substrate, the physical size and location of all electronic and mechanical components and the routing of conductors which interconnect the electronic parts. In practice, the layout of the PCB must incorporate all the information on the board before one can go on to the artwork preparation. Thus, the layout designer must be familiar with the design concept, details of the circuit and the philosophy behind the equipment.

The components and connections in the PCB layout are derived from the circuit diagram, and physically placed and routed by the designer to get the best result. The PCB layout defines the final physical form of the circuit and enclosure and labelling details of the equipment can then be finalized as the layout is completed.

In order to ensure that a good layout is made out, the circuit designer must provide the layout (PCB) designer information on the following points:

- A well-drawn schematic with minimum number of cross-overs and loops;
- Areas of circuit incompatibility, i.e. those where isolation of one circuit from another is required;
- The number and position of the external connections to be used with the interconnection system, as this plays an important role in determining the position of input, output, and ground terminations;
- The board contacts should be designated numerically/alphabetically on the schematic for easy and error-free recognition; and
- Shields and grounds to be clearly indicated, particularly if they are used for reducing interference or noise from couplings.

Such type of problems do not arise in small companies wherein the same engineer designs the circuit and carries it through the production stage.

Layout design is a complex operation as it has to take into account different requirements and constraints. The general considerations for a good layout design are:

- Type of product (required quality, reliability and safety considerations, applicable standards, approvals required etc.);
- Expected production volume;
- Assembly facilities and techniques along with desirable degree of automation;
- Maintenance set-up (factory premises, disbursed maintenance facilities and at customer's place);

- Working environment (storage, shock and vibration);
- Transportation arrangement;
- Electrical considerations/constraints such as electromagnetic shield, cross-talk between conductors;
- Components/connected modules requiring easy access;
- Heat removal considerations; and
- Requirement of insulation between the PCB or its mounted components and any other metallic part on the assembly.

While working out the layout design, there may be many reasons for setting a section of a circuit apart, such as heat dissipation, electromagnetic shielding, type of assembly technology, convenience of operation of final product, maintainability, etc. In such situations, it is always advisable to involve quality engineers in the project while working on these aspects of the design.

Making a layout was a very important and specialized task in the past as it was performed manually. It had a significant influence on the development time and cost of the printed circuit board. Extensive use of computer-aided design (CAD) systems and the developments in hardware and software for CAD have reduced layout relevance to only a few special situations such as care in dimensioning for current carrying capacity, thermal design, etc.

3.10.1   Grid Systems

The grid system is commonly used in practice in PCB layout design. The grid paper is available in either roll or standard sheet sizes. The standard grid spacing is 8×8 or 10×10 squares to the inch, and each one inch division is printed in heavier lines. Use of the grid paper has many time-saving advantages. For example, components are manufactured with their leads into grid intersections and numerically controlled drilling equipment can easily be programmed accordingly. Also, it is quite easy to segregate major sections of the circuit on grid paper and to carry out minor modifications on the same.

Grid systems are always basic and provide for no tolerance. When printed board features are required to be off a grid, they are individually dimensioned. The grid increment has to be specified on the master drawing. The choice of grid increment is based on the component terminal location for through-hole components, and on the component centre for surface mounted components.

3.10.2   Layout Scale

Layout scale is chosen depending on the accuracy required. It could be 1:1, 2:1 or even 4:1. It is best prepared on the same scale as the artwork, which minimizes the problems that might be caused by re-drawing of the layout to the artwork scale and also facilitates checking of the final layout.

The commonly applied layout/artwork scale is 2:1. This offers a good compromise between the accuracy achieved and convenience of handling. It may be remembered that a 2:1 artwork has 4x the actual PCB area. The 4:1 scale which gives 16x the PCB area is applied in special cases where very high precision is required.

If the layout is carried out on a 2:1 scale, one grid unit will be 5 mm. When the grid system based on 2.5 mm is considered as too coarse, grid units of 1mm should be adopted. Figure 3.17a shows a DIP (Dual-in-line-Package) drawn in the layout.

3.10.3 Layout Sketch/Design

The printed circuit layout sketch is the end-product of the layout design depicting components and the interconnecting conductors. It provides all information for the preparation of the final artwork. Besides this, the layout sketch also includes information on component holes, conductor width, minimum spacing between the conductors, etc.

Before the designer starts working on the layout for the design of a printed circuit board, it is advisable to prepare a trial layout drawing keeping the following factors in mind:

- Board size — dictated by equipment enclosure or the modular design concepts;
- Component outlines — available from data books;
- Component mounting data — in case of special mountings, data books may have to be consulted, thermal limitation may require heat sinking or large size de-coupling capacitors;
- Interconnecting patterns;
- Conductor width and spacing, depending upon the functional requirement of the conductor;
- Border lines — generally a 0.5 cm margin is left on all sides;
- Connectors — size and type of connectors based on external connections like power input and output signal, ground and device requirements; and
- Fixing arrangement — screws, clamps, etc.

Layout design is always a trial and error process with many iterations, and several trial layout drawings may have to be made before the final design specifications are met and the design is ready for taping for preparation of the artwork. When preparing the layout, the procedure is to assign arbitrary spaces to the symbols on the basis of grid spacing. However, a common basis is established for all symbols of that variety. Figure 3.17b shows components laid out on a grid.

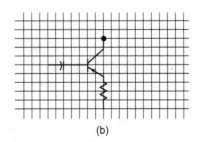

(b)

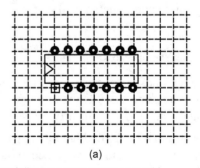

(a)

Fig. 3.17 **(a)** dip Package on Grid **(b)** layout of discrete components on a grid

3.10.4 Layout Considerations

All holes and pads in the board must be aligned according to a datum system. This is not only important but critical in case of multi-layer boards because during manufacture, every layer must align with all others. A typical datum system is shown in Figure 3.18. It makes use of two alignment patterns and a datum or a reference line, special alignment/datum marks are used and the layout of all the components is based on these marks. The following factors must be taken into consideration:

- Identify location of 0/0 datum features so that all or most of the board characteristics are in positive Cartesian coordinate zone.
- Establish majority of parts dimensional characteristics and choose that concept as the master grid. Identify all parts that do not fit that description (caused by intermixing of metric and inch-based parts) as off-grid and code as such in the CAD library definition.
- Consider tooling hole location requirements for both fabrication and assembly. If board size and/or density prohibit internal tooling holes, panelize boards as part of design, with external tooling holes.

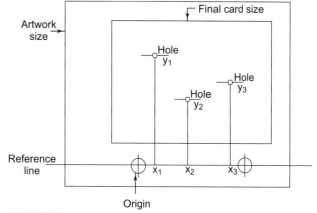

Fig. 3.18 Datum and reference line for a printed circuit board

3.10.5 Materials and Aids

The following materials and designing aids are required for manual layout design:

- Grid sheet;
- Pencil;
- Eraser;
- Scale;
- Adhesive tape;
- Puppets;
- Template set;
- Tracing sheet; and
- Indian ink pen (drafting pen).

3.10.6 Land Requirements

Land requirements designed so as to have a minimum diameter of at least 0.5 mm greater than the maximum diameter of the projection of the eyelet of solder terminal flange.

Test lands must be provided for all nodes as per ANSI/IPC 2221. A node is defined as an electrical junction between two or more components. A test land requires a signal name (node signal name), the X-Y position axis with respect to the printed board datum point, and a location, describing which side of the board the test land is located. The data is required to build a fixture for SMT and mixed technology printed board assembly layout to promote in-circuit testability with the help of "in circuit test fixtures" or commonly called "bed-of-nails fixtures". For this purpose:

- The diameter of test-lands used specifically for probing should be not smaller than 0.9 mm.
- Test lands should be located with clearances around a minimum of 0.6 mm and a maximum of 5 mm. If the component height is larger than 6.7 mm, the test lands should be located 5 mm away from tall components.
- No parts or test lands are to be located within 3 mm of the board edges.
- Test land should be in a grid of 2.5 mm hole centres. If possible, allow the use of standard probes and a more reliable fixture.
- Do not rely on edge connector fingers for test lands. Test probes easily damage gold plate fingers.
- Avoid probing of both sides of the PTH-PCB. Use vias, to bring test points to the non-component/solder side of the board. This allows for a reliable and less expensive fixture.

The number of different hole sizes shall be kept to a minimum. Table 3.7 gives the minimum holes tolerance range.

Table 3.7 Minimum Unsupported Holes Tolerance Range

Hole Diameter	Level A	Level B	Level C
0.10–0.8 mm	0.15 mm	0.10 mm	0.05 mm
0.81 mm	0.20 mm	0.15 mm	0.10 mm
1.61–5.0 mm	0.30 mm	0.20 mm	0.15 mm

The *Aspect Ratio* of plated through-holes plays an important part in the ability of the manufacturer to provide sufficient plating within the plated through—hole, as well as in the reliability of the PTH/PTV structure. When hole size is less than one-fourth the basic board thickness, the tolerance shall be increased by 0.05 mm. For drilled hole diameter 0.35 mm or less and aspect ratios of 4:1 or larger, the fabricator should mask or plug by a suitable method the plated through vias to prevent entry of solder. Generally, the ratio of board thickness to plated hole should be smaller than 5:1.

3.10.7 Manual Layout Procedure

The fundamental requirement of a good layout design is that it should reflect the concept of the final equipment. Although the circuit complexity and packaging density dictate the planning of the layout, there are general steps which need to be taken into consideration for the development of a good printed board design layout. These steps are listed below.

- Begin with a careful design of the electrical circuit and prepare a schematic or logic diagram.

- Make an initial evaluation of the schematic diagram, parts list and special circuit requirements and rules, if any, and carry out a rough comparison of this data with the physical limits of the usable board size.

- Compile a bill of materials, together with complete information on special environmental and performance requirements.

- Study the components carefully to understand their function in the circuit, requirement of heat sinks, ground and voltage connections, special width requirements and critical short conductor lengths. The circuit design engineer must define critical circuit design considerations such as capacitance coupling, feedback, current and clock signal grouping for the convenience of the layout designer.

- Understand the signal input and output connector interconnections and logical organization of different functional parts.

- Consider the general practices followed in the company/organization concerning component matrix location parameters, mechanical hardware and the automatic component insertion requirements.

- Select the shape and size of the board to accommodate all components and to fit the available space within the enclosure or the available area.

- Prepare a PCB layout as viewed from the component side. Double-sided designs may be represented on a single sheet by a coding system. Alternatively, it can also be done on two sheets, each representing the individual sides of the design. This method can sometimes cause alignment errors of the two sides. Multi-layer conductor layouts are also laid out singularly in this manner.

- Develop the layout in the direction of the signal flow as far as possible. This shall help to achieve the shortest possible interconnections.

- First place all components that need to be in specific locations. This includes connectors, switches, LEDs, mounting holes, heat sinks or any other item that mounts to an external location.

- The larger components are then placed and the space in between is filled with smaller ones. Heavy components should be located over or near the supported areas.

- Find out the method of fastening or mounting to be used, i.e. card guides, screws, stand-offs, etc.

- Components requiring input/output connectors are arranged near the connectors.
- All components are placed in such a manner that de-soldering of other components is not necessary if they have to be replaced.
- Divide the circuit into functional subunits. Each of these sub-units should be realized on a defined portion of the board. This shall ensure functional reliability, easier testing and quicker servicing of the board.
- Find out specific information on the widths of conductors, spacing of conductors and terminal areas.
- Obtain information on:
  - Component case sizes and shapes;
  - Distance between components;
  - Method of component lead termination;
  - Type of electrical interconnections;
  - Most suitable or critical routing of conductors; and
  - Holes sizes and locations.
- All packages, viz. transistors and ICs should be placed in a single-axis mode i.e. they should have the same orientation, either vertically or horizontally. This would make assembly and inspection of boards much easier. Uniform orientation minimizes mistakes in deciding routing of conductors.
- Position polarized parts (i.e. diodes and electrolytic caps) with the positive leads all having the same orientation. Also use a square pad to mark the positive leads of these components.
- After the components are in place, the next step is to lay the power and ground traces. It is essential when working with ICs to have solid power and ground lines, using wide traces that connect to common rails for each supply. It is very important to avoid snaking or daisy chaining the power lines from part-to-part.
- Draw the conductor lines only on grid lines. This will make spacing between the conductors uniform. All component hole locations and mounting holes should be laid on a grid system. The basic modular units of location are based on 0.1, 0.05 and 0.025" system applied along both the X and Y axes.
- Draw the initial layout by plotting a rough sketch/layout of the circuitry to establish the most practical placement of components and wiring.
- Give careful thought when placing components to minimize trace lengths. Put parts that connect with each other next to each other. Doing a good job here will make laying of the traces much easier.
- Convert the rough sketch/layout to a grid system. Establish the conductor paths, components, holes, test points and all other features. The layout should be viewed from the component side.

- Locate all holes preferably on a modular grid system.
- Provide proper fabrication indexing hole locations for each layout. Preferably, two or three indexing holes are recommended.
- Components, terminals, polarities and connector contacts should be labelled on the layout as they appear on the schematic. Marking of polarity is necessary in case of diodes, polarized capacitors and power terminals. The marking and identification marks should be as small as possible to be able to survive etching and still be legible. If possible, component reference designation markings should be considered to link the assembly with the documentation. On a DIP (Dual-in-line-package) and TO-5 type integrated circuits, pin 1 should always be marked.
- In addition, printed board assembly number serial number of assembly and company logo should be indicated on the component side. It is a good idea to place text such as a product or company name on the top layer of the board.
- After completion of the layout drawing, an art master is prepared either by manually taping it or with a computer. The art master is made to an enlarged scale, generally 2×1 or 4×1. It is then photo-reduced to provide a 1×1 scale film negative and positive (photomaster), which is a tool for manufacturing the PC board.

3.10.8   Layout Methodology

Approach with Sketching of Components

The process of layout starts by marking the board outlines and connectors, followed by sketching the component outlines with connecting points and conducting pattern. Since designing of the layout is based on trial and error, it necessitates continuous erasing.

An improved method of developing is to use a transparent tracing sheet over the grid sheet, making the erasing clean and convenient.

Layout Sketching with Puppets

A standard component template or puppet is one of the most important items the designer can use in printed circuit design. Puppets are individually die-cut, with transparent layout patterns for commonly used electronic components. They come coated with adhesive so that they can be pasted on non-matte drafting film. The use of templates not only saves valuable time, but also ensures that standard component lead spacing and body sizes are properly considered.

In order to develop the layout, a tracing sheet is placed over the grid sheet. The component layout can be done by using the respective puppets for each component. For interconnections, a sheet of tracing paper is placed over the component layout and the conductors can be pencilled on the tracing paper. The position of puppets can be simply changed by removing them and placing them on the new and preferred location. After all the components are allocated and all the interconnections are drawn, the component lead holes and outlines are also marked on the tracing sheet to complete the layout.

3.11   Layout Design Checklist

After the PC board layout design is completed, a check is recommended to ensure that all requirements have been taken care of. Often the requirements may be specific, but the following checklist covers the general areas of concern in the design cycle.

3.11.1   General Considerations

The following questions need to be asked:
- Has the circuit been analysed and divided into logical parts for a smooth signal flow?
- Are the components suitably distributed to give a uniform package density over the entire board?
- Is the placement of components is such that they result in short or isolated critical leads?
- Are the components easily accessible for easy replacement?
- Is some free space available for possible circuit extension or modification?
- Is the board size optimum?
- Are preferred conductor widths and spacing used; wherever possible?
- Are preferred pad and hole sizes used?
- Are heat-sensitive components kept at adequate distance from heat-producing components?
- Has heat sinking been provided, wherever necessary?
- Does a free air flow pass the heat sinks?
- Is jumper use kept to a minimum?
- Do the outermost conductors have enough distance from the edge of the board?
- Is the access to adjustable components (trimmers, pre-sets etc.) possible?
- Have test points been provided?
- Can the test equipment easily be connected to the board, e.g. clipping on of logic probe?
- Are the longer areas of copper broken up to prevent blistering?
- Has shielding been effectively provided, where necessary?
- Has proper type and size of lettering been provided for easy visibility after assembly?

3.11.2   Electrical Considerations

- Is there full compatibility between the circuit diagram and the layout?
- Has the circuit been divided into functional sub-units on the board?
- Have conductor resistance, capacitance and inductance effects been analysed, especially the critical voltage drops and ground?

- Is the signal flow smooth with interconnections being as short as possible?
- Have optimum precautions been taken such as minimum length, guidelines, clear separation of input and output lines?
- Has adequate conductor spacing been provided?
- Is there sufficient ground line width?
- Has a close coupling between supply and ground lines been realized?
- Have the analog and digital circuit parts independent ground lines?
- Are polarities adequately identified?

3.11.3   Mechanical Considerations

- Is the board size optimum?
- Is the board size compatible with the PCB manufacturing process?
- Are unstable or heavy parts adequately mounted?
- Will the mounted board meet shock and vibration requirements?
- Has the number of different hole diameters been restricted to the minimum?
- Are terminals and control locations compatible with the total assembly?
- Are specified standard components lead spacing used?
- Have tool locating holes provided?

 ## 3.12   Documentation

Documentation is an essential part of the printed circuit board design and fabrication process. Usually, companies have established norms, forms, lists and files for the purpose. However, certain drawings are considered essential for the maintenance of reference records. These are detailed below.

Schematic Diagram

Schematic diagram is the most important component of the documentation which goes along with the layout design as it represents interconnection of various components of the printed circuit board. The schematic diagram should clearly spell out the special requirements which need to be considered while making the layout design. Such requirements include: (i) heat sources, i.e. components which are likely to produce considerable heat and are required to be placed away from the heat-sensitive components; (ii) critical signal characteristics in terms of minimizing their length, conductor width consideration and provision of guards by ground lines around the signal paths; and (iii) input-output terminations or connectors which should have pre-determined locations and the mounting arrangements of the PCB in the equipment. All such requirements must be clearly marked with the necessary instructions.

Parts List

For simple circuits, it is not uncommon to incorporate the component specifications in the circuit diagram itself. However, for most of the cases, the parts forming an assembly are listed on a parts list which may include, depending on the requirement, the following information:

- Reference designation (R1, C1, C2; etc);
- Manufacturer's code or part number;
- Description of the item;
- Brief specifications; and
- Total quantity.

Fabrication Drawing

The fabrication drawing shows the dimensional configuration of the board, size and location of the holes, and material and process specification. The fabrication drawing should show the non-component side of the PC board and should include:

- Board outline and thickness;
- Material (board, conductors and plating);
- Hole location and size;
- Registration and mounting holes with locations and diameters;
- Test, inspection and qualification requirement; and
- Change/modification control information.

Assembly Drawing

An assembly drawing depicts an assembled printed circuit board with all the electrical components and mechanical parts contained on a particular assembly. It shows the board and instructions. For prototype boards, the assembly drawing may consist of a blueprint of the PCB artwork into which the component outlines and codes have been added.

Artwork

Artwork is an accurately scaled configuration of the printed circuit which is used to make the master pattern photographically. This shows only those items which have to be retained as a copper pattern in the manufacture of the board. It includes all solder pads and conductors in respect of both their dimensions and clearances; and their location on the board. It also carries identifying symbols and test patterns which may be required.

Artwork Scale: Planning a layout is very important in case of a manual layout design, because the modification, if required, is cumbersome and may even necessitate a total re-start of the artwork. In order to divide the artwork scale, it is useful to understand the requirements of the circuit design and to evaluate the same systematically in the form of a checklist. The checklist may include:

- Board size in mm (taking into consideration its final place in the equipment);
- Board shape (if dictated by the final product shape);
- Number and type of components;

- Mechanical mounting arrangement;
- Power supply details in terms of current and supply types;
- Signal line currents; and
- Component drill sizes.

Several sources may contribute to the inaccuracy of an artwork. The most obvious are the human limitation in placing solder pads at exact locations, parallelism between conductors drawn or printed on the artwork and the availability of self-adhesive materials of the required size. Therefore, printed circuit boards, especially those using integrated circuits for plated through-holes, the 1:1 scale is usually not adequate to meet the dimensional accuracy required for the reliable production of PCBs. Most of the artwork is, therefore, generated at a 2:1 scale, which gives an artwork four times the actual PCB area. For high accuracy fine line PCBs, the 4:1 scale is chosen which gives 16 times the actual PCB area.

3.12.1   Documentation File

It is recommended that a complete documentation file should have the following necessary information:

- *Front cover*: Title, date, version number, customer details, project features;
- *Schematic*: To be supplied in CAD format on a diskette;
- *Bill of material:* The parts list;
- *Parts key*: A glossary of the part number abbreviations, with package sizes, lead spacing, tolerance notes and preferred types;
- *Manufacturing notes*: Contains the notes relating to previous production runs-for instance problems encountered, methods of testing, etc.;
- *Drilling diagram:* A diagram showing the positioning and size of every hole on the PCB;
- *Printed circuit board layout*: To be supplied in CAD format on a diskette; and
- *Actual size PCB overlay:* A diagram showing the positioning and identification of the PCB components. The plan is printed in actual size to allow components to be placed against it for checking.

 ## 3.13   Useful Standards

IPC-2220: *Design Standards Series*: Includes all IPC current design standards in the IPC-2220 family.

IPC-2226: *Section Design Standard for High Density Interconnect (HDI) Printed Boards:* Establishes requirements and considerations for the design of high density interconnect (HDI) printed boards including component mounting and interconnecting structures.

IPC-JPCA-2315: Design Guide for High Density Interconnects and Microvias: A tutorial on the selection of HDI and microvia design rules and structures; includes design examples and processes, selection of materials, general descriptions and various microvia technologies.

IPC-2221A: Generic Standard on Printed Board Design: Establishes generic requirements for the design of printed boards and component mounting or interconnecting structures for single-sided, double-sided or multi-layer.

IPC-2222: Sectional Standard on Rigid Organic Printed Boards: Covers the design requirements of rigid organic printed boards, and component mounting and interconnecting structures.

IPC-2224: Sectional Standard on Design of PCB for PC Cards: Covers design requirements of printed board of PC card form factors including concepts on bow and twist constraints, heat dissipation considerations and component placement requirements.

IPC-2225: Section Design Standard for Organic Multi-chip Modules (MCM-L) and MCM-L Assemblies: Covers requirements and considerations with respect to thermal, electrical, electromechanical and mechanical for the design of single chip module (SCM-L), MCM OR MCM-L assemblies.

IPC-2615: Printed Board Dimensions and Tolerances: Includes fundamental dimensioning and tolerance rules, positional, profile, orientation and form tolerances and detailed geometric symbology.

IPC-D-322: Guidelines for Selecting Printed Wiring Board Sizes Using Standard Panel Sizes: Defines guidelines for choosing sizes of printed circuit boards using standard fabrication panel sizes.

IPC-M-105: Rigid Printed Board Manual: Latest standards addressing the dimensioning, tolerance, qualifying and performance aspects of rigid printed boards.It also includes solderability testing, polymer thick film printed boards, land pattern and high speed circuitry design.

IPC-1902: IPC/IEC Grid Systems for Printed Circuits: An international standard adopted by IPCA that ensures compatibility between printed circuits and the components to be mounted at the grid intersections.

IPC-D-325A: Documentation Requirements for Printed Boards: Specifies general requirements for documentation necessary to fully describe end-product printed boards; includes master drawing requirements board definition and artwork/photo-tooling.

4

Design Considerations for Special Circuits

4.1 Design Rules for Analog Circuits

In today's world, people can easily assume that they live in an all-digital world, yet analog signals are still found within more than 60 per cent of present-day electronic designs. The three important considerations which form the basis for design rules for analog circuit PCBs are:

- Component placement;
- Signal conductors; and
- Supply and ground line conductors.

4.1.1 Component Placement

Component placement plays a crucial role, especially in analog circuits PCB design. The important guidelines to be followed in this regard are:

- Components which need to be accessed from the front panel must be placed exactly according to the requirements of the equipment designer.
- Components for internal adjustments such as potentiometers, trimmers, switches, etc. should be arranged near the board edge and placed in the proper direction for easy operation.
- Components with metal cases should not be placed very near to potentiometers, trimmers and switches etc. otherwise while adjusting, the screwdriver may cause a short-circuit between the component and the equipment chassis.

- The placing of heat-producing and heat-sensitive components must be carefully planned. Heat-producing components should be placed away from the heat-sensitive components.
- Heat-producing components should be uniformly distributed over the entire board area as far as possible. This will avoid local over-heating of the board.
- Components likely to get heated must be separated from the board surface by suitable spacers. Provision for space for these spacers should be made on the board.
- Where mounting screws need to be provided, the requisite space for nut and washer must be planned for, and no conductive track should be run underneath.

4.1.2 Signal Conductors

Signal conductors in analog circuit PCBs have to perform a variety of different tasks including input, reference level, feedback, output, etc. Therefore, a signal line for one application has to be optimized in a different manner than for another application. But a common consideration in all analog circuit PCB designs is to keep the signal conductor as short as possible. This is because the magnitude of the undesirable inductive and capacitive coupling effects increases almost proportionally to the length of the signal conductor. It may not always be possible to keep all signal conductors as short as possible. A practical approach in such a case is to identify the most critical signal conductor and to put it first in the layout.

The signal conductor layout has to be made carefully, particularly for the following types of circuits:

- High frequency amplifiers/oscillators;
- Multi-stage amplifiers especially with high power output stage;
- High gain dc amplifiers;
- Low level signal amplifiers; and
- Differential amplifiers.

High Gain dc Amplifiers

High gain dc amplifiers are generally used to amplify low level signals. When a device like a transistor or dc amplifier is soldered on to the PCB, a thermocouple junction can be formed between copper and the lead of the device. This will create different voltages, which in turn, will generate a noise signal to the amplifier. In order to minimize the temperature gradient at the input stage of the dc amplifier and to maintain a stable temperature gradient, it is advisable to put the input stage in a separate enclosure which does not allow a free movement of the surrounding air.

Differential Amplifier

A differential amplifier amplifies the voltage difference between two signals and rejects the common voltage on both signals. When the signal level is low, the common voltage will interfere and create small difference signals if the differential amplifier and its PCB are not properly designed. The differential amplifier inputs have high impedance to ground and any unbalance in them will bring down the circuit performance to an unacceptable level. Therefore, the physical geometrical symmetry of the amplifier on the PCB must be ensured during layout design.

A finite leakage resistance exists at the input of the differential amplifier, which can cause an unbalanced offset voltage. This problem can be solved by providing *guarding* at the input circuit. This arrangement is shown in Figure 4.1. The guard encloses the signal conductors and if it is kept at the same potential as the low line of the two signal conductors, it would result in an increase in the effective resistance. This type of arrangement ensures that the source end and the guard line are at the same potential as the low end of the signal source. Figure 4.2 shows how guarding can be done on a PCB. The guard conductor in the form of a loop encloses the signal conductors from the input connector upto the amplifier input solder joints and is connected with the guard of the equipment. This method facilitates an efficient technique of handling low level differential signals. In addition, the PCB base material used for low level differential amplifiers should preferably be of glass epoxy type, which aids in reducing leakage currents.

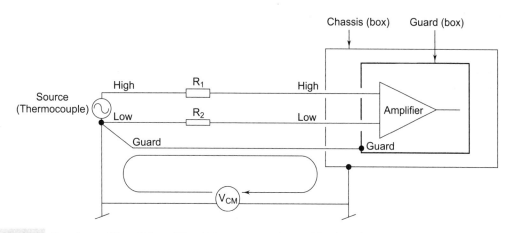

Fig. 4.1   Guarding a differential amplifier. At the source end, guard line is at the same potential as low end

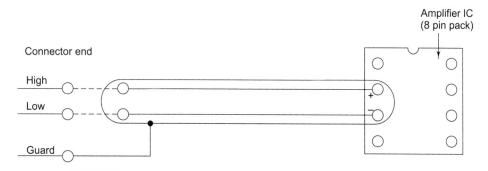

Fig. 4.2   Guarding signal conductors on a PCB (after Bosshart, 1983)

Low-level Signal Amplifiers: The amplifiers handling low-level signals are of two types.

High-impedance (low current) Amplifier: In these amplifiers, the capacitive coupling between two neighbouring signal conductors can seriously affect the circuit performance, even leading to masking of the low level signal.

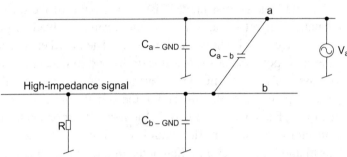

Capacitive coupling between the two conductors in high impedance circuit is shown in Figure 4.3. In order to minimize the coupling, it is advisable to provide a sufficient

Fig. 4.3 Capacitive coupling between two conductors in high impedance circuits (after Bosshart, 1983)

distance between the high impedance conductors and the other interfering signal. As a rule of thumb, the separation distance may be kept at least 40 times the signal conductor width.

However, the capacitance of low level signal conductors to ground should be high so that the coupled voltage is low. This implies that low level signal conductors should be close to ground conductors. If a wide separation is not possible, the coupling can be reduced by putting a ground conductor in between as shown in Figure 4.4.

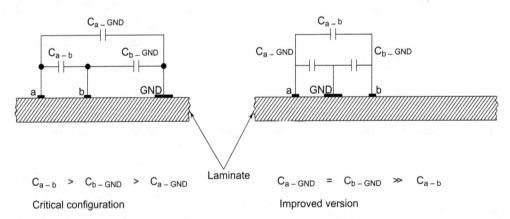

Fig. 4.4 Cross-talk reduction in parallel running signal lines by having a ground conductor between them (after Bosshart, 1983)

When amplifying signals from photocells and electrochemical cells, source impedance may be many millions or billions of ohms. If PCBs are inadequately cleaned after etching, the residual electrolytes on the board surface may result in comparable resistances between nearby conductors. Even with properly cleaned boards, leakage resistances of no more than 10^{12} ohms can be expected. These resistances, moreover, are unlikely to be isotropic so that the resistance between two adjacent tracks may be higher than that between two tracks separated by a much larger gap. For this reason, the inputs to low-level I/V converters should be protected by guard rings on both sides of the PCB (Figure 4.5) connected to a point at the same potential as the summing junction. If this is done, the exact value of the leakage resistance is unimportant since the potential difference across it will be small.

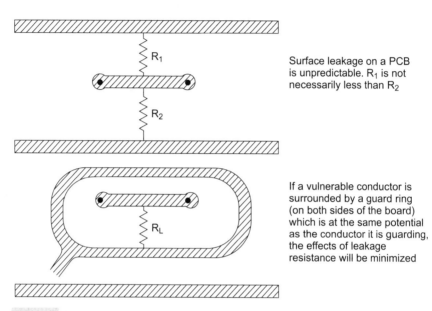

Surface leakage on a PCB is unpredictable. R_1 is not necessarily less than R_2

If a vulnerable conductor is surrounded by a guard ring (on both sides of the board) which is at the same potential as the conductor it is guarding, the effects of leakage resistance will be minimized

Fig. 4.5 Reduction of leakage resistance on printed circuit boards by using a guard ring

In applications of this type, the use of plated through-holes (PTH) is inadvisable. The bulk resistivity of PCB material is much lower than the sheet resistivity of its surface and it is very difficult to fabricate a guard ring in the bulk of a board. The best approach is to connect such high impedance amplifier terminals to a Teflon insulator rather than a PCB track. This is shown in Figure 4.6.

Low-impedance (low voltage) Amplifier: In case of low impedance circuits, there is a likelihood of having induced voltages due to inductive coupling or magnetic fields. This interference can be reduced to some extent by

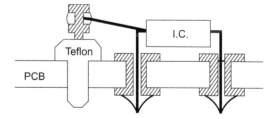

Fig. 4.6 Use of a Teflon stand-off insulator which has a much lower leakage than a PCB track

- Placing conductors carrying higher level ac signals sufficient away from low-level signal conductors;
- Providing ground conductors near the signal conductors; and

- Avoiding ground loops to disable the external magnetic field from disturbing low level signals.

High Frequency Amplifiers/Oscillators

An improper PCB layout of a high frequency amplifier results in a reduced bandwidth of the amplifier. Such a situation is shown in Figure 4.7. This is because the proximity of the ground conductors and signal conductors results in a high capacitance, which, along with the output resistance, acts as a low pass filter.

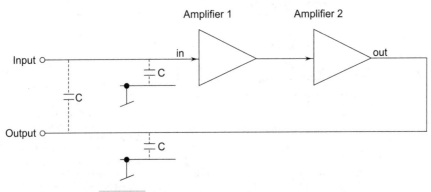

Fig. 4.7 High-frequency amplifier configuration

This action degrades the bandwidth of the amplifier. Also, if the input and output conductors are close to each other, there can be a feedback resulting in oscillations. In order to solve this problem, sufficient spacing must be provided between such conductors to avoid this effect (Lindsey, 1985).

It is a common experience of electronic circuit designers that at high frequencies (>10 MHz), that you design an amplifier, but in practice, it oscillates. Similar problems are encountered in designing the layout of an oscillator; it does not oscillate at the desired frequency. Such problems arise due to the presence of a capacitive coupling effect between the signal lines. One important precaution while making the PCB layout in such cases is to reduce the capacitive coupling between signal lines.

Multi-stage Amplifiers with High Power Output Stage

Multi-stage amplifiers are prone to low frequency oscillations, if supply and ground conductors are too long. The large current drawn by the high power stage will flow through the conductors with their own resistivity. This problem can be solved by de-coupling of the power supply conductors with sufficient large capacitors between supply and ground (Figure 4.8). Alternatively, separate power supply and ground conductors can be provided for the two different stages so that there is no common supply or ground path.

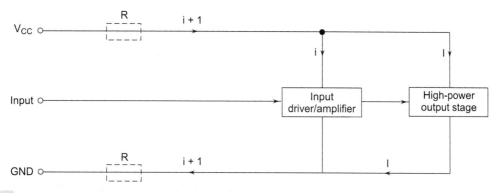

Fig. 4.8 Multi-stage amplifier with high-power output stage: providing separate power supply and ground conductors to avoid common supply guard path

4.1.3 Supply and Ground Conductors

Power supply lines should be of sufficient width to keep the resistance and inductance to a low value. However, the capacitive coupling to ground increases with more width.

Analog and digital circuits on the same PCB should strictly have independent ground network conductors. Similarly, reference voltage circuits, which are normally sensitive to ground potential fluctuations, should tap the supply lines directly at the input to the PCB and its ground line should be connected separately to the stable ground reference point of the equipment. Such an arrangement is shown in Figure 4.9.

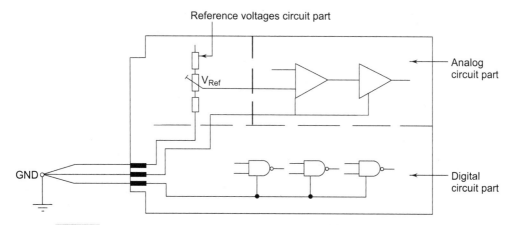

Fig. 4.9 Provision of separate ground conductors for reference, analog and digital circuit parts

In real life, ground conductors have both resistance and inductance, and may also be carrying unpredictable currents, which will have voltage drops when they flow in the ground impedances. CAD PCB programs are particularly bad at ground design because they tend to keep all conductors as thin as possible to conserve copper and the board area, and this, of course, results in high ground resistance. There is an obvious alternative to thin ground leads — a continuous "ground plane" of copper covering one side of a PCB to which all ground connections are made. The resistance of 0.001" (0.025 mm) copper is approximately 0.67 mΩ/square inch so that this solution is frequently adequate — but not always.

4.1.4 General Rules for Design of Analog PCBs

A few general rules concerning design of PCBs for analog circuits are:
- Keep the signal path as short as possible. This will help to minimize both voltage drops through the conductors as well as electromagnetic interference by controlling loop areas.
- Provide separate analog and digital grounds and tie the two together only once.
- Provide one connection from the system ground to the actual earth ground.
- Connect capacitive shields once to provide a return path to the noise source.
- Magnetic shields must be made out of a highly permeable material to be effective.
- Metal should not be left electrically floating.
- Maintain the balance of a system to prevent common mode signals from becoming differential.
- Limit the bandwidth of the system to the required signal bandwidth.
- Keep loop areas small and always think as to where the currents will flow.
- Between the two PCBs, use twisted pair cable to improve the noise rejection of a system.

The use of software packages for the design of high speed analog PCBs, typically containing through-hole and SMT components, shielding and signals running at 2 GHz are illustrated by Meyer et. al., (1991)

 ## 4.2 Design Rules for Digital Circuits

Until recently, the only task printed circuit boards were expected to perform was to provide electrical connectivity between various components and the conductors had to be of sufficient cross-sectional area to tolerate the necessary current without excessive over-heating. The conductor separation was so arranged that it should prevent voltage breakdown. The widespread use of digital integrated circuits has now resulted in devices with extremely fast switching speeds and rise times. Electromagnetic wave propagation characteristics have become important and need to be considered carefully. Under these circumstances, the printed circuit boards may act as transmission lines if the rise or fall time of the driving device is less than twice the propagation delay.

It is essential to understand that it is the rise/fall time that is critical and not the operating frequency.

However, the frequency is dependent on rise/fall times, since the lower the value of the rise/fall time, the faster the operating frequency of the device. Under these situations, the transmission line effects become applicable and knowledge of the electrical characteristics associated with the conductors acting as transmission lines is essential. Then, the characteristic impedance must be matched to that of the receiving device to prevent reflection.

4.2.1 Transmission Lines

In order to understand the concept of transmission lines, consider a long straight wire or trace with its return wire or trace nearby. The wire has some inductance along its length. There is also some capacitive coupling between the wire and its return. Figure 4.10 shows what is commonly called a *lumped* model of the wire pair, because the capacitors and inductors are shown as lumped components. In reality, the inductance and capacitance are spread continuously along the wires. In concept, the wire is infinitely long and the figure only shows an infinitely small part of the total length.

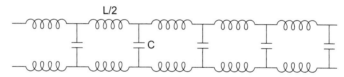

Fig. 4.10 Transmission line: made up of an infinitely long network of capacitors and inductors

If these wires are infinitely long, it is obvious that there will be no reflection at all from its far end. Also, if the wires are absolutely uniform, even then there will be no reflections. Therefore, one way to avoid reflections is to use an infinitely long, absolutely uniform wire or trace pair. Such a wire or trace pair has been given the special name *Transmission Line* (Brooks, 2002).

If we look into the front of this wire pair, there is input impedance which can be mathematically calculated. It is represented by a symbol Z_0, and is called the *intrinsic or characteristic* impedance of the line. By calculating the lumped values of inductance (L) and capacitance (C), the impedance would be then calculated as:

$$Z_0 = \sqrt{L/C}$$

Referring to Figure 4.11, let us take the infinitely long transmission line and break it into two parts. If we look at the second part, it also looks like an infinitely long transmission line with an impedance of Z_0. It thus turns out that a transmission line of finite length, terminated in its characteristic impedance Z_0 looks like an infinitely long transmission line. Therefore, even though it has a finite length, it will still have no reflections as all the energy traveling down the line is exactly absorbed or dissipated in the termination and there is no energy left to reflect back and no reflection to worry about.

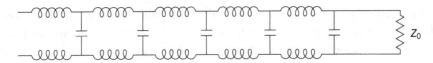

Fig. 4.11   Termination of an infinitely long transmission line to its characteristic impedance, oz

It is exactly this technique which is used to control reflections on PCB. It is required to make the traces look like transmission lines and to terminate them in their characteristic impedance Z_o. Certain types of transmission lines are commonly encountered. The coaxial cable leading to the cable TV is a 70 ohm transmission line. The 10 base 2 coax cable for networking is a 50 ohm transmission line.

From the above, it is clear that PCB traces can take on the characteristic of transmission lines. The point at which this happens (the critical length) is usually defined as when "the two-way delay of the line is more than the rise time of the pulse". For example, the critical length is approximately 3" for a signal with a 1ns rise time in FR4.

The characteristic impedance of a transmission line is a function of its geometry. In PCB applications, it is a function of several variables, two of which are the width of the line and the height of the line above the plane. If the signal trace length is greater than the critical length, and if there is no claim for the trace to reference to, it is likely that there could be no control over Z_o, no way to terminate the trace, and therefore, no way to control reflections. The reflections which are likely to occur in such cases can cause noise voltages and false signals that will cause the circuit to fail.

4.2.2   Problems in Design of PCBs for Digital Circuits

High frequency performance of printed circuit boards is becoming increasingly important in digital circuits and knowledge of electrical characteristics associated with conductors acting as transmission lines is essential (Jeffery, 1997). The main problems that can affect digital PCBs, if they are not properly designed; are:
- Reflections (causing signal delays and double pulsing, i.e. conversion of one pulse into two or more pulses);
- Cross-talk (interference between neighbouring signal lines);
- Ground and supply signal noise; and
- Electromagnetic interference from pulse type electromagnetic fields.

4.2.2.1   Reflections
Digital circuits are characterized by fast rise and fall times. Consequently, the conductors on the PCB cannot be considered as short-circuits, but as pieces of transmission lines. These transmission lines are normally mismatched with respect to source or load impedance, resulting in multiple reflections. These reflections have a deteriorating effect on the circuit performance which can be in the form of signal delays, and even double-pulsing, i.e. conversion of one pulse into two pulses. Therefore, the important point to consider while designing a digital circuit PCB is that signal conductors should have the proper value of wave impedance Z_o, so that its value gives the least reflection problems.

The desired value of the wave impedance can be obtained by properly choosing the width of the signal lines as well as the distance between signal lines and the ground line. A relatively large value of wave impedance is needed for designing TTL and CMOS logic circuits, which can be obtained from thin signal conductors. On the other hand, ECL circuits require broader signal conductors. It is important that reflections are kept small even if the digital circuit has a low operating frequency in order to avoid double-pulsing.

TTL Integrated Circuits   A basic digital circuit in which two gates are connected over a signal line with a wave impedance Z is shown in Figure 4.12. The wave impedance values plotted for a typical PCB are shown in Figure 4.13.

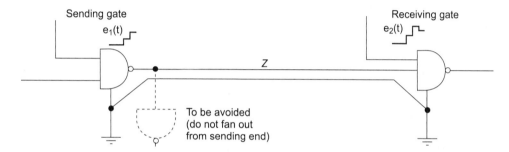

Fig. 4.12   Sending gate driving a receiving gate over a transmission line in a TTL circuit

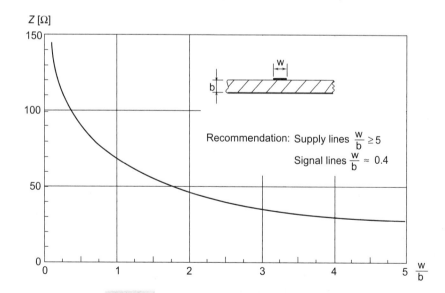

Fig. 4.13   Wave impedance Z for PCB conductors

The wave impedance desirable for TTL integrated circuit conductors has been found to range from $100\,\Omega$ to $150\,\Omega$. This value can be obtained by having conductor width of 0.5 to 1.0 mm. The use of broad signal conductors which may result in a wave impedance of 50 Ω or less should be avoided in TTL circuits, as the IC may be damaged due to large negative voltage spikes.

Also, very high impedance lines (≥ 200 Ω) should also be avoided as they may cause trouble in the form of harmful over-voltages and double-pulsing. Although very high impedance lines are not common in PCB layouts, they could occur with loose-ended open wiring. Therefore, it should be ensured that loose wiring is avoided and signal line connections between PCBs should always run near the ground line and preferably be twisted with the latter.

CMOS Integrated Circuits   The wave impedance required for CMOS integrated circuit conductors is 150-300 Ω. This large value of wave impedance can be obtained by keeping the signal conductor width as low as possible. Normally signal conductors with ≤0.5 mm width will have a wave impedance of 150 Ω to 300 Ω. So, avoid broad PCB conductors for the signal lines of CMOS circuits. Also, CMOS is not so critical with regards to cross-talk and ground and supply line noise and therefore does not require broad ground and supply conductors. In the wiring between PCBs, 50 Ω cables should be avoided. Also, for the wave impedance Z_o to remain high, the ground should not run too near the signal lines.

ECL Integrated Circuits: The wave impedance required for ECL integrated circuit signal conductors is 50 Ω-100 Ω. The wave impedance required for ECL integrated circuits is less as compared to TTL and CMOS integrated circuits. This small value of wave impedance can be obtained from broad conductors. The conductor width recommended for ECL integrated circuits is 1-3 mm.

Even with such wave impedances, reflections are likely to occur, which affect their performance especially at the rising edge wherein a lot of additional delay is caused; which is generally not acceptable for ECL circuits. So external circuit elements in the form of line driver/line receiver integrated circuits are used in ECL systems, which, at least, provide partial matching of the impedance.

4.2.2.2   Cross-talk

Cross-talk is nothing but interference of two neighbouring signals. If two signal conductors run parallel to each other for a length that exceeds roughly 10 cm (for ECL), 20 cm (for TTL) and 50 cm (for CMOS), it will induce a short spike or even a train of pulses on the neighbouring conductor. It is obvious that in the case of CMOS, cross-talk is much less dangerous because of the higher noise immunity of the CMOS family. On the other hand, ECL has a lower noise immunity and will be more sensitive to cross-talk than TTL.

In order to understand the basics of cross-talk, consider the trace A-B shown in Figure 4.14. Let us assume that a pulse is travelling down the line from A to B and is now at the point "X". There is a nearby trace, C-D. A signal may couple between A-B and C-D traces at the point X. It is likely that some capacitive coupling, though small, develops between the two traces. Since parallel traces look like a transformer, there is probably some inductive coupling between the traces also. In most PCB

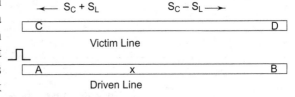

Fig. 4.14   A signal on the driven line, A-B, at point X couples a noise signal on the" victim" line, C-D, at the same point

applications, however, the material between the traces is probably a very good insulator, so there is probably no resistive coupling between the traces. Since capacitive and inductive effects are reduced with distance, any coupling reduces as the separation between the traces increases. Therefore, greater separation between traces is recommended to reduce cross-talk, (Scaminaci Jr., 1994).

It may be noted that the coupling between the traces A-B and C-D takes place only in the case of ac phenomenon and not with steady state dc signal. For an ac signal, the coupling will be greater for higher frequency or frequency components of the harmonics of the signal. So, to reduce cross-talk, lower frequency harmonics and slower rise times are recommended. The effects of coupling can be summarized as follows (Brooks 1997a):

- *Mutual Capacity Coupling*—a signal S_C caused by capacity coupling between the two traces, which travels along the victim trace in both the forward and backward direction with the same polarity.

- *Mutual Inductive Coupling*—a signal S_L caused by inductive coupling between the two traces, which travels along the victim trace in both the forward and backward direction with opposite polarity.

- *Directionality*—Cross-talk goes in both the forward and backward directions. Mutual capacitive and inductive forward cross-talk are approximately equal and opposite and tend to cancel. They are approximately equal and reinforcing in the reverse direction, and therefore tend to be additive.

- *Magnitude*—Forward cross-talk tends to look like the driven signal, and (at least in theory) continues to grow larger, the longer the coupled length A-B (and C-D). Reverse cross-talk tends to have a rectangular shape (in response to a step function) that reaches a maximum and then does not increase further regardless of the coupled length.

- *Environment*—If the two traces are contained within a homogeneous material, the inductive and capacitive forward cross-talk components are almost exactly equal and cancel. Therefore, we generally don't worry about forward cross-talk in such environments. If the surrounding material is NOT homogeneous, the inductive component tends to be larger than the capacitive component. Therefore, to reduce cross-talk, the sensitive traces should be kept in the strip line environment (traces contained within a homogeneous material).

For keeping the cross-talk low, the wave impedance between the signal conductors and ground should be low. Obviously, a close-by ground plate will definitely cut down cross-talk significantly. In more critical cases of TTL and ECL circuits, cross-talk problems can be solved by running a ground line between the two signal lines and maintaining proper wave impedance.

In high speed circuits, the cross-talk problem between parallel traces becomes acute. In Figure 4.15, cross-talk is proportional

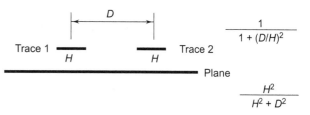

Fig. 4.15 Typical cross-talk configuration

to $H^2/(H^2 + D^2)$, where H is the distance between the trace and the plane, and D is the separation between traces. Intuitively, cross-talk diminishes as the separation between traces increases. But, all other things being equal, cross-talk will decrease as the distance between the trace and the plane decreases. So, planes are helpful in reducing and controlling cross-talk (Brooks, 1997b).

If the parallel traces are at different heights, the term H^2 really becomes the product of the two heights. As shown in Figure 4.16, the equation for cross-talk for this configuration becomes $1/1+(D^2 / H_1 \times H_2)$.

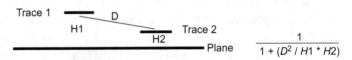

Fig. 4.16 Typical cross-talk configuration with different trace heights

Estimating cross-talk can often be difficult. The approach given depends on several simplifying assumptions and can lead to results that are closed but not necessarily precise. A practical approach is to calibrate the technique by applying it to boards that have been designed in the past and are known to be either good or bad with respect to cross-talk performance. That will provide insight into how future boards will perform based on these calculations.

4.2.2.3 Ground and Supply Line Noise

The main effects of the supply and ground signal are current spikes in the ground line. These current spikes will create transmission problems when digital ICs start switching. Internal and external current spikes are superimposed and must be carried by the same V_{CC} (supply) and ground lines. If many similar gates or flip-flops are connected to the same point, the current spikes will become excessive and the situation becomes worse.

This is a serious problem with TTL and in reduced form in ECL circuits; and in a highly reduced form in CMOS circuits. A designer should adhere to the following design rules while designing ground and supply line conductors in PCB design:

i. The wave impedance between supply and ground lines should be low, of the order of 20 Ω or lower. This will stabilize the voltage difference between Vcc and ground. This can be obtained by having broad supply and ground conductors sitting right across each other on opposite sides of a double-sided PCB. The ideal conductor width for supply and ground lines is 5 to 10 mm.

ii. An electro-magnetically highly stable ground conductor can be obtained by providing a large copper surface for ground. This is achieved by having a full ground board in case of multi-layer boards or leaving the copper in all unused parts of the PCB such as corners, etc., and connecting it to ground. The large copper area will make it difficult for ground to move up with a voltage spike. Alternatively, a closely knit grid of broad ground conductors is ideal for a digital PCB.

iii. Always avoid using the same ground lines for digital circuits and for sensitive analog circuits, because a digital ground line always has some ground noise, which may be in the range of millivolts.

iv. Use de-coupling capacitors in the power supply line: for every 2 to 3 TTL ICs, de-coupling capacitors of 10 nF are normally used whereas for C-MOS and ECL, 5 nF capacitors are usually employed. It is preferable to use ceramic chip capacitors and not electrolytic capacitors for de-coupling purposes.

The above rules must be followed strictly for TTL PCBs because the current spikes are very high in these digital circuits. Therefore, a large ground surface is absolutely essential in such cases. However, in C-MOS circuits, narrower ground lines are acceptable.

4.2.2.4 Electromagnetic Interference from Pulsed Noise

Electromagnetic radiation that adversely affects circuit performance is generally termed *EMI*, or *electromagnetic interference*. Normally digital electronic equipments are well shielded to avoid electromagnetic interference. *Shielding* is the use of conductive materials to reduce radiated EMI by reflection and/or absorption. Shielding can be applied to different areas of the electronic package, from equipment enclosures to individual circuit boards or devices. However, some interference may still get into the digital system, for one reason or other. A very common effect is that the mains and supply cables are infected with very high frequency pulse type noise due to electromechanical switches, commutators, motors, etc. This pulse type noise is carried into the casing by the mains cable and disturbs the whole system, influencing both the power supply and signal conductors. These problems can be solved by providing an EMI filter at the mains supply side. Robinson (1990) explains the types of shielding required to protect electronic equipment against electromagnetic and radio-frequency interference.

The use of polymer thick film for cost-effective EMC protection on PCBs for automotive applications, wherein EMC requirements are quite stringent, is described by Saltzberg et al. (1996). Markstein (1995) explains the theory of shielding and the shielding products available in the market to implement EMI protection for devices, PCBs and systems.

However, in the case of TTL and any other faster logic family, it is often difficult to filter all disturbances by shielding and the mains filter. Therefore, if TTL is used in a high EMI noise environment, it is essential to keep the distance between the logical signal lines and the ground line small. This means that the signal lines and the ground should be run close to each other on the PCBs and also when interconnecting PCBs.

As CMOS is much less sensitive to disturbances than TTL, it is therefore not usually necessary to keep the signal lines so close to a ground line. However, CMOS ICs get easily destroyed by over-voltages, and special protection circuits or measures, especially against electrostatic voltages are called for.

 ## 4.3 Design Rules for High Frequency Circuits

The signal which has a frequency above 300 MHz (in digital circuits) and 100 MHz (in analog circuits) can be referred as a high frequency signal. At these frequencies, even short pieces of conductors on a PCB act as transmission lines.

A conductor or a printed circuit board is considered as a transmission line if its length 'l' (in meters) is:

$$l > \frac{3\,\text{MHz}}{f_{upper}}$$

where f_{upper} = highest frequency in the signal in MHz.

Such a transmission line has a certain impedance, called the 'Wave Impedance'. A broad conductor has a smaller wave impedance than a narrow conductor. Similarly, a conductor which is near the ground plate will have a smaller wave impedance than the one which is far away. In a transmission line, if the wave impedance is not matched with the source and/or load impedance, reflection takes place. Such reflections are not desirable as they cause loss of bandwidth and an increase in rise and fall times of the pulses. Therefore, a mismatch on both sides must be avoided through proper design of the PCB when dealing with high frequency and fast pulses.

The rise-time increase due to a mismatched line will be a multiple of its transmission delay of approximately 5–10 nsec/m. As a rough estimate, it can be taken as 10–100 nsec/m or 0.1–1.0 nsec/cm

If R_S, $R_L \ll Z_O$, the line behaves as an inductor,

and if R_S, $R_L \gg Z_O$, the line behaves as a capacitor.

Where

R_S = Source impedance

R_L = Load impedance

Z_O = Wave impedance of the transmission line.

It is, therefore, necessary for the conductor impedance to be matched to the source and to the load. With such a matching, the line will hardly introduce any noticeable loss of amplitude in the frequency range of interest. However, for shorter conductors, matching is often very difficult and the conductors will then behave either capacitively or inductively. The PCB designer has to then decide which of the two, the capacitive or inductive conductor is preferable.

The ground and power supply lines also play an important role in high frequency applications. This is because the current drawn from the power supply line is fed back into the ground, where it appears as high frequency components, such as current spikes. Therefore, the dc potential of the power supply does not remain constant, leading to a significant deterioration in the behaviour of the circuit. Therefore, as a ground rule, power supply lines must be kept as short as possible.

The following guidelines are useful for high frequency circuit PCB design:

- Use a ground plate or very large ground surface for ground conductors.
- Use broad power supply conductors.
- Ground and power supply lines should run close to each other and they should be parallel.
- Provide a de-coupling capacitor between the ground and the power supply.

- The conductor length for a fast pulse system should be short as skin effect and dielectric losses increase in proportion to the length.
- For large-sized PCBs, dielectric losses play an important role. In such cases, use PCB with suitable high frequency directions.
- Decide which parasitic elements (capacitive and inductive) are more harmful and design the conductor layout accordingly.
- Keep all lines which are not matched very short, otherwise the rise-time increase could be as high as 1 nsec/cm.
- Provide ground lines (grounded or connected via a capacitor to the ground) when even a parasitic capacitance is likely to have a deteriorating effect.

4.4   Design Rules for Fast Pulse Circuits

A conductor on a printed circuit will behave as a transmission line if its length l (in meters) is:

$$l > \frac{t_r}{100\,\text{nsec}}$$

where t_r = rise-time of the pulse in nsec.

If a fast pulse travels over a line, the following situations can arise:

- If the transmission line or interconnection is matched with both the source and the load, and if the transmission line is not very long, a simple delay of approximately 5 nsec/m is introduced.
- If the transmission is matched at one end, either at source or load, there may be a single reflection, either at source or at the load.
- If the transmission line is mismatched at both ends, multiple reflections will take place, which will create disturbances and considerably slow down the rise and fall times of the pulses. Therefore, a mismatch on both sides must be avoided when dealing with fast pulses.

If the mismatch is upto about ±20 per cent, the resulting pulse will be distorted and could become rounded instead of having sharp edges.

When connecting two PCBs involving connections longer than 10–20 cm, the connection must be made either with coaxial cables or at least with twisted pairs (consisting of a signal wire twisted closely with the ground wire), otherwise complex multiple reflections will take place, causing unpredictable behaviour in the circuit. The wave impedance of commonly used coaxial cables are 50 or 75 Ω. Twisted pairs have impedances between 100 and 150 Ω.

In fast pulse circuits, various types of losses and especially skin effect and dielectric losses increase their effect on the rise-time as the conductor length is increased. Therefore, the most important rule in high frequency and pulse circuits is to keep conductors as short as possible. The specific measures taken to reduce such type of losses are detailed below.

- *Skin-effect losses:*   Increase PCB thickness and keep line length small.
- *Radiation losses:*   Use a ground plate on one side of the PCB, decrease PCB thickness and avoid discontinuities. Run signal lines near the ground plane.
- *Dielectric losses:*   Use a good high frequency dielectric PCB laminate.

Because of these losses, the design guidelines are: For very fast pulses ($t_r < 1$ nsec), even matched lines have to be kept very short. Rise-times of matched lines increase by 1 to 10 psec/cm or 100 to 1000 psec/m due to these losses.

The design challenges in high frequency and fast pulse circuits can be summarized as:

- *System timing:*   While designing high frequency PCBs, system timing plays an important role. System timing means checking whether the data get reliably transferred or not? The conductors from one component to another component should be as short as possible to achieve system timing. For example, a high frequency signal conductor with a large length may result in an unintended logic. What was supposed to be read as a 1 will be read as a 0.
- *Waveform integrity:*   Waveform integrity means that the signal should meet electrical requirements. One should get required waveforms similar to simulation results. Waveform integrity can be achieved by calculating the proper wave impedance of signal conductors. Wave impedance of a signal conductor depends upon the conductor width.
- *Cross-talk:*   Like reflection, cross-talk is a transmission line phenomenon. Cross-talk can be kept low if a ground line or a ground plane is nearby. These cross-talk problems can also be solved by maintaining proper wave impedance of the signal conductor. The wave impedance of a signal conductor depends upon the conductor width. Normally the conductor width is inversely proportional to the wave impedance of a signal conductor.

4.4.1   Controlled Impedance Considerations

It has been explained above that today's fast switching speeds or high clock rate PCB traces are treated as transmission lines whose electrical characteristics must be controlled by the PCB designer. Obviously, the critical parameter is the characteristic impedance of the PCB trace. In practice, the trace impedance has to be controlled when designing for digital edge speeds faster than 1ns or analog frequencies greater than 300 MHz. In general, the controlled impedance has to be considered when the electrical length of the signal line exceeds 30 per cent of the signal rise time (Polar Instruments, 2001).

The devices mounted on a PCB themselves possess characteristic impedance and the impedance of the interconnecting PCB traces must be chosen to match the characteristic impedance of the logic family in use. Referring to Figure 4.17, in order to maximize signal transfer from the source (device A) to the load (device B), the trace impedance must match the output impedance of the sending device (device A) and the input impedance of the receiving device (device B). For CMOS and TTL, this will be in the region of 80 to 110 ohms. If the impedance of the PCB trace connecting two devices does not match the devices' characteristic impedance, multiple reflections will occur on the signal line resulting in increased switching times or random errors in the high speed digital system.

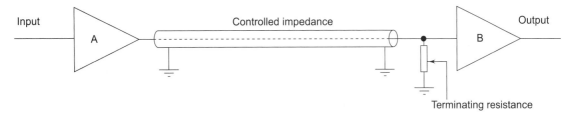

Fig. 4.17   Single-ended PCB trace

The single-ended transmission line as shown in the figure is probably the commonest way to connect two devices. In this case, a single conductor connects the source of one device to the load of another device. The reference (ground) plane provides the signal return path. This is an example of an unbalanced line. The signal and return lines differ in geometry—the cross-section of the signal conductor is different from that of the return ground plane conductor.

Controlled impedance PCBs are usually produced by using microstrip or stripline transmission lines in single-ended (unbalanced) or differential (balanced) configurations (Polar Instruments 2003C). The differential mode of operations is shown in Figure 4.18. The differential configuration is used when better noise immunity and improved timing are required in critical applications. This configuration is an example of a balanced line—the signal and return paths have similar geometry. The lines are driven as a pair with one line transmitting a signal waveform of the opposite polarity to the other. Fields generated in the two lines will tend to cancel each other, so EMI and RFI will be lower than in the case of the unbalanced line, and problems with external noise are reduced.

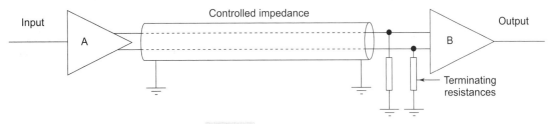

Fig. 4.18   Differential PCB trace

The impedance of a trace on a PCB can be controlled by carefully controlling the following geometrical dimensions (Bhardwaj, 2001), as shown in Figure 4.19:
- The width (w) and thickness (t) of the signal trace;
- The height (h) of core or pre-preg material on either side of the trace;
- The dielectric constant (D_k/E_r) of the core and pre-preg material; and
- The configuration of trace and planes.

The relation between the characteristic impedance of a trace and these physical factors is explained below.

Impedance (Z) is

- inversely proportional to trace width ($Z \propto 1/W$);
- inversely proportional to trace thickness ($Z \propto 1/t$);
- proportional to laminate height ($Z \propto h$); and
- inversely proportional to the square root of laminate Er ($Z \propto 1/\sqrt{Er}$).

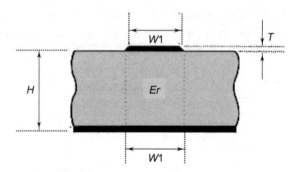

Fig. 4.19 Surface microstrip trace structure

The impedance of a PCB trace will be determined by its inductive and capacitive reactance, resistance and conductance. PCB impedances will typically range from 25 to 120 ohms. In practice, board designers will specify impedance values and tolerances for board traces and rely on the PCB manufacturer to conform to the specification.

Impedance calculations are usually very complex and depend upon variables that are difficult to control (UltraCAD Design, 2000). However, most applications require constant impedance traces, but not necessarily traces with the exact absolute value of the impedance, except when interfaces with backplanes and other circuit cards are required. Turn and vias generally have a minimal effect on impedance, except that moving a controlled impedance trace from one layer to another can cause significant and subtle problems unless the two layers are on immediately opposite sides of a single power distribution (reference) plane.

Most controlled impedance PCBs undergo 100 per cent testing. Impedance measurements are usually made with a time domain reflectometer (TDR). The TDR applies a fast voltage step to the coupon via a controlled impedance cable and probe. Any reflections in the pulse waveform are displayed on the TDR and indicate a change in impedance value. This is known as a discontinuity (Polar Instruments, 2003). The TDR is able to indicate the location and scale of discontinuity. Using appropriate software, the TDR can be made to plot a graph of the impedance over the length of the test trace on the coupon. The resulting graphical representation of the trace characteristic impedance allows previously complex measurements to be performed in a production environment.

 ## 4.5 Design Rules for PCBs for Microwave Circuits

4.5.1 Basic Definitions

It was explained in the previous sections that in the case of high frequency circuits, the conductors on the PCB behave as transmission lines. Besides the characteristic impedance of the transmission line, there are some other useful terminologies which should be understood. These are listed below.

Reflection Coefficient

When a power source is connected to a transmission line which is not terminated by its characteristic impedance, the energy gets reflected from the termination. The reflection is usually given as a fraction of the incident wave as follows:

$$\rho = \text{Reflection coefficient} = \frac{\text{Reflected wave amplitude}}{\text{Incident wave amplitude}}$$

Voltage Standing Wave Ratio (VSWR)

When a sinusoidal voltage is applied to a transmission line, the voltage at any point along it can be obtained by adding the incident and reflected voltages. In such a case, if the length of the line is greater than half the wavelength corresponding to the input frequency, voltage maxima and minima are established along the line. This is shown in Figure 4.20. From this, the voltage standing wave ratio (VSWR) is defined as:

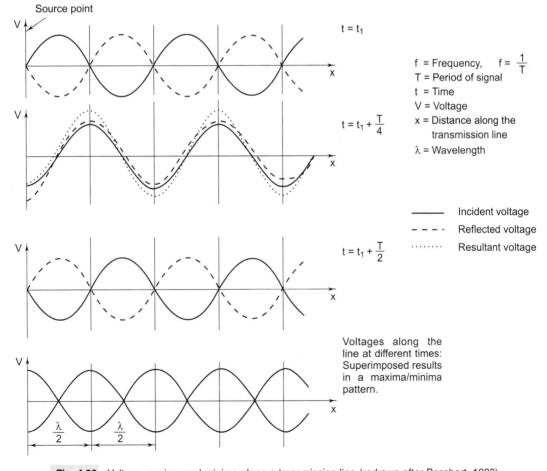

Fig. 4.20 Voltage maxima and minima along a transmission line (redrawn after Bosshart, 1983)

$$VSWR = \frac{Maximum\ value\ of\ the\ voltage\ along\ the\ line}{Minimum\ value\ of\ the\ voltage\ along\ the\ line}$$

VSWR is related to ρ by the relation

$VSWR = (1 + \rho)/(1 - \rho)$

Modes of Propagation

When a power source is connected to a transmission line, the wave propagation along the line can be described completely by describing the direction with respect to the propagation direction, magnitude and time variation of electric (E) and magnetic (H) waves. These two waves may have many possible orientations with respect to the propagation direction. Each possible orientation is referred to as a mode of propagation.

One of the most common types of propagation is the Transverse Electric and Magnetic (TEM) mode. In this mode, the electric and magnetic fields are perpendicular to the direction of wave propagation.

4.5.2 Strip Line and Microstrip Line

Today's high performance PCB traces are manufactured as transmission lines. In principle, several types of planar transmission lines can be fabricated. However, the strip line and the microstrip line are the most common type and are therefore described below.

4.5.2.1 Strip Line

A strip line is basically a sandwich of two PCBs: one double-sided PCB with the transmission line on one side, and a ground plane on the other, and one single-sided PCB with ground plane over its entire area. This is illustrated in Figure 4.21. The mode of propagation is TEM. In this mode, the electric and magnetic fields are perpendicular to the direction of wave propagation.

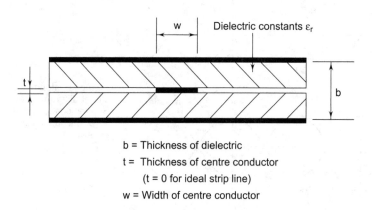

b = Thickness of dielectric
t = Thickness of centre conductor
(t = 0 for ideal strip line)
w = Width of centre conductor

Fig. 4.21 Practically realized strip line

There are typically two configurations of PCB stripline:

- *Centred or Symmetric Strip Line:* In this configuration, which is shown in Figure 4.22, the signal trace is sandwiched symmetrically, i.e. centred between the two reference planes. This is often difficult to achieve as the laminate above and below the trace will be either C-Stage or B-Stage (core or pre-preg) material.
- *Offset or Asymmetric Strip Line:* In this configuration, shown in Figure 4.23, the trace is sandwiched between the two reference planes but is closer to one plane than the other.

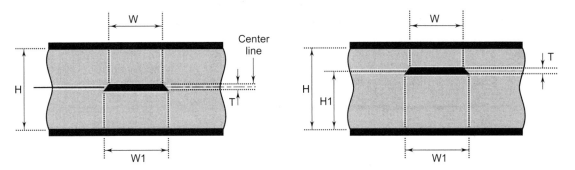

Fig. 4.22 Symmetrical strip line configuration   **Fig. 4.23** Offset or asymmetrical strip line

- *Dual Strip Line:* The structure of a dual stripline is shown in Figure 4.24, which has a second mirror trace positioned at distance $H1$ from the top ground plane. In this case, the two signal conductors are sandwiched between the two reference planes on adjacent layers. These two signal layers will be routed orthogonally to minimize inter-layer cross-talk; i.e. the signal layers are made to cross at right angles so as to minimize the crossing area. The structure then behaves as two independent offset striplines. (Polar Instruments, 2003)

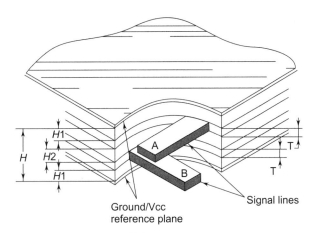

Fig. 4.24 Dual strip line

The value of the impedance of the stripline will be determined by its physical construction and electrical characteristic of the dielectric material. These factors are the width and thickness of the signal trace, the dielectric constant and height of the core or pre-preg material on either side of the trace and the configuration of trace and planes.

4.5.2.2 Microstrip Line

A microstrip line is nothing but a double-sided PCB with a conductor line on one side and a ground plane on the other side. This is illustrated in Figure 4.25. In other words, a microstrip transmission line consists of a conductive trace of controlled width on a low loss dielectric mounted on a conducting ground plane. The dielectric is usually made of glass-reinforced epoxy such as G10 or FR-4, or PTFE for very high frequency.

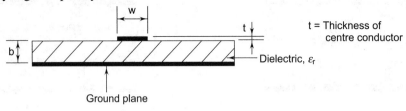

Fig. 4.25 Microstrip line construction

The mode of wave propagation in a microstrip line is not strictly TEM, but quasi-TEM. This is because of the discontinuity in the dielectric and the absence of symmetry of the ground plane with respect to the line conductor.

The important dimensions for the microstrip line are W (the upper track width), H (the laminate thickness) and G (the width of the ground plane). Ideally G should be infinite, but in practice, 10 W can be acceptable. For very low impedances, i.e. less than 30 ohms, it can be even reduced to 5 W.

Although straight lines are preferable when making a microstrip line, it is often necessary to use bends. For frequencies upto several GHz, low VSWRs are achieved if significant bends are trimmed at 45° as shown in Figure 4.26. Coaxial connectors to the boards are best inserted through the ground plane rather than at the board edges, as illustrated in Figure 4.27.

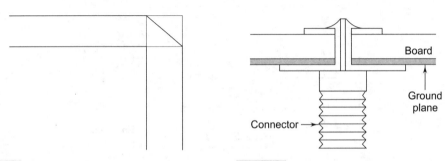

Fig. 4.26 For maintaining good VSWR, the stripline is trimmed at 45° corner

Fig. 4.27 Use of coaxial connectors inserted through the ground plane on a PCB

The following are the various configurations of a PCB microstrip (Polar Instruments, 2003b):

- *Surface Microstrip:* This is the simplest configuration and is shown in Figure 4.28. It consists of a signal line, with the top and the sides exposed to air, on the surface of a board of dielectric constant *Er* and reference to a power or ground plane. Surface microstrip can be implemented by etching one surface of double-sided PCB material.

- *Embedded Microstrip:* Also known as buried microstrip it is similar to the surface microstrip. However, the signal line is embedded (Figure 4.29) in a dielectric and located at a known distance *H1* from the reference.

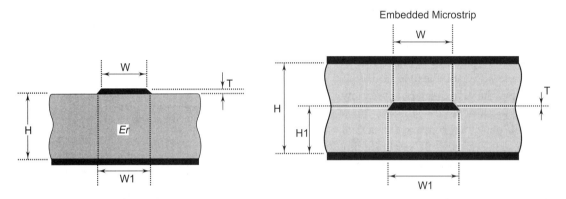

Fig. 4.28 Surface microstrip structure   **Fig. 4.29** Embedded microstrip with edge-coupled differential traces

- *Coated Microstrip:* This (Figure 4.30) is similar to the surface version. However, the signal line is covered by a solder mask. The solder mask coating can lower the impedance by up to a few ohms depending on the type and thickness of the solder mask.

- The equations for characteristic impedance for the microstrip require complex mathematics, usually using field solving methods including boundary element analysis.

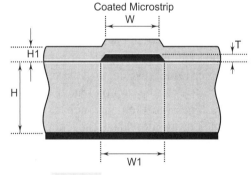

Fig. 4.30 Coated microstrip

4.5.3 Transmission Lines as Passive Components

Many electronic circuits such as tuned amplifiers, filters, etc., operating at microwave frequencies, make use of passive components like inductors and capacitors. However, their values are too small to be constructed by conventional means such as wound inductance and parallel plate capacitors. Not only is the physical size too small but the parasitic effects may disturb the circuit function. In such situations, the only method is to use transmission lines of suitable length with suitable termination as inductors and capacitors. Figure 4.31 shows a typical configuration of a transmission line connecting a source to a load.

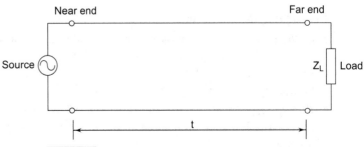

Fig. 4.31   Transmission line connecting a source to a load

By choosing a proper length of the transmission line, either an inductance or a capacitor of desired value at any frequency can be realized. It may be noted that for a given length of transmission line, the value of inductance or capacitance is a function of frequency. Therefore, microwave circuitry can be realized on a PCB by having transmission line elements with suitable length and termination. With developments in PCB technology and the availability of microwave dielectric materials, it is now possible to fabricate transmission lines in a single plane, wherein for all practical purposes, the width of the conductor controls the property of the transmission line. This has facilitated the formation of complex microwave circuits by interconnecting components to the transmission line elements.

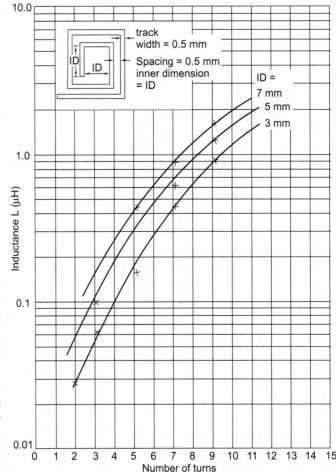

As explained earlier, the transmission lines can be used as passive elements. A line which is short in length as compared to the wavelength of the signal transmitted can be approximated as a *lumped* passive element such as an inductor or an individual capacitor. A thin line gives an inductance and a thick line, a capacitance. Of the various possible geometries, the flat spiral type is the most common because it provides the greatest inductance per unit area. With a double-sided board, the inductance per unit area can be increased by placing turns on both sides of the boards. Figure 4.32 gives the inductance values and shape for a range of printed flat spiral inductors. Where a simple low Q

Fig. 4.32   Inductance values and shape of printed flat spiral inductors transmission line connecting a source to a load (redrawn after Haskard, 1998)

inductor is required, the form as shown in Figure 4.33 can be used. Its inductance value for a given area is nearly one-tenth that of a spiral inductor.

Fig. 4.33 Example of printed circuit meander inductor

These components can be used to make filters, impedance transformers, matching devices, etc., for example, Figure 4.34 shows a low pass filter, which can be built using transmission line segments such as an inductor and capacitor. This circuit can yield low pass filter with a cut-off frequency in the range of a few megahertz.

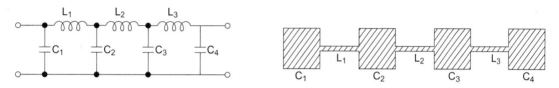

Fig. 4.34 Low pass filter circuit and its microstrip form of realization

Transformers can also be realized in printed circuit form. Figure 4.35 shows several configurations used for making transformers. It may be noted that coupling on the meander transformer is low, typically fewer than 10 per cent whereas it can approach 90 per cent with spiral inductors.

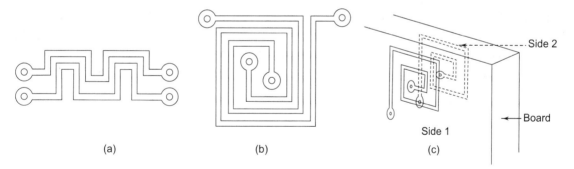

(a)   (b)   (c)

Fig. 4.35 Examples of printed circuit transformers: (a) meander transformer (b) spiral transformer on one side of a board, and (c) using two sides of a board (after Haskard, 1998)

4.5.4 General Design Considerations for Microwave Circuits

The requirement of high accuracy for line width for microwave applications is much higher; otherwise it seriously affects the VSWR (voltage standing wave ratio). For example, consider a 75Ω source connected to a 75Ω load through a line whose characteristic impedance is exactly 75Ω. In such a case, no reflection takes place and the VSWR is exactly one.

Consider now a line whose characteristic impedance is 80 Ω, caused due to reduced line width and connected to the same load and source. Then the reflection coefficient $\rho = (80 - 75) / (80 + 75) = 0.032$

and therefore VSWR is = $(1 + \rho) / (1 - \rho) = 1.032 / 0.968 = 1.066$. This shows that VSWR becomes poorer if the line width is not exactly what has been calculated.

To obtain high accuracy of line width, the artwork, should therefore, be made 4 to 16 times larger than the actual size while designing microwave PCBs.

The wave which propagates along the strip and microstrip transmission line gets attenuated due to: (i) dielectric loss (ii) loss in the conductor of the line, and (iii) radiation loss (mainly in the microstrip line). Therefore, PCB materials for microwave applications should be selected so as to yield minimum loss. This is achieved by choosing a material with high dielectric constant which reduces radiation and dielectric loss and results in reduction of the size of the microwave circuit, thus extending the usefulness of microwave PCBs to higher frequencies. The dielectric materials used at high frequencies, including microwave frequencies are Rexolite 1422 polystyrene, Silicon resin with ceramic powder filling and Teflon fibre-glass.

4.6   Design Rules for Power Electronic Circuits

If the circuit is characterized by low power requirements and signals with slow rise times and large signal levels, the power distribution system may not be too critical. But as many of these factors change, power conditioning requirements increase, and effective methods for distributing power around a PCB and handling large heat dissipation devices and conductors need careful design effort.

The design of power electronic circuits is more critical than common electronic circuits. This is because comparatively high power is flowing on these PCBs and a failure occurring under the operational conditions on such boards can easily lead to far more serious consequences, including danger to personnel. Several factors need to be considered in the designing of power electronic circuit PCBs. These are discussed below.

4.6.1   Separating Power Circuits in High and Low Power Parts

In power electronic circuits, the circuit which carries less than 3 amp current can be considered as a low power circuit. The circuits which carry more than 3 amps are to be considered as high power circuits. Generally, a control circuit of a considerably low power level controls an active high power electronic component. For example, it is common that a TTL circuit drawing less than 1 amp at a voltage of only 5V may be controlling a thyristor through which the current flow may be as high as 50 amp. One may normally plan to have both the power conditioning and its control circuit on one PCB. However, Figure 4.36 illustrates, in a simplified way, an SCR control circuit. It may be noted that here even the pulse transformer, which provides isolation, is mounted on the high power part of the PCB and not a control PCB, as its secondary winding is driving the high power SCR circuit. If we design both low and high power circuits on one PCB, it will result in capacitive and inductive coupling between power circuits and control circuits, leading to malfunctioning of the equipment. Therefore, the low and high power circuits should be designed on separate PCBs.

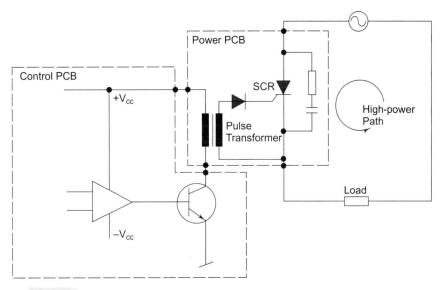

Fig. 4.36 Division of circuit into high- and low-power parts (after Bosshart, 1983)

4.6.2 Base Material Thickness

Power electronic devices dissipate a certain amount of heat which usually needs suitable heat sinks. If the heat sink is directly mounted on the PCB, the whole board will be heated up to the same temperature. Therefore, the base material selected must withstand the continuous operation of the equipment. A very common choice is glass epoxy laminates. The most used laminate has a thickness of 1.6 mm; however, 2.4 mm and 3.2 mm will thicknesses meet the mechanical property requirements for mounting of heavier components such as pulse transformers, heat sinks, chokes, etc. Heat sinks are now available as pastes for printing.

4.6.3 Copper Foil Thickness

A copper clad laminate with a 35 μm standard thickness is preferable for low power circuits. For high power circuits, normally a copper clad laminate with a 70 μm thick copper foil is commonly used. For special cases, even 105 μm thickness of copper foil may be required.

4.6.4 Conductor Width

In the design of power electronic PCBs, the copper available on the board surface should be fully utilized for the larger currents. The procedure is to first determine the required spacing between the conductors and then allot the remaining copper area to the conductors. Conductors carrying large current should be designed with large conductor width. It is also necessary to analyse the circuit to determine the most probable circuit failures and the conductors likely to be affected on the PCB. A check must be carried out to ascertain that they can carry the fault current. If, not, the conductor width may be increased as far as possible.

4.6.5 Resistive Drop of Voltage

In power electronic circuits, high currents flowing through the PCB conductors can cause a considerable voltage drop. Wherever possible, these heavy load currents should be avoided on the PCB. In case it is unavoidable to bypass the load current and it has to be carried through the PCB, the conductor should be so designed that the voltage drop caused thereby should not have any influence on the functional ability of the circuit.

4.6.6 Thermal Considerations

Heat gets generated on the PCB from two sources: the board itself and the components mounted on it. Since each system (and components) has a maximum temperature of operation, care must be taken to ensure that this temperature is not exceeded. Use of heat sinks, forced air cooling, placement of components as well as the mounting of the board in horizontal or vertical position will affect the temperature of the board and the components mounted on it.

Considering the maximum allowed temperature rise, caused by the copper track, Table 4.1 gives the minimum allowable track width which can be used in order to ensure temperature rise less than 10°C, 20°C and 40°C for various dc currents. It may be remembered that 1mm track width has a safe current rating of a little more than 2 A and this will not cause an excessive voltage drop in the conductors.

Table 4.1 Minimum Cu-track Width (mm) for Temperature Rise of Less than 10°, 20° and 40° C for Various DC Currents

DC Current Amps	Temperature Rise °C (35µm copper-foil)		
	10°C	**20°C**	**40°C**
0.5	0.15 mm	0.10 mm	0.06 mm
1.0	0.40 mm	0.25 mm	0.15 mm
2.0	0.80 mm	0.50 mm	0.30 mm
5.0	3.25 mm	1.75 mm	1.00 mm
10.0	8.00 mm	4.70 mm	3.00 mm

These values only determine the increase in board temperature due to static currents. To this must be added any large direct, alternating or switched currents causing significant component heat dissipation. Fortunately, modern EDA tools allow thermal analysis to be undertaken quickly and accurately by using the analogy between the current flow and the heat flow. Any analogue electrical simulation program (like SPICE) should be able to model the static and dynamic thermal considerations.

 ## 4.7 High-density Interconnection Structures

The increasing use of fine pitch ball grid array (BGA), chip scale packaging (CSP) and other evolving technology factors means that new design and fabrication techniques must be adopted for PCBs to accommodate components with extremely tight pitches and small geometries. Besides, extremely fast clock speed and high signal bandwidths necessitate board configurations that overcome the negative effects exerted by radio frequency (RF) and electromagnetic interferences (EMI) on a product performance. In addition, increasingly restrictive cost targets limit the use of traditional methods for fabricating smaller, denser, lighter and faster interconnect systems. The use of PCBs incorporating microvia circuit interconnects represents a viable way of addressing solutions to these problems (Brist et al; 1997).

Microvias are vias of less than or equal to 6 mils (150 micron) in diameter. Their most typical use today is in blind and buried vias used to create interconnections through one dielectric layer within a PCB. Microvias are commonly used in blind via constructions where the outer layers of a multi-layer PCB are connected to the next adjacent signal layer. Used in all forms of electronic products, they effectively allow for the cost-effective fabrication of high-density assemblies. The IPC has selected High-Density Interconnection Structures (HDIS) as a term to refer to all the various microvia technologies (Holden, 2003a).

By using microvias, components can be placed much closer to each other thereby freeing up the trace routable area. Microvias can be placed directly in landing pads, eliminating the need for fan-out of short traces that connect pads to hole. Increasing the density of board components can sometimes make it possible to place all the components on one side of the board, thus eliminating the expense of double-sided component assembly. Figure 4.37 shows the reduction in size of the printed circuit board through the conventional and microvia way.

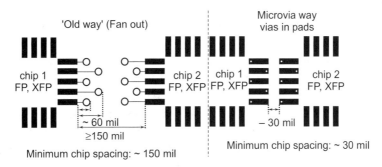

Fig. 4.37 Using microvias, components can be placed much closer to each other, freeing up the trace routable area (courtesy Merix Corporation)

The PCB density can be measured by its *wiring capacity* which is measured in inches of wiring per square inch of substrate. The total wiring capacity of a board depends on the *channel width* and the number of tracks per channel. Channel width is the distance between adjacent via or component pads, and tracks per channel refer to the number of traces that can be routed on one channel width. The number of tracks per channel depend, of course, on the trace and space widths, and the pad sizes. The reduction in via pad size with microvia design makes it possible to achieve much higher wiring density.

Blind and buried vias (Figure 4.38) further help to reduce the size of the board and the layer counts. They connect only those layers that require connection. By definition, *a blind* via is a copper plated hole, just like a regular via, except that it interconnects only one external layer of the PCB with one or more internal layers, but does not go all the way through the board. *A buried* via is a copper plated hole that interconnects one or more internal layers, but does not connect to an external layer, hence the hole is completely internal or buried within the board. Today's semiconductor packages demand more interconnections, and blind and buried vias provide a creative means of fitting those connections into less space.

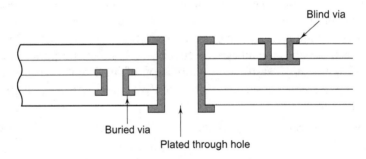

Fig. 4.38 Blind and buried vias enable to fit more interconnections in less space

4.7.1 Drivers for HDI

4.7.1.1 Density

Semiconductor complexity and increases in total gates have necessitated more pins for integrated circuits (ICs) as well as finer pin pitch. Over 2000 pins on a 1.0 mm pitch BGA is not unusual, nor is 296 pins on a 0.65 mm pitch device. As discrete components continue to get smaller and IC packages are increasingly becoming BGAs, the total number of connections on both sides of a board increases. When the average connection per square inch begins to exceed 100 pins (connections) per sq. inch, there is less room to wire these devices. The space occupied by the SMT land pattern, the through-hole via and the traces that connect them, begin to exceed what you can put in a single square inch. Beyond around 120 connections per square inch, additional layers have to be added to complete the interconnect. The layer count begins to go up exponentially as shown in Figure 4.39.

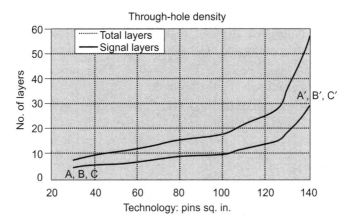

Fig. 4.39 Relationship between pin count and umber of layers. As the pins per square inch goes above 130, the total number of layers grow exponentially(after Holden, 2003a)

Faster rise-times as well as the need for signal integrity (*SI*) necessitate an increasing number of power and ground pins. Consequently, this creates the need for layers in multi-layer boards and the need for HDI with microvias (Holden, 2003a).

4.7.1.2 Fine Pitch Devices

1.0 mm pitch devices benefit from HDI, but the use of 0.8 mm pitch devices (Holden, 2003b) is where HDI really begins to provide advantages. The blind vias save room on inner layers and have reduced via lands, besides making via-in-lands possible. Typical of these devices is the 240 pin, 0.65 mm pitch, Digital Signal Processor (DSP).

4.7.1.3 High I/O Area Arrays

The other new components becoming more widespread are ones with very high pin counts of around 600 to 2500 pins, even at 1.27 mm and 1.0 mm pitches. While some of these are telecom digital switches, the vast majority are the new field programmable gate arrays (FPGAs). Current products have packages with 240 to 1200 pins.

4.7.2 Advantages of HDI

Microvias offer several distinct advantages over their mechanically created counterparts. These are listed below.

- Systems with higher circuit density with better electrical performance can be created by using the smallest and the most advanced components available. As a result, smaller, lighter and more robust products can be built. They enable greater track density, which, 'in turn', increases the potential for layer count reduction and reduced fabrication cost.
- Microvias reduce switching noise, which is attributed to the decreased inductance and capacitance of the microvia as its physical size becomes smaller and shorter. There is a reduction in signal reflection and cross-talk between traces.

- Due to increase in the routability area, it is possible to place more ground plane around components. By doing this, the size of the ground return loops decreases, resulting in a reduction of radio frequency and electromagnetic interferences.
- Microvias require the use of fewer materials and fewer processing steps, both of which reduce the cost of manufacturing the product.
- Microvias are made with photo definition, laser ablation, or plasma-etching and offer distinct advantages over their mechanically drilled counterparts.

4.7.3   Designing for HDI

IPC has developed standards with which the board designers must be thoroughly familiar. The basic information required for this purpose is as follows:

- IPC HDI Standards (IPC-2315, IPC-4104, IPC-6016, IPC-9151 [2]);
- Material selection (IPC-4104); and
- Stack-up and design rules (IPC-2315).

IPC HDI Design Standards (IPC-2315):   This would be helpful to select the minimum and simplest technology/architectures for the design. Figure 4.40 shows the most common microvias structures from the simplest (type 1) to the most complex (type 3) with stacked vias.

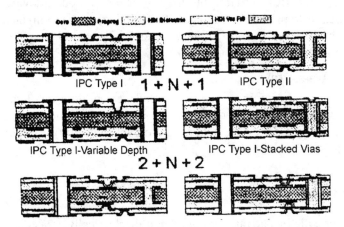

Fig. 4.40   The IPC HDI Type I to Type III are the most common microvia structures used from the simplest (Type I) to the most complex (Type III with stacked vias).

Material Selection: The most important step in HDI design is the selection of materials as they determine performance and fabrication technology. When designing for HDI, one can choose from an increasing range of new materials available that are not commonly used for conventional multi-layer boards. Glass-reinforced laminates and resin-coated copper foils are the most popular HDI materials (Holden, 1997).

Stack-up and Design Rules: The selection of signal layer stack-up and design rules determines the maximum wiring capability (*Wc*) for a design. The schematic and total component parts list, along with their connections, can be used to estimate the total wiring lengths required to connect this design. This is the wiring demand (*Wd*). The actual wring capacity is the maximum wiring capacity multiplied by the design layout efficiency (*LE*). The actual wiring capacity must always be larger than the wiring demand, $Wd < = LE * Wc$ (Holden, 2003b)

The layout efficiency (*LE*) is the ability to deliver the design rules to the final product. The higher the layout efficiency, the smaller or fewer signal layers will be required.

Holden (2003b) suggests that if you plan to design a HDI board, start by setting up your CAD system and create a test vehicle as your first HDI board. A simple test vehicle of fine-pitch BGAs land patterns, via-in-pads, in-circuit assembly test schemes, high-frequency test structures and reliability daisy-chains will all help to answer nagging questions and provide an insight on HDI without critically obstructing some new project. The help of a typical CAD menu, which defines the HDI microvia structure in a PCB design, is always available in the packages such as the one from Mentor Graphics.

One very useful HDI design technique is to use the blind vias to open up more routing space on the inner layer. This is shown in Figure 4.41. By using blind vias, the routing space effectively doubles on the inner layers and many more traces can be used to connect pins on the inner rows of a BGA. With this technique, half to one-third the number of signal layers is required to connect a complex, high-I/O BGA.

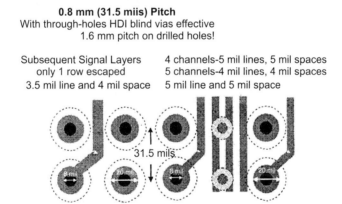

0.8 mm (31.5 miis) Pitch
With through-holes HDI blind vias effective
1.6 mm pitch on drilled holes!

Subsequent Signal Layers   4 channels-5 mil lines, 5 mil spaces
only 1 row escaped   5 channels-4 mil lines, 4 mil spaces
3.5 mil line and 4 mil space   5 mil line and 5 mil space

Fig. 4.41 Channel routing utilizes blind vias to create an inner-layer channel to route out interior pins on devices (courtesy Merix Corporation)

Many HDI boards have controlled impedance requirements as in conventional PCBs. When edge speeds are fast and traces are long in comparison, then impedance needs to be taken into account. In case of boards having a design for both low voltage and high speed, the noise margin will be lower and hence more susceptible to changes in impedance -causing reflections (Gaudion, 2000),

4.8   Electromagnetic Interference/Compatibility (EMI/EMC)

High speed digital circuits are a major source of electromagnetic radiation. The level of emissions from a PCB is significantly affected by high clock rates, fast signal rise/fall times, long track length and poor ground grid. At high frequencies, traces on a PCB act as a mono-pole or loop antennas. The transmission line effects become an important design consideration when the trace length approaches one-seventh of the wavelength of the signal being transported. If the system clock frequency is 200 MHz, the wavelength in FR4 is about 0.5 m.

Onley (2003) explains the EMC design considerations for high-speed PCBs. In digital circuits, the pulse information of "1" or "0" is carried on the leading edge of the pulse. The frequency and the rise time of the signal are related as follows:

Tr (rise time in nS) = 0.35 / frequency in GHz

Table 4.2 shows the rise times and wavelengths for common high-speed ICs.

Table 4.2   Rise Times and Wavelengths of Commonly Used Digital Devices

Parameter	TTL	Schottky TTL	ECL	GaAs
Output Rise Time (ns)	6-9	2-3	0.45-0.75	0.05-0.20
Wavelength in free space (m)	6.8	2.5	0.52	0.086
Wavelength in FR4 (m)	3.1	1.2	0.24	0.04

In case of ECL, the frequency would be 0.35/0.45 = 777 MHz which translates into a wavelength of about 375 mm in free air or 175 mm in FR4 and 100 mm in ceramic. Therefore, if the trace length is more than 25 mm for PCBs fabricated from FR4, then the electromagnetic properties of the ECL signal and the transmission line effects should be considered.

Another design consideration concerns characteristic impedance. Fifty to eighty ohm characteristic impedance is often used in high-speed designs. Lower impedance values cause excessive di/dt cross-talk and can double the power consumed to create a heat dissipation problem. Higher impedances not only produce high cross-talk, but also produce circuits with greater EMI sensitivity and emission.

An important concept in PCB design for EMI compliance is the loop area concept. A close circuit of tracks as shown in Figure 4.42 comprises a loop. Here, interference can both exit and enter the PCB into areas inside the loops. The smaller the dimensions of each loop, the smaller will be the magnitude of interference to be dealt with. The circuitry on the board may be contained inside several small loops, through a gridded power supply distribution. The best distribution of the lot is one or two continuous planes or sheets of copper.

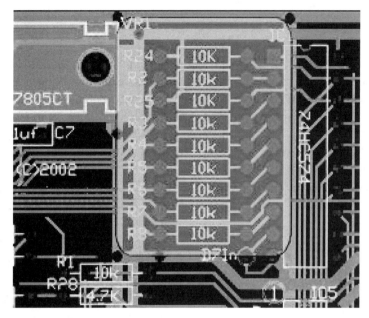

Fig. 4.42   The concept of loop area for EMI compliance in PCB Design (courtesy M/s AirBorn Electronics)

A four-layer circuit board, with the two centre layers being power and ground, is often used to minimize EMI emissions and susceptibility. One way this works is by making the loop area very small. When using double-sided PCBs, without the opportunity for ground planes, special attention must be paid to the loop area. High frequency circuits, such as crystals, should be enclosed by an overall ground. If there is an opportunity to earth the metal case of such devices, it is usually worthwhile to do so.

Enclosing signal traces between the ground and power planes provides a shield which reduces both radiation (by up to 45 dB) and susceptibility to radiation, besides providing ESD protection. It is a good practice to route high-speed, fast -rise time signals between those planes to eliminate radiation. If a large capacitance exists between the rails, both ground and power planes may be used as reference planes.

The clock generation components must be located near the centre of the PCB or adjacent to a chassis ground location, rather than along the perimeter of the board or near I/O section (Montrose, 2003). The clock lines should always be kept short. The longer the trace, the greater will be the probability of the RF currents being produced, and more spectral will be the distribution of the RF energy created. Clock traces must be terminated to reduce ringing (enhance signal integrity), and to prevent the creation of avoidable RF currents. In order to minimize electromagnetic interference and susceptibility, and to verify their design, software packages are available which can be used to analyse the physical layout and estimate the radiated emission potential of signal tracks on a PCB (Lum and Waddell, 1996). These tools range from fast, interactive tools that provide conservative analyses to full 3-D analysis tools designed to provide detailed results.

4.9 Useful Standards

- *IPC-SM-782-A: Surface Mount Design and Land Pattern Standard-Includes Amendments 1 & 2:* Includes land patterns for all types of passive and active surface mount components; also covers EIA/JEDEC registered components, land pattern guidelines for wave or re-flow soldering, dimensioning system, via location guidelines, etc.
- *IPC-EM-782: Surface Mount Design and Land Pattern Standard Spreadsheet:* Includes industry consensus land pattern sizes; providing appropriate size, shape and manufacturing tolerance of lands.
- *IPC-HDI-1: High Density Interconnect Microvia Technology Compendium:* Includes standards, specifications and guidelines to produce high density interconnect (HDI) and microvia boards. The document comes with other multiple documents, defined separately.
- *IPC-6016: Qualification and Performance Specification for High-density Interconnect (HDI) Layers or Boards:* Covers specific electrical, mechanical and environmental requirements for organic HDI layers with microvia technology.
- *IPC-2141: Controlled Impedance Circuit Boards and High-speed Logic Design:* Provides design guidelines for printed circuit board designers, packaging engineers, printed board fabricators and procurement specialists.
- *IPC-2252: Design Guide for RF/Microwave Circuit Boards:* Covers information required to design practical, functional and cost-effective microwave circuit boards in the frequency range of 100 MHz to 30 GHz.

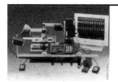

Artwork Generation

5.1 What is Artwork?

Artwork is basically a manufacturing tool used in fabricating printed wiring because it uniquely defines the pattern to be placed on the board. Artwork displays only those items that have to be generated as copper patterns in the manufacture of the PCB. Therefore, the artwork will necessarily include solder pads, lands and conductors true to scale in respect of their dimensions, but shown at the scaled level. In addition, the artwork will also show lines that represent the boundary of the board.

Since the artwork is the first step of the PCB manufacturing process, it has a great influence on the final product. Obviously, the final product can only approach the quality and accuracy of that of the artwork. The generation of artwork to the requisite sophistication is highly dependent upon the area (board size) and accuracy. Therefore, the importance of a perfect artwork should not be underestimated. Problems like inaccurate registration, broken annular rings or too critical a spacing observed on the PCBs often occur due to bad artwork. Developing a clean and exact artwork necessitates skills and patience on the part of the personnel entrusted with the task.

5.2 Basic Approach to Manual Artwork

Artwork generation for single-sided boards is a straightforward exercise, whereas preparing artwork for double-sided printed circuit boards, leads to a problem in securing accurate registration between the two sides.

Correct and accurate registration is of great importance in multi-layer boards also where shorting between layers may occur because of improper alignment when the board assembly is drilled for plated-through holes.

The various methods used for artwork preparation are discussed below.

5.2.1   Ink Drawing on White Card Board Sheets

This is the simplest and earliest method used for PCB artwork design. The materials required are a white cardboard paper, good quality Indian ink and an ink pen. A paper with blue lines grid is usually preferred as these lines do not get reproduced in the photographic process for film master production. Suitable polyester foil, which is dimensionally more stable than cardboard paper, can also be used for drawing the ink pattern. However, the one side of the foil should have a rough or mat surface so that ink will stick to it.

In this technique, circumferences of the solder pads and the centre holes are drawn with a drawing compass and the spaces in between are filled with ink. Conductors are drawn with an ink pen either by directly providing the desired width or as double lines, which are thereafter filled with ink.

Templates in the scale of 1:1, 2:1 and 4:1 are available, which serve as useful aids for drawing of the solder pads. Since this is a totally manual method, it suffers from drawing inaccuracies. For example, conductor widths may vary at least by 0.1-0.2 mm and solder pad locations and conductors can easily by displaced by 0.3-0.5 mm. Also the dimensional instability of the cardboard, 0.01 % °C with respect to temperature and 0.005 %RH (relative humidity) with respect to changes in relative humidity, could be a source of error in high accuracy boards. The method is now rarely used because of the poor stability and precision results.

5.2.2   Black Taping on Transparent Base Foil

Manual artwork generation has become very convenient since the advent of self-adhesive or transfer type pads and precision tapes, and no longer relies on the high drawing skills and patience of the draftsman.

Self-Adhesive Tapes

They are available in a wide range of widths, which can be selected depending upon the application. They are supplied in rolls with standard widths of 0.5 mm, 1 mm, 2 mm, 4 mm and 8 mm; etc. (Figure 5.1a). The width tolerance can be 0.05-0.1 mm depending upon the tape width and the manufacturer.

Fig. 5.1(a)   Sizes of tapes

Self-Adhesive Pads

Pads are available in standard shapes such as donut, square, hex, oval, and tear drop and triangular. These are shown in Figure 5.1b. The two different types of pads which are commonly used are: Self-adhesive type and the Transfer Type.

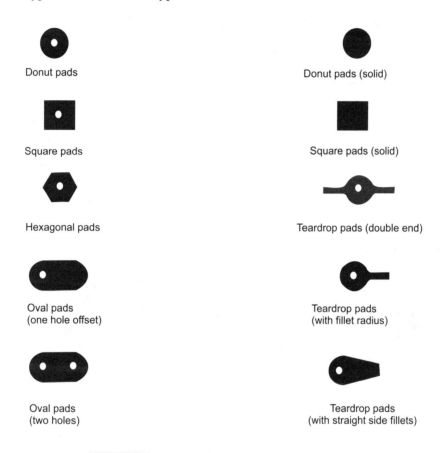

Donut pads	Donut pads (solid)
Square pads	Square pads (solid)
Hexagonal pads	Teardrop pads (double end)
Oval pads (one hole offset)	Teardrop pads (with fillet radius)
Oval pads (two holes)	Teardrop pads (with straight side fillets)

Fig. 5.1 (b)   Standard pad shapes, all pads are 1.27 mm

Transfer Pads

They are printed on a thin adhesive film of typically 10 μm thickness, which is mounted on the top side against a transparent carrier strip. The pad can be transferred from the carrier strip onto the artwork base (polyester base sheet) by just rubbing with a wooden stick end or pencil. The carrier strip can then be lifted from the artwork base leaving behind the pad.

In order to minimize the artwork preparation time and to achieve accuracy in positioning, pads in the form of solder pads, pre-arranged patterns like integrated circuit footprints, contiguous patters for connectors and special patterns for TO casing components can be used. These pre-arranged patterns are usually for the multi-lead components and are generally referred to as footprints: for example, the 16 pin IC pattern is called DIP 16 footprint and so on. Some typical footprints are shown in Figure 5.2.

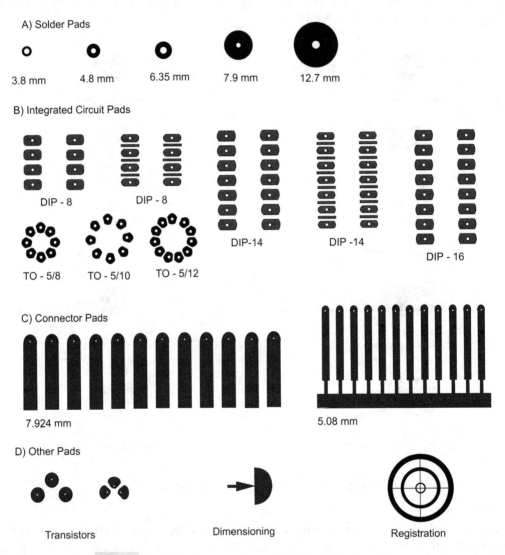

A) Solder Pads

3.8 mm   4.8 mm   6.35 mm   7.9 mm   12.7 mm

B) Integrated Circuit Pads

DIP - 8   DIP - 8   DIP-14   DIP -14   DIP - 16

TO - 5/8   TO - 5/10   TO - 5/12

C) Connector Pads

7.924 mm   5.08 mm

D) Other Pads

Transistors   Dimensioning   Registration

Fig. 5.2   Special pads commonly used in 2:1 artwork generation

Polyester Films

They are usually employed as the artwork base foil as they provide an excellent dimensional stability. The typical values are 17 ppm/°C with respect to temperature changes and 11 ppm/%RH with respect to changes in relative humidity (ppm = parts per million). Polyester films are available in many thicknesses. However, the minimum thickness should be 100 μm to facilitate sufficient mechanical stability against wrinkles. Polyester films are available with pre-printed grids.

The artwork for double-sided PCBs may be generated by using either the 'Two Layer Artwork' or the 'Three Layer Artwork'.

Two Layer Artwork

In this method (Figure 5.3), separate artworks are prepared for the component side and the solder side of the PCB. The method is tedious as it is quite difficult to achieve accurate registration between the two layers.

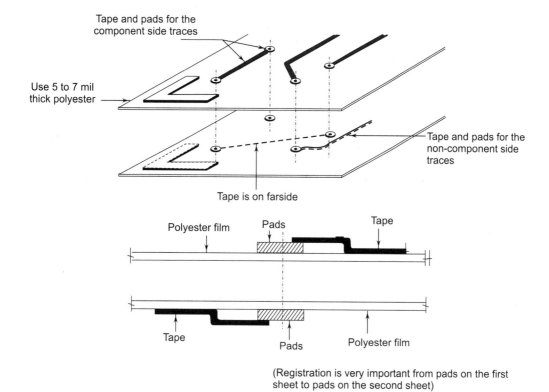

Fig. 5.3   Two-layer Method for artwork generation

Three Layer Artwork

This method solves the registration problem for the common pads by preparing a separate artwork for the: (i) component side conductors, (ii) solder side conductors (solder pads, pads or ICs, via holes, etc), and (iii) the pads, which are common to both sides and form the third layer. Special registration marks need to be provided for each of the layers of the artwork.These are required to align the final master films at the time of fabrication.

A major disadvantage of the three-layer method is that for any artwork modification, a simultaneous rectification of two or even three layers is required which is sometimes a very cumbersome procedure.

5.2.3 Red and Blue Tape on Transparent Polyester Base Foil

By using special tapes, it is possible to draw the two sides of the artwork on the two sides of the same transparent film. The component pads, which have to appear on both sides of the PCB are generated by using black self-adhesive pads. The conductor pattern is, however, done by using red transparent tape for one conductor side and with blue transparent tape for the other conductor side.

While producing the two film masters, colour separation is achieved by using special filters. For example, a red filter used with the camera will filter off the red-taped conductors (normally solder side) while only the blue tapes and black pattern will appear. Similarly, a blue filter will filter out the blue taped conductors (normally component side) leaving only the red tapes and black pattern as being visible. The end result is absolute registration for double-sided PCBs and is the most convenient method for their preparation.

 ## 5.3 General Design Guidelines for Artwork Preparation

Irrespective of the method used for the preparation of the artwork, some basic guidelines, need to be followed. Although these guidelines are specific to the manual technique of artwork preparation, some of the important criteria like optimization of the number of pad sizes, drill diameters, route length, etc., are also relevant even in the case of automated artwork preparation procedures.

In the PCB libraries for components, major parts such as ICs and connectors are still set in an *Inch*-based raster. However, there is an increasing trend nowadays to lay on a *millimeter* raster, specially for new connectors. Therefore, there is a need to convert the information in inches to millimeters, specifically for pads, holes, tracks and clearances. Table 5.1 presents a calculator to assist in making the necessary conversion.

1mil = 1/1000 Inch => 0.0254 mm

Table 5.1 Calculator to Convert mil in mm

PADs		Holes		Tracks		Clearances	
mil	mm	mil	mm	mil	mm	mil	mm
50	1,3	28	0,7	8	0,2	8	0,2
62	1,6	32	0,8	10	0,25	10	0,25
80	2,0	40	1	15	0,4	15	0,4
100	2,5	48	1,2	20	0,5	--	--

5.3.1   Conductor Orientation

It is a usual practice to run the conductors basically on one side in the direction of the X-coordinate and on the other side in the direction of the Y-coordinate. This provides a fairly regularly distributed pattern with a minimum number of via holes.

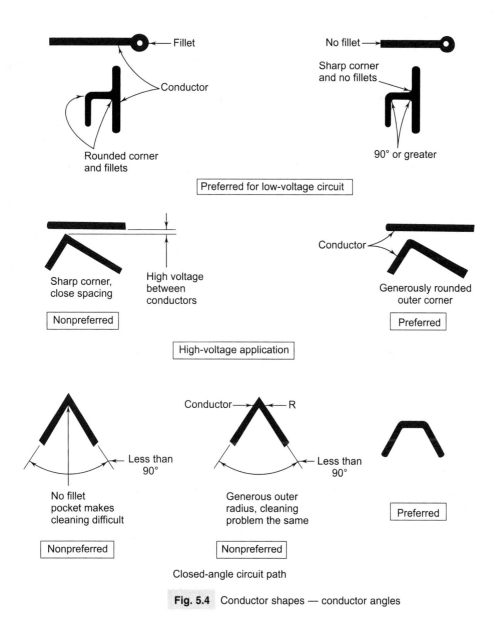

Fig. 5.4   Conductor shapes — conductor angles

Conductor angles should be made at 45 or 90 degrees or 30/60 degrees. This is mainly adopted for the layout design of digital circuits. For analog boards, the bends are usually made in the form of arcs. Figure 5.4 shows preferred angles of taping an artwork. Although the traces could be directed at almost any angle, it is important that all traces running parallel lie at the same angle for uniformity. It would also ensure optimum utilization of the space available and give a good appearance.

5.3.2 Conductor Routing

The following guidelines are suggested in respect of conductor routing:

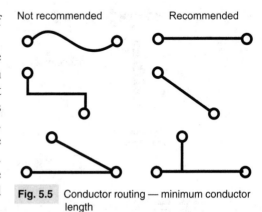

Fig. 5.5 Conductor routing — minimum conductor length

- Even in circuit patterns where plenty of space is available and the conductors can be run in any direction, it is essential to give the shortest (Figure 5.5) interconnection length. This is especially true in case of high frequency PCBs. Normally, it is a good rule to begin and end the conductors in a solder pad or in another pad. But if this practice results in an increase in the length of the conductor, it can also be terminated by joining it to another conductor.

- The minimum angle that any trace should be placed at is 60 degrees. Angles of less than 60 degrees create a situation during the manufacturing process which could allow the etching solution to build upon the inside angle and etch away excess material as illustrated in Figure 5.6.

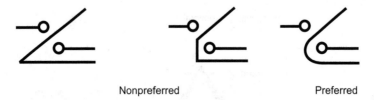

Fig. 5.6 Conductor routing—avoid sharp internal angles

- Conductors running parallel must preferably be at the same angle (Figure 5.7) to ensure uniformity. This eliminates the variance in the conductor to conductor spacing.
- Where one or several conductors have to pass between pads or other conductive areas, the spacing has to be equally distributed. Maximum spacing is obtained if conductors are put perpendicular to a narrow passage. Figure 5.8 shows how the available space is optimally utilized.

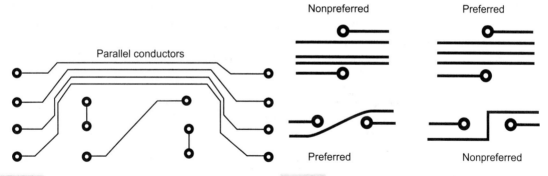

Fig. 5.7 Conductor routing — parallel pattern

Fig. 5.8 Conductor routing — utilize available space to optimum extent

- Closely-spaced parallel running conductors can result in manufacturing problems. In order to avoid this, the conductor should be as w idely spread as possible over the available area.

- Conductor routing must ensure that there is no unwanted bunching (Figure 5.9) of paths at the same point. The routing should also avoid grouping number of conductors at the same solder pad.

- Generally, the conductor widths are chosen based on the current carrying capacity required by specific signal or line. However, it is preferable to adopt standardization, and also keep the number of conductors widths used in

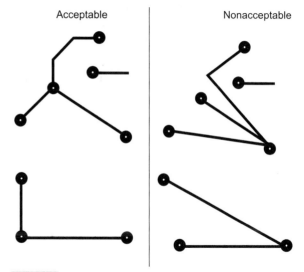

Fig. 5.9 Conductor routing — no bunching is acceptable

any layout to the minimum. The recommended widths of conductors are:
- Signals   : 0.2 mm to 0.3 mm
- Power Lines : 0.762 mm to 1.5 mm
           (depends on current)
- Ground Lines : 1,0 mm to 2.0 mm
           (depends on current)
- The ground conductor width should always be greater than the power line widths. The conductor width is normally chosen while assuming temperature rise of a maximum of 20°C.

- The conductor routing must generally avoid passing in between pads resulting in narrow gaps between the pad and the trace, if an alternative path can be worked out. This minimizes the rejection at the manufacturing stage. Figure 5.10 shows some of the preferred routing patterns. The aim of any routing method should be to minimize the route length.

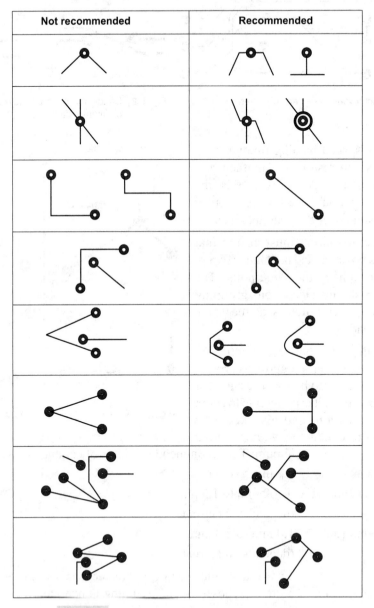

Fig. 5.10 Preferred routing pattern of conductors

- The routing pattern should preferably be distributed equally between the various layers of the PCB. For example, in a double-sided PCB, this applies to the solder and component side. This helps in the manufacturing process by way of ensuring uniform plating in case route densities are equally distributed.

Scaminaci (1994) states that the major causes of system noise begin during the artwork design stage. By incorporating the following guidelines, it is possible to keep system interconnect noise to a minimum:

- Route critical signal nets on a common signal layer. Keep line distances as short as possible. Avoid changing layers for a single net. Layer-to-layer signal runs will cause reflections and lower line impedance.
- Run adjacent signal layers orthogonally (mutually perpendicular) to each other.
- Isolate signal layers from each other by placing ground or voltage planes between them.
- Running ground isolation traces between signal lines will reduce noise but will also slightly lower the trace impedance.
- Route parallel signal lines as far apart as possible.
- Differential pair traces must have the same length.
- Space signal lines equally in routing channels, keeping a maximum distance from connector pads.
- Run a trace directly to a connector pad without line branching. This will prevent reflections and impedance changes. Common connector points can then be made.
- Use curves or two 45-degree turns to avoid minor line reflections.
- Minimize vias or through-holes as far as possible. Every plated through-hole in a net will add to the distributed capacitance of the line. If through-holes are unavoidable, via drill size should be the same as that of the connector; this will lower fabrication costs and maintain the same aspect ratio.
- Avoid line width changes through connector pin fields, and use single trace routing if possible.
- Use the widest line width possible to reduce the dc resistance.
- Signal surface layers are the most difficult to control with respect to line width and copper thickness. Only non-critical signals should run on the outside layers.
- Pads with soldered signal traces should be tear-dropped at the pad junction.

For reference planes, the following points should be kept in mind:

- Maintain a solid ground or voltage plane for signal layers that are impedance-controlled and referenced to those layers
- Keep the ground returns common to the logic family.
- Multiple ground and voltage planes provide the current handling support and low inductance ground return to minimize ground bounce.

- Use minimum diameter clearance pads for maximum copper web within the connector fields.
- Full ground and voltage planes located back-to-back with minimum dielectric separation, develop a capacitor which helps in filtering high-frequency noise in the voltage supply.
- Soldered pads should be thermally relieved.

5.3.3 Conductor Spacing

The fundamental principle in determining the minimum spacing requirements between conductors is that spacing should be provided only when it cannot be avoided. The PCB manufacturing process specifications play an important role in spacing considerations. Mainly, the plating process must be indicated. For example, in pattern plating, a conductor width increase which can be as much as 125 μm, can take place whereas in panel plating, width reduction of the same magnitude can result due to under-etching. The width change also depends upon the thickness of the copper foil in subtractive PCB processes and also on the image transfer method used, i.e. wet-film resist, dry film resist, screen printing, etc.

In order to rule out the chances of a voltage flash-over between conductors due to insulation failure, minimum spacing requirements with respect to voltage are usefully specified. The specified value must be maintained under all circumstances including the worst case tolerances of artwork generation and PCB processing.

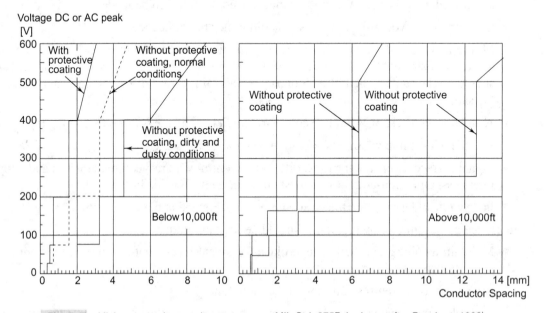

Fig. 5.11 Minimum spacing requirements as per MIL Std. 275B (redrawn after Bosshart, 1983)

Minimum spacing specifications are available in the standards issued by various international agencies such as IPC (Institute for Interconnecting and Packaging Electronic Circuits), IEC (International Electrotechnical Commission), MIL (US Military Standard), and UL (Under - writers Laboratories). As a guide, spacing specifications are divided for PCB applications in altitudes below and above 10,000 ft. (3048 m) and whether or not a protective coating is applied on the PCB. Figure 5.11 shows minimum spacing requirements as per MIL 275 B.

5.3.4 Hole Diameter and Solder Pad Diameter

5.3.4.1 Hole Diameter

It is essential that the component lead should be fitted only into the hole of an appropriate diameter. The hole must allow easy insertion of components without excessive pressure. At the same time, it must be large enough to allow gases generated at the time of wave soldering to escape, otherwise blow holes may develop at the solder joints and inside the through-holes, thereby reducing reliability. At the same time, the number of different hole diameters on a PCB must be kept at a minimum. Therefore, it is essential to optimize the number of drill sizes, otherwise it adds to the cost of manufacturing. For satisfactory soldering, the hole diameter of finished and plated holes should give about 0.2-0.5 mm clearance as compared with the nominal diameter of the component lead. For the hole diameter drilling, tolerances should also be taken into consideration, which are:

> For nominal drill dia < 0.8.........+ 0.10 mm
> For nominal drill dia > 0.8..........+ 0.13 mm

Using this as a guideline, the nominal drill size is normally slated to be about 0.2-0.5 mm larger than the component lead diameter. Drill holes have been standardized as 0.8, 1.0, 1.3, 1.5 and 1.6 mm.

The degree of complexity of the circuit mainly decides the density of the conductor patterns. Each company has its own standards of this kind. For example, Table 5.2 shows PCB classification according to complexity.

Table 5.2 PCB Classification according to Complexity*
(all figures in mm)

Item	Single-sided		Double-sided PTH	
	Consumer	Professional	Low Density	High Density
D=most used Hole diamter	0.8 - 1.5	0.8 - 1.5	0.8 - 1.2	0.6 - 1.0
Copper and diameter	D plus 1.0 - 2.0	D plus 0.8 - 1.3	D plus 0.4 - 1.0	D plus 0.3 - 0.6

(Contd.)

Table 5.2   *(Contd.)*

Minimum path width (excluding limited lengths)	0.8	0.5	0.35	0.15
Minimum clearance between adjacent conductors	0.6	0.4	0.3	0.2
Overall tolerance on hole positions	0.1 - 0.2	0.1 - 0.15	0.05 - 0.1	0.05 - 0.075

Note *Redrawn after Leonida, 1981.

It may be noted that even if four different standards are shown, yet there is no clear division between classes and that a single PCB can belong to different classes according to different factors.

Holes drilled for mounted leaded components and vias must be of the current size. If the drilled hole is plated for through-hole mounting of components, allowance for the plating thickness should be made. The preferred drill sizes such as 0.4, 0.5, 0.6, 0.8. 0.9, 1.0, 1.3, 1.6 and 2.0 mm in diameter are often specified in some standards. As a general rule, the minimum drilled hole is about one-third the board thickness. For example, for a board thickness of 1.6 mm, the hole diameter is 0.6 mm. For standard dual-in-line packages, a drill size of 0.8 mm is common. Some basic guidelines regarding holes are:

- The number of hole diameters on a PCB has to be kept at a minimum. One may vary the hole sizes within the range given to minimize the different sizes used, but remember that a hole which is larger than the ranges shown may be difficult to solder.
- Satisfactory soldering results are usually obtained, if the diameter of the finalized and plated holes gives about 0.2-0.5 mm clearance as compared with the nominal diameter of the component lead.
- Hole diameter = effective lead diameter + hole location tolerance (PTH) + 0.2 mm.

The pad or land size in relation to the hole depends upon whether the land has to support a leaded component or whether it will be plated. If D is the land or pad diameter and d is the drilled hole diameter, then the recommendations are:

D/d ≥ 40 mil (1 mm) for non-plated holes

≥ 20 mil (0.5 mm) for plated through-hole.

The board material is important in the determination of the pad and hole sizes.

$D/d = 2.5$ to 3.0 for non-plated holes in phenolic boards

$= 1.8$ to 3.0 for non-plated holes in epoxy boards

$= 1.5$ to 2.0 for plated through-holes.

As illustrated in Figure 5.12, the solder mask for pads or lands used to mount through-hole leaded component must allow at least a 0.25 mm clearance around the pad. The edges of the pad are to be covered by the solder mask.

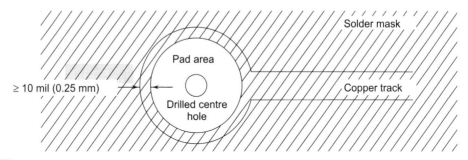

Fig. 5.12 Solder mask minimum dimension overlap for a leaded component land (redrawn after Haskard, 1997)

Table 5.3 gives the lead thicknesses and spacing for the commonly used components in electronic products. Also, given in the table are minimum standard hole sizes that are required to be used for these components.

Table 5.3 Typical Component Lead Sizes (After George, 1999)

Component type	Lead spacing	Lead thickness	Standard hole Size
1/4 W Resistor	0.400"	0.023"	0.028"
1/4 W Carbon Comp Resistor	0.400"	0.025"	0.028"
1/2 W Carbon Comp Resistor	0.600"	0.032"	0.035"
1 W Carbon Comp Resistor	0.900"	0.041"	0.052"
2 W Carbon Comp Resistor	1.000"	0.045"	0.052"
Small Ceramic Capacitor	0.100"	0.020"	0.028"
Large Ceramic Capacitor (>0.2 uf)	0.200"	0.020"	0.028"
Small Silver Mica Capacitor	0.150"	0.015"	0.028"
Small Transistors (TO-92)	0.050"	0.018"	0.028"
Power Transistors (TO-220)	0.100"	0.036"	0.042"
T-1 3/4 LED	0.100"	0.028"	0.035"
Small Crystal	0.200"	0.018"	0.028"

(Contd.)

Table 5.3   *(Contd.)*

IC	0.100"	0.023"	0.028"
IC Machine Pin Socket	0.100"	0.020"	0.028"
IC Solder Socket	0.100"	0.025"	0.028"
Headers/Jumpers	0.100"	0.035"	0.042"
Large Headers	0.156"	0.063"	0.086"
D Connector	0.109"	0.035"	0.042"
#4 Clear Hole	N/A	0.124"	0.125"
#6 Clear Hole	N/A	0.150"	0.156"

Note *For conversion to millimeters: 1 inch = 25.3994 millimeters.

5.3.4.2.   Solder Pad Diameter

Pads are the entities that interface the part pins to the copper traces of the board. The hole in the pad must be big enough to allow for variations in the pin size, in the hole size, in the hole location and in the pin location. The pad must be big enough to ensure that the hole always has some copper around it on the surfaces of the board. Therefore, the diameter of the solder pad with respect to the finished hole diameter is very important for reliable solder joints. Generally, in PCBs with plated through-holes, the widths of the annular ring should be between 0.3 to 0.6 mm. For non-plated through-hole PCBs, the solder pad size must be bigger because there is no through-plating to give mechanical strength to the solder pads. It is however, essential to provide a sufficient solder pad size in order to avoid broken annular rings because of drill position tolerances. In addition, another important consideration is the size of the solder pad and width of the joining conductor. The conductor width should always be less than the solder pad diameter, preferably about one-third as shown in Figure 5.13.

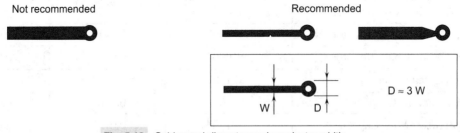

Fig. 5.13   Solder pad diameter and conductor width

Figure 5.14 gives an idea of the preferred and minimum pad sizes applicable for a range of drill diameters. For example, for a drill size of 0.85 mm to 1.3 mm, a solder pad size of 2.54 mm can be used. However, for this range of drill dia, the solder pad size of 1.98 mm meets the minimum criterion only. The solder pad size is specific to manual artwork generation only. The following are important solder pad diameter rules:

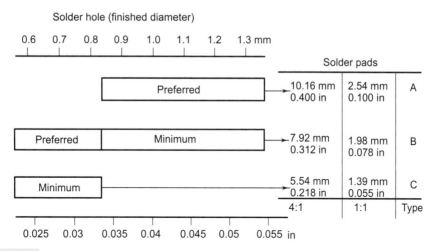

Fig. 5.14   Preferred maximum and minimum pad sizes applicable to a range of drill diameter

- In PCBs with PTH, the widths of the annular ring (Masaoka, et al, 1993) should be at least 0.5 mm, but without PTH, the annular rings must be more because there is no through-hole plating to impart mechanical strength to the solder pads.
- As a rule, the solder pad diameter is approximately three times the component lead diameter.
- The pads are always placed exactly and properly centred on grid intersections.
- The conductor width should always be less than the solder pad diameter.
- The pattern around the hole should be maintained as uniformly as possible to enable symmetrical solder points.

The main issues pertaining to pad sizes are solderability and manufacturability. Solderability is a matter of skill whereas manufacturability is concerned with the issue that the pad will not be broken when the hole is drilled in it. If a drill hole is slightly off centre, the pad may be broken at one edge, possibly leading to an open in the circuit. A standard requirement for pad sizes is a 5 mil annulus. This means there must be 0.005" all around the hole (i.e. a 28 mil hole would require a 38 mil pad). Something a little larger than this (maybe 10 mils) is recommended for solderability. In exceptional cases, a 2.5 mil annulus, (i.e. a 20 mil hole with only a 25 mil pad), can be used but is not usually recommended (www.leonardo.caltech.edu).

Ball grid array (BGA), fine pitch ball grid array (FBGA), and other chip-scale packages (ACSP) require special consideration for successful assembly. As package geometry continues to shrink, processing margins decrease when compared to leaded packages. Therefore, proper pad layout is essential to achieve reliable solder joint structure in such cases. A traditional BGA package has a ball grid pitch of 1.0 mm pitch or greater. An FBGA package is defined as a ball grid array (BGA) with a ball-to-ball pitch of 0.80 mm or less. In both cases, a substrate interposer is used to re-distribute the IC I/Os to the grid of solder balls. For PCB pad layout, it is recommended to follow the IPC-SM-782A, "Surface Mount Design and Land Pattern Standard" requirements. Table 5.4 gives examples of pad sizes and tolerances recommended for various components.

Table 5.4   Recommended Nominal Values (mm) for BGA (Courtesy Micron Technology Inc.)

BGA pitch	BGA pad size	Ball diameter*	PCB pad size	Foil thickness	Square aperture
1.25	0.60	0.75	0.60	0.127	0.60`
1.00	0.33	0.40	0.33	0.102	0.33
0.80	0.33	0.40	0.33	0.102	0.33
0.75	0.27	0.35	0.27	0.102	0.27

Note *Ball diameter measured prior to mounting and re-flow.

5.3.5   The Square Land/Pad

The square land/pad is a common method to designate the polarity or orientation of a component. For polarized components that have positive or anode lead marked, the square land typically indicates where the positive lead should be placed. For components which have the negative or cathode lead marked, such as diodes or LEDs, the square land indicates where the marked negative lead should be placed.

Similarly, the square land is often used to indicate where the marked lead or pin-1 of a multi-pinned component should be placed. This facilitates proper orientation of the components on the surface board.

In case of integrated circuits, the orientation symbol is usually Pin-1 of the IC. In such cases matching orientation marks are made on the printed circuit board with silk screened symbols and markings.

5.4   Artwork Generation Guidelines

5.4.1   No Conductor Zone

Traces should not come any closer to the edge of the board than 0.50 inches (12.5mm) at a scale of 1:1 and if possible, at least 0.10 inches (2.5 mm) should be left. This gives the necessary tolerance that is required to shear the board to the specified size; otherwise a portion or all the trace could get sheared off or shorted out at the next assembly.

If metal or electrically conductive card guides are used, then the gap from the edge of the card guide to the edge of the nearest conductor should be the standard required air gap 1.25 mm (0.05 in. or greater). This is shown in Figure 5.15.

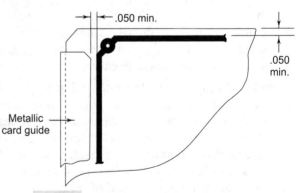

Fig. 5.15   Spacing for conductive card guides

5.4.2   Pad Centre Holes

The pad centre holes are always kept free, i.e. they should not be covered by tapes. Only this can enable them to act as registration marks and in artwork design and as a centring aid in the drilling operation.

5.4.3   Conductor and Solder Pad Joints

For ensuring reliability of the solder joint, it is important that the pattern around hole should be maintained as uniform and as small as possible to enable symmetrical solder joints. The overlap should be such that even if the tape creeps, there is no gap between the pad and the condition. Figure 5.16 shows examples of some bad solder joints and the suggested improvements.

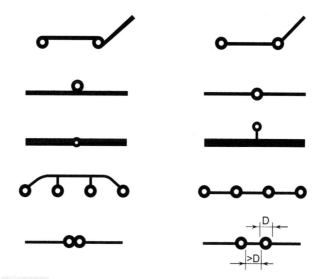

Fig. 5.16   Guidelines for the design of reliable solder joints

 ## 5.5   Film Master Preparation

The components and the various connections according to circuit design are done in the artwork. Film master is prepared from the artwork. It is the film negative or film positive, which is finally used for the direct exposure of the photo-resist coated PCB or the light-sensitized screen. In order to ensure PCBs of high quality, the film master must have high dimensional accuracy, sharpness and wear-out resistance. The imperfections of the film master get materially reflected in each PCB made with it afterwards. The final quality of the film master, in general, depends upon the chemicals, the film emulsion and exposure units such as cameras.

5.5.1 Photographic Film

The photographic film basically consists of two layers: the emulsion and the base (Figure 5.17). The emulsion has an approximate thickness of 5-10 μm and is responsible for the photographic properties of the film. When unexposed, it has light-sensitive silver halides that are stabilized in halide or in a synthetic suspension.

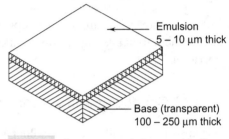

Fig. 5.17 Components of photographic film

The material of the base can be acetate, polyester or glass and has thickness of 100-250 μm. The dimensional stability of the film is primarily decided by the material of the base. The photographic films are available in rolls, sheets, strips or cards, mounted or unmounted without dimensional limitation.

Polyester-based films are popularly used in PCB technology as they offer the best compromise between dimensional stability and convenience in handling and processing. For highest dimensional stability, such as in microelectronics applications, glass base film is used. If the dimensional stability is not much of a concern, acetate-based film can be employed.

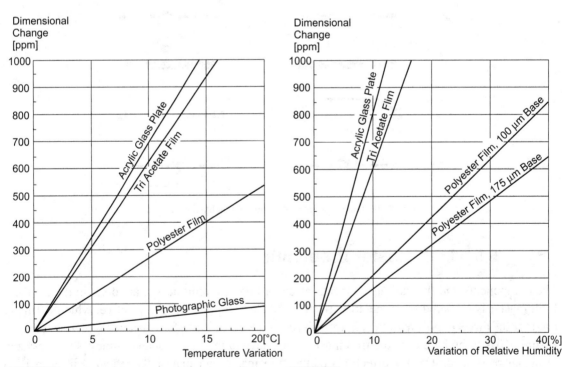

Fig. 5.18 Dimensional changes in the film due to temperature variation (redrawn after Bosshart, 1983)

Fig. 5.19 Typical values for dimensional changes caused by a change in the relative humidity (redrawn after Bisshart, 1983)

The dimensional stability of the film masters varies with temperature variations, changes in relative humidity, processing effects and aging, The dimensional changes are usually expressed in ppm (parts per million). Figure 5.18 shows dimensional changes occuring due to temperature variation whereas Figure 5.19 gives typical values for dimensional changes caused by a change in the relative humidity.

A moisture content of 60 per cent and above destroys the image. If the relative humidity is below 30 per cent, it affects the emulsion's brittleness. Hence, the optimum level of humidity is about 30 to 40 per cent in the surrounding air. Table 5.5 gives the recommended values of relative humidity range for different type of films.

Table 5.5 Recommended Relative Humidity Range for Different Type of Films

Sensitive layer	Base type	Recommended relative humidity range (per cent)
Microfilm		
Silver-gelatin	Cellulose ester	15 to 40
Silver-gelatin	Polyethylene terephthalate	30 to 40
General		
Silver-gelatin	Cellulose ester	15 to 50
Silver-gelatin	Polyethylene terephthalate	30 to 50
Colour	Cellulose ester	15 to 30
Colour	Polyethylene terephthalate	25 to 30
Diazo	Cellulose ester polyethylene	15 to 30
Vesicular	Ethylene terephthalate	15 to 50

The master artwork cannot be directly used in PCB image transferring processes as the artworks are normally prepared to an enlarged scale such as 2:1 or 4:1. Hence, the exposure unit should be able to reduce it to its actual size.

The reproduction film is used to obtain a negative or positive photo tool from the artwork, with or without photo reduction through camera exposure and is available in a 4 to 8 micron thick ester base and also in thin ester base.

Duplicating film is used to obtain either a negative from another negative or to obtain a positive from a positive master. The image is transferred by contact exposure or by camera exposure without reduction. The base thickness and the process are the same as for the reproduction films.

Diazo film is entirely different from the silver gelatine films. It is opaque to ultraviolet light but transparent to normal light after exposure and developing. This does not require a dark room for exposing. Films are exposed using a camera or contact exposure units. This film is developed by passing through a vapour of ammonia. It does not need any chemical bath for developing and the film dries very fast.

The dimensional stability of the diazo film is the same as that of the silver gelatine films. It is available in transparent amber form or burgundy in colour with the base of thick or thin ester. Due to its transparency to normal vision, precise registration of holes and pads can be ensured in film registration.

The *resolving power* of the photographic material is its ability to maintain in the developed image the separate identity of parallel bars which are very close together and is expressed as lines per mm. A typical test pattern to test the resolving power is given in Figure 5.20.

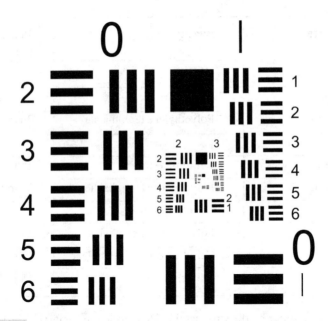

Fig. 5.20　Typical test pattern to test resolving power of the photographic film

The variation of temperature in the surroundings or the storage area affects the dimensional stability of the film. If stored above 40 °C for a long period, it will degrade the image produced on the film. The temperature range acceptable for storage is about 15 to 20 °C. Exposure to direct sunlight will lead to a poorly processed print. This results in the formation of stains and the film may fade. Tungsten and UV-free fluorescent lamps are recommended for periodic examination of the film.

Gaseous impurities like nitrogen oxide, sulphur dioxide, hydrogen sulphide and ozone react with the silver in the emulsion resulting in faded images, which also tend to decompose the base material. Hence the film should be stored away from the above impurities.

Handling of films is very important as improper handling damages the film emulsion. Good housekeeping and cleanliness are very important. Films should be held only at the edges. Thin cotton hand gloves are also recommended for handling.

5.5.2 Exposure through Camera

The camera is used for exposing photographic films, to achieve colour separation and photo reduction. So the camera should have flashlight units, reduction lens systems, a vacuum system as also colour filters.

The camera should be mounted on vibration-free mounts in a processing room or dark room. The image formed on the photographic film depends upon the camera resolving power of the lens and light intensity. The exposure time depends upon the light source. Table 5.6 gives the time required for reproduction of line work along with exposure requirements.

Table 5.6 Examples of Camera and Contact Exposures

Light Source	Lighting	Time for 1:1 Reproduction of Line Work
Camera exposures		
Pulsed-xenon	Two 1500 watt lamps at 3ft.	10 seconds or more
White flame carbon arcs	Four 30 amp. Arcs at 4ft.	20 seconds or more
Tungsten/Tungsten Halogen	Four 750 watt lamps at 4ft.	28 seconds or more
Contact Exposure		
Tungsten	6 volt, 30 watt point source lamp at 6ft	12 seconds or more

The above table indicates the exposure time for lith ortho films type-3 reproduction film. The distance indicated in the table is the distance between the light source and the artwork.

Cameras are of two types. They are:

- Vertical reduction camera, and
- Horizontal reduction camera.

The working principle of the camera is shown in Fig. 5.21. The horizontal camera basically permits bigger artwork and film sizes, whereas the vertical camera needs comparatively lesser floor space, but the copy board size is limited to enable a normal standing person to operate the camera conveniently. Usually, a maximum reduction ratio of four and a useful copy board size of 45×60 cm are needed in PCB technology.

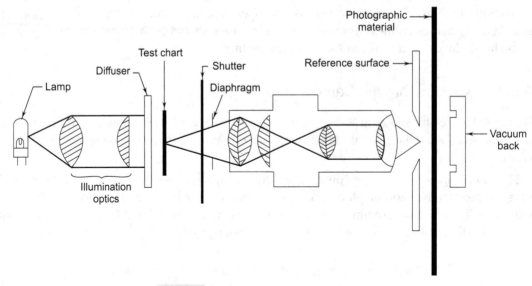

Fig. 5.21   Working principle of the camera

The cameras must have accurate exposure time control, with conveniently accessible focus and operative control. They should have flat field type of lens, free of spherical aberration, colour-corrected and essentially giving a distortion-free image.

5.5.3   Dark Room

Film processing requires a dark room. All photographic films are light-sensitive and hence the dark room should be light-proof and air-conditioned. In order to ensure that the dimensional stability requirements of the film are met, the temperature and humidity inside the dark room should be kept within the permissible limits along with safe lighting. Generally, the temperature of the dark room is maintained at $20° \pm 2°C$ and the relative humidity around $55\% \pm 5\%$. A slight atmospheric over-pressure inside the dark room prevents dust from entering through the doors and windows. The occurrence of vibrations in and around the dark room should be avoided and this has to be considered right at the planning stage of the layout of the building.

Dark rooms should have separate provision for the storage of films and chemicals. The processing room and the film dryer unit should be provided with running water facilities. It is preferable to have two sinks with running water, one for rinsing the film and the other for washing of hands, etc. The layout of a typical dark room is shown in Figure 5.22.

The film processing room should be illuminated with safe light with filter. The type of safe light depends upon the type of film used. The safe light should not sensitize the film emulsion. Normally, yellow, orange or red light of 15 watt bulb is used. It must be mounted at least 120 cm away from the film processing area.

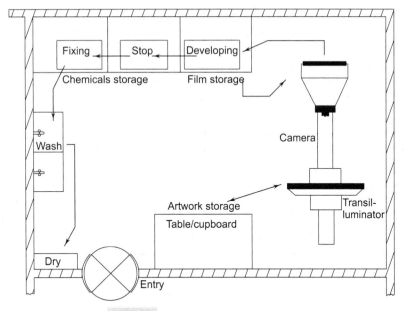

Fig. 5.22 Typical layot of darkroom

The photographic films consist basically of two layers namely the emulsion and the base. The emulsion is light-sensitive silver halide. The light-sensitive chemicals are reduced or oxidized by light with a wavelength of 360 nm to 680 nm falling on the surface. This wavelength region determines the sensitivity of the photographic film. In the visible region, yellow, orange and red spectrums do not affect the film emulsion. Therefore, these coloured lights are used as safe lights in the dark room.

Some useful items needed in the dark room work are:

- Trays of suitable size for developer, stop bath and fixing — they could be of plastic or stainless steel;
- Film forceps to handle film during processing;
- Bath thermometer;
- Dark room alarm clock;
- Refrigerator for storage of unused film; and
- Storage space for artwork.

5.5.4 Film Development

Since films are affected by any natural or bright light source, film processing should be carried out only in the dark room, which is lighted with safe lamp of low intensity lighting.

The artwork is placed on the trans-illuminated copy board of the camera. The camera is adjusted to obtain a sharp distortion-free image of the artwork in a 1:1 standard scale. If the artwork is colour-taped (for example, using the red and blue method), colour filters are used to separate the two layers. The photographic film is indexed on the film plane of the camera and held under vacuum. The film is then exposed for the required time depending upon the type of film used. The exposed photographic film is removed and developed by following the steps given below:

- Silver bromide in gelatine is sensitized by light radiation, which produces metallic silver and bromide ion that will dissolve in the fixer.

$$Ag\ Br \rightarrow Ag^+ + Br^-$$

- After exposing, the film is immersed in a solution of mild reducing agents like pyrogallic acid mixture called 'developer'. The parts that have been exposed to light are reduced to metallic silver more readily than the unexposed parts.
- After developing, the unexposed grains are dissolved in the fixing bath which contains a sodium thiosulphate mixture. The bright part of the object develops dark spots on the film and the film is called a negative film. Then the film is washed with solid free distilled or de-ionized water and dried by using a film drier.

$$\text{Developer} \rightarrow \text{Stopper} \rightarrow \text{Fixer} \rightarrow \text{Washing} \rightarrow \text{Drying}$$

Developer Bath

The developer is a mixture of chemicals, dissolved in distilled or de-ionized water. The developer (chemicals) used has to be of a type recommended by the film manufacturer for the particular film being used. The developing time depends upon the concentration of the chemicals and the agitation of the bath. However, it is advisable to go in for the recommended developing time at the beginning and modify it only after sufficient experience.

Stopper Bath

The function of the Stopper Bath is to effectively stop the development action. After the development is over, the film is lifted above the developing tray for a maximum period of 3sec. which enables the excess developer to drop, followed by immersion in the stopper bath.

The film is immersed for about 30 seconds in the stopper bath, consisting of dilute acetic acid maintained at a temperature of $20 \pm 2°C$. The stopper bath also needs moderate agitation.

Fixing Bath

For fixing the emulsion properly, the film is immersed in the fixing bath. The following chemicals are used in the fixing bath:

a) Sodium thiosulphate, anhydrous;
b) Sodium metasulphate, anhydrous;
c) Boric acid (crystals);
d) Sodium acetate, anhydrous;
e) Aluminium potassium sulphate (alum); and
f) Distilled water.

The temperature of the bath should be maintained at $20 \pm 2°C$. The fixing time depends upon the bath concentration and agitation. Normal light in the dark room may be switched on after disappearance of the milky emulsion.

Washing and Drying

After fixing is done, the film is washed in running water at $20 \pm 2°C$ till all the suspended solids are dissolved. The film is preferably dried at room temperature and takes about one hour to dry. The use of water filter with a porosity of 50 µm with the tap water is recommended to avoid damage to the film from solid matter that may be present in the tap water.

 ## 5.6 Automated Artwork Generation

The challenges posed by modern technologies and the resulting complexity of interconnection networks make the use of the computer a preferred design tool. Computer-based PCB design systems facilitate the generation of the required artwork and documentation for the PCB manufacturer such as a set of master films that represent the circuit connectivity; and drilling information which gives different types of drill sizes used in the PCB, soldermask films and component marking master films, which are present-day standard requirements.

The computer-based design process is both faster and more accurate than the manual process. It also provides the flexibility that is imperative in the development phase of the product when changes in the PCB sometimes become frequent in the circuit till it is stabilized.

If we make a time-analysis of the manual layout design and artwork preparation, the total time spent for these two operations is typically 40 per cent for the layout sketch design; while 60 per cent is spent on the artwork preparation. It is therefore, natural that the first step in automation is the elimination of the manual artwork preparation. The layout sketch is digitized and the information fed to a plotter which directly produces the 1:1 artwork on film. The next step in automation is the use of a computerized layout design which may still need the active involvement of the designer. The last step in automation is to make even the layout design fully automatic and independent of any major involvement of the designer. The availability of high-performance personal computers today has changed the scenario drastically. Low-cost PCB design software has made the manual methods of artwork generation almost obsolete.

When automatic artwork facilities are used, there is a significant reduction in the turn-around time in PCB design. The artwork preparation step is eliminated and replaced by the considerably faster method of digitizing of the circuit pattern. In the earlier systems, a digitizer was used to provide the paper tape or the punched cards (Figure 5.23) to the photoplotter as well as the paper tapes required for NC drilling and NC routing (NC stands for Numerically Controlled). A digitizer usually consists of a cursor-operated digitizer, a control console, a disc storage system and a fast pen plotter, which are all linked to a computer. Digitizing is a graphic-to-digital conversion of all the relevant information incorporated in the layout sketch such as component location, solder pad size, conductor configuration, etc.

With the hand-held cursor, all the relevant positions on the layout sketch are sensed while the cursor movement gets resolved into x/y co-ordinate pulse trains. These pulse trains are stored in the storage system. Additional information on hole diameters, conductor widths, etc. can be entered via the keyboard on the control console. Figure 5.24 shows an automated artwork generation method.

The pulse trains from the digitizer are temporarily stored in the storage system of the computer. The data are then processed by the computer in OFF-line mode. The output is usually on paper tape as required for the photoplotter and the NC routing equipment. Design tools for high performance PCB design and for controlling the design process are illustrated by Isaac (1995).

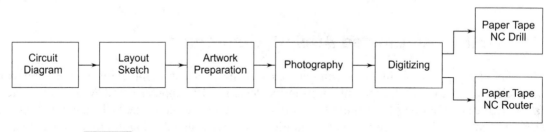

Fig. 5.23   Layout and artwork generation processes: manual artwork process

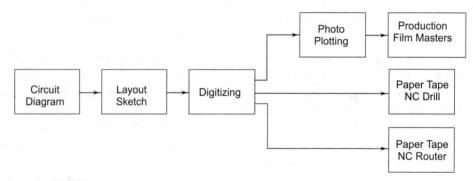

Fig. 5.24   Layout and artwork generation process: automated artwork process

The resolution capability of the digitizers is mostly in the range of 0.025–0.1 mm. The absolute positional accuracy varies between 0.075–0.25 mm.

 ## 5.7   Computer- Aided Design (CAD)

Computer-aided design provides an interface between the PCB designer and the computer. The combination of a graphic terminal (video display unit), an input device and a functional keyboard gives the designer an automated drawing board, which brings about a significant improvement in productivity. In recent years, there has been a phenomenal growth in the availability of software for the design of printed circuit boards.

Early software programs were simple geometric editors allowing only the placement and routing of tracks. However, they were interactive and it was, therefore, easy to erase, shift and replicate the components and blocks of the circuitry. They had some severe limitations as they did not allow grids other than 100 mil or components on both sides of the board. With the developments in software, schematic entry was added to the geometric editor, which, in turn, allowed automatic routing. Today, automatic placement and routing is possible, as the software library now contains not only standard footprints of individual devices and integrated circuits, but full electrical data on standard product lines (such as TTL and CMOS packages), so that simulation can be undertaken easily on digital circuits. In addition, net lists, drill sizes and other relevant data are automatically generated.

A CAD system with various possibilities offers tremendous advantages over manual methods of designing. An important advantage is the reduced time for the layout procedure. Also, in many cases, the capability to make circuit modifications simple simultaneously provides a completely updated production documentation.

With the assistance of CAD, higher package densities can be achieved and complex circuitry with a larger number of ICs per board are realized, which can hardly be arranged by a manual design. The resulting patterns are constantly of the same high precision and of a consistent quality. In a multi-layer board design, especially, interactive CAD plays an important role in the design process.

5.7.1 System Requirements

Figure 5.25 shows a generic block diagram of a CAD system. The various components of the CAD system are discussed below.

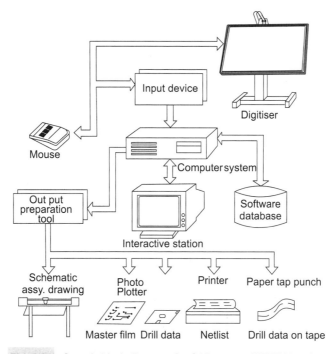

Fig. 5.25   Generic block diagram of a CAD system (NTTF Notes)

5.7.1.1 Hardware

A CAD system comprises both hardware and software, wherein the hardware is the most visible component. Depending upon the user requirement, the hardware may range from a personal computer (PC) to a workstation. A workstation, in general, has more computing power and offers a network environment. In a networked system, all users can share the common database to component libraries and peripheral devices like pen plotters and printers, etc. The PC or workstation configuration is usually governed by the CAD software. The following PC hardware is, however, adequate for work of the PCB design packages:

CPU	:	Intel Pentium IV Processor 2.0 GHz or better with 512 KB cache Memory or better;
Bus architecture	:	Integrated Graphics, 3 PCI and AGP;
Memory	:	128 MB 266 MHz DDR RAM upgradable upto 2.0 GB Master
Hard Disk Drive	:	40 GB Ultra DMA 100 HDD PCI Dual Channel Bus Master
FDD	:	1.44 Floppy Disk Drive (3.5") Internal
Monitor	:	15" SVGA Digital Colour Monitor (Support 1024 × 768 NI Resolution)
Video Controller	:	on board 4 MB or better Dynamic Video Memory
Keyboard	:	104 Keys Keyboard cherry type
Ports	:	2 USB Ports, 2 Serial Ports, 1 Parallel Port,
CD ROM	:	48X or better CD ROM DRIVE

The above configuration provides a good PCB design CAD workstation environment.

Data Entry Devices: The keyboard is the most commonly used data entry device. In addition, some graphic input devices such as light pens, touch screens, joysticks, track balls, mice and digitizers allow operators to enter data such as lines and points in the graphic form. Graphic input devices are also used to select items from a menu.

Output Devices: Hard copies are required for various purposes, including preview, file storage, reports, presentations, finished drawings and precision photo-tools. Normally, an A-1 size plotter can meet most of the requirements, but it is expensive. Therefore, an A3/A4 plotter is adequate.

Two basic types of electromechanical pen plotters are used in CAD systems. In the flat bed X-Y plotter, servo-controlled pens or stylus are moved in two axes over flat stationary sheets of paper. In the *drum* or roll type of plotters, the pen or stylus moves in one axis while the paper moves in another axis on a revolving drum. In *photoplotters*, a moving light head transmits a focused beam of light through an aperture onto a photo-sensitive film or paper. The aperture selection and motion of the light head are under the control of software. Photoplotters generate high resolution images and are a primary source of high quality PCB artworks from the CAD system.

The printer is used to generate the netlists, and bill of materials. The most commonly used printers are inkjet or laser printers.

5.7.1.2 Software

The selection of an appropriate computer-aided design software is often difficult and many articles keep appearing in professional journals to assist people in their selection. There are several considerations for selecting a suitable software for the intended applications, which include the capabilities of the software and its cost. In general, PCB design needs, and factors such as user interface, learning time, help provisions, speed of use, etc., should be taken into account.

Since making a schematic is the first step in designing a PCB, it is necessary to analyse the attributes of the schematic editor, which may include schematic capture programme, libraries supplied with the software, scaling of symbols, netlist generation, online packaging, automatic bus connection, etc.

Since a good placement forms the backbone of a well-designed PCB, it is important that the placement tools offered by the software should facilitate placement of the components in the most optimal manner and in the shortest possible time. Routing percentage, definition of spacing parameters, placement control, matrix placement, re-entrant autoplace, rotation controls, tools for ascertaining optimal placement, independent viewing of bus connections, and component search during placement form the basics of these tools. The time needed to finish a board with a combination of auto and manual routing, throughput, routing percentage, degree of control, and support of post-routing optimization form the guidelines for evaluation of an auto-routing tool.

 ## 5.8   Basic CAD Operation

The CAD design process is usually started with a schematic or logic diagram. This can be either in the sketch form or an electronic transfer from a Capture system. It is followed by the merging of the netlist with the physical layout design. The board outline is then created in accordance with the input requirements. The placement technique is then selected for placing the components. Once the placement process is complete, the routing phase of the design is applied.

During the design process and on completion of the layout, the system can check design errors like space violations, land-to-hole size ratios, and clearance for automated insertions, among other things. This is perhaps the most significant benefit of CAD systems in their ability to check the design in real-time.

The currently available software packages do considerably much more than artwork generation. Figure 5.26 shows the various steps in the design and manufacturing process. These steps are detailed below.

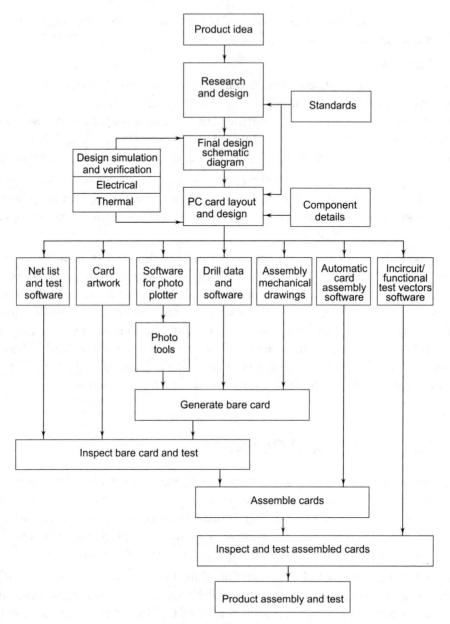

Fig. 5.26 Steps in the design and production of a board showing the wide ranging information required from a designer to manufacture and assemble a board

Schematic Diagram: This provides the functional flow and graphical representation of an electronic circuit and consists of electrical connections (nets) and junctions. The schematic diagram consists of:

- Symbols representing discrete components like resistors, capacitors, transistors and integrated circuits; symbols to follow international standards;
- Input/output connectors;
- Buses;
- Power and ground symbols;
- Component reference names; and
- Text supporting the diagram.

Layout: This involves decision-making pertaining to:

- Surface mounts or mixed technology;
- Single, double or multi-layer boards;
- PTH or non-PTH;
- Placement of components, vias, lands, test pads and device footprints; and
- Generation of interconnection tracks, etc.

Design: Design considerations should be decided in parallel with the layout. The following factors must be taken care of:

- Track widths to ensure that the current densities are not excessive;
- Adequate track separation so that there is no electrical breakdown;
- Proper thermal design to avoid any hot spots on the board; and
- Effect of stray parasites, particularly stray capacitances in case of high frequency circuits, etc.

Netlist: This includes:

- Generation of software list to show all interconnection paths; and
- Comparison with the netlist generated by the schematic diagram to check the accuracy of interconnection lines.

Card Artwork: This involves:

- Test plot of the final card layout;
- Manual inspection of components and visual inspection of the card; and
- Generation of bare board testing information from via information and the netlist.

Photo-tooling: This implies:

- Generation of photo negative or positive.

Drill data: This involves:

- Specifying drill sizes;
- Ensuring that the drills are in the correct position; and
- Software to drive numerically controlled (NC) drilling machine.

Assembly Drawing: This is required to:

- Provide mechanical design information, specifically card material, thickness, overall dimension and shape;

- Check the datum position; and
- Fix the component sizes and their positions.

Automatic Card Assembly: This is needed to:

- Drive radial, axial and integrated circuit insertion machines; and
- Pick and place machines.

In circuit/function Testing: This involves:

- Test vector generation; and
- Final testing and verification.

Figure 5.27 shows the traditional design flow in CAD systems.

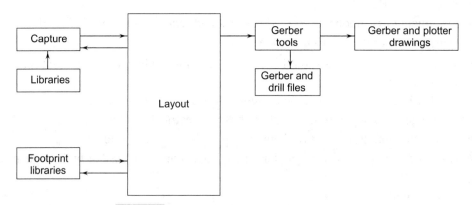

Fig. 5.27 Typical design flow in CAD systems

5.8.1 Layout Procedure

Before the layout procedure is started, complete and detailed specifications on the electronic circuit are required. This includes the following material:

- Schematic diagram with component details, interconnections, and edge connector specifications;
- Component list including component name, type specifications, type number, manufacturer;
- Mechanical specifications such as board size and shape, mounting holes, identification of restricted areas with respect to component height, edge connector location, etc;
- PCB specifications, whether single-sided; double-sided or multi-layer board, plated-through holes;
- Pattern specifications on pad type and size, conductor widths, spacing;
- Electrical specifications like restrictions on component placement because of heat production, capacitive or inductive coupling, ground planes, critical-length interconnections, etc. ; and
- Data operation.

First of all, a component library is prepared describing each component package type used with its outline, pad type, and size and pad position. In the complete component list, the package type, component name orientation and side of interconnection on the PCB are identified for each component used. The connection list gives an exact account of all the interconnections in point-to-point form. The list of board details finally contains the board information and the x/y co-ordinates of the board corners. The correct entry of these data and, in particular, of the connection list, is of utmost importance. In many CAD systems, there is a provision to enter two connection lists for the same circuit but prepared by two independent operators. Any further processing of the data is possible only after the discrepancies between the two connections lists are sorted out.

5.8.1.1 Entry of Schematic Diagram

As explained in Chapter 3, a schematic diagram provides the functional flow and the graphical representation of an electronic circuit. The entry of the schematic diagram is the first step in PCB design using a CAD system. A schematic diagram consists of:

- Electrical connections (nets);
- Junctions;
- Integrated circuits symbols;
- Discrete component symbols like resistors, capacitors and transistors, etc.;
- Input/output connectors;
- Power and ground symbols;
- Buses;
- No-connection symbols;
- Component reference names; and
- Text.

It is desirable that the symbols used in a schematic diagram follow the international standards.

In schematic capture, each design that you open is in a separate *project manager window*. If you need to work simultaneously with several designs, you can open them all, and each will have its own project manager window. The project manager is used to collect and organize all the resources you need for your project. These resources include schematic folders, schematic pages, part libraries, parts, and output reports such as bills of materials and netlists. A project manager doesn't actually contain all the resources. It merely "points to" the various files that the project uses. For this reason, be sure that you don't move or delete any files referenced by a project. If you do, the project won't be able to find them.

The project file is saved as an ASCII-file (American Standard Code for Information Interchange), and can be viewed in any text editor.

5.8.2 Library Manager

The library manager contains all the files for the component footprints. These footprints are used to design PCB. The layout provides the facility to create a footprint library for component footprints. Libraries also contain a variety of symbols that you can use in your boards. The layout has more than 30,000 footprints.

Footprints:

Footprints refer to the physical description of components. A footprint generally consists of the following three types of objects:

- Pad stacks;
- Obstacles, representing the physical outline of the component, silkscreen outline, assembly outline, and placement and insertion outline; and
- Text.

A library footprint can be viewed as a graphical display wherein you can perform various actions on the footprint, such as editing, saving, copying and deleting it.

Creating New Library Modules

Using the library manager, you can create a new library module by saving new or existing footprints to a library that you name. You can then add other footprints by selecting them in the footprints list and saving them to the newly created library module. You can also create new footprints and add them to the libraries.

Similarly, you can add pins to a footprint. Pins can be numeric or alphanumeric, and placed in any order. By default, layout names the pins in numerical order beginning with the number 1. You must change the pin names in the layout to match the pin numbers in the schematic, or change them in the schematic library.

Assigning Pad Stacks to Footprint Pins

Pad stacks define the pins on each layer of the footprint. Pad stacks show the shape and size of each pin. You can assign the same pad stack to all the pins in the footprint using the edit footprint dialog box.

5.8.3 Component Placement

After the input data have been fully accepted by the computer, component placement is carried out. Those components for which the position has earlier been determined by the designer will get their place accordingly. The other components are usually placed in a manner that ensures minimum overall conductor length. Two placement procedures are possible: either the complete placement can be fully guided by the operator, or it can just be attempted by the automatic placement routine with the intervention of the operator, wherever necessary.

This part of the software enables one to place as well as route the components and set unit of measurements, grids, and spacing. For the placement and routing of the components, the Auto-placement and Auto-routing facility is normally used. However, in many softwares, some critical signals have to be routed manually before Auto-routing. The following steps are taken for board design:

- Create a netlist from the schematic diagram by using capture. Netlist file is a document file which contains information about the logical interconnections between signals and pins. A Netlist file consists of nets, pins, components, connectors, junctions, no-connection symbols, and power and ground symbols. Before you create a netlist file, be sure that your project is completed, annotated and free from electrical rule violations.

- Software includes design rules in order to guide logical placement and routing, which means, that you must load the netlist into layout to create the board by specifying the board parameters.

- Specify board parameters. Specify the global setting for the board, including units of measurements, grids, and spacing, create a board outline and define the layer stacks, pad stacks and vias.

- Place components. Use the component tool in order to manually place the components which are fixed by the system designer on the board or otherwise use Auto-placement.

- Route the board. Use different routing technologies to route the board and take advantage of push-and-shove (a routing technology), which moves tracks to make room for the track that you are currently routing, and you can also auto-route the board.

- Provide finishing of the board. The system supplies an ordered progression of commands on the Auto menu for finishing your design. These commands include Design Rule Check, Clean-up design, Re-name Components, Back Annotate, Run Post Processor, and Create Reports.

5.8.3.1 Manual Placement

The graphic display shows the board contour as specified in the board list. At the same time, all the components are shown with their outlines in one corner of the display including the straight pin-to-pin interconnections.

With the help of the light pen, components can be called up one by one, and moved across the display to the position found to be the most suitable. The light pen is a hand-held electronic device. If held close to the display, its position can be detected by the computer with a special program. When components are moved, the interconnections simultaneously follow in straight lines like elastic bands.

5.8.3.2 Automatic Placement with Operator's Interaction

With the computer's placement routine, it is possible to attempt the optimum component placement by automatic means. The components having the most interconnections are placed together in groups while the components with pre-determined locations are placed accordingly. The component allocation occurs now by minimizing the overall conductor length and can be observed on the display. The huge amount of data to be processed and considered by the automatic placement routine (Figure 5.28) makes it very difficult to find the optimum solution. Since the designer has the option to interact, an acceptable solution can be found with combined efforts.

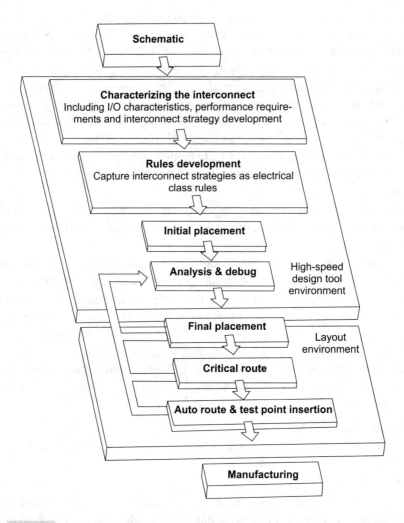

Fig. 5.28   High-speed design tools: physical design process (after Isaac, 1995)

Most automatic placement routines or algorithms automatically place the components at strictly defined locations on the printed board. Other systems allow random positioning of components, restricting only component body interference. All systems attempt to minimize the length of the connection paths (Ginsberg, 1992a). The most commonly used automatic placement techniques are:

- *Swapping*: A possible feature of CAD systems is the ability to swap the positions of similar components so the possibility of 100 per cent interconnection is enhanced. Some auto-placement packages provide the capability to swap parts, gates and pins as shown in Figure 5.29a.

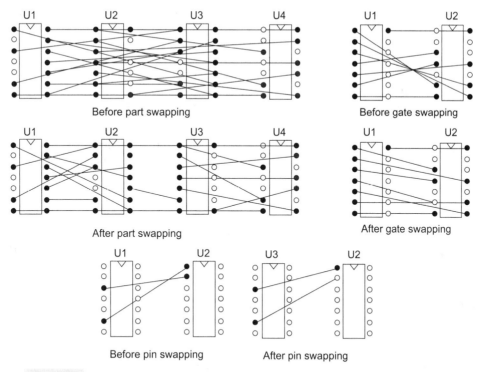

Fig. 5.29(a) Swapping of parts, gates and pins in CAD systems (after Ginsberg, 1992 a)

Components and pins can be swapped automatically to obtain the shortest possible overall wire length; components, gates and pins can be swapped interactively to minimize routing complexity and to shorten overall wire length.

- *Moving/rotating:* This is usually an interactive design tool accessed from a menu during the design placement phase. It enables the designer to move and/or rotate any individual part in the design. Smarter software packages also have the ability to move and rotate a complete group, or to rotate a complete component type.

- *Partitioning:* In this technique, the design process is reduced to a sequence of partitioned problems. First, a vertical boundary line splits the set of component locations into left and right halves. The set of components is then partitioned into halves; the partitioning procedure swaps component groups to minimize the crossing count for the vertical boundary line.

Next, the set of locations is split into upper and lower halves separated by a horizontal boundary line, and the same component swapping method is used. Each component now has been placed in either the left-upper, right-upper, left-lower or right-lower quadrant. The partitioning technique is applied to each of the quadrants in turn, and then to the resulting octant until each region contains one board location.

placement: The most basic approach to automatic placement is to begin with a ...nt, apply the metric of choice to obtain a "score" for the current placement, and ...ate the operations of:

- ...osing a new, legal arrangement of devices;
- ...aining a score for the new arrangement; and
- If the new score is "better" than the current score, making the new arrangement the current arrangement and the new score the current score.

Through this technique the designer is able to get a better component placement arrangement. However, the process is very slow because whole choices for a new arrangement are random.

5.8.4   Conductor Routing

After the component placement, the interconnections are still shown as straight pin-to-pin connections in all directions across the display. The automatic routine now has the task of finding an interconnection pattern which is feasible on the PCB. This procedure is executed by laying stress on the following priorities:

- Interconnections are sorted out according to length.
- Conductor routing begins with the shortest interconnections.
- All the existing obstacles like pads and copper areas are sorted out.
- Via holes are minimized by transferring conductors or portions thereof to the optimum board side. (A via hole is a plated through-hole of usually a smaller diameter which serves as an interconnection between the circuit pattern on the two sides of the board, but it is not used for component lead mounting
- All the information in the computer memory and on the display is continuously updated.

On the display, the patterns of the two different board sides are distinguishable according to different light intensity. The software will attempt to place all the conductors in x direction on one board side while conductors in y direction are placed on the opposite side.

The computer as such will hardly be able to place all the conductors onto the board or it might also find solutions with very long tracks which are very risky because of the possibility of an unwanted coupling effect. Unfinished interconnections (usually about 20 per cent) are displayed as dotted lines. Since the operator has a better overview on the pattern, he can interfere and finish the conductor routing.

5.8.4.1   Auto-Routing

Nowadays CAD systems are able to handle fine line technology, variable trace widths, buried and blind vias as well as multiple land sizes on the same layer with a variable grid. A good routing program attempts to make all the required connections by using the shortest grid. A good routing program attempts to make all the required connections by using the shortest total connection paths and at the same time reducing the number of vias. Many of the Auto-routers are restricted to route between pairs of layers only. For efficient routing, all layers must be taken into consideration simultaneously. The Auto-router of the particular software should be capable of constructing an intelligent routing pattern while constantly checking clearances and connectivity. All this must be achieved while keeping both routing and circuit costs to a minimum.

The different categories/variations of automatic routers for PCB design in use include:

- Maze-running routers;
- Line-probe routers;
- Channel routers;
- Rip-up routers; and
- Push-and-shove routers, as in 'OrCAD-Layout Plus'.

Maze-running Routers: In this technique, the router makes interconnections by expanding out from the primary grid connection point until the secondary point is found, and connected. The maze runner attempts to complete every possible path, compare all the successful connections and pick the best (shortest path and/or least vias) available. The method usually requires longer processing time.

Line-probe Routers: Here, the router connects a pair of points by constructing simultaneously or sequence of line segments (probes) out from each point to be connected. When the two-line sequences intersect, the connection path is complete. The method gives over 95 per cent completion rates.

Channel Routers: This method involves placing circuit components in such a way as to leave vertical and horizontal channels between the components through which the connection paths may be run. Any two points to be connected must lie in the same channel, column or row, for the router to make the connection. The router proceeds by connecting the shortest paths first and progresses until it has completed all the connections within the channel. It will then move to the next channel and eventually to the other side. This is because the basic router algorithm operates on two-circuit layers, making vertical connections on one side and horizontal connections on the other layer.

Rip-up Routers: The algorithm used in this method divides the circuit board into a grid with each connection point residing in a grid cell. The cells containing connections are called *obstacles* and those without connections are known as *free cells*. The algorithm keeps a track of available cells and finds the shortest path between two connections. The process continues until all connections have been routed. (Ginsberg, 1992c).

Push-and-shove Routers: With this algorithm, the completed paths which form obstacles to subsequent paths are not removed but are re-routed. Instead, the path is completed by pushing or shoving small sections of previously routed conductors in order to make additional routing channels. This technique is commonly used by human designers and is shown in Figure 5.29b.

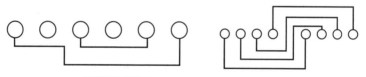

Fig. 5.29 (b) Push and shove router

The performance of an auto-routing algorithm is usually measured against the percentage of routing completed or in other words, the amount of manual effort needed to complete a job. The execution time and system resources required are also important considerations. Figure 5.30 illustrates completion rates based on density and routing completion time. The best algorithms are designed to balance both economic and technical factors. The system usually allows the user to complete the unfinished connections, while it monitors his action against the ground rules for interconnections and signals him when any rules are violated. The re-entry capability helps the designer to make changes on previously laid out designs in terms of movements, and the deletion and addition of conductors.

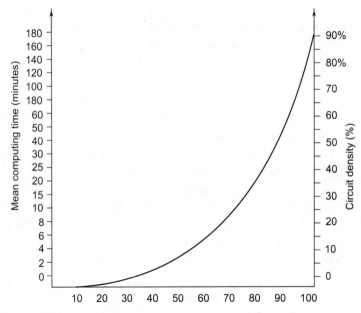

Fig. 5.30 Completion rates based on circuit density and routing completion time (redrawn after Herrmann, 1992)

As no router gives 100 per cent auto-routing on every board, this needs to be completed by manual means. Generally, the last few routes are the most difficult to connect and require changes in earlier placement. Figure 5.31 illustrates the manual work required as opposed to the number of open connections after auto-routing. It may however, be noted that faster and more powerful microcomputers are bringing artwork routing systems into an affordable price range, which will eventually result in higher and higher levels of auto-routing possibilities. Mantay, et al. (1991) describe a simple model for optimizing auto-routing PCB design for manufacturability and routability.

The wiring space used by the routers is connected by the various types of space organization methods used in auto-routing. These are the grided, gridless, plastic grid, and channel methods and combinations of these (Ginsberg, 1992a).

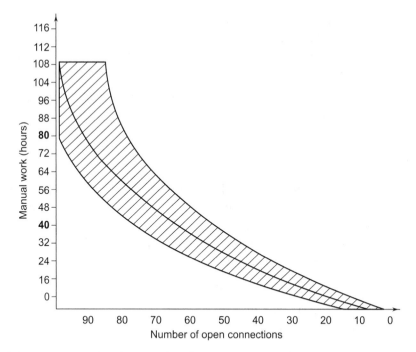

Fig. 5.31   Completion curve: manual work vs. number of open connections after auto-routing (redrawn after Herrmann, 1992)

Grid arrangement: In this method, the conductors can be placed on any gridline that does not cause interference with a land, component lead, via or keep-out area. Grided routers work best with designs consisting mainly of small-scale (SSI) and medium scale (MSI) integrated circuits. The higher the component density becomes, the slower the router runs because it has to manipulate a larger number of co-ordinate vectors.

Gridless arrangement: Gridless routers also rely on design rules, but are not constrained by fixed grid or cell dimensions. Instead, they assess conductor width, conductor spacing and via size for the net currently being routed, in order to vary the grid size on the fly. This technique allows the stacking of conductors and spaces of different widths across a span surface while dodging obstacles such as pins and vias. The advantage of gridless routers is that they adapt easily to changes in topological and packaging technology such as the requirements of surface mount devices. However, the method is inherently slow.

Plastic grid arrangement: Plastic grid or flexible field surface organization permits each routing grid to be of a different size. Run before the actual routing, the plastic grid algorithm establishes a grid pattern of various sized rectangles, customized to the components and other restrictions on the printed circuit board. As such, dense component placements and extraordinary land spacing, such as off-grid, are easily accommodated, with connections to the centre of all lands. This process is also inherently slower than a grided solution.

Channel routers: Channel surface organization creates routing surfaces made up of a series of channels that pass between the pins of the devices placed on the routing surface. Vias are placed in between the channels using the same pattern as the component leads. This allows for pre-reserved space for both wire and vias.

There are also possibilities of combining the above-mentioned techniques for proper space utilization on the PCB surface.

5.8.5 Checking

After completion of the conductor routing, checking routines are used to check that the design is fully in accordance with specified standards. Such check parameters can be rules on spacing, minimization of the number of plated through-holes, special design rules, the current-carrying capacity of the conductor, opened or disconnected pins, positional errors, etc. The errors are indicated on the display by flashing and can be corrected by the designer's interaction.

After completion of the design, artwork generation is carried out.

5.8.5.1 Verification/Design Rule Check

In manual designs, everything was required to be checked as a possible source of error. Component sizes, hole sizes, conductor widths and clearance, land-to-hole ratio, board areas to be free of components, clearance to the edges, positional accuracy and, of course, electrical interconnections had to be personally reviewed with a great deal of care. Automated design also requires no less attention as many types of checking are performed by the software during the design cycle.

After completing the design of the printed circuit board with the help of CAD Tool, a designer again has to verify the PCB in order to find out errors. Such type of verification/design rule check entails:

- General verification;
- Physical verification; and
- Electrical verification.

General Verification: General verification checks are broad checks carried out to establish that the basic requirements have been met by the design. The checks seek answers to the following questions:

- Has the circuit been analysed and divided into basic areas for a smooth signal flow?
- Has a basic grid pattern been fully utilized?
- Are the board size and the number of layers selected for low cost manufacturing?
- Are preferred conductor widths and spacing used wherever possible?
- Are preferred pad and hole sizes used?
- If jumpers are used, are they kept to a minimum?
- Are the larger areas of copper broken up to prevent blistering?

- Have mounting holes been provided or not?
- Are dummy areas inserted where necessary?
- Has shielding been effectively used where necessary?

Physical Verification: In physical verification, checks are carried out on the component placement and various other violations are identified. This verification attempts to answer the following questions:

- Are terminals and control locations compatible with the total assembly; i.e., does the concerned part work in conjunction with other parts?
- Will the mounted board meet shock and vibration requirements?
- Are specified standards (like: IPC 2221, IPC 2222) used?
- Are unstable or heavy parts adequately retained?

CAD systems provide for verifications of placement violations, including component-to-component spacing violations and other placement errors like insertion outline or grid restrictions.

Electrical Verification

The assignment of signal paths generally follows a priority system which starts with the highest priority signal such as of low level and descends to those signals which can tolerate greater movement on the layout. The ground signal is usually placed at the outer extremity of the printed circuit format because it allows for easy connection of internal ground buses to the main bus. The ground and power signals are the first to be assigned to ensure a good, clean path for the ground signals. The ground signal has to be of sufficient width to more than adequately carry the intended current for the board. Mostly, ground and power signals are put between the rows of pins of the ICs on the component side of the board. This will allow for all cross traces to be on the wiring side of the board where they are clearly visible after assembling. This allows the service engineer to easily follow a signal path during troubleshooting.

It is important to check the following from the point of view of electrical considerations:

- Have conductor resistance, capacitance and inductance effects been analysed? This is especially important for the critical voltage drops and grounds.
- Are the conductor and hardware spacing and shape compatible with insulation requirements considered?
- Have controlled impedance for frequencies above 100 MHz digital and 300 MHz analog signals been provided?
- Are polarities adequately identified?
- Have the dielectric changes related to surface coating been evaluated?

The tools available for checking electrical violations are designed to verify adherence to spacing criteria as listed in the route spacing spreadsheet. Any problem found by the route spacing violations is thus identified.

5.8.5.2 Design Rule Violations Check

The CAD tool scans schematic designs and checks for conformance to basic design and electrical rules. These checks usually are easiest for continuity verification and clearance violation checking. Clearance checking has assumed great importance because of tight tolerances on technologically advanced devices. The following checks are made:

- Conductor-to-conductor spacing;
- Conductor-to-land spacing;
- Land-to-land spacing;
- Via hole-to-via hole; and
- Mechanical clearance (to edge of board, around on-board connectors, around mounting holes).

The basic design rule checks involve checking of pad-to-pad, pad-to-trace, trace-to-trace and pad-to-via gaps, as shown in Figure 5.32.

5.8.5.3 General Guide Lines for Avoiding Design Errors

The following factors must be considered to avoid design errors:

- The design must have a complete, closed board outline. A design without a board outline or with only a partial outline is not valid and acceptable. There cannot be more than one board outline in the design.
- When placing components, enough clearance must be left between the connected pads and the board outline to avoid spacing violations during routing. Edge finger connections are a frequent offender in this category.
- Component pads outside the board outline to which connections are attached, will be rejected.
- Do not try to sub-divide plane layers with copper, outlines or details until after routing.

5.8.5.4 Identification Checks

For the sake of proper documentation and identification, the following checks are made on the artwork:

- The PCB number;
- Assembly title and number;
- Revision information and number; and
- Serial number.

Most of the CAD systems are capable of performing many of the design checks including connection and dimensional checks. The netlist driver CAD systems usually provide for checking circuit continuity versus the layout and vice versa.

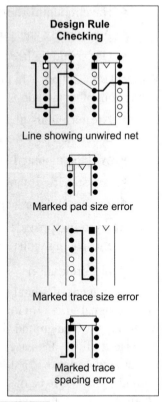

Design Rule Checking

Line showing unwired net

Marked pad size error

Marked trace size error

Marked trace spacing error

Fig. 5.32 Examples of design rule violations (IPC-D-390)

The most pressing problems faced by CAD software developers and users today are those of board size and density. With most systems, the packaging design engineer can find ways to work around these problems by fooling the system, changing the grid or splitting the board. Also, higher technology designs involving extremely dense analog circuitry typical of high speed RF designs, fine line technology, surface mount technology, flexible circuitry and thick film design, pose a considerable challenge to CAD software developers, and to the packaging designers that must work with existing CAD systems.

 ## 5.9   Design Automation

A design automation system is one in which the computer acts as the designer rather than just as an aid to the design engineer. Because of the large amount of data to be processed, high capability computers are utilized for this purpose.

While it is typical in designing with CAD systems that only about 80-90 per cent of the conductor routing is automatically accomplished, design automation results in nearly 100 per cent complete conductor routing. This is achieved by iterative routing programs which also replace previously routed conductor tracks and via holes. Iterative routing requires considerable processing time since a new step can only start after the previous run has been completed. Hence, only computers with high processing capacity are used in design automation.

Many of the digital IC packages house two to eight similar gates within the same package. The software does not only optimize the location of the package on board but can also move gates from one package to another in order to keep the overall length of interconnections and via holes at a minimum. An automatic check certifies that the fan-out capacity of the gates is taken care of.

Design automation software also provides conductor routing and component placement programs, which develop the board design simultaneously on all the conductive layers of the board. This eliminates the possibility of a dense pattern on the first layer and a far more sparse pattern on the final layer. This requirement is of special importance to the multi-layer design.

Design automation is an extremely efficient tool, especially for the design of highly complex double-sided and multi-layer boards with a large number of integrated circuits. If compared with CAD, the overall system flexibility in design automation is limited since it does not utilize human ingenuity which is still superior to computer algorithms. Design automation can be successfully applied where this high flexibility is not required and the tasks follow a certain standardization, as for instance in a digital system design. Standardization in this context is related to mechanical board outlines and the types of components/packages used. For standard designs, design automation can usually complete the design job without human involvement. But now and then, the need for an interaction by the designer is still required, particularly for more complex tasks.

Design modifications are also possible with design automation; as compared to CAD, they need more time and are costlier because the whole design gets involved. In CAD, the modifications are carried out manually on the display which works more in a straightforward way.

The advantages of CAD in providing production documentation can be further enhanced in design automation. It may even include engineering report facilities; giving an account of spare gates and unutilized pins or thermal placement.

The most challenging problems faced by the developers of EDA (Electronic Design Automation) systems are diminishing board sizes and increasing packaging density. Also, the need for higher technology designs involving extremely dense analogue circuitry typical of high speed high frequency designs, fine line technology, surface mount technology and flexible circuitry are some of the pressing problems being faced by the system developers and PCB designers. The ingenuity of the designer lies in funding ways to work around and find solutions by changing the grid or splitting the board, etc.

A number of companies are now commercially selling PCB-EDA software systems. They have similar features, with small variations here and there.

5.9.1 How to Judge CAD Systems?

A number of CAD systems are available and it is quite difficult to select the right tool you need. The early simple programs that ran on personal computers have now become quite sophisticated. Examples of such programs are SMARTWORK, PROTEL, P-CAD and CAD-STAR. Some of the latest CAD Systems are quite commendable. They include:

- Mentor Expedition 2000;
- Cadence Allegro V13.6;
- Innoveda Power PCB V3.6; or
- even much cheaper system like EDW 2000.

The standard elements are schematic capture, simulation, library, placement, auto-routing, post-processing and documentation. Evaluate each of these areas.

Demo Board: Avoid the demo board. It is too easy and doesn't represent the real world and will always look good. Use a board you have already designed as a comparison (take your toughest board).

Demo Licence: This is the best way to evaluate the learning curve, but will cause you to spend the maximum time to evaluate the tool and see how it integrates into your environment.

EMI, Thermal, Geometric and Other Technological Restrictions: Certain issues such as RF, EMI, thermal-, geometric and other technological restrictions are often better done on systems that specialize in these topics.

Auto-router: The percentage of completion and amount of time taken to complete the routing can be used as a criterion to judge which tool has the better router, provided that the quality is similar.

Metric-system: Is the software truly 100 per cent metric compatible? Designers and component manufacturers need to drop the English measurement system and embrace the metric system.

Library: Does the component library, provided by the software vendor, mostly contain old and outdated parts with no support documentation?

What kind of information is included in the library? A good library should meet all manufacturing requirements like:

- Component Families
- Documentation
- Silkscreen outline
- Naming convention
- Polarity Marking
- Transition to Metric Units

- Component Geometry
- Tolerances
- Links or Search Engines
- Creation of components
- Mounting and Tooling Holes
- Land patterns

Standards Interface: There is a time when we need a standard interface such as "GENCAM". Check if the CAD software is GENCAM import and export compatible.

 ## 5.10 Manual Versus Automation in PCB Design

The decision to go in for automation in PCB design and artwork generation, and to what extent, depends upon a number of factors. Each method has its area wherein it can be justified as a method of preference.

Manual Design and Artwork Generation

This is a preferred method for simpler boards like single-sided or double-sided PCBs. It can also be successfully employed for higher circuit complexity in single or low-volume production quantities. Manual designing offers high flexibility and gives all possibilities for human ingenuity. However, digital boards of high complexity, especially those involving more than 100 ICs are difficult to design manually. Other limitations of manual methods relate to quality, time and the need for trained personnel.

Worldwide, a large percentage of the PCB design and artwork generation is still done manually. The fully manual method does not require any investment and may therefore, be used for some more time though its share gets continuously reduced especially for digital PCB designs.

Design Automation

The full automation of PCB design and artwork generation is a valuable tool wherein the input constraints can be standardized and reduced to a smaller number of straightforward approach rules. It is a desirable tool for the design of the highly sophisticated digital boards with more than 150 ICs, and for complicated multi-layer boards wherein it is extremely difficult, even for an experienced design engineer, to keep an exact overview on the design. Here the total design time can be reduced from several weeks to a few days while giving near optimum results. A tight delivery schedule is important for a considerable number of PCB designs while simultaneously the need for de-bugging and rectifications should be minimum, CAD is often a preferred choice. Automated artwork draughting also provides a more precise artwork than manual draugting or taping.

Design automation is normally not used for analog PCBs because it is difficult to reduce the various design constraints for a wider range of analog circuit boards to a list of a view straightforward-approach rules as it is normally done with digital circuit boards.

The large capital investment into CAD facilities will always call for a full utilization of the system. CAD can hardly be justified for boards with less than 20 digital ICs, with discrete component contents of more than 50 per cent or where only a very small number of PCBs is normally required of such designs.

5.11 Photoplotter

The processed digital data from CAD systems are converted back to graphics by photographic means with the help of an equipment called a 'photoplotter'. In this equipment, the drawing head which is a light spot projector changes its relative position towards the drawing medium which is the photographic film. Depending upon the design of the photoplotter, the drawing head or drawing table or both of them are moving parts.

A photoplotter, as the name implies, is a plotter that writes by using light and accordingly has to be programmed for the tool to be used and the path to be followed for drawing a trace. For a photoplotter, tool means especially shaped apertures through which light passes to create a given shape on film. An aperture can be used without movement to make a shape or with movement to make a line or an arc.

There are two major types of photoplotters, 'vector' and 'raster' (or laser). Each of these types handles the apertures differently. The vector photoplotter has become almost obsolete and is being replaced by the laser photoplotter. Laser plotters operate much quicker than vector machines. For example, a complex plot that required hours. on a vector machine can usually be performed in ten minutes or less on a laser photoplotter. This decrease in turnaround time has also brought down the photoplotting cost considerably. The photoplotter includes three main features

- A photoplotting head, which comes with a set of apertures;
- Film plotter; and
- Software.

The main advantages of photoplotting as compared to pen plotting; are that the line quality is sharp, consistent and accurate. There are no pens to skip or dry up. The photoplotting head has more than 24 apertures thereby allowing the user to have several line widths and pad shapes. The details of the two types of photoplotters are available (Ward, 1992):

5.11.1   Vector Photoplotter

The computer numeric control (CNC) vector machines produce photoplots by moving a projection head across a sheet of photographic film. When stationary, the projection head could "flash" the light source to image a pad or target of any fixed shape. When moving, the projection head could "draw" circuit lines by leaving the light source on.

The projection heads that the vector plotters use are mechanically sophisticated and so the size and shape of the projected light is controlled by using apertures. These apertures could either be fixed or variable.

Aperture Wheels: Traditionally, the photoplotters counterpart to a pen plotter's pen rack has been the aperture wheel. The aperture wheel is a disk with 24 or 70 apertures arrayed radially along its circumference. When the photoplotter selects an aperture, the aperture wheel is rotated to place the desired aperture between the light source and the film. Apertures are themselves pieces of film and can be made to any shape required, though in practice, this is a time-consuming process and there is a physical limitation in size.

Flash and Draw Apertures: In order to achieve constant exposure on a vector photoplotter, apertures used for flashing pads are filtered differently than those used for drawing traces. Therefore, Flash and Draw apertures cannot be used interchangeably without the risk of localized over-exposure and under-exposure.

The set-up of an aperture wheel is an exacting and time-consuming process since each aperture in the wheel must be hand-mounted and aligned. In order to avoid repeated set-up costs, designers have the photoplotting vendor keep a wheel in file and are forced to always use that same set of apertures. This has obvious drawbacks, both in terms of design flexibility and the ease of migration to other vendors.

5.11.2   Raster (Laser) Plotters

In the laser raster plotting technique, the instructions which are to be executed are completely read into and interpreted by a raster image processor (RIP) before the machine plots. The RIP usually sorts the commands into some intermediate form which is then processed in real time to produce a stream of on/off commands to a laser modulator.

A raster plotter produces images as a series to tiny dots clustered together to form the lines and pads called from a vector plotter file. Generally the smaller the dots (pixels), the better the image. Also, it is a general rule that the smaller the pixels, the longer it takes the plotter to plot.

The SMDs require a wide variety of pad geometries and orientations. There are two methods of dealing with this requirement on CAD systems. When many different pad shapes are needed, one choice is to define a symbol for each shape. The easier and more economical way of producing many pad shapes is to define an aperture for each shape.

In conclusion, two factors led to the demise of the vector photoplotters. Increased design densities and the photoplotting of stepped multiple images increased photoplotting demands to the point where plotters had to be tremendously fast. Under these conditions, it's no surprise that the laser raster machines succeeded. It is a lot easier to make a beam of laser light sweep than to make a CNC machine tool move.

Aperture Lists: While the use of the term "aperture" to describe a pad or trace shape continues, the term aperture wheel is now replaced by "aperture list". The following are important advantages with aperture lists on raster plotters:

- Aperture shapes can be easily generated in software, thus eliminating the need to design a physical wheel.
- The aperture shapes can be described in the Gerber file, if certain extended-Gerber formats are supported. This reduces chances of getting the wrong list.
- More apertures can be defined on a list.
- Allowable apertures sizes are typically (but not always) greater than those imposed by the physical dimensions of an aperture wheel.

5.11.3 Talking to Photoplotters

The de facto standard for photoplotter data is the Gerber format, more properly known as RS-274D. The term Gerber refers to the Gerber Scientific Instrument company, a pioneer and leader in photoplotter manufacturing. Popular variants include Extended Gerber ("RS274X" and MDA FIRE AutoPlot, both of which embed the aperture list information in the file). RS-274D is a variation on traditional Numerical Control (NC) machine tool languages. It differs from traditional NC formats (i.e. drill data), as far as its use of tool selection codes goes but is otherwise compatible.

D Codes: D codes have multiple functions. The first is to control the state of the light being on or off. Valid codes for light state are D01, D02 and D03.

> D01 — Light on for next move.
>
> D02 — Light off for next move.
>
> D03 — Flash (light On, light Off) after move.

D codes with values of 10 or greater represent the aperture's position on the list or wheel. D 10 and higher values have aperture shapes and dimensions assigned to them by each individual user. For example, in one job, D10 could be a 10mil round, while another D10 job could be a 40 mil square.

X and Y Codes: The X and Y values in the Gerber file determine where the aperture shape and dimension will be positioned and drawn. X and Y values are used as co-ordinate pairs to determine where the light will be exposed, using the D codes shapes (i.e. D10) and light exposure status (i.e. D01, D02, D03) for drawing lines and arcs, as well as moving between drawing entities.

G Codes: G codes are used to configure the photoplotter. The commonly implemented codes include:

G01 — Future X, Y commands are straight-line moves;

G02 — Future X, Y commands are clockwise arcs;

G03 — Future X, Y commands are counterclockwise arcs;

G04 — Ignore the rest of this block (used for comments);

G54 — Prepare to change apertures (not necessary in laser photoplotters);

G74 — Future arcs are quadrant arcs;

G75 — Future arcs are full 360 arcs, allowing for a single command to draw a complete circle;

G90 — Absolute data: the controller will move to the absolute value given by the X and Y value; and

G91 — Incremental data: the machine will move the data by the amount of the X and Y value, whether then to the absolute coordinate point.

M Codes: M codes are used for machine control. The most commonly used codes are:

M00 — Full machine stop, commonly ignored by many plotters;

M01 — Temporary machine stop, commonly ignored by many plotters; and

M02 — End of plot.

In order to implement the CAD designs which are directly useable by CAM systems, the drawings must be supplied in either Gerber, HPGL, DXF or hard copy form. This drawing is used as a reference during the manufacturing process and should, therefore provide the dimensions for the board profile as well as the size and location of any internal routs. Drill hole locations along with a hole chart should also be included. It is important that any non-plate through-holes (NPTH) are clearly identified.

5.12 Computer-Aided Manufacturing (CAM)

Computer-aided manufacturing takes the output of the design system and applies it to the manufacturing process. Broadly, the hardware and its operation are very similar to the CAD system. While the CAD system has placement and routing programs, the CAM station has programs suitable for the manufacturing process. CAD/CAM systems can be integrated through a common database (Ginsberg, 1992b) as shown in Figure 5.33 .

All the necessary data to operate numerically controlled (N/C) Printed Circuit Board fabrication machinery can be derived entirely from the common database after converting the same to the equipment's required data format. The data can also be optimized by the computer to take advantage of the actual machine characteristics, such as drill speed, routing capabilities, tool selection speeds etc. In addition, the data obtained from the CAD/CAM database can be used for the control of component assembly equipment in sequence that provides the maximum use of automation. CAD/CAM systems can also provide the necessary inputs for test equipment. With the appropriate software, the test data can be provided in the correct format for bare-board testers, in-circuit testers; etc. This has led to an increasing interest in computer-aided testing (CAT) technology by printed circuit manufacturers.

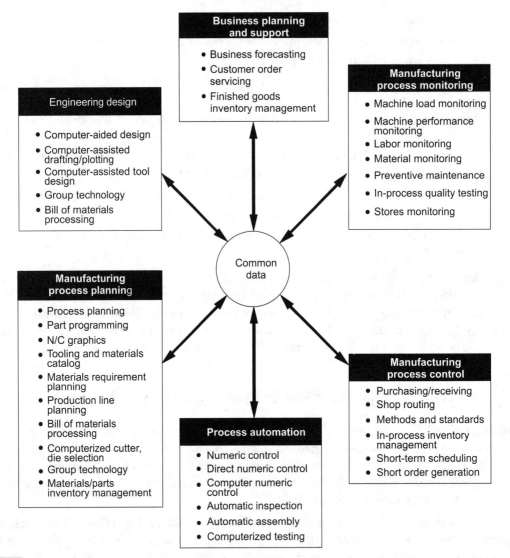

Fig. 5.33   Common CAD/CAM database: CAD system with a common database can be accessed to satisfy CAM requirements (adapted from Ginsberg, 1992 c)

A typical sequence of activities and operation, which are followed to take a printed circuit board from concept to final delivery is shown in Figure 5.34. The diagram illustrates as to how the computer-aided design, manufacturing and testing functions can be interfaced with each other during the process flow.

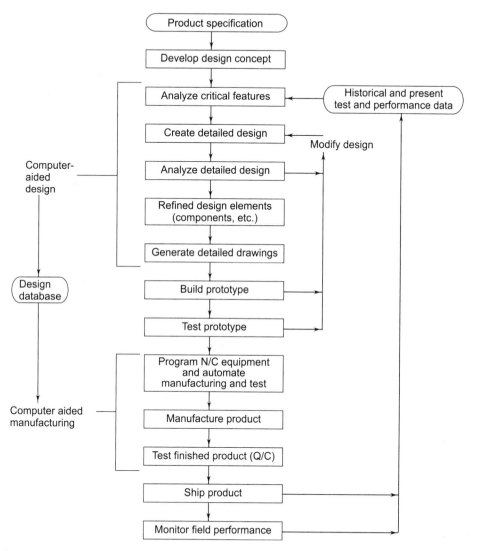

Fig. 5.34 Typical CAD/CAM System showing sequence of activities and operations required to take a printed circuit board from concept to final delivery (adapted from Ginsberg, 1992 b)

Integrated computer-aided engineering, design and manufacturing (CAE/CAD/CAM) with a common database and networking are now becoming popular as the users can link front-end CAE and back-end CAM. This is shown in Figure5.35. For example, test patterns created with the aid of CAE's simulation capability can offer the possibility of production testing, especially as the increase of fine pitch surface mounting may drive electronic manufacturers from in-circuit to functional testing.

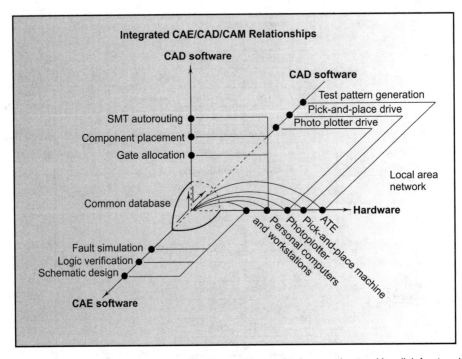

Fig. 5.35   Integrated CAE/AD/CAM relationships, based on a common database and networking, link front-end CAE with back-end CAM (redrawn after Ginsberg, 1992 a)

The use of CAD/CAM systems also helps to optimize printed circuit board manufacturability and can also enhance the electrical characteristic of the product. In addition to minimizing the use of vias, CAD/CAM software can be used to achieve enhancement features such as:
- Converting sharp 90 degree bends in conductors to 45 degree bends or curves;
- Increasing spacing between conductors and vias;
- Increasing the clearance between conductors;
- Increasing land and hole sizes.
- Centering conductors routed between lands.
- Generally increase conductor feature size, (e.g., more copper).

In a CAD system, the designer selects the basic routes that determine the board's complexity. The rules include the number of layers, minimal track and via sizes, and minimal spacing between objects. In the placement stage, the designer selects the physical location of the components on the board. When the placement process is complete, the physical netlist can be derived. A netlist is composed of groups of pins, each of which is expressed by a reference designator, a side, and an X-

Y location. Each group of pins represents a physical net. In addition, the placement of the components determines the location of the footprints, which are composed of all the pads in all the layers (toeprints) needed to connect the component to other parts of the design. Most footprints contain information about auxiliary features to be added to support layers and drawing. A typical example is the silk screen and soldermask layers derived from the placement process. It is an unwelcome reality that these layers are rarely viewed by the designer, who is primarily interested in the copper layers.

The algorithmically challenging routing process can take several hours or days to complete. Its intent is to implement the physical netlist using physical features on the board. These features are typically composed of three types of entities: traces, which are chains of lines carrying the signal between various locations on the same layer; vias, composed of pads in all or some of the layers and a hole that creates the physical connection, and which can be viewed as vertical lines carrying the signal between layers; and planes, the solid or hatched copper areas carrying power or ground signals between multiple toeprints.

The last stage of the design process is the manufacturing output. Through the previous stages, the designer worked mostly with logical entities. Components were added as needed, and as a result, all kinds of pads were automatically attached to multiple layers. Routing was initiated with a set of technology rules, which automatically implied the usage of certain line widths and via hole sizes. Through the manufacturing output stage, the designer has to generate files in various formats including:

- Gerber files representing the plotted layers;
- Drill files, typically in Excellon format, representing toeprint holes, via holes, and mechanical holes;
- HPGL/DXF files representing the mechanical drawings;
- A netlist file (IPC-356D or various CAD formats) to represent the physical connectivity; and
- A bill of material with a list of components and packages.

For an efficient CAD/CAM system, standardization of the output data is desirable. Therefore, PCB manufacturers increasingly use CAD data to assist in generating production ready photo-tooling. For this purpose, the following drawings are, therefore, normally plotted:

  i. photomaster for component side;
 ii. photomaster for solder side (mirror image);
iii. negative for component-side ground plane;
 iv. negative for circuit-side ground plane (mirror image);
  v. board assembly drawing including board outline, component outlines, and reference designators;
 vi. board assembly reference designator tables (optional — none or one drawing per table);
vii. dot pattern; and
viii. drill edits.

The master drawing is usually prepared from a component side view. However, to avoid misunderstanding, it is always advisable to print 'component side view' clearly on each drawing.

As the conductor tracks become finer and clearances smaller, front-end automation appears to be the only real solution to the problems of PCB manu-facture.Figure 5.36 shows a scheme using a front-end system which helps to produce accurate PCB tooling. The input to the manufacturer who uses front-end automation is the output from the CAD systems in the form of photoplotter steering data. The most commonly used data structure is the Gerber format. There are also companies with libraries of artworks for which Gerber plot files are not available. In this case, each art-work layer can be scanned and converted into a Gerber plot file, which can be loaded into the front-end system.

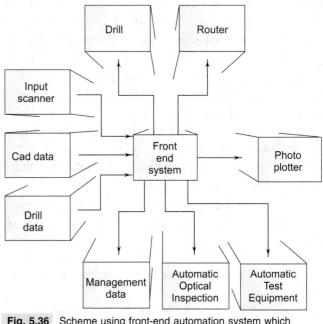

Fig. 5.36   Scheme using front-end automation system which helps to produce accurate PCB tooling (redrawn after Williamson, 1990)

The data for each layer is loaded into the workstation and electronically registered in X,Y and theta. Once loaded, the operator has the opportunity to execute automatic design rule checks to verify clearances between tracks, pads, track and pad, hole to copper edge; etc. The workstation operator can activate the system to step and repeat and rotate and/or mirror the design to obtain the maximum raw material utilization by appropriate planning of the panel. The operator can then add the test coupons which conform to the relative approval or to the customer's own coupon, the reference numbers, the resin venting pattern, test pattern, plating border and tooling holes. The operator can also compensate for mash stretch in a silk screen process.

Data is the output to a laser photoplotter and it is these first generation silver halide masters that are used directly in production. These masters are more accurate than contacts and reduce the amount of visual inspection required. In fact, tooling pins can be inserted into the film platen or flatbed of the laser photoplotter, enabling the imaging of the panel outwork directly onto pre-punched film. The CNC drill data is produced with 100 per cent accuracy. The routing data can also be the output as a post-processing exercise. Similarly, data can be the output to the bare-board test equipment and automatic optical inspection machine. This procedure enables the manufacturers to set up the inspection tolerances for the process against the database feature dimensions, generating the whole range of tooling in a short time.The high degree of accuracy and the quality of outputs enable the

PCB manufacturers to produce the board with a higher yield. Williamson (1990) explains the concept of front-end automation system for the bare board manufacturing process.

Murray (1996), while discussing the issue of CAD to CAM data transfer, brings out the importance of Design for Manufacturability (DFM) and states that it will now break the wall between the designer and the fabricator of the printed circuit boards. Cost-effective manufacturing can and should be the normal output from design, even when the fabrication time is compressed (Baumgartner, 1996).

 ## 5.13 Data Transfer Mechanisms

Data transfer from design systems has always been a ticklish job for the PCB fabrication industry (Dolberg and Kovarsky, 1997). It took many years for the industry to convert from physical pieces of film to digital data, but the data is still based on archaic formats for driving photoplotters and drill machines. The standardization of operational methods used to transfer data between CAE, CAD and CAM is absolutely essential if the intended time-to-market goals of the customers are to be met.

The 1995 IPC (The Institute for Interconnecting and Packaging Electronic Circuits) Technology Roadmap identified several solutions for data transfer from CAD to CAM. This led to the evolution of the IPC-2510 series, Generic Computer-aided Manufacturing (GenCAM) standard, with its first release in November 1998. The generic requirements of GenCAM are documented in IPC-2511. The whole series is available on the GenCAM website (www.gencam.org), which can be downloaded freely.

GenCAM is an ASCII format driven by domain experts that represent printed board fabrication and test; list of materials, assembly, inspection and in-circuit test as well as documentation, administration and configuration control management issues. It is a single file, able to completely describe a printed board, a printed board assembly, an assembly array, multiple assemblies on a sub-panel, a board fabrication panel, quality assessment coupons, and assembly/test fixtures. It establishes a level playing field software implementation concept for all CAD/CAM tool developers and is fully supported as an ANSI standard.

The IPC-2511 is the generic standard and describes in basic text all the requirements of the Gen-CAM data model. In addition, it has a computer code which enables us to read and understand while certain things are characterized in a certain manner. The computer code was developed by two German scientists, Backus and Nauer. The Backus and Nauer format (BNF) has been used for many years to test the GenCAM format for applicability and computer processing characterizations.

The data transfer, be it product description, equipment messages or supply chain communication will use the Internet in order to communicate information to the trading partners. We are witnessing a constant change in the way we do business, e-commerce, the Web, and many other factors which are forcing the industries to re-think the way information is moved from one source to another. There is no doubt that using the Web as a transfer mechanism not only has a great appeal but becomes a must in any future transaction.

The format which has been developed to transfer data across the Web is known as XML. The BNF and XML formats are very similar and several individuals/companies have written tools that can convert a BNF file into an XML file without loss of intelligence.

Many tools are now being developed to help in the implementation of XML. It is expected that as the industry moves ahead to XML, many tools will be developed that can be used to enhance the characterizations and descriptions of the components of the data transfer files.

Most electronics companies have links to an Internet facility. The connection can be dial-up (a modem link to a host computer at the service provider site) or a direct digital link. The speed of the link ranges from 14.4 kbps to 1.544 mbps or higher. Prices vary with the amount of bandwidth purchased by the company. Creighton (1996) points out that they have found the Internet to be a reliable, economical means to transfer PCB designs to the vendors. However, he advises that even if you chose to drive your work through the Internet, a modem should always be kept as a back-up as Internet links have been known to fail on rare occasions.

Internet file transfer has become a standard part of the PCB production cycle in many companies. But, as technology and the Internet evolve, file transfer methods will continue to improve. The Web supports forms and file transfers, many vendors and customers integrate PCB file uploads and downloads into their Web homepages. As high speed digital links have become available to more users, on-line checking and display of drawings has become possible.

 ## 5.14 PCB Design Checklist

The following checklist shall help to ensure the completeness of the design process:

- **Initial Board Features**
  - Place the board outline
  - Edit the board name, assembly number, and the part number for the board text.
- **Import Netlist**
  - Check for single pin nets, duplicate net names, alphanumeric pins and the netlist format.
- **Place Keep-Outs**
  - Include placement, component height, component drill, routing, copper and plane area, via and jumper, test point, and layer restrictions.
- **Place and Fix Constrained Parts**
  - Place mounting holes, tooling holes, connectors, potentiometers, switches, and everything that is a constrained part.
  - After placing the mounting holes, set the origin to the centre of the mounting or tooling hole that is closest to the bottom left corner.
  - After setting the origin, dimension the board from 0,0.

- **Global Fiducials**
  - Place the global fiducials in three corners of each side of the board that will contain parts. The fiducials should be 5 mm from the corner of the board, if possible, and keep other parts at least 2 mm away from fiducials.
- **Placement of Components**
  - Place the main ICs to establish the flow of the layout.
  - Place all critical components that have EMI, hi-speed, impedance control, thermal, and high voltage considerations.
  - Place the remainder of the non-critical ICs.
  - Place any local fiducials that are needed for fine pitch parts.
  - Place the remaining parts. Keep in mind not to let ensure that the components do not get within 2 mm of the board edge.
  - Verify design clearance check for overlapping parts.
- **Place All Reference Designators**
  - Text height should be as large as possible given the board density — no more than 4 mm / 3.5 mm and no less than 1.5 mm/0.15 mm.
  - Place all reference designators outside of parts at 0.3 mm nominal clearance.
  - Orientate reference designators at either 0° or 90° rotation.
  - Check that the assembly/part number appears in a clear area of the board along with the company name and board name, etc.
  - Polarized components must have at least one or a combination of the following that is visible both before and after assembly: a polarization mark; pin 1 text or dot; keying device.
- **Check Part List and Netlist**
  - Compare the PCB netlist to the schematic netlist and correct any differences before continuing.
- **Set-up Design Rules**
  - Review default design rules, set-up design rules for bus groups, set-up design rules for power nets and set-up design rules for the remainder of nets and classes.
- **Define Vias**
  - Define all vias and test pads that will be used in the design.
  - Generally shared vias are acceptable on voltage nets.
  - Single vias are preferred on ground.
- **Fan-Out Signals**
  - Manually fan-out remaining signals before routing or auto-routing the rest of the board.
  - Route clocks, matched line length and differential pairs
  - Fan-out and rout buses.
  - Route the rest of the board.

- **Add Test Points**
  - Set-up design for test audit rules and preferences.
- **Perform design rules checks**
  - Check for clearance, connectivity, hi-speed, planes, test points and fabrication.
- **Process Gerber, Drill and Assembly Data**
  - Numerically sort out the drill table.
  - Organize the drill sizes from smallest to largest by synchronizing the lowest drill symbol to the smallest drill size. Continue until all drill sizes have been defined.
- **Generate Gerber Data**
  - Select the Gerber files except the top and bottom assembly drawings.
  - Compare the ASCII netlist to the design netlist.
- **Generate Assembly and Drill Drawings**
  - Select pen, assembly top, bottom, and drill drawings and press.
  - Create the zipped file needed for fabrication and assembly.
  - Archive all data in the source safe and send the appropriate files to the fabrication and assembly houses.

5.15 Useful Standards

- *IPC-2531: Standard Recipe File Format (SRFF) Specification:* Outlines the requirements that an SRFF file must meet; describes the file format, outlines the file sections and indicates how data should be represented through objects which can either be vendor-independent or specific.
- *IPC-2547: Sectional Requirements for Shop Floor Equipment Communication Messages (CAMX) for Printed Circuit Board Test:* Describes event message content and an XML encoding scheme, which enables a detailed definition of messages in the domain of electronics inspection, test and repair/rework i.e. product and process quality.
- *IPC-2571: Generic Requirements for Electronic Manufacturing Supply Chain:* Defines an XML encoding scheme to facilitate supply chain interactions; the scheme is defined for bill of materials, approved manufacturer list, changes in engineering, manufacturing or product.
- *IPC-2511-A: Generic Requirements for Implementation of Product Manufacturing Description Data and Transfer Methodology:* Identifies the generic requirements for implementation of product manufacturing description data and transfer; helps users to transfer design requirements and manufacturing expectations from computer-aided design systems to computer-aided manufacturing systems for printed board fabrication, assembly and test.
- *IPC-2511B: Generic Requirements for Implementation of Product Manufacturing Description Data and Transfer XML Schema Methodology:* Specifies the XML scheme that represents the data file format to describe printed boards for tooling, manufacturing, assembly inspection and testing requirements.

Copper Clad Laminates

 ## 6.1 Anatomy of Laminates

The basic function of the laminate is to provide mechanical support for electronic components and to interconnect them electrically. Laminates for PCBs are composite materials. They can be simply described as products obtained by pressing layers of a filler material impregnated with resin under heat and pressure. The resulting thin insulating material, which is the mixture of filler (reinforcement) and resin on which all conductors and components are mounted, is called *base material*. This can be either rigid or flexible material.

6.1.1 Fillers (Reinforcements)

Fillers are meant to provide mechanical strength, stability and rigidity to the laminate. The commonly used fillers are a variety of papers, cotton fabric, asbestos sheet, glass in various forms such as cloth and continuous filament mat, ceramic material, molybdenum, etc. However, the most common materials used are paper and glass fibre.

Paper has been used as reinforcement in a vast majority of printed circuit boards. They are low priced and are easily machinable. However, they have a tendency to absorb a lot of moisture.

Fibre-glass as reinforcement has gained popularity because of its high tensile strength and dimensional stability. It offers a high resistance to temperature variation and has a low moisture absorption property.

6.1.2 Resins

Resins are used to impregnate the selected fillers. The commonly used resins in the manufacture of base materials are phenol, polyester, cyanate ester, epoxy and polyimide. Of these, the epoxies and phenolics are used for about 90 per cent of all laminates. They are mostly synthetic types of materials, either thermoplastic or thermosetting, formed by the polymerization process. The selection of a resin takes into account electrical, mechanical, chemical and thermal characteristics. All these characteristics have varying degrees of importance depending upon the specific application of the PCB.

Epoxy resins, which are the most commonly used, are sometimes modified with additives to achieve higher thermal properties or improved chemical resistance. In a composite, the properties of the laminate depend upon the type and quantity of raw materials used, their curing schedules and the procedures used to produce the printed boards. It has to be ensured that there is enough resin to fill all the spaces between the fibres as internal voids may lead to premature mechanical failure of the laminate.

Polyimide is the material of choice when extreme thermal condition exists, such as extended time at high temperature during assembly or use. Another consideration is the need to replace defective components on expensive assemblies. Polyimide maintains its bond to the foil during excursions in order to solder temperature extremely well. Its Tg is greater than 220 °C and is responsible for polyimide's excellent high-temperature performance.

Goosey (2003) describes the characteristics expected from the laminates to meet the present-day requirements of increasing the packaging and interconnected densities of electronic assemblies. New types of laminates have been developed in recent years that have Tg values stretching well above 200 °C and in some special cases, up to nearly 300 °C. Examples of these laminates include those based on cyanate esters, alleylated polyphenylene ethers, and the so-called BT-epoxy and tetrafunctional epoxy systems. Table 6.1 shows the properties of some of the new laminate types.

Table 6.1 Properties of High Tg Laminates (after Goosy, 2003)

Laminate material	Tg (°C)	Dielectric constant (10 GHz)	Dissipation factor (10 GHz)
FR4	130–150	4.5	0.022
Tetrafunctional epoxy	175	4.4	0.020
Polyphenylene ether	175	3.4	0.009
Epoxy/polyphenylene oxide	180	3.9	0.013
Bismaleimidetriazine	180	4.1	0.013
Thermount	220	4.1	0.022
Cyanate ester	240	3.8	0.009

(Contd.)

Table 6.1   *(Contd.)*

Laminate material	Tg (°C)	Dielectric constant (10 GHz)	Dissipation factor (10 GHz)
Polyimide	280	4.3	0.020
Liquid crystal polymer	280	2.8	0.002

A large variety of copper clad laminates are commercially available these days. They have been standardized at national and international levels in terms of specifications which have been laid down for each copper clad laminate grade and the minimum/maximum limits of important properties. In general, the laminates should have good electrical, mechanical and environmental characteristics and meet the standard specifications of the Institute for Interconnecting and Packaging Electronic Circuits (IPC), National Electrical Manufacturers Association (NEMA), Department of Defence Design Standard or Military Standard (MIL), International Electrotechnical Commission (IEC) and the American National Standard Institute (ANSI), among others.

6.1.3   Copper Foil

The conductive layer on a laminate can be made of copper, nickel, stainless steel or beryllium copper. However, the most widely used substance is copper due its easy availability, cost and functionality. Copper cladding can be on one side or on both sides of the composite, depending upon the need and use.

The quality of PCB depends, to a large extent, on the properties of the copper foil. Therefore, the quality requirements of copper foil are very demanding. The thickness of copper foil is usually expressed in ounces per square foot ($oz./ft^2$), which corresponds to about 3.052 gram/square cm (g/cm^2) or 305.2 grams/square m (g/m^2).

Table 6.2   Standard Thickness of Copper Foil and Weight

Thickness	Weight			Resistivity
	Basic weight		Tolerance	M Ω -max
	Oz/ft^2	g/m^2		
12 μm	3/8	107	± 10 %	9.3
18 μm	1/2	153	± 10 %	7.0
35 μm	1	305	± 10 %	3.5
70 μm	2	610	± 10 %	1.8

Usually, a layer of very thin copper foil of thickness 17.5 microns, 35 microns or 70 microns is bonded to one or both sides of the base material. 17.5 micron copper foil is also denoted as half ounce since half ounce of copper is used to get 1sq.ft. of copper clad sheet with 17.5 micron thickness of copper. The copper foil is normally available in large rolls weighing 136 to 181 kg (300-400 pounds). Table 6.2 shows the standard thickness of commonly used copper foil. The vast majority of laminates used are with foils of 305 gr./m^2 or lower. The tolerance of weight is usually ±10 %.

Copper foil must satisfy strict quality requirements. Its resistivity should not exceed 0.1594 ohm-gram/m^2 at 20 °C. The foil should be free from pin holes, pits, scratches and nodules. Copper foil is available in two forms: rolled annealed copper foil and electrolytic copper foil. These are detailed below.

6.1.3.1   Rolled Annealed Copper Foil

This is manufactured by melting electrolytically formed copper cathodes into large ingots. The ingots are hot rolled in specially designed rolling mills and annealed to get large copper foil rolls. These foils are available in widths ranging from a minimum of 635 mm to a maximum of 965 mm.

The rolled copper is 99.9 per cent pure and has a good horizontal grain structure. The rolled copper is mainly used in the flexible PCB manufacturing process. Even though foil manufacturing is easier, it has some disadvantages like limited width, poor solderability, and adhesion and ductility problems which are created due to grain structure deformation. The rolled copper foils produced by annealing electrodeposited copper at a high temperature are also called High Temperature Elongator (HTE) foils. HTE is advantageous as it is the most ductile foil available and thus has a niche in some microwave applications.

6.1.3.2   Electrolytic Copper Foil

This is manufactured by the electroplating method. The tank has either a lead or polished stainless steel rotating drum which is used as cathode and pure copper as the anode. Both are immersed into the copper sulphate electrolyte as shown in Figure 6.1.

The deposited copper is easily peeled off because of poor adhesion on the polished drum. The peeled copper has a very smooth shiny finish on one side and dull finish on the other side. The dull side increases its adhesion with suitable adhesives. The grain structure size is vertical in nature and gives excellent bond strength. The dull side is again subjected to further process to enhance roughness by chemical oxidation in order to improve adhesion. The electrodeposited copper foil rolls are available in widths of up to 1970 mm.

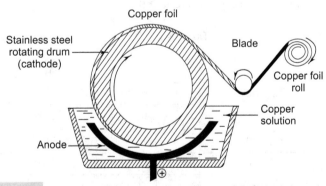

Fig. 6.1   Cell for manufacturing electrolytic copper foil. Cathode is made of stainless steel and anode of copper

The purity of the foil is around 99.5 per cent and its resistivity should be 0.1594 ohms gram/m^2 at 20 °C. Now, thin copper foils with the thickness of 5 microns and 9 microns are available for manufacturing of multi-layers and rigid PCBs. The advantages of thin foils include very rapid etching time, less etchant waste, elimination of bonding treatment prior to photo-resist application and better photo-resist adhesion.

 ## 6.2 Manufacture of Laminates

Copper clad laminates are manufactured by pressing layers of filler material impregnated with resin under heat and pressure with copper foil. A hydraulic press is used for this purpose.

Although the following manufacturing procedures can be extrapolated to include any type of laminate available today, the process detailed herein pertains to the industry standard epoxy-glass FR-4 materials.

6.2.1 Materials

The materials needed for the manufacture of laminates are glass fabric (filler), epoxies (resin), solvent and copper foil.

6.2.1.1 Fibre-glass Cloth

Glass cloth acts as the main structural reinforcement in most laminates. The rigidity and strength offered by glass compliments the binding, encapsulating and insulative properties of the epoxy resin. The singular fibre-glass filament is the building block with which glass clothes are constructed. These thread-like fibres are put together to form a yarn or bundle. Subsequently, like weaving of any other type of cloth, numerous yarns are woven together in the manufacture of cloth. Various combination of filament and bundle diameters, filament counts and weave density, among other variables, will resist in a multiplicity of glass cloth thicknesses and weights. Finally, glass clothes are coated with a finish that facilities resin impregnation of and bonding to the cloth.

6.2.1.2 Epoxy Resins

The function of the resin is to act as a 'glue' to hold the laminate together. Epoxy resins can be purchased from various vendors at various steps of manufacture. Epoxy resin can be had in a liquid form so that it can be concocted to upstaged resin using proprietary recipes and processes. It can also be purchased in the advanced or upstaged state, wherein the solid resin, complete with hardness and catalysts, is ready for use in treating.

6.2.1.3 Copper Foils

Most foils used in FR-4 manufacture are electrodeposited type foils. These are manufactured by plating copper onto slowly revolving drum-shaped cathodes that are partially immersed in the plating solution. As the drum revolves, the plated copper deposit is removed from the cathode drum at one continuous speed. Varying the drum speed and current density helps to vary the copper deposit and consequently, the resulting foil thickness. At this stage, the 'raw' foil becomes available, which is then subjected to various processes designed to increase the roughness of the matte side, thereby increasing its mechanical adhesion to the substrate. In addition, the foil is coated with a micro-thin film of protective coating to prevent oxidation of the copper during lamination and storage.

6.2.2   Process

The three primary raw materials — glass, resin and copper-are pressed at the laminator to give a fully cured final product that is dimensionally stable and resistant to moisture, chemicals and thermal excursions occurring in the PCB manufacturing process. The process utilized in bringing the thin major raw materials together is shown in Figure 6.2.

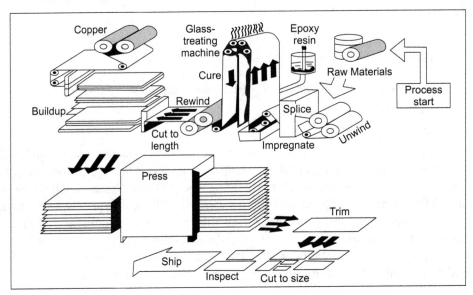

Fig. 6.2   Process for manufacturing laminates (Courtesy GE Electromaterials, 2001)

6.2.2.1   Treating

Treating is the process whereby the liquefied resin is applied to the glass cloth, usually via a combination of immersion and metering rollers. The treated cloth is then subjected to a controlled heat source to semi-cure the resin. The heat source is a drying oven, which is air-circulating or infra-red type and can be up to 40 m long. Most of the volatiles are driven-impregnated glass sheet is now dry to the touch, and in this stage, it is generally referred to as prepreg or 'B' stage.

Rigid process control is necessary during treating as the immersion and metering process are critical. Thorough wetting of the glass cloth by the resin, as well as precise control of the resin quantity absorbed are of utmost importance to the laminate consistency and quality. Practically, the ratio of resin to base material, the final thickness of the prepeg and the degree of resin polymerization need to be monitored.

6.2.2.2 Lay-up

Lay-up is the process wherein treated prepegs and copper foils are assembled for pressing. In this operation, the copper foil is first laid against a large polished stainless steel press plate. Then, a number of sheets of prepegs are laid on top of the copper. The number of layers depends upon the desired thickness of the laminate. The final sheet of copper foil is placed on top of the prepeg if the material is to have copper on both sides. If copper is desired only on one side, a release film is used to replace one of the sheets of copper.

6.2.2.3 Pressing

Pressing is the process wherein simultaneous heat and pressure are applied to the packs or books (prepegs, copper foils and release film, if any) to produce fully cured laminates. This operation is carried out in a press which is hydraulically operated and capable of developing pressure up to 1000 psi. Steam is a typical heat source. Packs or books are loaded into each press, with the typical process being capable of moulding 80 sheets 36×48 inches or 250 sheets of 48×144 inches, 1/16 inch thick.

During pressing, the semi-cured epoxy resin liquefies and flows, expelling any entrapped air or gases. This flow acts to encapsulate the treatment side of the foil(s), facilitating foil adhesion, and also to homogenize the resins in each laminate. After a certain period of time, the epoxy groups in the liquefied resin begin to form cross-links, leading to curing of the resin. Thermocouples are placed in several sheets to monitor and control temperature, while a timer automatically records time against a pre-set cure cycle.

When the curing is completed, the steam is automatically cut off, press cool down begins and the press books are cooled to a temperature (80 °F) at which they may be handled. After removing the material from the press, the edges are trimmed from the sheet to remove the irregular excess resin flow areas. At this stage, the laminate sheets are sheared down to the desired sheet or panel size.

During the manufacturing process, several quality control checks need to be implemented to ensure uniformity in thickness of the laminate, lamination integrity (endurance of extreme thermal, mechanical and chemical abuses) bow and twist, surface quality and dielectric variations. Knowledge of the laminate manufacturing process is helpful for designers, fabricators and assemblers in understanding the capabilities and laminations of this critical building block of PCB manufacture.

6.2.2.4 Quality Control

After the laminate has been formed, it undergoes various tests. They are conducted to check for the following:

- Cleanliness;
- Dents;
- Scratches;
- Thickness;
- Water absorption
- Solder float test (Solder resistance);
- Bonding strength;
- Warp and twist;
- Flame resistance;
- Dimensional stability;
- Resin content;
- Volatile content;
- Resin flow and gel time;
- Printability;

- Flexural strength;
- Peel-off test;

- Drillability; and
- Punching and shearing qualities.

6.3 Properties of Laminates

The properties of laminates vary from grade to grade, depending upon resins and fillers. The electrical, mechanical, chemical and thermal characteristics are of laminates mainly depend upon the selection of the resin. The properties of laminates are:

- Dielectric constant;
- Dielectric breakdown strength;
- Dielectric strength;
- Dissipation factor;
- Arc resistance;
- Loss factor;
- Absorption of water;
- Tensile strength;
- Compression;
- Shear;
- Flexural strength;

- Impact strength;
- Environmental resistance;
- Fungus resistance;
- Flammability characteristics;
- Self-extinguishing characteristics;
- Laminating difficulty;
- Copper adhesion;
- Heat resistance;
- Machinability; and
- Dimensional stability.

The electrical and mechanical properties of laminates are affected by environmental factors such as humidity, temperature, corrosive atmosphere, etc. Table 6.3 lists the important properties of laminates commonly used for PCB construction.

Table 6.3 Important Properties of Base Materials

Grade	Composition	Remarks
XXXPC	Paper/phenolic	High moisture resistance
FR-2	Paper/phenolic	Similar to XXXPC but flame retardant
XXXPC	Paper/phenolic	Best mechanical characteristics of paper/phenolic grades
FR-3	Paper/epoxy	High mechanical and electrical characteristics, flame retardant
FR-4	Glass/epoxy	Flame retardant, chemical resistant, low water absorption
G-3	Glass/phenolic	High flexural strength and dimensional stability
G-5	Glass/melamine	High resistance, high impact strength
G-9	Glass/melamine	Same as G-5 but better electrical characteristics
G-10	Glass/epoxy	Same as FR-4 but not flame retardant
G-11	Glass/epoxy	Same as G-10 but higher flexural strength under heat

(Contd.)

Table 6.3 *(Contd.)*

Grade	Composition	Remarks
G-30	Glass/polyimide	High dimensional stability under heat, flame retardant
FR-5	Glass/epoxy	Same as G-11 but flame retardant
GPO-1	Glass/polyester	General purpose mechanical and electrical grade
GPO-2	Glass/polyester	Similar to GPO-1 but lower flammability

6.3.1   Electrical Properties

The electrical properties of a laminate depend upon the electrical properties of the filler, cured resin and the by-products of the curing reaction.

Laminate absorbs moisture to some extent when exposed to high humidity conditions. Consequently, this absorbed moisture adversely affects the electrical properties. For 1.6 mm thick laminates, the appropriate water absorption values are as follows:

a) Papmer phenolic
   Example: NEMA grades X, XX, XXX, etc.   0.75 to 6 %
b) Glass epoxy
   Example: NEMA grades G10, G11   0.23 %
c) Glass PTFE (polytetrafluoroethylene)
   Example: NEMA grades GTE   0-0.68 %

6.3.2   Dielectric Strength

This is the ability of an insulating material to resist the passage of electric current of a disruptive discharge produced by an electrical stress. It depends upon a large number of factors pertaining to the material such as chemical composition, molecule structure, degree of moisture, thickness cleanliness and roughness of surface and material ageing.

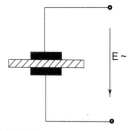

Fig.6.3 Schematic electrode arrangement for testing dielectric strength

The test is performed by applying 50 Hz ac voltage on a piece of laminate whose copper surface is etched off before it is placed between two electrodes as shown in Figure 6.3. The test is carried out under oil in the following two ways:

- *Short-time Test:* In this test, the voltage is increased at a uniform rate of 0.5 KV/s.
- *Step-by-step Test:* Initially, 50 per cent of the short-term breakdown voltage is applied. The voltage is then increased in increments according to a pre-determined schedule at 1-mm interval. The test values of dielectric strength vary with the form and size of the electrodes, the frequency and shape of the voltage waveform and the surrounding material.

6.3.3 Dielectric Constant

Dielectric constant is the ratio of the capacitance of a capacitor with a given dielectric to the capacitance of the same capacitor with air as dielectric (Figure 6.4). It is calculated from the capacitance as read on a capacitance bridge, the thickness of the sample and the area of the electrode.

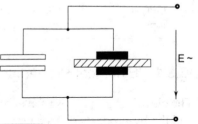

Fig. 6.4 Electrode arrangement for testing dielectric constant

The dielectric constant is also referred to as *Permittivity* and being a ratio, is a dimension-less entity.

The dielectric constant measures the ability of an insulating material to store electrostatic energy. It varies with the thickness, temperature, humidity and frequency and chemical composition of the material. The effects of temperature and frequency variations on the dielectric constant vary for different materials.

6.3.4 Dissipation Factor

The dissipation factor of an insulating material is the ratio of the total power loss (in watts) in the material to the product of the voltage and current in the capacitor in which the material is the dielectric. It varies with frequency, moisture, temperature, etc. and is a dimension-less entity.

Expressed in another way, the dissipation factor is the ratio of parallel reactance to parallel resistance. It is measured with the electrode arrangement as shown in Figure 6.5 whereas Figure 6.6 shows the vector diagram of the equivalent parallel circuit.

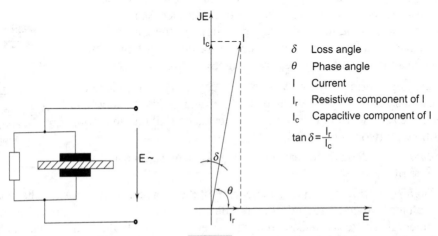

δ Loss angle
θ Phase angle
I Current
I_r Resistive component of I
I_c Capacitive component of I

$\tan \delta = \dfrac{I_r}{I_c}$

Fig. 6.5 Electrode arrangement for testing dissipation factor

Fig. 6.6 Vector diagram of equivalent parallel circuit for a laminate

The dissipation factor is expressed as tan δ (the tangent value of loss angle δ). The dissipation is directly related to the resistive power loss in a laminate. Therefore, for an electronic circuit operating at a high power loss, it is desirable to use laminates with a low dissipation factor. The value of the dissipation factors for various combinations of fillers and resins are:

Paper phenolic laminates	0.02–0.08
Glass epoxy laminates	0.01–0.03
Glass PTFE laminates	0.0008–0.005

The dissipation factor of a laminate varies with the frequency, temperature and moisture absorbed as follows:

- Increase in frequency results in decrease in the dissipation factor.
- Increase in temperature results in increase in the dissipation factor.
- Increase in moisture constant results in proportional rise in the dissipation factor.

This implies that the dissipation factor given in the data sheet must be related to the condition under which it has been determined.

6.3.5 Insulation Resistance

This is the ratio of the voltage applied to the current flowing in the base laminate. Although the absolute value of insulation resistance is initially important, the change in resistance under a specified environmental condition is usually more significant. Insulator resistance varies with the environmental conditions and the process techniques of the test samples. Therefore, during testing for the insulation resistance, the test specimen should be subjected to the same environment as would be available in the final application. Insulation resistance is composed of both the volume and surface resistance in a copper clad laminate.

6.3.6 Surface Resistivity

This is the resistance to electrical leakage current along the surface of an insulating material. This depends upon surface humidity, cleanliness, finish, temperature, and environmental conditions, among other things.

The surface resistance between two points on the surface on any insulation material is the ratio of the dc potential applied between the two points to the total current between them. This is measured by using a three electrode circular system as shown in Figure 6.7. The electrodes are formed by completely removing the metal foil by etching but leaving the foil outlines on the specimen to form the edges of the electrodes. The guard

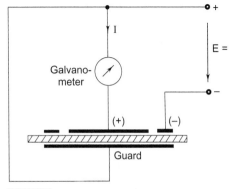

Fig. 6.7 Schematic electrode arrangement for measuring surface resistivity

electrodes on the opposite, kept at positive (+) potential, help to avoid stray currents passing through the laminate. Measurements are made by applying 500 V dc voltage with a Megohm meter.

Surface resistivity also indirectly depends upon the chemical composition of the dielectric material. The material which has a high moisture absorption property results in a reduction of surface resistivity as is evident from Figure. 6.8. The effect of humidity on the surface resistivity of glass epoxy have been found to decrease logarithmically with an increase in humidity at approximately the rate of one decade per 20 per cent humidity change.

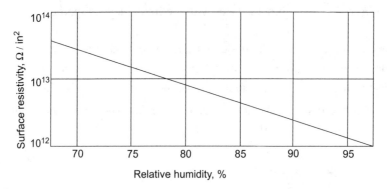

Fig. 6.8 Variation of surface resistivity with relative humidity for glass epoxy surface

6.3.7 Volume Resistivity

This is the measured resistance to leakage current through the body of an insulating material. In other words, volume resistance is the ratio of the dc potential applied to electrodes embedded in a material to the current between them. It is usually expressed in ohm-centimeter.

The volume resistance is again measured by using the three electrode method as shown in Figure 6.9. The volume resistivity is governed by:

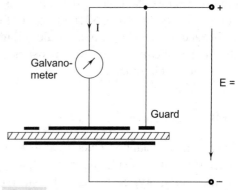

Fig. 6.9 Schematic electrode arrangement for measuring volume resistivity

$$\text{Volume Resistivity} = \frac{R \times A}{t} \ (\Omega \ \text{cm}),$$

where

R = resistance measured (Ω)

A = area of guarded electrode (cm^2)

t = thickness of sample (cm)

Volume resistivity is influenced by factors such as the chemical nature of the material, temperature and moisture absorbed in the sample. In general, it falls rapidly with increasing temperature. Volume resistivity is one of the most important electrical properties of a laminate as it is related to its moisture absorption property. Laminates in which the volume resistivity does not change with the moisture, like PTFE are preferred for high frequency applications.

6.3.8   Dielectric Breakdown

Dielectric breakdown is the disruptive discharge measured between two electrodes inserted in the laminate on 25.4 mm centres perpendicular to the lamination. The test is carried out in oil in a similar manner as is done for measurement of the dielectric strength. The data on 1.59 mm thick material is given in Table 6.4.

Table 6.4   Dielectric Breakdown Data

Sr. No.	Material	KV
1.	XXX PC and FR-2	15
2.	FR-3 and FR-6	30
3.	CEM-1, CEM-3, G-10, G-11 Fr-4, FR-5	40
4.	GT and GX	20

 ## 6.4   Types of Laminates

6.4.1   Phenolic Laminates

Phenolic laminates make use of phenolic resins, which consist of a solution of the reaction product of phenol and formaldehyde in a solvent. The phenolic resins are reinforced with paper fillers and the copper foil is pressed to the base material. Hence, they are also called as 'paper phenolic copper clad laminates'.

The first laminate based on phenolic resins was probably the well-known bakelized fabric. It is made from cloth and resin which is dissolved in a suitable solvent to form a solution used to impregnate the cloth. The cloth is cut into sheets, stacked in a pile and placed between two heated plates of a laminating process. The combined action of the temperature and pressure causes the resin to melt and flow to form a single mass. At the end of the lamination cycle, a strong rigid laminate is obtained.

Phenolic cloth has now been virtually replaced by phenolic paper. The laminate grade classification as per NEMA is X, XX and XXX, which means a content of phenolic resin of about 35, 45 or 55 per cent respectively, the remainder being paper and other fillers and additives. Out of these laminates, X and XX have rather poor characteristics and are seldom used. Further modification of the XXX materials gives XXP and XXXPC grades, where P means punchable and PC is cold punchable. Their properties are summarized below.

- *Grade XX*: Paper-based phenolic, fair electrical properties, good mechanical properties, not recommended for punching and little abrasive action to drills.
- *Grade XXP:* Similar to XX, but can be punched at high temperatures (200 to 250 degrees F)
- *Grade XXX*: More resin than XX, better electrical and mechanical properties and recommended for use in radio frequencies.
- *Grade XXXP:* Same as XXP, but with more resin, recommended for most general purpose applications.
- *Grade XXXPC*: Same as XXXP but can be cold punched (70-120 degrees F), has higher insulation resistance, has lower water absorption and is good for high-humidity applications.

Phenolic laminates offer good punchability and ease of fabrication. They are also have relatively low cost and therefore, have the largest commercial use. One of the drawbacks of the paper phenolics is their poor arc resistance and higher water absorption property as compared to epoxy laminates. Figure 6.10 shows the impact of moisture on the surface resistivity of paper phenolic laminates.

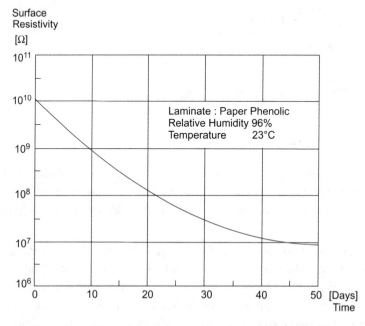

Fig. 6.10 Impact of moisture on surface resistivity of paper phenolic laminates (redrawn after Bosshart, 1983)

6.4.2   Epoxy Laminates

The base material is reinforced with glass fibre or paper fibre as filler and epoxy resin and the copper foil is pressed over it to get copper clad laminates. Hence they are called 'glass epoxy copper clad laminates' or 'paper epoxy copper clad laminates' depending upon the type of filler used. During lamination, the epoxy resin is cured by means of heat, pressure and the action of an added catalyst. The final product is a thermosetting resin which is neither feasible nor soluble. They withstand all chemicals, with the exception of oxidizing acids and are almost indestructible, except at high temperatures. Some of the special properties of epoxy resins are that they:

- Show very low shrinkage during cure;
- Are tough materials;
- Form an outstanding adhesion bond;
- Exhibit high mechanical strength;
- Have extremely high alkali resistance or even excellent acid and solvent resistance;
- Exhibit good electrical properties over a wide range of temperatures and frequencies; they are excellent insulating materials and have high dielectric strength; and
- Demonstrate excellent moisture barriers and low water absorption.

Epoxy resins are usually combined with high quality reinforcing material such as glass cloth and composite material, and exhibit exceptionally high mechanical strength and excellent electrical properties.

6.4.3   Glass Cloth Laminates

The glass cloth is used for the manufacture of epoxy resin reinforced glass cloth laminates. G-10, G-11, FR4 and FR5 types are some of the glass cloth laminates available. The glass fibres or filaments are usually 9.6 microns (0.38 mil) in diameter, are bunched together and twisted to form thread and are woven from silk fibre in the silk industry. The threads are of different types. The mesh of the cloth is indicated by the number of threads provided per linear inch. 1.6 mm thick laminates contain eight glass cloth layers and if a hole of 1mm diameter is drilled, it cuts about 10,000 filaments. The glass fibre in the holewall will have about 20,000 ends. Spiak and Valiquette (1994) describe the trends in the laminate industry for better and faster materials. Table 6.5 gives the physical characteristics of laminate materials.

The most widely used and industries' standard base material for PCB fabrication is woven-glass-cloth-reinforced epoxy or FR-4 with a 135 °C Tg. The dielectric constant is approximately 4.4 and propagation delay is approximately 178 ps/in. These laminates are constructed on one or multiple piles of epoxy resin impregnated woven glass cloth. FR-4 is the most widely used material because its properties satisfy the electrical and mechanical needs of most applications. They exhibit high mechanical strength and machinability, consistent drilling properties, thermal stability and V-O fire resistant. Designed for single- and double-sided boards, they provide a combination of processing flexibility and finished board performance needed for many SMT applications. However, demands for the following high performance (Guiles, 1998) requirements has necessitated the development of new materials:

- Boards with special thermal requirements, i.e. those handling high concentration of power in small areas;
- Large or high layer count boards that are complex to manufacture; includes boards with complex etched patterns requiring precise registration;
- Boards requiring controlled thermal expansion characteristics due to special assembly technologies (flip chip, chip-on-board) or enhanced reliability needs;
- Boards with special or tightened electrical requirements, especially those requiring a low dielectric constant for low propagation delay, lower cross-talk and higher clock rates, and a low dissipation factor for low attenuation, better signal integrity, higher clock rates, and lower power consumption in portable electronics.

Table 6.5 Physical Characteristics of Laminate Materials
(after Lucas, 1993)

Material type	Tg °C	Dielectric constant		Propagation delay ps\in	Per cent of H_2O	Copper peel strength	
		Resin	w/E-glass		absorption	lb/in.^{2} @ 25°C	Lb/in.^{2}@ 200°C
Standard FR-4	135	3.6	4.4	178	0.11	11.0	5.7
High-performance FR-4	180	2.9–3.6	3.9–4.4	168	0.04–0.20	9.0–11.0	7.5–7.9
Polyimide	220	3.4	4.3	176	0.35	8.5	8.0
BT	195	3.1	4.0	170	0.40	8.7	5.2
Cyanate ester	240	2.8	3.7	163	0.39	8.0	6.3
Silicon Carbon	190	2.6	3.4	156	0.02	5.0	5.0
PTFE	16	2.1	2.5	134	0.01	10.0	8.0

An important physical characteristic of the laminates is the amount of Z expansion occurring when a circuit board, and subsequently an assembly, are exposed first to a series of excursions to solder temperature and then to life cycles from ambient to operating temperature. The detrimental effect of Z dimension expansion is the work-hardening of the copper cylinder (plated through-hole).

From a laminate perspective, the lower the Z expansion, the lower the overall work-hardening and the greater the life expectancy of the electronic assembly (Lucas, 1993).

Ehrler (2002) examines the properties of various epoxy materials and reinforcements, which are responsible for the development of new materials with improved electrical and thermo-mechanical performance of printed circuit boards. As shown in Figure 6.11, it usually results in improved through-hole reliability due to the enlarged range below the Tg with lower Z-axis expansion in the range upto the soldering

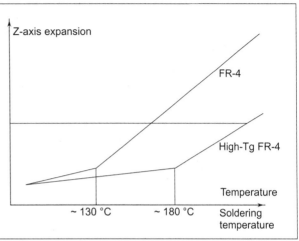

Fig. 6.11   Z-axis expansion of FR-4 and high Tg FR-4 as soldering temperature increases (after Ehrler, 2002)

temperature. This is usually achieved by increasing cross-linking of the material. High-Tg (180 °C) FR-4 materials with associated reduction in Z expansion have better dimensional stability, which is important for complex PCBs because the cores do not require heating much above the Tg, as conventional FR-4 would require. This i smproves the laminate's resistance to stress-related defects and increases product life expectancy. This high performance FR-4 laminates with standard E-glass reinforcement have a dielectric constant as low as 3.9 and propagation delay as low as 168 ps/in. However, the thermal stability of some high-Tg FR-4 resins does not match that of conventional FR-4 and they have lower copper peel strength.

Lin (2003) describes the properties of the NPLD series material which posses the desirable characteristic of high Tg (200), low Dk (3.5-3.9) and low Df (0.038-0.01). The material provides with improved thermal dielectric and moisture resistance performance and is best suited for use in the Giga-hertz frequency range application.

6.4.4   Prepreg Material [B-Stage]

Epoxy resins are widely used in the manufacture of B stage or prepreg material. When two monomers of epichlorohydrin and bisphenol-A are mixed together, a polymer is obtained. This polymer is called B Stage resin, which melts and is soluble in some solvents. The resin is mixed with fillers like paper or glass cloth to get B stage epoxy resin material and are used in manufacturing multi-layer laminates.

They have some characteristics like chemical and solvent resistance and low Z-axis expansion at high temperatures.

6.4.5 PTFE (Polytetrafluoroethylene) Laminates

Polytetrafluoroethylene is a thermoplastic material. When reinforced with glass, it results in a laminate which offers several advantages, particularly when used for high frequency (RF) and microwave applications. The thickness distribution in a composite material consisting of resin and glass fabric affects the uniformity of the dielectric constant and the dielectric loss factor. With microwave PCBs being mainly active components, variations of the dielectric characteristics inevitably lead to performance losses. The PTFE woven glass fabric base materials have an exact thickness distribution across the entire manufacturing sheet 1220 mm × 914 mm (48" × 36") as 0.020" ± 0.0015" without any taper towards the edges of a sheet. In addition, PTFE laminates have a low dielectric constant and a low dissipation factor under a wide temperature, humidity and frequency range.

The molecular structure of PTFE is responsible for an extremely low moisture absorption of < 0.02 % of the base material. Thermoplastics undergo a chemical as well as a phase change when they are heated. Their molecules form a three-dimensional cross-linked network. Once they are heated and formed, they cannot be re-heated and re-formed. Microwave laminates made from thermoset resin systems exhibit significantly higher values. The influence of moisture in degrading the electrical behaviour of the laminates, particularly on the impedance, is well understood. PTFE therefore offers very high electrical strength having electrical insulation under severe environmental conditions. Therefore, besides their use in RF and microwave applications, PTFE laminates are often used where high insulation resistance under humid conditions has to be maintained. Typical examples are input stages of high impedance / low current measuring instruments and amplifiers.

A further advantage of PTFE is the higher inter-laminar bond strength and the resultant copper peel strength. Higher peel strength is required during re-work of the assemblies, when conductors are partially exposed to very high temperatures. PTFE/woven glass materials exhibit a minimum peel strength value of 1.8 N/mm (10 lbs/in) even with 0.5 oz/ft^2 (17.5 μm) copper foil.

PTFE laminates meet the required class of the flammability UL-94, V-0 without using any flame retardant. Microwave base materials based on thermoset resins mostly require flame retardants degrading other performance parameters such as peel strength or do not meet V-0 for the thinnest material thickness. This is a definite advantage of PTFE-based laminates.

Besides using woven glass as the filler with PTFE, ceramic-filled PTFE laminate has also been developed which is not only low in cost but its use as RF microwave PCB substrate is now well established. The laminate is used for double-sided strip line PCB construction and multi-layer application. The various types of epoxy laminates are:

- NEMA-FR3-flame resistant paper epoxy;
- General purpose glass epoxy G-10, temperature-resistant glass epoxy G-11;
- Fire retardant glass epoxy FR-4; and
- Fire-and temperature-resistant glass epoxy Fr-5.

Flame-retardancy is achieved by substituting some functional groups of the resin with halogens like chlorine or bromine.

6.4.6   Polyester Laminates (Mylar Lamination)

Polyesters are solutions of unsaturated polyester resin in copolymerizable monomers such as styrene. Glass fibre as filler is used to reinforce the polyester. This is pressed with copper foil to get mylar laminations. The commercially available grade of laminates is FR-6 of NEMA. They are usually of glossy colours such as white or red and are relatively cheap. Their electrical properties resemble those of XXXPC laminates and mechanical properties are fair. Their dimensional stability and water resistance are also good. After soldering, the warp and twist can be considerable.

6.4.7   Silicone Laminates

These are made of silicone resins with glass reinforcements. They have good resistance to chemicals and to heat; and are used in the range of 175 °C to 400 °C depending upon their type. It is difficult to obtain a good bond between copper foil and base material in the silicone resin system. Even though the electrical properties of these laminates are very good, their usage is limited because of high cost.

6.4.8   Melamine Laminates

Melamine resins can be combined with a variety of reinforced fillers such as glass fibre to produce melamine laminates. They offer very high surface hardness and high arc resistance. However, their main disadvantage is their poor dimensional stability, particularly with cyclic variations of humidity.

6.4.9   Polyamide Laminates

Polyamide is reinforced with filler like woven glass fabric, or aramide fibre or quartz fibre to form laminates with good electrical and mechanical properties, higher copper bond strength, good heat resistance and low Z-axis expansion. They find use in demanding military and aerospace applications and in special multi-layer circuits. They, however, have poor peel strength, which can result in lifted copper conductors after soldering at high temperatures.

6.4.10   Teflon Laminates

Teflon is reinforced with glass fibre to get a laminate of low dielectric constant and it is used in Radio Frequency (RF) applications with small leadless components. Due to the high co-efficient of thermal expansion, teflon laminates have limited use.

6.4.11   Mixed Dielectric Laminates

Mixed dielectric laminates are becoming popular in today's microwave and high-speed digital PCBs.

One such material is Speedboard C prepreg, which can be used as a prepreg or an outer-layer HDI dielectric for critical signal routing, while using low cost FR-4 for non-critical layers in multi-layer PCBs provides a good solution for these applications. Speedboard prepreg is used as a low loss HDI dielectric that improves the performance of the controlled impedance high speed digital, RF and microwave PCBs by making them thinner, lighter and faster. The product is comprised of expanded polyterafluoroethylene (ePTFE) that has been impregnated with a modified BT resin. The air space inside the ePTFE is replaced with resin and the ePTFE membrane becomes the carrier or delivery system for the resin. The conformable ePTFE toughens the dielectric, improving reliability and enabling excellent surface planarization for high density fine lines and space.

Speedboard C material features a low dielectric constant of 2.6 and a low loss tangent of 0.00036 that is stable over frequency and temperature. It utilizes standard thermoset processing as opposed to fusion processing. The material is manufactured by M/s W.L Gore and Associates Inc. Newark, DE. Additional information may be obtained from the company's website at www.gore.com.

 ## 6.5 Evaluation of Laminates

6.5.1 Laminate Testing

Tests for laminates are described in many publications such as IEC Publication 249, NEMA and ASTM standards MIL-P-13949, BS4584 [UK] and DIN 40802 [Germany]. Generally, electrical and mechanical design, and the particular fabrication process will demand that the final material has certain features and controls. Some of the methods most commonly used today to evaluate the laminates are described in the following sections.

6.5.2 Surface and Appearance

Surface and appearance standards are probably the most difficult tests to define adequately. Laminates generally get rejected for pits and dents in the copper surface. Most laminators advocate that surface and appearance standards must be applied only on the finished boards. Some users of laminate require copper surface standards only on certain critical areas, such as the areas for tips inserted into edge connectors. In those cases, the user usually specifies the laminators with an overlay of the critical copper areas to be inspected on each sheet of material before it is shipped. Thus, the surface standard is applied only to areas that are pertinent to the finished board. With more than 90 per cent of the copper being ultimately removed, the change of pit or dent affecting a critical area is quite small.

Dimensional Stability

Copper pits and dents are defined in the surface standards wherein the specification defines the longest permissible dimension of a pit or dent and supplies point values for rating all pits and dents. As per the MIL-P-13949 standard, the defect dimensions and the point values are given as follows:

Defect dimension (mm)	Point-Value
0.13 --- 0.25	1
0.28 --- 0.51	2
0.53 --- 0.76	4
0.79 --- 1.02	7
over 1.02	30

The sum of points for all defects within the inspected area must be less than 30 per 645 mm^2. Scratches are permitted that have a depth of less than 140 μm or a maximum of 20 per cent of the foil thickness.

Colour

Colour variation of the laminate can take place from lot to lot due to variation in colour of the batches of resin, types of paper used or due to variation of alloy coating on the copper. Generally, a set of samples that illustrate the preferred colour extremes are established.

6.5.3 Water Absorption

Water absorption of a laminate must be as low as possible. If it is high, the electrical characteristics of the laminate will change considerably as a result of humidity or of the absorbed liquid during PCB manufacture. Absorbed water can also cause outgassing during soldering or promote blistering of a heated laminate.

The amount of water absorbed by a sample of specified size, when immersed in distilled water for a specified period at a specified temperature, is taken as a measure of water absorption. Usually, three specimens of 50 mm × 50 mm are taken and subjected to immersion in distilled water at 20 °C for twenty-four hours. The average value of water absorbed in milligrams represents the water absorption.

Alternatively, the gain in weight expressed as a percentage of increase over the initial weight can also be taken as a measure of water absorption. A 76.2 × 25.4 mm piece of the laminate is chemically etched to remove the copper and then dried by heating it at 107 °C for one hour. After cooling, it is weighed and immersed in distilled water at 25 °C for twenty-four hours. It is then surface dried and weighed again. The increase in weight as a percentage of the initial weight must not exceed the valves shown in Table 6.6.

Table 6.6 Water Absorption Limits

Material	Thickness		
	0.79 mm	1.59 mm	3.17 mm
XXXP,XXXPC,FR-2 and FR-3	1.00	0.65	0.50
G-10, G-11, FR-4 and FR-5	0.80	0.35	0.20

6.5.4   Punchability and Machinability

Punchability is the most desirable mechanical property of copper clad laminates. Good punchability, in simple terms, means that when the laminate is punched, there is no cracking of the board, there is no lifting of the copper around the punched hole, the edges are smooth and there is smoothness inside the holes. These properties are possible if the laminates are dimensionally stable and free of warp and twist.

Manufacturers should be consulted on the recommended drilling speeds and feeds prior to performing any machining operations. Sectioning the board often helps in evaluating the type of the hole being obtained and is particularly important on plated through-applications. Often, sectioning will show that the material has been heated by drilling to such an extent that the surface is smooth and not palatable, is smeared with resin, or is so roughened that glass fibres protrude and will inhibit continuous plated through-holes.

Punchability can be measured in a die that simulates the conditions used in the fabrication process. Various hole sizes, spacing and configurations are incorporated in the test die. Careful physical inspection and sectioning of the holes will indicate the types of punching being obtained. The material must be carefully inspected to ensure that no cracking occurs and that there is no lifting of copper around the hole. Many paper base grade laminates will tend to vary in punchability from lot to lot. So care must be taken to measure a wide range of sample panels.

A preferred test for measuring the punchability of a laminate is defined in the German standard DIN 53488. The test entails punching square holes at exact spacing on a strip of laminate measuring 120 mm × 15 mm and visually inspecting to determine the minimum spacing between two adjacent holes which do not crack or chip. The punchability of the sample is expressed in points from 1 (optimum punchability) to 4 (low punchability) as a function of the minimum unbroken spacing and the laminate thickness. The test is particularly useful in comparing different laminates.

6.5.5   Peel Strength

Peel strength indicates the measure of adhesion of copper to the base material. The basic test pattern for testing peel strength or copper bond strength specified in MIL-P-13949 and by NEMA standards is illustrated in Figure 6.12. The pattern is processed by the same fabrication technique as in the user's final process, with the exception of exposure to various plating solutions or solder which will also be tested. When testing for peel strength, the specimen should be mounted on a flat, horizontal surface. The wide copper end of each trace is peeled back approximately 25 mm so that the line of peel is perpendicular to the edge of the specimen. The end of the peeled strip is then gripped by a clamp, which is attached to a force indicator or tensile tester adjusted to compensate for the weight of the clamp, and connecting chain. The minimum load of the force indicator is recorded. The ratio of the recorded force F to the conductor width is the *peel strength*. This is usually expressed in pounds per inch of width or in grams per millimeter of width.

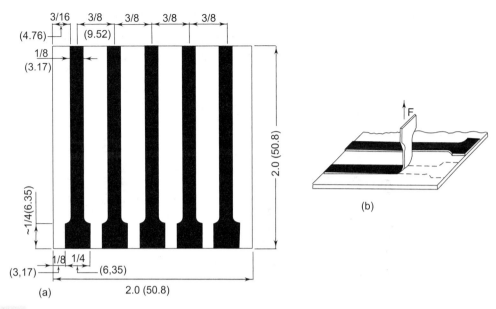

Fig. 6.12 Peel strength test (a) NEMA test sample (b) the test is performed by applying a force 'F' to a conductor

For the most commonly used laminates, the minimum peel strength for 1 oz/ft^2 copper thickness is 0.143 kg/mm. The peel strength value varies with the thickness of copper. For example, for a 2 oz/ft^2 copper foil, the minimum peel strength required is 0.1756/mm for XXXP, XXXPC, FR-3, FR-4, FR-5 and 0.1964 kg/mm for G-10 and G-11.

6.5.6 Bond Strength

Bond strength is a measure of the adhesion of a conductor pad to the base material. It is an important parameter as the conductor de-lamination in most cases starts from a pad. Bond strength can be measured on a finished PCB or on a test pattern. For doing so, a specimen of the laminate is drilled as recommended, with at least 10 holes of 50 mil (1.27 mm) diameter, spaced about 10 mm. Around each hole, a 2.54 mm conductor pad is made by printing and etching the sample. It should be seen that land centres are not offset by more than 4 mils (0.1 mm) from the hole centre. The specimen is then assembled by using staples of AWG 20 tinned copper wire. The specimen is soldered according to standard operating parameters on a wave soldering machine. The lead projections on the solder side are pulled vertically to the laminate and the values which cause pad removal are recorded. Figure 6.13 shows an arrangement for carrying out the bond strength test. The average value from at least eight tests should not be less than:

4.0 kg [8.8 lb] for XXXP, XXXPC and FR-2; and

8.0 kg [17.7 lb] for G-10, G-11, FR-4 and FR-5

and the minimum value should not be less than 70 per cent of the average.

The same test can be used [pulling the lead from the component side] for determining the strength of the solder, provided that the tinned copper wire is highly solderable.

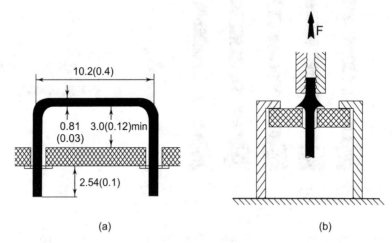

(a)   (b)

Fig. 6.13   Arrangement for test for bond strength (a) dimension of the sample (b) test arrangement (redrawn after Leonida, 1983)

6.5.7   Solder Resistance

The resistance of the laminate to soldering can be determined by floating a specimen on the surface of molten solder at 260 ± 3 °C (500 ± 5 °F) for a given time, usually from 10 to 60 seconds. The specimen must have conductors and holes (non-plated or PTH) of the same type used on the PCBs to be manufactured. The sample should not be fluxed. In order to prevent wetting by solder, a thin layer of silicone grease or oil is used for this purpose. After floating, the specimen is visually inspected for blistering, haloing, measling, etc.

6.5.8   Warp and Twist

Warp is the warpage along the edge of a sheet whereas the twist is the warpage along the diagonal. Excessive warpage or twist can lead to problems at various stages of PCB manufacturing and assembly. The measurement of warp and twist is carried out by taking a square specimen of laminate having a side of 914.4 mm or on a rectangle with the shorter side not less than 610 mm.

By definition, the sample is warped, if the sample, when placed on a flat area, has all its four corners in contact with the supporting plane. For measuring warp, the sample is placed so that all corners contact the supporting plane. Along each side, the maximum deviation from the plane is measured and expressed as a percentage of the length of the side.

Similarly, if at least one corner of the sample does not contact the supporting plane, it is twisted. For measuring twist, the specimen is supported on these corners and the maximum deviation is measured on the triangle internal to the corners. One support is then measured to the unsupported corner and the maximum deviation is measured on the other half of the panel. Twist value is usually expressed as a percentage of the diagonal of the specimen. Warp and twist values are governed by the MIL-P-13949 standard.

6.5.9 Flexural Strength

Flexural strength is a measure of the force per unit area, that a laminate strip will be able to withstand without fracture, when supported at ends and the force is applied at the centre. Figure 6.14 shows the testing arrangement for flexural strength.

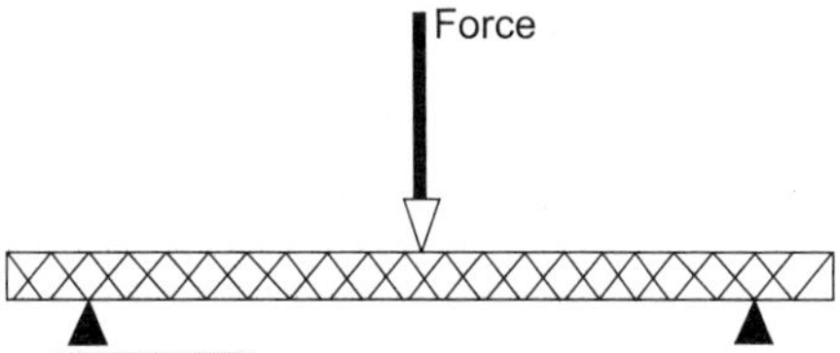

Fig. 6.14 Testing of flexural strength

The values of flexural strength are not the same in the two directions, i.e. parallel to the length of the filler (lengthwise) and perpendicular to it (cross-wise). For measurement of flexural strength, the specimen is 25 mm wide and at least twenty times the thickness of its length. The span of support is generally sixteen times the thickness. Five specimens, each corresponding to the two directions (lengthwise and cross-wise) are studied. The minimum average of the values in the two directions is taken as the flexural strength.

6.5.10 Flammability

The problem of inflammability arising out of the materials used in the industry is a cause for great concern. So, there is a need to make and use more flame-resistant laminates. The standards for testing flammability have been laid out by IEC, NEMA and Underwriters Laboratories (UL).

UL 94 is a general standard for the flammability testing of plastic materials. According to this standard, materials tested for flammability are classified as 94 V-D, 94V-1 94V-2 and 94 HB. The tests are conducted either in horizontal or vertical position. These tests are conducted in a chamber which is free from draught. The sample taken is of 127 mm length and 12.7 mm wide. Before the tests are conducted, the samples are conditioned for forty-eight hours at 23 °C and 50 % RH. The equipment used and test set-up are described in detail in the standard.

The classifications of the various tests are:

Vertical Burning Test (94V-O, 94V-1, 94V-2).

- Specimen must extinguish within
     10 seconds (94V-O)
     30 seconds (94-1, 94V-2)

- For each set of five specimens, the specimen should not have a total combustion time exceeding:

  50 seconds (94V-O)

  250 seconds (94V-1, 94 V-2)

- There should be no samples to dip flaming particles that may ignite the dry absorbent surgical cotton below (94V-O, 94V-1). Samples may dip flame particles, while burning briefly.

- After the second flame test, specimen with flaming combustion should not last beyond

  30 seconds (94V-0)

  60 seconds (94-1).

Horizontal Burning Test: (94 HB):

- The burning should not exceed 76.2 mm per minutes over a 76.2 mm span.

- Samples must cease to burn before the flame reaches the 102 mm reference mark.

6.5.11   Glass Transition Temperature

The glass transition temperature (usually expressed as Tg) is an established indicator of as to how well a laminate resin system would resist softening from heat. At Tg temperature, the resin changes from its glossy state and its molecular bonds begin to weaken enough to cause a change in physical properties (dimensional stability, flexural strength, etc.). FR-4 epoxy shows a Tg of 115-125 °C and polyimide 260-300 °C. Tg, in a way, is an indicator of the amount of expansion that will take place from ambient to solder.

6.5.12   Dimensional Stability

The increased circuit density results in the increased need for circuit board dimensional stability. The dimensional stability in X,Y directions is a function of the laminate reinforcement (glass or paper), whereas the thickness expansion 'Z' is generally function of the system or resin matrix.

Z-expansion: This is the amount of expansion which occurs in the Z-dimension when a circuit board, and subsequently an assembly, are exposed first to a series of excursions to solder temperature and then to lifecycles from ambient to operating temperature. The detrimental effect of Z-dimension expansion is the work hardening of the copper cylinder (plated through-hole). From a laminate perspective, the lower the Z-expansion, the lower is the overall work-hardening and the greater the life expectancy of the electric assembly. Ideally, if a laminate with a Z-expansion equal to that of the plated through-hole could be created, the stress and the associated work-hardening could be virtually eliminated. On a practical basis, the laminates that meet specified performance requirements are selected. Most selections are made on the basis of the glass transition temperature.

6.5.13   Copper Adhesion

This is usually tested by the 'peel strength test'. As the test is destructive, it is not done on the production board. Rather, the supplier is asked to prepare some test specimens with each production batch.

6.6   Useful Standards

- *IPC/JPCA4104: Specification for High Density Interconnect (HDI) and Microvia Materials:* Covers various conductive and dielectric materials that can be used for the fabrication of HDI and microvias, also includes qualification and conformance requirements for such materials as photo-imageable dielectric dry films and liquids, epoxy blends and coated foils.

- *IPC-4103: Specification for Base Materials for High Speed/High Frequency Applications*: Includes requirements for high speed/high frequency laminate or bonding layers to be used primarily for the fabrication of rigid or multi-layer printed boards for high speed/high frequency electrical and electronic circuits.

- *IPC-M-107: Standards for Printed Board Materials Manual*: Contains the requirements for the various reinforcements, foils, laminates and prepregs.

- *IPC-4101A: Specification for Base Materials for Rigid and Multi-layer Printed Boards:* Covers the requirements for base material (laminate and prepreg) to be used primarily for electrical and electronic circuits.

- *IPC-4562: Metal Foil for Printed Wiring Applications:* Includes nomenclature and requirements for metal foils used in laminate and PCB fabrication.

- *IPC-CF-148A: Resin-coated Metal for Printed Boards:* Covers the requirements for metal foils coated with a resin or composite of resins on one side, to be used for the fabrication of PCBs, including specification sheets that outline engineering and performance data for resin-coated metal foil, indicating foil material type and resin type.

- IPC-CF-152B: *Composite Metallic Materials Specification for Printed Circuit Boards:* Includes requirements for copper/invar/copper (CIC), copper/molybdenum/copper (CMC) and three-layer composites for use in electronic applications.

- *IPC-TR-482: New Developments in Thin Copper Foils:* A compendium of technical methods and techniques used for evaluating the quality aspects of present and future interconnection product and electronic assemblies; addresses base materials, conductor physical requirements, internal planes, construction, registration, plated-through-holes, component mounting areas, cleaning evaluation, soldermask and printed board electrical requirements.

- *IPC-TR-484: Results of IPC Copper Foil Ductility Round Robin Study:* The test report evaluates ductility of the foils provided by the industry's copper foil vendors.

- *IPC-TR-485: Results of Copper Foil Rupture Strength Test Round Robin Study:* The test report evaluates rupture strength testing as a means of determining the mechanical properties of electrodeposited and rolled copper foil.

- *IPC-4412: Specification for Finished Fabric Woven from "E" Glass for Printed Boards:* Covers the classification and requirements for finished fabrics woven from E glass fiber yarns.

- *IPC-4130: Specification and Characterization Methods for Non-woven E Glass Mat:* Defines the nomenclature, definitions and requirements for materials made from non-woven 'E' glass fibres; includes specification sheets for selecting and purchasing these materials.

- *IPC-4110: Specification and Characterization Methods for Non-woven Cellulose Base Paper for Printed Boards:* Defines the nomenclature, chemical and physical requirements of paper made from cellulose fibres for PCB fabrication.

- *IPC-4411-K: Specification and Characterization Methods for Non-woven Para-aramid Reinforcement:* Includes the nomenclature, definitions and requirements for reinforcement made from non-woven para-aramid fibres.

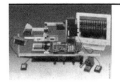
Image Transfer Techniques

 ## 7.1   What is Image Transfer?

Image transfer basically involves the transfer of the conductor pattern from the film master on to the copper clad base material or any other metal clad laminate. In the fabrication of the PCB, the two methods common for image transfer are:

- Photo printing method; and
- Screen printing method.
  - Photo Printing: This is an extremely accurate process, which is generally applied to the fabrication of semiconductors and integrated circuits wherein the conductor widths are typically in the region of a few microns. Although such a precision technique is not required in the production of general purpose PCBs, yet where conductor widths of 100 μm are required and for PCBs for professional applications, the photo printing process is resorted to.
  - Screen Printing: Although less precise than the photo printing process, screen printing is a comparatively cheap and simple method. The majority of PCBs produced worldwide are screen printed.

 ## 7.2   Laminate Surface Preparation

Copper surface plays a major role in the success or yield of the image transfer process. It demands that the surface should be carefully inspected for pits, drilling burns and any other types of irregularities. If unacceptable defects are observed, the image transfer process should not be carried out further and the defective material should be rejected straightaway. So, for the image transfer to take place, the cleaning of the copper surface prior to resist application is an essential step for any type of PCB process. The difficulties most often encountered in PCB fabrication arise due to insufficient cleaning of the laminate surface. Therefore, the laminate should be free from oil, grease, dust, fingerprints and foreign particles. Possible sources causing contamination could be the equipment used for shearing, drilling, punching or air from the air compressor. Any contamination on the surface of the laminate may impair the adhesion of the photopolymer or decrease the bonding of the electro-deposited copper. Hence, very good cleaning methods are required to prepare the laminate surface. The methods commonly used are:

(i) Manual Cleaning Process — This includes:
 • Chemical Cleaning; and
 • De-greasing (vapour or aqueous)

(ii) Mechanical Cleaning.

7.2.1 Manual Cleaning Process

7.2.1.1 Chemical Cleaning or Cold Cleaning

Chemical cleaning entails the use of concentrated alkali chemicals to remove the oil, grease and soil particles on the surface of the laminate. The concentration of the alkali chemical which is between 80-100 per cent at a temperature range between 60 °C and 70 °C is used for twenty to thirty minutes for cleaning the laminates. After this process of alkaline soak, the laminate is effectively rinsed with filtered tap water which is oil-free. Water immersion, followed by strong water spray, ensures complete removal of cleaners. Neutral or acidic cleaners are sometimes preferred because of the attack of hot alkaline solutions on exposed epoxy or polyamide substrates.

The steps followed for the chemical cleaning process are shown in Fig. 7.1. These are:

 • De-grease with hot soak cleaner;
 • Water rinse (using pressurized water of more than 4 bar or 60psi);
 • Water spray;
 • Micro-etch copper (optional);
 • Water rinse;
 • Inspection (of oil and grease complete removal);
 • Acid dip (neutralization); and
 • Water rinse.

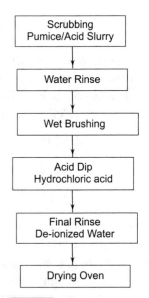

Fig. 7.1 Simple manual cleaning process

7.2.1.2   Vapour De-greasing

This process involves cleaning of laminates by condensing pure solvent vapour. Here, a non-flammable solvent such as fluorochlorocarbons (trichloroethylene or perchloroethylene) is brought to its boiling point within a vapour de-greaser. This solvent vapour de-greases the surface contaminants of the laminate. Figure 7.2 shows the cleaning process with solvent de-greasing. The first step is to gently rub a clean cloth, soaked in a solvent, over the entire PCB area. The solvents act as efficient de-greasers and do not react chemically with the materials they dissolve. This step is followed by *scrubbing* with a pumice or salt solution, which is intended to remove inorganic matter like particulates and oxides, and helps in de-greasing to a certain extent. The next step is to *water rinse* the PCBs and then to remove the fine particles of the pumice with brushing. From this stage onwards, the cleaned PCB should be held only at the edges and rubber hand gloves should be used as far as possible.

For removing residual alkali and metallic oxides, and preparing the surface for the image transfer, the board is acid-dipped in hydrochloric acid (10 volumes per cent). The *final rinse* process is best carried out by using de-ionized water. Rinsing with tap water may lead to the risk of introducing troubles caused by water impurities. The final step in the cleaning process is *drying* which is generally done by blowing compressed air over the laminate surface. The compressed system should have a filter in the air pipe to avoid contamination through oil from the compressor. It is often desirable to keep the PCB in an oven for about fifteen minutes at a temperature of 90 °C for complete drying.

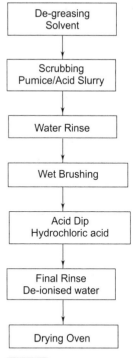

Fig. 7.2   Manual cleaning process with solvent de-greasing

The vapours of the solvents are toxic, and are even known to be air pollutants. Therefore, when working with them, care must be taken not to breathe in the vapours. Sufficient air circulation should be ensured so that the maximum vapour concentration does not exceed 100 ppm.

De-greasing can also be achieved by using a soak cleaning solution which reacts chemically with the organic soils. It removes oils and greases quite efficiently, without causing the threat of any severe air pollution.

The vapour de-greasing process is an improvement over the cold solvent cleaning process because the laminates are always washed with pure solvent. The de-greasing parts are heated to the boiling point of the de-greasing solvent. So the laminates are dried faster than the cold solvent cleaned laminates. Vapour de-greasing equipments are also available with ultrasonic agitation system in which agitation is done by sound waves.

7.2.2   Mechanical Cleaning

In the mechanical cleaning process, the laminates are typically cleaned by abrasive brush cleaning units employing abrasives such as emery, corundum, aluminium oxide and silicone carbide. These abrasive materials are impregnated into a nylon or similar plastic matrix. The resulting brush construction can be either a compressed lamella or a filament type. The silicon carbide 320-grid abrasive filament brush is generally used for cleaning the copper surface prior to lamination of the dry film resist. There are different types of mechanical cleaning operations. They are:

- Polishing;
- Brushing;
- Buffing or sanding;
- Deburring or scrubbing; and
- Scrubbing

Polishing

This process involves smoothening of the metal surface to improve its appearance. To do this, polishing wheels (muslin) are generally used.

Brushing

This process is a metal smoothening operation which involves rotary non-metallic wire brushes such as nylon brushes. The wire brushes are generally run wet. For very dull finishes, pumice slurry of grade 3F or 4F with water is used. Although the cleaning performance with respect to heavy oxides and other metallic soils as well as the removal of particulate matters is excellent, the removal of organic soils such as oils and greases is not complete. In fact, some of these soils are retained in the brushes. Therefore, the de-greasing step with a solvent is recommended before brushing. Figure 7.3 shows the scheme of a typical cleaning process with abrasive brushing for high volume PCB production with plated through-holes.

Buffing or Sanding

Buffing implies that the metal surface is buffed with the muslin wheel, to obtain low gloss finish to high gloss finish. The method removes heavy inorganic soils and particulate matter, and also helps to remove the burns around the drilled holes. Buffing can be carried out with powered hand vibrators.

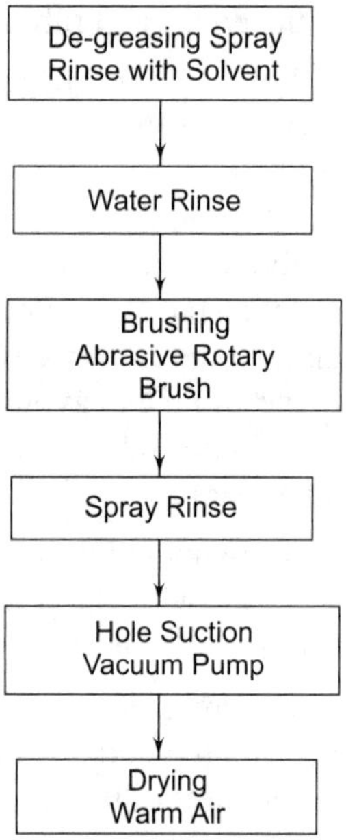

Fig. 7.3 Cleaning process with abrasive brushing for high volume PCB production with plated through-holes

Deburring:

After the board is drilled or punched, burr formation is noticed. Burr is the extra projection of metal when a hole is drilled through the PCB.

Even with the best drill bit and drilling equipment, the formation of certain amount of burrs cannot be avoided, especially on the side where the drill bit comes out from the base material. For reliable plated through-holes, these burrs must be completely removed. This is usually done by abrasive cleaning. The machine used for deburring consists of a deburring wheel (fused silicone carbide), which is used to remove the burr on the drilled laminate. This process is carried out only in the presence of wet conditions. The sanding process (wet deburring) involves a water-resistant abrasive of 280 to 600 grid.

In all the above methods, after the laminate is cleaned, the boards are rinsed with water at high pressure (> 60psi) and dried with hot blow air. Normally, all mechanical processes involve automatic or semi-automatic machines. During these processes, the base material may be removed which, in turn, degrades the electrical properties of the boards. Hence, strict process monitoring is required.

Scrubbing

Scrubbing is carried out with fast rotating brushes made of fine plastic brushes which are continuously fed with abrasive slurry. The slurry is prepared by adding pumice powder to water and is generally re-circulated. The method includes a final water brush – rinse cycle to ensure the complete removal of particulate matter. Such machines provide an excellent surface finish on the PCB surface.

7.2.3   Test for Cleanliness

In order to test the adequacy of the cleaning process, an in-process water break test can be used. The cleanliness of the copper surface can be expressed in terms of the contact angle and wettability.

- *Contact Angle*: A clean copper surface should have a very low contact angle.
- *Wettability:* The clean copper surface should hold a continuous film of clean water of 15 to 30s. The water film should not break into droplets nor show areas of de-wetting during this test. If water breaks, it indicates that the cleaning process is not adequate.

 ## 7.3   Screen Printing

The photographic image is transferred to copper clad laminate by the screen printing technique. This technique has been used for a long time for printing cloth, panels and so on, and reaches its maximum degree of accuracy when applied to PCB manufacture. The technique is particularly adopted for low cost print and etch and plated printed boards, when the ultimate resolution and definition are not very exacting.

A screen comprises an aluminium frame, mesh, emulsion and adhesive bonding. The emulsion is removed by a photochemical process where deposition is required. A specification of 1:1 ratio of open area to board pad area is typical, but ±10 % variations are not uncommon. The flexibility of the emulsion creates a good gasket against the PCB and aids print definition.

It is basically a stencil operation, which depends on the transfer of resist to the copper clad laminate surface using a stencil image of the circuit design. The stencil is firmly attached to the surface of a silk, nylon or stainless steel screen. A liquid resist material is forced through the open areas of the screen mesh that are not protected by the stencil onto the copper clad substrate by the pressure of a squeegee wiped across the top surface of the screen. The squeegee pressure deflects the screen downward in point contact with the substrate. As the squeegee passes a given point, screen fabric tension snaps the screen back, leaving screen ink behind. Figure 7.4 illustrates the screen printing process.

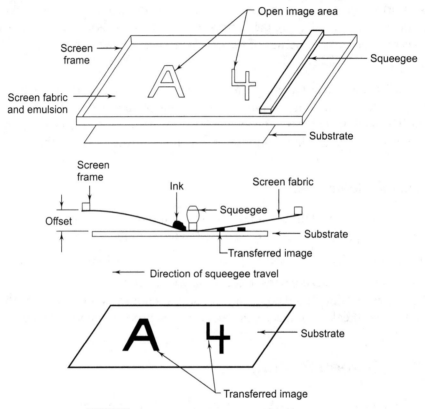

Fig. 7.4 Screen printing process (NTTF Notes)*

7.3.1 Screen Frame

The frame provides a support on which a screen fabric of uniform mesh or opening is stretched. Normally two types of flat screen frames are used: 'rigid type' and 'adjustable type' using floating bar frame. These are shown in Figure 7.5.

The frames are made of strong materials like wood, aluminium alloy, stainless steel or plastics. Aluminium alloys and stainless steel frames are the most widely used frames in the PCB industry. Even though the weight of a wooden frame is comparatively lesser, it is not used due to its absorption of water during the process, leading to a bend tendency. Normally, aluminium alloy is used for rigid type and stainless steel frame for stretchable type (variable tension) frames. The frame size is usually 400 m × 400 m. However, only 10-25 per cent of the screen area should be utilized.

---

* NTTF: NTTF Electronics Centre, Bangalore, India.

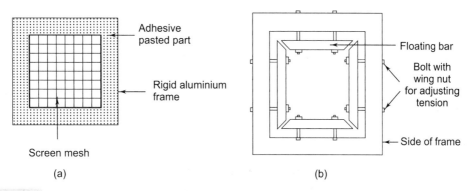

Fig. 7.5 Types frames for screen printing (a) rigid type frame (b) stretched type floating bar frame

7.3.2 Screen Cloth

The woven mesh fabric used in screen printing may consist of stainless steel, silk, polyester, nylon or similar material.

The threads of the above fabrics are classified into the following two types:

- *Mono-filament:* This is woven with single thread fibres. Its examples are stainless steel and phosphor bronze, nylon and polyester.
- *Multi-filament*: A multi-filament cloth is woven with more than one fibre or many separate fibres. Its examples are silk and polyester.

Screen fabrics are classified on mesh counts, mesh opening and the percentage of open area or ink coverage area. The screen fabric mesh is classified according to the number of threads or wires (openings) per linear cm thread or wire diameter. The weave style and mono-multi-filament are important factors to be considered along with the mesh classification. The larger the mesh classification number, the more wires per linear cm and therefore, the smaller the openings, the finer will be the resulting circuit image. However, it may be sometimes be more difficult to push the screen ink through the small screen openings onto the substrate surface.

Multi-filament polyester is not used in circuit printing. It is used basically in ceramic printing. Mono-filaments like polyester and stainless steel are generally used since it gives more passage for inks to penetrate deposits better and to give uniformity of colour or paste. It produces less drag during the squeezing process and so mono-filaments are used for screen printing with direct emulsion and indirect emulsion methods.

Mono-filament polyester fabrics have a mesh number in the range of 100-140 mesh /cm. However, the applications where a maximum precision, such as registration error to be less than 0.1 mm is required, stainless steel fabrics are recommended for best results. With stainless fabrics, the mesh number is chosen within the range of 120-140 mesh/cm. Typically maximum particle size should be no larger than one-third of the mesh opening to prevent jamming. For example, an 80 mesh screen has openings of about 224 μm, so the particle size should not exceed 75 μm. Finer mesh screens tend to be used for thinner deposits. A 180-mesh and a fine powder paste could be used to produce a deposit thickness of 100-150 μm, using an 80-mesh.

The thickness of ink deposit is controlled by the fabric material. For determining the correct mesh count of the fabric, theoretical colour volume is used. The theoretical colour volume is the amount of ink held in the fabric prior to the passage of the squeegee during the printing stroke, and it is determined by the thread count, thread thickness and mesh opening (Figure 7.6).

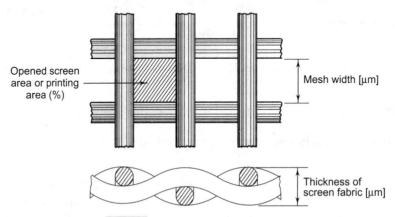

Fig. 7.6 Geometry of a screen fabric

The edge definition of the image produced by the printing screen varies according to the thread thickness, (d) and mesh opening, Mo. The printing fabrics differ according to the quality of the threads (HD, T, S — thread types). The most commonly used thread types for screen printing in the PCB industry are T-type and S-type, as shown in Figure 7.7.

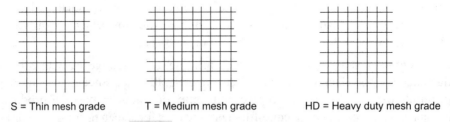

S = Thin mesh grade   T = Medium mesh grade   HD = Heavy duty mesh grade

Fig. 7.7 S, T and HD mesh grades

7.3.3 Screen Preparation

The screen is prepared by considering various factors such as the image transfer line and space requirement, panel size, fabrication process, run length, etc. It is followed by properly selecting the fabric type, mesh size, frame stencil and ink. Once this is done, the fabric is cut several cm longer than the frame moulded into the rails of the frame with staples and adhesives, under maximum tension uniformly applied by hand.

The correct stretching of the fabric is very important in the process of circuit printing. Insufficient stretching of fabric will cause poor registration of the printed detail, colour, pattern, off-contact printing, loss of open mesh and loss of ink passage. During the stretching, the threads should be parallel or at an angle of 45° to the sides of the frame, and the frame should be rigid. The sides should not bend during stretching. The tension of the screen should be maintained between 10 to 15 N/cm (Newtons per centimetre).

The screen is stretched over frame in the following ways:

- The fabric is stretched permanently using special adhesive or nail on the frame and in this case, the tension is not adjustable. For stretching, equipment like a pneumatic system (using air pressure) containing pneumatic clamps can be used.
- The fabric can also be stretched by using a floating bar frame. The screen tension is adjustable in this case.

Once the stretching is complete, the tension of the fabric is measured using the tension gauge. The meter is placed on any part of the screen fabric and the reading becomes available on the gauge dial. The gauge measures tensions from 7 to 50 and the dial is calibrated in Newtons per centimetre (N/cm).

The stretched frame is removed and the fabric is thoroughly cleaned and de-greased with a proper de-greasing agent. The de-greasing agent should not affect the screen fabric. It can be 20 per cent sodium hydroxide or disodium phosphate or any other de-greasing solution recommended by the screen fabric manufacturer.

After the screen is de-greased, by the screen is rinsed with warm or cold water. The screen is then dried by using a fan or a blower which gives hot air. The screen is now ready for printing.

7.3.4 Squeegees

The squeegee used in the screen printing process consists of an elastomeric material such as urethane or a strip of rubber which is mounted on a wooden or an aluminium holder (Figure 7.8a). The material is available in various thicknesses consisting of different compositions of natural and synthetic (polyurethane) rubber. The flexibility of the rubber is measured in terms of durometers (hardness). Usually:

- 40 to 50 durometer rubber is used in textile industries for printing; and
- 70 to 80 durometer rubber is used in the PCB printing process.

The sharpness of the squeegee is very important for printing. Line sharpness, uniform ink coating thickness and the edge definition depend upon the squeegee's sharpness. The sharpness is reduced after continuous operations as the edges become rounded. Hence, it is required that the sharpness be maintained. This is done by sandpaper or any other abrasive material. Over-sharpness also damages the screen. The application of a squeegee is shown in Figure 7.8b.

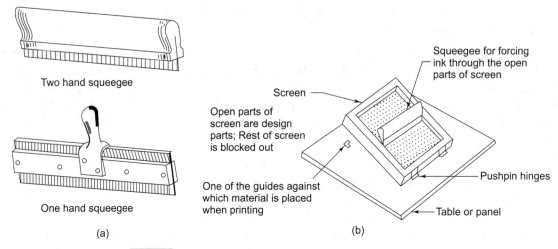

Fig. 7.8   (a) Types of squeegee (b) application of a squeegee

 ## 7.4   Pattern Transferring Techniques

7.4.1   Screen Stencil Method

In this method, a photographic emulsion is coated directly on to the screen fabric and is exposed to light, thereby establishing a direct contact with the film master. When the screen is developed, the mesh opens at the pattern areas. The photographic emulsion used is polyvinyl alcohol, polyvinyl acetate or polyvinyl chloride, and they contain some dye to make the pattern visible while processing.

The emulsion is not in its sensitized form. In order to sensitize it, the emulsion is mixed with a sensitizer before use. Some of the sensitizers used are potassium dichromate, ammonium dichromate or sodium dichromate.

The sensitized emulsion-coated screen is exposed to ultraviolet or other fluorescent light sources. The exposure time depends upon the light intensity and the distance between the bulb and the screen. Normally the distance between the bulb and the screen is 1.5 feet.

During the process of exposing, the positive or negative film is placed on the printing side of the emulsion-coated screen. The exposure brings about a polymerization in the exposed areas, which makes them insoluble in the developer. If any air gap is present between the photo-tool and screen, it must be removed by using a vacuum system. The air gaps can lead to incomplete pattern forming and insufficient half-tone work. Vacuum printing facilitates sharp details of line, half-tone work and no pin holes. Vacuum printing machines with light source and all the other regulating facilities are available in the market. The suppliers of the emulsion also specify exposure calculations that ensure the correct exposing time.

Developing is carried out immediately after the exposure, keeping the screen with the emulsion side up. The screen is rinsed by using a strong water supply of hot or cold water till the mesh at the pattern areas opens and the sharp pattern with all details is visible. The screen is then dried well with cold or warm air of 40 °C maximum. The screen is then thoroughly inspected and is then ready for printing.

The direct method provides very durable screen stencils with a high dimensional accuracy. However, the finest details are not reproduced. There may be porosity in the washout stencil, which could be due to thin coating, too short exposure, too old emulsion, too long developing or improper sensitizing.

7.4.2 Indirect Method [Transfer Type Screen Method]

In this method, a separate screen process film supported on the backing sheet is used. The film master is exposed on to this film, followed by film developing which dissolves the unexposed areas of the film and leaves the rest on the backing sheet. The film on its backing sheet is thereafter pressed on to the screen fabric and stuck. Finally the backing sheet is peeled off, opening all those screen meshes which are not covered by the film pattern. Thus the screen is prepared with photographic screen printing films and this is known as the indirect or transfer type screen method. This process is carried out only in low intensity illuminated rooms with yellow, orange or red lights.

Emulsions are available as unsensitized or pre-sensitized products. The unsensitized emulsions are readily mixed with the correct amount of sensitizer and are applied directly from the original container on to the fabric. Pre-sensitized emulsions eliminate possible mixing errors and save working time, and are sold with a prescribed shelf life. In the transfer type screen method, the following commercially available films are used: (a) Chromoline film, (b) Five-star film, and (c) Capillary film.

The emulsion container must be kept under controlled humidity conditions, well closed and kept in low intensity lighted surroundings. However, the shelf life is seriously affected.

The indirect method is more suitable when small quantities of PCBs are to be manufactured. The method is faster but dimensionally less accurate. The screen stencils are less durable, and more sensitive to mechanical damages and interruption of the printing process.

The films should be handled in yellow rooms only. They should not be subjected to bends as this will affect adhesion to the screens. Any contacts of the moisture with the films must be avoided.

7.4.3 Knife-cut or Hand-cut Film Process

For electronic circuit printing, the manually prepared screens are usually of the knife-cut film types. This consists of a transparent or translucent film coating. The emulsions on the knife-cut film are either coated or adhered semi-permanently with an adhesive or by means of the electrostatic process to a transparent or translucent plastic or paper backing sheet. The film may be either of the lacquer or of water soluble type; water soluble film may be softened or dissolved, or adhered with water, water and alcohol, or with an adhering liquid recommended by the manufacturer of the film. The film or coating used for the printing screen must be one that will not get dissolved by the inks or solvents used in the inks or resists that are being printed. Water soluble films will resist all type of inks except those having a water base. Printing inks such as vinyls, lacquers, epoxy, and oil vehicle inks are suitable for printing. Lacquer films will resist all inks except those having a lacquer base.

Knife-cut films are available in various colours with different film thicknesses. Knife-cut films may be adhered to such screen fabrics as metal cloth, nylon, polyester and silk. The preparation of knife-cut films for circuit printing is similar to that done in general screen printing and the films used in both are also similar.

7.4.4 Photographic Techniques

For electronic circuit printing, the most commonly used and practical method is the one that has plastic type backing sheet. Since this type holds the pattern better and is least affected by atmospheric conditions, there is a tendency to eliminate shrinkage or expansion of film, especially larger screens.

The knife-cut film is adhered to the screen fabric with the correct adhering liquid only. Each type of film uses a different type of adhering liquid. But water alone may be used for adhering water soluble film. A mixture of water and alcohol in the ratio of 3:1 is recommended for some soluble films. While adhering film to the fabric, it is advisable to make a build-up layer on which the film is placed so that perfect contact is obtained with fabric. The build-up layer may be glass or cardboard.

After placing the film on the fabric, the screen frame should be held down by placing some weight on the frame side. The adhering liquid is applied by using either a piece of cloth or a squeegee. If a cloth is used, care should be taken not to saturate the cloth. The top of the fabric over the film is wet for about nine square inches at a time and the wetted area is wiped immediately with a dry cloth. The adhering liquid should not be allowed to remain too long on the fabric. After the film has been allowed to dry for about ten to fifteen minutes, the packing sheet may be peeled off. Then the screen is ready for the pattern transferring process as shown in Figure 7.9.

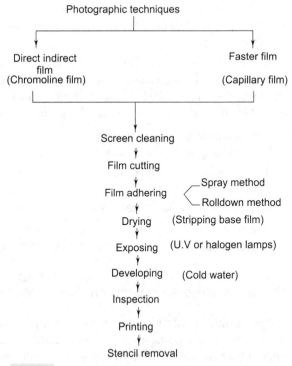

Fig. 7.9 Flow chart for photographic technique for image transfer

7.5 Printing Inks

Screen printing inks are characterized by the chemicals used to strip them or by curing method.

They are classified as:

- Alkaline or solvent-strippable resists; and
- Thermal, uv or air-cured.

The inks used in the screen printing process for etching and plate resists are mostly vinyl-based and can be broadly classified into two groups: (a) Solvent soluble ink, and (b) Alkali soluble inks.

- *Solvent Soluble Resists*: They are highly resistant against chemical attacks in both the acidic and alkaline range. Trichloroethylene or trichloroethane solvent is used for stripping.
- *Alkaline Soluble Resists*: They are acid resists and alkali soluble. For stripping, caustic soda solution (2.5 per cent) at a temperature of 30° to 40 °C is used.

Printing ink contains pigment and vehicle. Vehicle is a mixture of binder and solvent. Binder is based on vinyl resin, phenolic resin, ethyl cellulose lacquer, nitrocellulose lacquer, linseed oil, cotton seed oil, castor oil or commercially trade named binders. The function of the binder is to hold the ink permanently on a surface. Solvents are ketones, acetones, alcohols, aromatics, chlorinated solvents and commercially trade-named solvents.

The selection of an ink depends upon many factors. The ink should dry rapidly on the PCB, but should dry slowly on the screen. It should be highly resistant against all the chemicals, but should be easy to be stripped. It should be easily visible, but at the same time, it should not contaminate the cleaners. The final selection of the ink is a compromise among the above mentioned desirable qualities.

7.5.1 Ultraviolet Curing Inks

The ultraviolet curing inks cure by exposure to high intensity mercury vapour lamps, generally in a conveyorized mode. It is a single-sided system, and if the second side image is required, it is applied after the first side is cured.

Ultraviolet (UV) inks consists of three major ingredients: (1) Photo-sensitizable monomers, polymers or photo-initiators, (2) Inhibitors, and (3) Pigments and dyes.

The photo-initiators in the ink absorb the ultraviolet energy and produce polymerization. The inhibitor allows the inks to remain stable and prevents the compound from polymerization when kept in the container prior to the actual use. Curing of the ink is dependent upon its chemical composition, amount or type of pigment or filling materials, thickness of the applied coating and energy of the ultraviolet light.

UV curing inks are 100 per cent solids, and offer rapid cure and high productivity. However, there are problems of getting batch-to-batch consistency, and they may become brittle if over-cured and through adhesion to copper. It is also difficult to cure completely when the wet film thickness increases.

 ## 7.6 Printing Process

The circuit patterns are screen printed on the substrate by two ways: (1) Manual screen printing process, and (2) Automatic or semi-automatic screen printing process.

7.6.1 Manual Screen Printing Process

Manual screen printing is carried out on a sturdy workbench. The patterned screen is fixed in a frame holder held by a high mechanism. A squeegee is selected with width of 2-4 cm more than the pattern width. The squeegee speed and squeegee pressure are controlled manually. The quality of the printing depends upon the operator skill. This type of process is used in the manufacture of low cost non-professional boards.

7.6.2 Automatic or Semi-automatic Screen Printing Process

In the automatic screen printing technique, the machine operates with a hydraulically controlled squeegee for constant speed and a pneumatically controlled squeegee for constant pressure. The squeegee pressure and speed are constant in all strokes, so the quality of printing is more uniform in this case than in the manual printing process. Normally all inner layers of the multi-layer boards are printed by this method.

Advances in the design and operation of screen printing machines have resulted in a progressive reduction in the use of manual machines. Semi-automatic and fully automatic machines are now available with automatic feeders, racks and conveyorized curing units to make them very efficient. Standard panel sizes, 0.25 to 0.4 mm line and space and long run lengths, are most suitable for machine printing. However, the highest resolution in screen printing is only obtained by manual means as skilled persons are able to adjust their technique to changes in ink viscosity, humidity, screen tension and conditions of the substrate. But the manual method is generally too slow for most PCB operations.

 ## 7.7 Photo Printing

Photo printing of PCBs basically means applying photo-sensitive material having the ability to form a continuous film, which is sensitive to light or other radiation so that the exposed (or unexposed) areas of the film can be further processed without affecting the unexposed (or exposed) areas. This photo-sensitive material is called *'photo-resist'*. The essential property of a photo-resist is that an exposure to proper radiation must produce a change in it to enable a clear distinction in the later operations between the exposed and unexposed areas. Usually, in all the photo-resists, a light-induced change in solution forms the basis of their action.

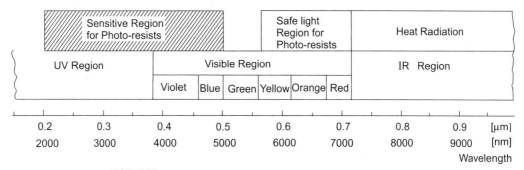

Fig. 7.10   Photo spectrum and sensitive region for photo-resists

The photo-resists are sensitive to ultraviolet light with a wavelength of 200-500 nm. Like the photo-footing film, the photo-resist has a safe visible light area in the yellow to red wavelength (560 to 700 nm). This is shown in Figure 7.10.

Photo-resists can be classified as wet film (liquid film) and dry film (solid film) resists. Irrespective of whether wet or dry resists are used, the following steps are applied for image transfer with the photo printing technique. The coated board is exposed through the appropriate negative or positive film to ultraviolet light. If the board is double-sided, a double-sided exposure is made by ensuring that the patterns on both sides are aligned by matching up, special punch holes on the boards and photographic film. Exposure times can range from a few tens of seconds to several minutes. The resist is then developed leaving those portions of the copper to be retained on the board covered by the resist.

In the case of liquid resists, however, two additional steps are employed. These are: (1) Applying a dye to the resist so that it is visible on the board, and (2) Baking the resist at about 80-100 °C for ten minutes so that the resist is hardened before the etching process.

7.7.1 Liquid Photo-resist (Wet Film Resist)

The wet film resists are organic liquids which, when exposed to light of a proper wavelength, chemically change their solubility to certain solvents (developers). They are available in unsensitised (two-component system) and pre-sensitized (single-component system) forms. The chemical constituents, the mixing ratios, the residual solvents and/or monomer levels are all important in determining the correct chemical and physical performance of the wet film resist. The wet film resist is coated on the laminate by several techniques. These are detailed below.

- *Dip:* The board is dipped into the resist and pulled out slowly. The coating is usually thicker at the bottom of the board than the top.

- *Spray:* Thin coatings of desired thickness can be made by this technique. The use of multiple spraying can build up the coating thickness. Double-sided systems can coat both sides with an appropriate carriage system, which can continue to carry panels through a drying oven. The system loses material through over-spray. This method is not currently used.
- *Roller Coating:* This is done in a special two-sided roller coating equipment, which is directly coupled to a drawing system without releasing the panel. Multiple panel thickness presents no problems and surface uniformity over a large panel is good.
- *Flow:* The resist is placed on the board and is gently tilted; till the resist covers the whole board. The method is useful for large area boards.

After applying, the resist is normally baked at 80 °C for a few minutes so that it is hardened.

Liquid photo-resists, in general, have been difficult to work with. The coating thickness is uneven and plated through-holes have plugs of resist that are difficult to remove. Pin holes (dirt) in the resist image necessitate inspection and quite often require a touch-up operation. Liquid film photo-resists are classified as:

- Negative acting resists; and
- Positive acting resists.

Negative Acting Resists

Negative acting photo-resist is initially (before exposure to ultraviolet light) soluble in the developer or the solvent. This is before the exposure. But after exposing, it is polymerized and becomes insoluble in the developer. Here, the artwork must therefore be in the form of a negative. This type of photo-resist is widely used in PCB manufacture.

Positive Acting Resists

The characteristic of the positive acting photo-resist is opposite to that of the negative acting resist. It is initially not soluble in the developer. But after exposing, the polymerized part becomes soluble in the developer. Here, the artwork must be in the form of a positive. The use of this type of photo-resist is highly limited.

Liquid film resists are inherently more fragile. They are relatively thin and subject to damage if handled roughly (Gurian and Ivory, 1995). Stacking must be done carefully as any contact with the working surface may generate defects, besides leading to entrapment of foreign dust and fibres. Liquids always demand clean working conditions to prevent attraction and inclusion of defect-producing contaminants.

7.7.2   Dry Film Photo-resists

Dry film photo-resists are widely used in PCB manufacture, especially for the production of professional grade PCBs. This image transfer technology was introduced as an alternative to screen printing and liquid photo-resists by Du Pont in 1968. The technology has not only found wide acceptance, but has revolutionized the fabrication of professional grade PCBs with plated through-holes. Increased thickness, uniformity and ease of application are the strong points for the dry film photo-resists. The dry film resist is available as a composite material consisting of three different layers as shown in Figure 7.11. The photo polymer layer of 17 to 75 μm thickness is sandwiched between a layer of polyester (mylar) film on one side, and a polyolefin film, on the other side. The PCB manufacturer therefore has a wide choice for selecting the optimum resist film thickness for a particular application. The photo polymer is applied to the board surface by using heat and pressure as the board passes through the laminator. In the laminating process, the polyolefin sheet is removed just prior to lamination. The polyester cover sheet (film) serves as a protection from fingerprints and foreign particles. The protective layer is removed only before the development of the exposed resist. The polyester and polyolefin layers allow the dry film resist to be wound into rolls and packaged in lengths of 122 to 305 m (400 to 1000 ft.).

Dry film resist have been widely used because they have largely simplified application, cleanliness and handling problems. There is a reported evidence that the technical resolution capability of dry film resist approaches 1mil (25 microns), while the current state of production capability drops dramatically below the 3 to 4 mil (75 to 100 μm) range (Gurian and Ivory, 1995).

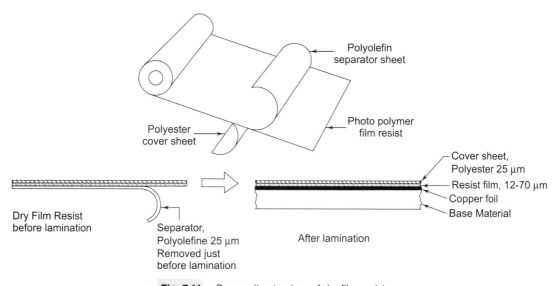

Fig. 7.11   Composite structure of dry film resist

7.7.2.1   Processing of Dry Film Resist

Figure 7.12 shows a typical process flow for the image transfer process utilizing dry film resist. The dry films are applied to the surface of the laminate, which can be either a single side or double side type. It is essential that the surface of the laminate should be cleaned properly, so as to avoid contamination on the surface. This ensures that a good adhesive property exists between the film and the laminate to achieve reliable chemical properties.

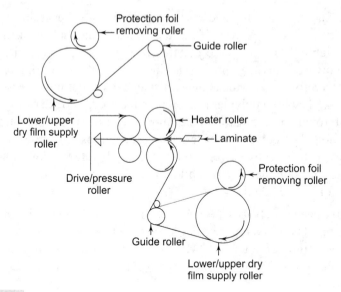

Fig. 7.12 Typical process flow for image transfer process based on dry film resist

The flow chart for dry film resist processing is as given below:

- Prelamination cleaning and drying (prebaking)
- Dry Film Lamination
- Exposure (Printing)
- Developing and Drying (post baking)
- Inspection and touch up
- Stripping

Prelamination Cleaning (Surface Preparation):   The laminate surface is cleaned, using any one of the methods mentioned under surface preparation techniques discussed earlier. Adhesion of dry-film on surfaces contaminated with traces of oil and greases is more critical than with wet films because such defects cannot be immediately detected. Similarly, surface defects like deep stretches are also harmful.

The laminate surface whether it is drilled or undrilled will absorb water or moisture during surface cleaning process. The absorbed moisture is removed by a 15–20 minutes bake in an oven set at 110 ± 5 °C.

Water must be physically removed or blown off from the surface and via holes before oven drying. A high volume air turbine dryer should be used at this stage. This process is called prebaking process. The drying time varies with the substrate grade and the processed laminates should be stored in a clean, non-corrosive environment.

Dry Film Lamination: The lamination is carried out in a laminator which has heated rubber rolls and mechanical or pneumatic pressure arrangement. The cleaned and dried laminates are coated with dry film photo-resists using hot roll laminator, the hot roller is preheated at 120 °C and the lamination is done at high pressure

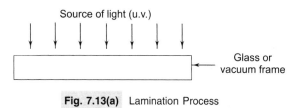

Source of light (u.v.)

Glass or vacuum frame

Fig. 7.13(a)   Lamination Process

(15 to 40 psi or 100 to 275 kpa) as shown in Figure 7.13 (a). The separator sheet is automatically removed just before the heating occurs. Coating speeds are typically around 1.5 m/min. The temperature and pressure vary from one laminator to the other based on the recommendation of the dry film photo-resist manufacturers. After lamination, the board is separated from the continuously coming dry film by a knife cut.

The distance between top chamber and the surface heating which is called the *free space* must be at the required level. Less free space may lead to air entrapment between resist film and the laminate surface. The entrapment may lead to metal chippage during plating or under-cut during etching.

In the case of double-sided boards, both the board sides are laminated simultaneously. For single-sided boards, lamination can be done by putting two boards back to back or by removing one of the rolls with the dry-film resist, the drilled holes remain absolutely clean.

The hold time between lamination and exposure should be controlled carefully for optimum results. Dry film photo-resist after lamination is cooled for 10 to 15 minutes for better dimensional stability.

If no time is allowed for cooling, and the photo-resist is exposed, the resolution of the circuit image is likely to vary due to varying dimensional properties.

Exposure (Printing):   The most commonly used exposure units for image transfer in case of dry film resists produce UV (ultraviolet) light, around a wavelength of 365 nm. For this, the recommended light source is mercury-vapour lamp, though carbon-arc lamps and UV fluorescent tubes can also be used. In production environment, use of a bank of high-pressure mercury arc lamps providing 6000 watts of irradiation are provided. Such an arrangement provides intense, collimated UV light at levels conducive to high board throughput and reliable image transfer. For laboratories, a much smaller system is usually adequate.

Source collimation, or the degree of parallelism with the UV light illuminates the artwork/sensitized substrate, ultimately determines the fidelity of image transfer and the minimum size feature that can be reliably resolved. In addition, uniformity of illumination determines the consistency of exposure from one point to another and affects the trace-width uniformity in the developed image. From a uniformity point of view, the ideal source would be an isotropic emitter whose emission area was as large as, or larger than the substrate you are imaging. Any light source intended for use in PCB production must strike an acceptable balance between these two factors and the cost of implementation.

Exposure is carried out by registration of the emulsion side of the photographic film like diazo film or silver halide film—on the polyester sheet. The two are then placed in a vacuum frame for intimate contact between surfaces. The proper exposing time depends on light intensity, temperature, thickness of photopolymer, and type of equipment being used (Figure 7.13b).

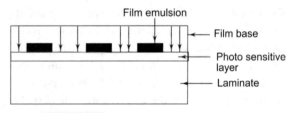

Fig. 7.13 (b)   Exposing photo sensitive resist

To determine the required exposure time, hold time and the suitable development conditions, a trial production work may be carried out. The required exposure level varies with the type of film and photopolymer manufacturer normally supplies suitable data for the same.

Selection of ultraviolet bulb or mercury vapour bulb plays an important role in the uniformity of exposure. The aged bulb will affect the degree of polymerization of photopolymer. Optimum exposure is best controlled with the step-tablet method.

After exposing the resist coated laminate, a certain time is allowed for polymerization to take place, and to attain equilibrium or stabilization. The minimum and maximum time allowed for the same is specified in the polymer manufacturer product data sheet.

Developing: Developing the exposed laminate after stabilization starts with the peeling off the protective polyester cover sheet. The developing process washes away the unexposed or unpolymerised photopolymer without affecting the exposed polymerized portions of the film.

Developing time is dependant on the chemistry and temperature of the developer solution and the thickness of the photopolymer. Recommended chemistries and temperatures must be maintained before starting the developing process. If not controlled, the developer affects the polymerized resist causing degradation of the surface and the side walls of the film.

Thorough removal (development) of unexposed resist from the surface is necessary to ensure good etching and plating. The total time between lamination and development should not be more than 8 hours to minimize difficulties in developing process.

Normally the recommended chemistry accounts for 0.8 to 1.2 per cent solution of anhydrous sodium carbonate (soda ash of 99 per cent purity) dissolved in water. If other grades of sodium carbonate are used, the quantity required must be adjusted to compensate for the water crystallization.

Aqueous developing solutions tend to foam in spray developing machines. Antifoaming agents are usually used to control the same. These have to be added in their proper proportions to ensure that it does not affect the polymer. Normally butyl carbitol or n-octanal are used to control the foaming. This increases developing speed also.

Baking or drying after developing process is done at temperatures between 85–95 °C for 10 to 15 minutes. Moisture absorbed by the resist and base material during the development and rinsing steps must be removed completely. It also aids the pattern plating adhesion to a considerable extent. The chemical resistance is enhanced by baking.

Inspection and Touch up: Inspection should follow after the board is dried. The inspection should determine the quality of the board and also the major and minor defects. The defective boards are stripped and whole process is repeated.

Defects like small pin holes and various other minor defects can be corrected by certain lacquer coatings that will not affect further processes.

Stripping: If the boards are found defective after plating/etching, the photopolymer is stripped out using certain chemicals recommended by the supplier. The chemical is usually a 20 to 30 per cent solution of anhydrous sodium hydroxide dissolved in water. The temperature of the bath is maintained at about 50 to 60 °C. The boards are washed with water and blow dried with filtered air.

The processing area of the photopolymer dry film resist (laminator, exposure, developer) should be illuminated with yellow light of low intensity and must be separated from all the other processing areas.

7.7.2.2   Conditions for a Good Processing Area

The photopolymer film (liquid/dry film) is a photographic product. Hence working in a dust-free atmosphere reduces the need for touch-up or any other re-work and considerably improves the quality of the final board.

The following measures aid in attaining good quality:

- The processing areas should be separated and illuminated with yellow light.
- The area of work and the equipment require regular cleaning.
- Suitable protective clothes such as lint-free clothing, acid-alkali resist shoes, hat etc. are recommended.
- Temperature and humidity should be controlled like the photo-tool production area.
- Solvent developer must be separated from the ultraviolet processing equipment.

 ## 7.8   Laser Direct Imaging (LDI)

The conventional image transfer system makes use of artworks and photoplotters. Photoplotters are generally considered infallible in terms of dimensional accuracy, as they are periodically calibrated and routinely maintained. The best positional accuracy the photoplotter can offer is 28 microns, which is without the effects of varying temperature and humidity. Inaccuracies are observed as both linear and non-linear errors in each axis along with the associated rhombic distortion, wherein the deviation across one diagonal is greater than across the other.

Artworks used for PCB manufacture are made from extruded polyester and, therefore, have different characteristics in the direction of extrusion and across the extrusion. In addition, the humidity characteristics for artwork material are approximately 14 ppm %RH. The temperature characteristic for artwork material is approximately 16 ppm/°C.

Punched tooling systems also tend to add further errors due to the accuracy and repeatability inconsistencies of the punch. Some print frames heat up during use and can cause further distortion of upto 25 microns.

When these errors are compounded during the inner layer manufacturing process, the mean error can be upto 130 microns. This would require a design rule of 175 microns to guarantee alignment when the layers are stacked for bonding. This means that high yielding HDI designs would be difficult to achieve on the large panels required for high volume cost-effective manufacture using conventional imaging technology. The removal of artworks from the imaging process would obviously substitute all sources of errors associated with their use. The use of laser direct imaging technique offers an artwork-free manufacturing route, instantly removing the large errors associated with environmental control during both production and use of the artwork. The use of the LDI system has shown that the total mean error is reduced form 130 microns to 12 microns.

LDI is a process of imaging printed circuit boards directly without the use of photo-tool. The exposure of the photo-sensitive resist is done using a laser beam, i.e. scanned across the panel surface and switched on and off by means of a computer control system. The laser used in this process is in the UV spectrum region, as this tends to suit most of the commonly available photo-resists. However, systems exist that operate in both the visible and infra-red spectrum, working with specially formulated photo-resist.

LDI systems first started to find their way into the printed circuit manufacturing arena in the late 1980s. These early systems were much slower than the conventional contact printing, and it has only been in the last few years with a new generation of faster LDI systems that the process has become a viable threat to contact printing with a photo-tool. With the availability of high-powered lasers, giving 4 watts of power at the work piece, and with the introduction of a new generation of photo-resists requiring exposure levels of the order of 8-10 mJ/sq.cm along with faster computers for data rasterization, it is now possible to get LDI exposure and handling times down below 30 seconds per side for an 18" × 24" panel.

A Laser Direct Imaging system has the following important components (Sallan and Wiemers, 1999):

- The laser system;
- The optical system;
- The mechanical system; and
- Data processing system.

The Laser System

The laser system is used as the light source and consists of a water-cooled argon-ion laser with a capacity of 1.5 W. The exposure takes place in the UV range at 2 wavelengths of 351 nm and 358 nm because of the maximum sensitivity of the common photo-resist within this range. The pixel diameter is 28 μm. Since the intensity within the pixel is different, an exposible pixel diameter of 10 μm remains.

The resist exposure by a laser beam is effected in the horizontal direction with a velocity of 240 m/s. During the scan movement, an exposure clock pulse ensures the light/dark scan of the laser beam. The vertical movement of the material to be exposed towards the scanning causes a line per line image structure all over the whole surface. The exposure grid is 10 μm. In case of a maximum exposure area of 340 × 600 mm, a data quantity of about 2 billion pixels has to be processed.

The Optical System

Figure 7.14 shows the optical system of an LDI system. The system consists of an acousto-optical modulator, a quartz in which an acoustic wave rectangular to the laser beam is generated. It is situated in the optical axis of the laser beam. The incoming laser light is diffracted and its direction changes depending upon the frequency of the acoustic wave. The light reaches either a beam trap or the optical beam path. The scanning is achieved by deflection via a ten-face polygon whose mirror faces have an angle error to each other of less than 2". A line of 340 mm is exposed per each mirror face.

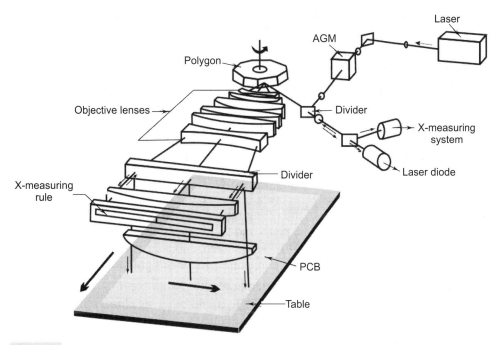

Fig. 7.14 The optical system of direct laser imaging system (redrawn after Sallan and Wiemers, 1999)

In order to achieve a precise rotation speed and smooth running, the polygon is placed on an air slot of 10 μm and is synchronized with the table drive. The table moves on 10 μm with each new mirror face. The optical system consists of eight lenses, which ensure a telemetrically beam path and therefore, a clear pixel image.

After the laser beam passes through the lens system, it is guided via a semi-reflecting mirror down to the resist-laminated PCBs or inner layer laminates. The X-measuring system ensures the exact positioning of the pixel. The system consists of a laser diode of 685 nm wavelength, a glass measuring rule and several deflection and filter units as well as a reference grid.

The red laser beam of the measuring system is guided parallel to the blue laser beam of the exposure system via the polygon through the lens system and through a 50 per cent-mirror on an etched X-measuring rule. There, the image is reflected back through the mirror and the lens system to the polygon where it is de-coupled and mapped on a reference grid.

During the rotation of the polygon, the image of the X-measuring rule passes the reference grid and generates a light signal from which the exposure clock pulse can be deduced. The exact position of the red laser beam can be determined by counting the impulses of the exposure clock. The position of the blue laser beam is determined at the same time due to the coupling of the red laser beam with the blue one.

Mechanical System

The main mechanical components of the LDI are the Y-measuring system, the Z-positioning system, the registration and the air ventilation system for clean room facility.

The work table is directly connected with the Y-measuring system. Since the table movement and the polygon are synchronized, the table receives a positioning pulse after each mirror face. This impulse is given to a stepper motor situated on a ball screw, which is also connected to the work table. The positioning is carried out by counting the increments referring to a reference mark.

Before the exposure process, the PCB must be lifted in the focus area and registered. The work table is equipped with three stepper motors for this Z-positioning. These motors lift the table until a proximity sensor detects the copper clad of the PCB. The table is therefore always situated within the focus area with a rectangular alignment to the laser beam.

For the registration of the PCBs, two service holes can be found on the production panel. This procedure enables proper adjustment of the fitting for the pattern structure to the drilling pattern to be produced. There are LEDs beneath the panel and there are also four quadrant sensors above the panel. The sensors measure the light quantity per quadrant and adjust via the positioning motors in X, Y and Z-direction as long as each quadrant captures the same light intensity.

The work table with the optical system and the laser unit has air bearings to eliminate the effect of external vibrations caused by vehicular circulation and other machinery in the vicinity. This vibration damping system ensures a correct exposure quality.

Data Processing System

CAD or CAM data usually exist as GERBER data which is normally used for photoplotting. Due to compatibility reasons and the need to simplify the processing, the LDI system also prefers to process these data formats.

Figure 7.15 illustrates the typical data flow arrangement in which the external data of the CAD system is received by ISDN or by e-mail. CAM reads out the data from the modem, processes it and makes it available via an internal LAN system on the LDI system.

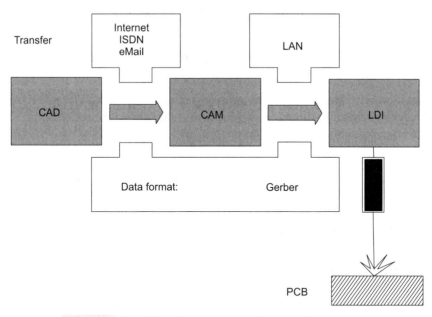

Fig. 7.15 Typical data flow arrangement for laser direct imaging

Since vectors are described in the GERBER format, which is inappropriate for a line per line presentation, the data have to be transformed into information for 10 μm grid pixels. The original amount of data is therefore multiplied. The compilation of the data is performed in real time by multiple simultaneously working transputers. The data are read line per line and stored on a 250 MB RAM. The RAM is sufficient to store exposure data for approximately five double-sided PCBs.

7.8.1 Benefits of LDI

A comparison between film and laser exposure clearly reveals the advantages that the laser system offers in terms of quality, production tolerances and savings in time and cost (Barclay and Morrell, 2001). The savings in cost of manufacture can obviously be seen from any one or all of the following factors:

- Elimination of photo-tools and the cost involved in their manufacture and storage;
- Reduced job set up time between prints and manufacture of PCB; Manufacture can start as soon as the data leaves the engineering department. The time saved (from 10 hours for films compared to three hours by laser) is very important during the production of prototypes;
- Possibility of adopting a flexible manufacturing route to meet the varied demands of production without impact on throughput; and

- Reduced manufacturing lead times by enabling manufacture to start as soon as the data leave the engineering department.

Similarly, quality improvements offer several benefits such as:

- Elimination of film- and printing-related defects; and
- Elimination or reduction of temperature-and humidity-induced effects on the product due to the controlled environment employed within the laser imaging systems.

Over and above the cost saving and quality improvement benefits, there are a number of technical advantages, which are detailed below.

- **Resolution:** LDI systems offer improved resolution due to the small laser spot size. Sub 50 μm features can be easily resolved. With process optimization it is possible to produce fine lines of the order of 35 μm in a 40 μm resist. With the future LDI systems improvements, 25 μm lines and spaces are likely to be realized.
- **Registration:** Improvements in registration are achieved by eliminating the photo-tool, which has always given alignment problems, especially as the tools move anisotropically with temperature and humidity changes. With the LDI system, it is possible to use a CCD (charge coupled device) camera system and target fiducials on the panel to align the print image and panel. It is also possible to use these target positions to calculate any panel or drilling movement, thus achieving an improved registration.
- **Tolerances:** A qualitatively different tolerance class is achieved by using a LDI system. The industrial standard of 0.1 mm permissible misalignment from drilling pattern to conductive pattern can be reduced to 0.03 mm in the ideal case.

Kelley and Jones (2002) illustrate the application of laser direct imaging. The LDI systems are bringing about fundamental changes in the organization of sequences, and in the logistics and data storage facilities in the PCB industry. One LDI system is replacing all exposure systems in PCB manufacture. Both the tasks that are being currently carried out by conventional film exposure systems and those of a photoplotter, can be taken over by LDI systems. The software control adaptation of the machine to the required task can easily be carried out and all CAD systems would work with a uniform table and every PCB would have the same material specifications. The use of LDI systems, in future, is likely to drastically reduce the development time for electronic assemblies, a sensational but possible solution. Vaucher and Jaquet (2002) provide an update on Laser Direct Imaging and Structuring.

 ## 7.9 Legend Printing

Legend printing is generally done on the component side of a PCB, which gives the details of the components and their assembly position on the board. It is done by the process of screen printing. Epoxy-based ink is used for this purpose. This ink, when once printed and cured, bonds permanently with the epoxy of the base material. This marking, which gives the component outline with their identification numbers, helps in easy assembly and also in the identification of the components during the re-work and troubleshooting of the PCBs.

Legend printing is generally the last process done in the fabrication of a printed circuit board. The colours generally used for legend printing are white, black and yellow.

 ## 7.10 Useful Standards

- *IPC-A-311: Process Controls for Photo-tool Generation and Use:* Covers the information and data to be collected during the generation and use of photo-tools to improve artwork quality, thereby improving yields downstream.
- *PC-D-310C: Guidelines for Photo-tool Generation and Measurement Techniques*: Covers manufacturing and design considerations, input data requirements, test coupons, process control, tape and perform artwork, cut and strip artwork, vector photoplotting, raster plotting, direct imaging, measurement and quality assurance.

Plating Processes

 8.1 Need for Plating

In printed circuit boards, copper is used for interconnecting the components on the substrate. Although it is a good conducting material on the PCB to form the conducting track pattern, it is liable to tarnish due to oxidation, if exposed to atmosphere over a long period of time. It undergoes corrosion and thus loses its solderability. Therefore, various techniques are used for the protection of the copper tracks, and via holes and printed through-holes (PTH). They are organic lacquer coating, oxide coating and plating.

Organic lacquer coating, though simple in application, is not suitable for long term usage due to variations in thickness, composition and curing cycles. It can also bring about unpredictable deviation in solderability. *Oxide coating* can be used to protect the circuit from corrosion, but it fails to preserve the solderability. *Plating* or the metal coating process is a standard practice to ensure solderability and protect the circuit from corrosion. So it plays an important role in the PCB manufacturing of single-sided, double-sided and multi-layer (PTH) boards. In particular, plating a solderable metal over the tracks has now become a standard practice to afford solderable protection to the copper tracks.

Edge connectors with spring contacts mating with suitably designed connector tabs on printed circuit boards, are used for the interconnection of various modules in electronic equipment. Such contacts should have a high degree of wear resistance and low contact resistance. This requirement has resulted in precious metal plating on these contacts and the most commonly used metal is gold. Other metals used for coating are tinning of the tracks, nickel plating and in some instances, copper plating to build up some of the track areas.

Another type of coating on the copper tracks is of the organic type, which is usually a solder mask, a screen printed epoxy coating to cover those areas that are not required to be soldered. The process of applying an organic surface protectant (OSP) coating does not require electron exchanges since the circuit board is coated upon submersion in a chemical bath. A nitrogen-bearing organic compound allows adhesion to the exposed metal surfaces and is not absorbed by the laminate.

The precise technical requirements of electronic products and the demands of stringent environment and safety compliance have resulted in highly advanced plating practices. These are evident in the technology to produce complex, high resolution multi-layer boards. In plating, such a level of precision has been achieved by the development of automatic, computer-controlled plating machines, highly sophisticated instrumental techniques for the chemical analysis of organic and metallic additives, and the possibility of precisely controlling the chemical process.

There are two standard methods to get metal build-up onto the circuit traces and the holes: pattern plating and panel plating. These are discussed below.

Pattern Plating

This is the process wherein only the desired circuit pattern and holes receive copper build-up and etch-resist metal plate. During pattern plating, the circuit lines and pads increase in width on each side about as much as the surface thickness during plating. For this purpose, allowance thus needs to be made on the master artwork.

Pattern plating basically involves masking off most of the copper surface and plating only the traces and pads of the circuit pattern. Due to the reduced surface area, a much smaller capacity current source is generally needed. Further, when using contrast reversing photopolymer dry film plating masks (the most common type), a positive image is produced on a relatively inexpensive laser printer or pen plotter. Pattern plating consumes less copper from the anode bank and requires the removal of less copper during etching, thereby reducing bath analysis and maintenance. The disadvantage of the technique is that the circuit pattern is required to be plated with either tin/lead or an electrophoretic resist material prior to etching and then stripped prior to soldermask application. This increases the complexity and adds another set of wet chemical baths to the process

Panel Plating

This is the process wherein the entire surface area and the drilled holes are copper plated, slopped off with resist on the unwanted copper surfaces and then plated with the etch-resist metal. While this requires a fairly large current source for even a modest size PCB, the end result is a smooth, bright copper surface that is easy to clean and prepare for later processing. If you do not have access to a photoplotter, there is a need to use negative artwork to expose the circuit pattern into the more common contrast reversing dry film photo-resists. When you etch a panel plated board, you end up removing most of the material that you plated, so the burden of extra erosion of the anode banks is exacerbated by an increased copper loading in the etchant. Figure 8.1 shows a printed circuit board plating flow chart for the subtractive process.

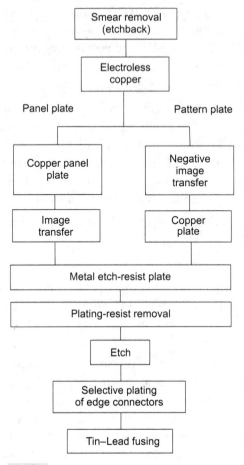

Fig. 8.1 Printed circuit board, plating flow chart

Pattern plating is the preferred method for manufacturing PCBs, with the standard thickness as follows:

- Copper : 1 mil
- Tin-lead : 0.5 mil
  (traces, pads, holes)
- Nickel : 0.2 mil
- Gold : 50 μm
  (connector tips)

The electroplating process parameters are so maintained that the metal deposits provide for high electrical conductivity, good solderability, and high mechanical strength and ductility to withstand panelling of component terminals and fill copper coverage from surface into PTH.

8.2 Electroplating

Electroplating is the process by which a metal is deposited on a conductive surface by passing a direct current through an electrolytic solution containing a soluble salt of the metal.

When a uni-directional current is passed through a solution, it results in movement of the charged particles through it. These particles are called ions and the terminals which are used to pass the current into the solution are called *electrodes*. The combination of the two electrodes and the solution form what is known as *electrolytic cell*. The other electrochemical terms associated with electroplating are listed below.

- *Anode*: The pole or electrode at which the chemical reaction of oxidation takes place is called the *anode*. It is a positively charged electrode. During electrolysis, positive ions are formed at this electrode.
- *Anion*: This is a negatively charged radical which, on electrolysis, is attracted towards the anode.
- *Cathode*: The electrode or pole at which the chemical reaction of reduction takes place is called the *cathode*. It is a negatively charged electrode. During electrolysis, negative ions are formed at this electrode.
- *Cation*: This is a positively charged radical which, on electrolysis, is attracted towards the cathode.
- *Electrolyte*: This is a conducting medium in which the flow of current is accompanied by the movement of ions. For electroplating, the electrolyte must contain dissolved salt of the metal that is to be deposited.

The metal to be plated is used as the cathode in the electrolytic cell. The anode in the cell can be the same as the metal to be coated on the cathode or any other inert metal having good conductivity. The reaction which takes place at the junction between the electrolyte and the electrodes is called electrolysis. It is accompanied by the transfer of electrons.

8.2.1 The Basic Electroplating Process

The article to be electroplated is first cleaned thoroughly to make it free from oil, grease and foreign particles. The cleaned article is used as cathode in an electrolytic solution. The electrolyte is kept in an electroplating tank. The two electrodes, anode and cathode, are dipped in the electrolyte.

When direct current is passed through the electrolyte, coating metal ions migrate towards the cathode and get deposited. Most plating solutions are similar in nature. Therefore, their use and the quality of the resulting deposits would depend upon the processing variables involved. Accordingly, the deposition depends on the temperature, current density, metal ion concentration, pH, solution movement and filtration. Figure 8.2 shows the basic principle of electroplating. Therefore, the conditions must be controlled so that uniform composition is maintained over a wide range of operating variables.

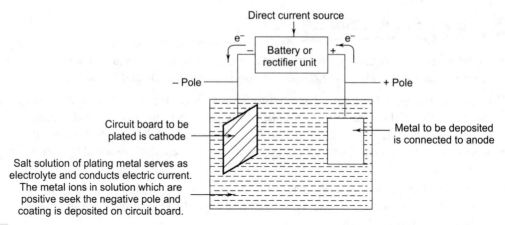

Fig. 8.2 Principle of electroplating. Metal to be deposited is connected to the anode whereas circuit board to be plated is the cathode

If the anode is made up of a metal which is the same as the metal to be deposited on the cathode, the concentration of electrolyte remains the same during electrolysis. The metal which is deposited on the cathode is removed from the electrolyte. This loss in the electrolyte is compensated since the anode dissolves in the electrolyte proportionately. If the deposition and the dissolution are not equal, it will affect both the metal concentration in the electrolyte and deposition rate.

8.2.2 Faraday's Laws of Electrolysis

The fundamental principle of electrolysis is governed by Faraday's laws, which are detailed below.

First Law

The mass of a substance produced or consumed at an electrode (either anode or cathode), is directly proportional to the quantity of electricity that passes through the solution.

$$M \propto Q \qquad Q = It \qquad \text{or } M \propto It$$

Therefore $M = ZIt$

where M = Mass of the substance in grams,

$\quad I$ = Current (in amperes),

$\quad Z$ = Electrochemical equivalent, and

$\quad t$ = time in seconds.

When $I = 1$ A, $t = 1$ sec. then $M = Z$

Second Law

When the same amount of electricity is passed through different electrolytes, the weight of the metal deposited is proportional to its chemical equivalent

$$M = \frac{ITA}{nF}$$

where A = Atomic weight (or equivalent weight × valency),

N = Valency (number of electrons involved in metal ion reduction), and

F = Faraday's constant.

Plating takes place at the cathode, i.e. the negative electrode. Therefore, deposit thickness on the cathode surface would depend upon the time and the current impressed on it. For example, 0.5 mil (0.0005 in) of tin-lead alloy is plated at $17 A/ft^2$ for fifteen minutes.

From Faraday's Law, it may be inferred that if 1 coulomb (1 A sec) deposits 1.118 mg of silver, then 8 coulombs will deposit 1.118×8.944 mg of silver for achieving this quantity of deposit. It is immaterial whether a current of 1 A flows for 8 secs. or 2 A for 4 secs., or any other current and time which may yield 8 coulombs.

8.2.3   Water Quality

Water supply for the plating solutions and for the cleaning process should have low level of impurities. If the water supply contains high levels of dissolved ionic minerals and impurities such as calcium, magnesium, silica, iron and chloride, they can cause copper oxidation, residues in PTH, peeling, staining, roughness and ionic contamination. These problems result in board rejects and equipment downtime, reduced bath life and difficulty with rinse water recovery. It is thus imperative to have good quality water with low hardness and total dissolved solids. Generally, the quantity of dissolved solids should not be more than 5 ppm.

The dissolved solids are usually measured by recording the conductivity of the solution. This is done with conductivity meter. The presence of 5 ppm of dissolved solids correspond to 1 µ mho cm.

For obtaining such high purity water needed for plating and PCB manufacturing, two processes are widely used: reverse osmosis and de-ionization process. These are discussed below.

Reverse Osmosis

In this process, raw water under pressure is forced through a semi-permeable membrane. The membrane has a selected porosity which allows rejection of dissolved salts, and organic and particulate matter. The membrane allows the passage of water through the membrane.

To understand the process, let us consider that there is pure water and a saline solution on opposite sides of a semi-permeable membrane. The pure water will diffuse through the membrane and dilute the saline water on the other side. This process is called 'osmosis', and the pressure under which this process takes place is called the osmotic pressure. Similarly, if pressure is applied on the saline solution, the osmosis pressure is reversed. This is called 'reverse osmosis'. Reverse osmosis is an effective method of water purification and allows the removal of 90-98 per cent of dissolved minerals and 100 per cent organics having molecular weights over 200.

De-ionization Process

De-ionized water is prepared by the ion exchange technique. This involves passing water containing dissolved ions through a bed of solid organic resins. These convert the ionic water contents to H^+

and OH^-. The process takes place in ion exchange columns, which are commercially available. The de-ionization water purification process is used when high purity water is required in applications such as bath make-ups and for rinsing in the plating process. The typical de-ionized water should have the following characteristics:

- Residual dissolved solids in ppm   < 1
- Electrical conductivity in micron/cm at 20 °C   < 1
- Turbidity   1.0 NTU
- Silica (SiO_2) in ppm   0.05
- pH Value at 25 °C   6.8 to 7.0
- Chloride in ppm   2.0
- Total organic carbon in ppm   2.0

8.2.4 pH of a Solution

Pure water contains equal concentration of H^+ ions and OH^- ions.

$$H_2O = H^+ + OH^-$$

The ionic product of water is expressed as

$$[H^+][OH^-] = 10^{-14}$$

When acid is added in water, the concentration of H^+ ions increases and that of OH^- ions decreases. When alkali is added, the concentration is vice versa. The acidity or alkalinity of a solution is expressed by its pH value. The hydrogen ion concentration forms the fundamental basis of the pH scale as shown in Figure 8.3. The actual value of the hydrogen ion concentration is extremely low, and for convenience it is expressed as a negative logarithm.

$$pH = \log \frac{1}{H^+} = -\log[H^+] = 10^{-pH} \quad OR \quad [H^+] = 10^{-pH}$$

pH is the most important factor in electroplating process control.

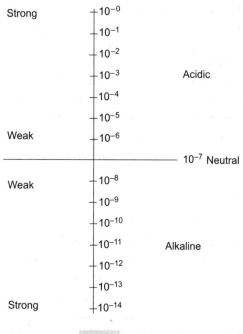

Fig. 8.3 pH scale

8.2.5 Buffer

When a solution contains a weak acid or alkali, its conjugate alkali or acid is known as a buffer solution.

Buffers are only partially dissociated in solution. They are added to a solution in order to reduce the effect of the addition of acid or alkali on their pH value and to keep the pH value constant. In electroplating, buffers have an important influence on the process parameters.

8.2.6 Anodes

In electroplating, the positive electrodes or anodes are used in the form of bars, plates, rods or pallets. The metal used may be the same as that of the electrolyte or it may be different, but it should be free from any impurities. The contaminated anode will affect the electroplating process. It should, therefore be 99.9 per cent pure. The size of the anode is also very important as it determines its current-carrying capacity.

8.2.7 Anode Bags

Anode bags are used to cover the anodes and filter the dissolved particles. The anode bags are made up of polypropylene materials. They are available in various grades from 1 micron to 10 microns.

8.2.8 Pre-treatment for Electroplating

A well cleaned and pre-treated copper metal surface is essential in circuit board plating. Many bonding failures on copper metal can be traced directly to poor cleaning of their surfaces before the application of metal plating. The following methods are used to activate the copper metal surfaces:

   a) Solvent cleaning or de-greasing;

   b) Alkali cleaning; and

   c) Mechanical cleaning.

The above processes are explained in Chapter 7 on Image Transfer Techniques. After the surface cleaning, the PCB is subjected to various protective coating or plating techniques to prevent corrosion and to increase solderability.

 ## 8.3 Plating Techniques

Printed circuit boards are plated or metal coated by three types of techniques. They are:

   • Immersion plating;

   • Electroless plating; and

   • Electroplating.

8.3.1   Immersion Plating

Immersion plating is the simplest technique in which the deposition of a metallic coating on a substrate takes place by chemical replacement from a solution of salt of the coating metal. The substrate metal reduces the atoms from their ionic state in the solution.

Although this method does not require much of a capital cost and has a good ability to deposit in recesses, it allows only a limited thickness of deposition. This is because as soon as the substrate metal is almost completely covered by the deposit, the reaction slows down to the rate at which substrate metal is available in discontinuities or pores in the coating. The limited thickness offers only short term protection. Therefore, immersion plating is done only where immediate assembly of components is planned. Immersion plating is usually carried out with only two coating metals. These are tin and its alloys, and gold.

The print and etched boards are tin-plated by immersing the board in tinning solution and tin metal is deposited on copper metal. The amount of copper dissolved in the solution is equal to the deposition of tin on the PCB. The typical thickness deposited per hour is 3 μm or less.

A typical tinning solution has the following composition:

- Stannous sulphate   5-8 g/l
- Thiourea [CS (NH$_2$)$_2$]   70-80 gms/l
- Sulphuric acid   10-15 ml/l
- Bath working temp.   30 °C
- Immersion time   5 to 20 mins

The copper on which tin is to be deposited forms an electrolytic cell. The potential at the electrode which exists due to ionization is called the *electrode potential*. The reaction which takes place in the immersion tank is

$$Cu + Sn^{++} \longrightarrow Sn^+ Cu^{++}$$

Cu^{++} and Sn^{++} are positively charged metal ions, due to the lack of two electrons. The electrode potential of copper can be made more negative by adding a complexing agent in the tinning solution. Thiourea is used as a complexing agent which also forms stable complexes.

Immersion Plating of Gold: This offers an advantage in that the gold coatings are relatively pores-free. A typical formulation for immersion gold can deposit about 0.025 μm in half an hour.

8.3.2   Electroless Plating

With the introduction of plated through-hole technology and developments in multi-layer boards, it is necessary to provide a copper layer on a drilled hole on its insulating area or to connect the different layers on a multi-layer printed circuit board to obtain electrical conductivity. The most popular method for achieving this is electroless copper plating. Once the resin surface has been made conductive, electrolytic plating can be used to build up the desired copper thickness.

The plating process is done after the through-holes have been drilled in the substrate and before the etching process is performed. Electroless copper plating depends upon many different processes, chemistry, equipment and materials (Hodson, 1991). The synergy of these elements determines the quality of the interconnection and the ultimate performance of the board.

The electroless plating mechanism is similar to electroplating but the electrons are obtained from the chemical reduction process. In electroplating, the electrons are obtained from an electrical current source to reduce metal ions to the metallic state. But in case of electroless plating, an external source of energy is not used. The electroless plating process is very simple. The chemical reducing agent reduces the metal ions in solution to neutral metal atoms for deposition. With this process, a continuous build-up of metal coating on a substrate takes place by simple immersion in an appropriate aqueous solution. The chemical reducing agent in the solution supplies electrons for the reaction. But the reaction takes place only on a catalytic surface.

Electroless copper is an autocatalytic immersion plating process in electrolysis or electroplating that dispenses with the need for providing electrical connectivity between the outer and inner layers of a printed circuit board. A catalytic surface is a surface on which a noble metal like palladium is seeded to trigger the chemical reduction process. In practice, the drilled printed circuit board is activated by coating with a thin film of palladium by an activator. It is then immersed in the copper bath. The autocatalytic process deposits the metal without using electrical current. The copper provides a conductive metal layer needed for the electroplating process. The chemistry of copper reduction in electroless plating is rather complex. However, the following equation sums up the deposition process using formaldehyde as the reducing agent.

$$CuSO_4 + 2\ HCHO + 4\ NaOH \xrightarrow{\ Pd\ } Cu + 2\ HCO_2Na + H_2 + 2H_2O + Na_2\ SO_4$$

This autocatalytic plating necessarily involves the use of an aqueous solution copper ions, copper complexants, formaldehyde, sodium hydroxide and stabilizers. These constituents and their functions are shown in Table 8.1.

Table 8.1 Constituents of Electroless Bath and their Function
(after Hodson, 1991)

	Constituent	**Function**
Copper salt	$CuSO_4. 5H_2O$	Supplies copper
Reducing agent	HCHO	$Cu^{2+} + 2e \rightarrow Cu^\circ$
Complexer	EDTA, titrates, Rochelle salt	Holds Cu^{2+} in solution at high pH, controls rate
pH controller	NaOH	Controls pH rate, 11.5–12.5 optimum for HCHO reduction
Additives	NaCN, metals, S, N, CN organics	Stabilize, brighten, speed rate strengthen

Keeler (1990a) explains that in subtractively processed double-sided printed circuit boards, electroless plating applies copper on the walls of drilled holes to provide electrical connectivity between the top and bottom of the PCB. In the case of multi-layers, the barrel of plated copper in the hole interconnects not only the two outer layers, but also circuits on inner layers. Electrolytic plating is done over this conductive primary layer.

On additively processed boards, in addition to plating up the walls of drilled holes and vias, the electroless copper serves to form the basic circuit patterns on a bare substrate. In semi-additive processing, by contrast, only the initial layer of copper is deposited from an electroless copper bath and the remainder of the copper metal is electrolytically deposited on the board.

8.3.2.1 Electroless Plating Process

The procedure for electroless copper plating is complex. The exact chemistry and electroless copper plating process will vary depending upon the size and type of the substrate and the production throughput demands. There are a number of chemical solutions and electroless plating equipment suppliers that manufacture different products to accomplish more or less the same process. Therefore, to ensure that electroless copper plating is done properly, the fabricator must understand the intricacies of the plating chemicals and equipment along with the substrate material used in the PCB. In general, the following operations are carried out in sequence to complete the electroless plating on a PCB:

- Laminate deburring;
- Laminate cleaning;
- Water rinse;
- Micro-etch or etch back multi-layer boards;
- Water rinse;
- Dipping in sulphuric acid of 20 per cent concentration;
- Water rinse;
- Sensitizer dipping;
- Activator/catalyst treatment;
- Water rinse;
- Accelerator dip;
- Water rinse;
- Electroless plating (copper electroless plating); and
- Acid rinse.

Deburring: The drilled board surfaces are cleaned by using a deburring machine and a 320 grade silicon carbide brush.

Laminate Cleaning: In this step, the drilled through-hole is cleaned in an alkaline cleaning solution, removing any dust or residue. This will also remove any fingerprints, oils, grease layers and epoxy smears. The bath contains mixed alkali chemicals. The temperature of the bath is maintained at 60–70 °C and the dipping time is about twenty minutes.

Water Rinse: After each process step, the substrate is rinsed to ensure that none of the chemicals is transferred to the next bath. Each step relies upon the chemical reaction within itself and can contaminate the reaction of another bath.

Micro-etching: This is done in persulphate acids, sulphuric acid and hydrogen peroxide and stabilizers. It is used for micro-etching or cleaning of copper surfaces already existing on the laminate. In the process, some amount of copper gets removed from copper surfaces and gets dissolved in the solution used to provide a uniform surface. The importance of micro-etching is that it ensures a uniform metal surface and increases the bonding between the copper layers by roughing up the surfaces. This also gets rid of any contaminants that may impinge oxidation. The etch-back method is used only for multi-layer PTH processing to clean the hole walls.

Acid Dip: The sulphuric acid dip rids the board of any remnants from the micro-etching process. If the substrate is put in the sulphuric acid dip, it must be rinsed before the pre-catalyst stage. Twenty per cent sulphuric acid is used for the dipping process.

Sensitizer Dipping: This is the pre-catalyst stage, which prepares the substrate for the next step, i.e. catalyzation. This bath is a combination of salt and acid, and is a mixture of stannous chloride and hydrochloric acid. The bath can also serve as a pre-heater for the activator stage.

Activator Catalyst-treatment: The activator/catalyst is a surface treatment process that makes the substrate surfaces and hole walls amenable to copper. In this, the sensitizer and activator are mixed in a bath to get a mixed catalyst solution. The mixed catalyst produces a catalytic film of palladium. The catalytic film is produced only on the surface of the insulator (resin and filler) but the copper foil of the laminate remains inactivated. The thin palladium film improves adhesion between the base copper and the plated copper.

The sensitizing step involves immersion of the board in an acidic solution (sensitizer) containing stannous ions, which are absorbed by the surface of the resin. The stannous ions reduce the palladium ion and the stannous ions are oxidized to stannic ions according to a reduction reaction. The tin is removed in the next step, leaving the palladium layer ready for copper deposition.

The temperature in the activator/catalyst tank is extremely important. An air pump is used in this bath to agitate the solution to create an even temperature throughout the bath.

Due to the volatility of this bath, the board must be rinsed thoroughly so that it does not contaminate the accelerator/post-accelerator bath. Generally, the PCB is subjected to two rinse cycles.

Accelerator Dip: Accelerator is a strong acid or alkali and its function is to remove the colloidal tin from the activator bath. Remnants from this bath can neutralize the electroless copper plating bath, and therefore rinsing is again very important at this stage.

Electroless Bath: The copper is plated on the drilled holes and the substrate surface in the electroless copper bath. This can be done as a subtractive or an additive process. The *subtractive process* involves plating a substrate with conductive copper and then exposing it to electrolytic plating. Then, the substrates circuit pattern can be created by masking and etching.

In the *additive process*, the substrate is exposed to a copper process only once. The additive process metalizes the substrates through holes and the circuitry pattern at the same time, thereby eliminating the need for electroplating. With the industry pushing for high-density PCBs, the use of additive plating is on the rise. This is because the additive process can produce the finer lines demanded by surface mount technology as compared to the boards produced by using the standard etching process. Another reason why additive plating is becoming more popular is its ability to metallize high aspect ratio holes uniformly.

Electroless copper plating baths are of two categories: (i) Room temperature, and (ii) Hot solutions.

A typical composition for a room temperature bath is as follows:

- Copper sulphate   5 to 10 g/l
- Sodium potassium tartrate   40 to 50 g/l
- Formaldehyde (37per cent)   10 to 20 ml/l
- Sodium hydroxide   10 to 15 gl/l
- Sodium carbonate   5 g/l
- Potassium cyanide   1 to 10 ppm
- Wetting agent   0.5 ml/l

The following are the typical operating parameters of the room temperature bath:

- Temperature   15 to 30 °C
- pH   11 to 13
- Deposition/hour   1 to 1.6 micron

These operating conditions help to attain the deposition of a thin copper layer, sufficient for subsequent electrolytic plating, in about 15-30 minutes.

The room temperature bath is not suitable for the additive process because its deposition rate is too slow and the loading volume is 1 to 1.2 dm^2/hour.

Hot solutions are based on EDTA stabilizer, which is a copper complexing or chelating agent. It thus stabilizes the bath even at temperatures well above the ambient temperature. The following is a typical formulation for hot electroless plating solution:

- $CuSO_4 - 5H_2O$ 5 to 10 g/l
- EDTA [ethylene diamine tetra acetate]   5 to 25 g/l
- Formaldehyde   10 to 20 ml/l
- Sodium hydroxide   10 to 20 g/l
- Copper cyanide   1 to 10 ppm
- Stabilizer [dimethyl poly siloxane]   200 ppm
- Wetting agent   0.5 ml/l

The typical working conditions for hot temperature baths are:

- Temperature   35 to 70 °C
- pH   11 to 12
- Deposition/hour   3 to 8 microns

In this process, the plating rate increases both with temperature and with the copper and formaldehyde content, whereas EDTA and copper cyanide have the opposite effect. High-speed baths are more sensitive to contaminants and are less stable. So, they require more frequent control and adjustment of their composition.

Hot baths are used mostly for the additive process. A 25 micron (1mil) conductor can usually be built-up in 6-12 hours.

As the bath is repeatedly used, the chemical parameters change with time. This changes the dynamics of the copper. The chemical parameters of the bath must be closely monitored because if they are out of balance, the plating quality will suffer. This can result in insufficient coverage of copper (voids), cracking of the plating material or peeling of the plating material away from the hole wall.

Baths are inspected and controlled by chemical analysis. The sodium hydroxide and formaldehyde are analysed and replenished. The bath control and replenishment procedure varies from manufacturer to manufacturer.

Acid Rinse: After the electroless copper bath, the board is rinsed in a sulphuric or phosphoric acid to neutralize any caustic solution that may remain on the panel. The board is then placed in an anti-tarnish bath in order to prevent the newly deposited copper from oxidizing. Next, the electrolessly plated substrates can be scrubbed and subjected to a copper flash plate process.

Successful electroless plating is determined by the thickness of copper and how well it adheres to the hole walls and substrate surfaces. The most important parameters which influence the plating quality are chemistry, bath agitation and temperature. Electroless copper plating is vital to the performance of a multi-layer printed circuit board. Regardless of whether the plating is done subtractively or additively, tight process control must be maintained to guarantee uniform plating thickness.

Automated plating lines are necessary for large volume production to ensure the integrity of the printed circuit boards. In these machines, it is possible to maintain uniformity by continuously monitoring and controlling parameters such as temperature, solution concentrations and the activity of the bath. In smaller operations, where high throughput is not required, manual plating lines can be used.

8.3.2.2 Electroless Nickel Plating

Just like in electroless copper plating, nickel is also deposited on the substrate by the electroless method. The temperature of the bath is high and can go upto 90 °C while the pH range will be from 8 to 10. Sodium hypophosphite is used as a reducing agent and ammonium hydroxide is added to adjust the pH. The typical bath composition is as given below:

Alkaline Nickel Electroless Bath [Typical Values]

- Nickel chloride   30 g/l
- Sodium hypophosphite   10 g/l
- Sodium citrate   100 g/l
- Ammonium chloride   50 g/l

- pH   8 to 10
- Temperature   90 +1 °C

Electroless nickel plating is used in the additive process.

8.3.2.3 Electroless Gold Plating

Electroless gold plating is better than immersion gold plating. As in electroless nickel plating, sodium hypophosphite is used as a reducing agent and sodium citrate as a complexing agent. The temperature of the bath is 90 to 95 °C with a pH range of 7 to 8. The typical bath composition is given below:

- Potassium gold cyanide   2 g/l
- Sodium citrate   50 g/l
- Ammonium chloride   75 g/l
- Sodium hypophosphite   10 g/l
- pH   7 to 7.5
- Temperature   90–95 °C

This process is used to plate the edge connector section of the PCB on a fully additive process.

Gold is an expensive alternative to other surface plating methods. While it has significant advantages in many applications, there are certain limitations to its overall usefulness. Not the least of gold's drawbacks is that if not properly controlled, gold metal in the solder joint can lead to embrittlement and early failures. Therefore, when gold is specified, careful consideration must be given to all trade-offs. Banks (1995) points out that when properly controlled, gold plating on printed circuit board pads offers highly effective solderability protection and is not detrimental to solder joints.

8.3.3 Electroplating

8.3.3.1 Copper Electroplating

In printed circuit boards, copper is electroplated to increase the thickness of the copper layer on the surface and drilled hole walls. The plating thickness on the surface is 70 μm and on through-hole it is 30 μm. Copper plating is mainly done to build plated through-holes in double-sided or multi-layer boards. Copper has many properties which make it suitable for widespread use in PCBs. However, the boards are sometimes subject to harsh environmental conditions during soldering and assembly, and therefore, must have good adhesion and thermal expansion properties. Several formulations are commercially available. However, in the preferred industrial process, an acid copper sulphate solution, containing copper sulphate, sulphuric acid, chloride ion and organic additives are used.

The important functions of the bath ingredients are:

- Copper sulphate   :   Gives copper metal ions
- Sulphuric acid   :   Improve the ionic movement or increases the conductivity of the solution

- Hydrochloric acid : Improves the uniform dissolution of the anode
- Additives : Increases the cathode wetting and gives wetting agents the uniform copper deposition on the cathode

Typical bath composition for acid copper plating is as follows:

- Copper sulphate ($CuSO_4$) 75 – 100 gm/l
- Sulphuric acid (H_2SO_4) 90 – 110 ml/l
- Hydrochloric acid 30 – 70 ppm
- Additives and wetting 4 – 8 ml/l Agents

The most common additives used are gelatin and potassium tartarate.

The optimum operating conditions for uniform plating are as given below:

- Temperature 25–30 °C
- Anode current density 150–300 A/m^2
- Cathode current density 300–600 A/m^2
- Agitation Vigorous oil-free air
  Cathode movement
  Continuous solution filtration
- Anode Phosphorus de-oxidized copper [0.03 per cent phosphorus]
- Anode Bags Polypropylene
- Hooks Stainless steel [316 grade or titanium]
- Anode to Cathode ratio 2:1 (minimum)
- Distance between anode and cathode 15 cm [maximum]
- Plating rate 1 micron in one minute
- Chloride content in the bath 40 ppm

The plating is generally about 25 μm copper in forty minutes at 300 A/m^2 to carry out copper electroplating. Panels are racked and immersed vertically in the electrolyte bath. The racks are connected to the cathode bar of the cell (-ve electrode of a current generator). Anode is usually a vertical copper bar.

Cupric ions in the solution are attracted towards the cathode, where they acquire two electrons and are deposited as metallic copper:

$$Cu^{2+} + 2e \longrightarrow Cu.$$

The anode dissolves in the solution according to the following reaction:

$$Cu \longrightarrow Cu^{2+} + 2e$$

Both reactions occur simultaneously. They are caused by the current which flows through the bath from anode to cathode and vice versa outside the bath.

Electrodeposited copper must satisfy strict requirements regarding ductility, tensile strength, fatigue resistance, low electrical resistivity, low porosity and strong adhesion to electroless copper. Most of these properties would depend upon both the composition of the bath and its operating conditions. In practice, it would be difficult to test the copper layer. The quality of plating is thus often judged from its appearance and on a microstructure examination as methods of indirect quality control.

One parameter, which is related to the thickness of the coating, is known as "Throwing Power". It is defined as the ratio of the thickness plated halfway into the holes to the average thickness plated on the panel surface. A bath will have a *throwing power* equal to 1, when it plates an even coating over the entire platable area. As the drilled holes are perpendicular to the board, a high *throwing power* bath enables adequate depreciation inside the hole without excessive plating on the flat board surface.

The PCBs are copper electroplated at different stages, which are:

- Panel Plating, and
- Pattern Plating.

The flow chart for the copper plating process is shown in Figure 8.4.

Panel Plating Process — This involves:

- Electroless copper plating;
- Inspection;
- Acid dip-10% H_2SO_4
- Water rinse (de-mineralized water);
- Copper electroplating;
- Water rinse;
- Drying; and
- Resist Coating [Screen Resist and Dry Film Resist].

Pattern Plating Process — This involves:

- Electroless copper or copper electroplating;
- Resist coating [exposing and developing];
- Water rinse using spray;
- Acid dip;
- Water dip [de-mineralized water];
- Copper electroplating;
- Water rinse;
- Acid dip;
- Water rinse [de-mineralized water];
- Resist metal coating;
- Water rinse;
- Stripping and etching.

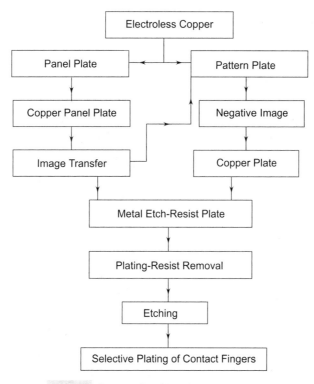

Fig. 8.4 Process flow chart for copper plating

8.3.3.2 Tin Electroplating

Pure tin is widely used for plating printed circuit boards because of its advantages of having good solderability, easy plating control, corrosion resistance and metal-resist properties. It is also recommended that tin must be fused on component leads.

Although a variety of processes are available for tin plating, the process using acid tin sulphate type electrolytes is the most widely used system. Tin is plated from acid sulphate type electrolytes similar to copper acid electroplating.

The bath ingredients for tin plating have the following functions:

- Stannous sulphate   :   Gives the tin metal ion
- Sulphuric acid   :   Medium for ionic movement
- Additives   :   Increase the levelling of deposition and increase the brightness

The bath compositions of various solution constituents are:

- Stannous sulphate ($SnSO_4$)   25 to 30 g/l
- Sulphuric acid (H_2SO_4)   100 ml/l
- Additives

- Phenol sulphonic acid   30 g/l
- Gelatin   28 g/l
- B-napthal   1 g/l

The typical operating conditions are:

Temperature	20 to 30 °C
Anode current density	0.5 to 1.5 A/dm^2
Cathode current density	1.5 to 3.0 A/dm^2
Voltage	1 to 3 volts
Agitation	Cathode rod movement, Air agitation not recommended
Filtration	Continuous
Anode to cathode ratio	2:1
Anode bags	Polypropylene (3 to 10 micron)
Plating rate	1 micron/minute at 2 A/dm^2
Anode	99.99% pure tin bars
Length	Rack length minus 5 cm
Hooks	Titanium or Stainless steel (316 grade)

Tin has a uniform lustrous finish and the plated surface should be smooth to the touch. The drill deposits are due to out-of-balance solution constituents and operating conditions. Control of additives levels and plating quality can be tested with Hull Cell.

8.3.3.3   Tin-Lead Alloy Plating or Solder Plating

Tin-lead alloy in the ratio of 63:37 is widely used as a finish plate for printed circuits. It is considered as a standard etch resist for professional PTH boards because it features excellent etch resistance to alkaline ammonia, excellent initial solderability and, large shelf life for the boards, and is relatively cheap.

The properties of lead tin alloy coating which play an important part are: (i) the thickness and uniformity of the coating, (ii) its porosity, and (iii) solderability. For all these factors, the composition of the deposited alloy plays an important role. It is possible to adjust, within certain limits, the tin-lead ratio in the deposit by adjusting the tin-lead ratio in the solution, adding specific additives and adjusting the current density.

Most specifications require the percentage of tin to be between 50 to 70 per cent, though with modern plating baths, a narrower range like 60 to 70 per cent of tin can be deposited on all parts of the board, including holes. Tin-lead alloy with 63 per cent tin and 37 per cent lead is the eutectic alloy, which melts at a temperature lower than the melting point of either tin or lead and this makes it easy to re-flow and solder. The thickness of tin-lead should be minimum 8 microns for resist etching. However, it is common practice to deposit 13 to 25 microns on the surface side.

Bath ingredients: Nearly all tin-lead alloy electrolytes currently available include the high-concentration fluoro boric acid-peptone system. The basic functions of these components are:

a) Tin fluoroborate and lead fluoroborate: give metallic ions.

b) Free fluoroboric acid: increases the bath conductivity and inhibits the decomposition of fluoroborate; produces a deposit with a linear grain structure.

c) Boric acid: increases the conductivity and inhibits the decomposition of lead fluoroborate into lead fluoride.

d) Peptone: inhibits the formation of dendrites or trees.

e) Additive agents: increase the grain deposit of tin.

f) Anti-oxidants: prevent oxidation of tin from stannous to stannic.

The typical bath composition is:

Tin fluoroborate equivalent to 15 to 20 g/l of metallic tin.

Lead fluoroborate equivalent to 9 to 10 g/l of metallic lead

Free fluoroboric acid	25 g/l
Boric acid	400 g/l
Additive and peptone	3 to 5 g/l

The optimum operating conditions are:

• Temperature	20 to 30 °C
• pH	0.5 or less
• Cathode current density	1.5 to 2 A/dm^2
• Anode to cathode ratio	1:2
• Agitation	Slow movement of the panels
• Anode	60 per cent tin and 40 per cent lead
• Anode length	minus 5 cm.

Copper is the most serious contaminant in the bath. It causes darkness at low current densities, specially in the plated through-holes, and may even coat the anode. The maximum levels of allowable copper are 15 ppm.

The overall plating can be checked by using Hull Cell. This test also shows the need for peptone, additives or carbon treatment, as well as the presence of dissolved copper in the solution.

Hot Air Levelling (HAL): The bare copper on the printed circuit board generally tends to deteriorate with time resulting in an inferior appearance and making soldering difficult. As such, bare copper boards are acceptable only for experimental and prototyping work. For high quality products, some type of coating is applied to prevent the occurrence of these problems. The most common type of coating is the tinning of the copper tracks. This is done either by roller tinning or electroplating. A modified process of roller coating is the hot air levelling process.

Hot air levelling is a process of applying a thin coating of eutectic tin-lead on the exposed portion on a printed circuit board. This deposition increases the solderability of the board during assembly operations. Solder has many advantages over any other conventional organic coatings. These solder coatings have very good shelf life, short solder wetting time during assembly and very high mechanical stability. The formation of an inter-metallic bond even before the PCB assembly is a unique characteristic of HAL. Due to the presence of solder on copper, the PCB can withstand multiple soldering and de-soldering operations during the assembly cycle, with no adverse effect on the solderability of the board. HAL also offers many advantages in the field of SMD and fine pitched quad packs.

The procedure involves dipping a board in pure molten solder and blowing off the excess air between air knives. The steps for hot air levelling are shown in Figure 8.5. The boards are dipped vertically into a solder bath. They are then subjected to hot air blasts as they are withdrawn from the bath. This process removes all excess solder, and clears vias and holes leaving a high quality flat surface.

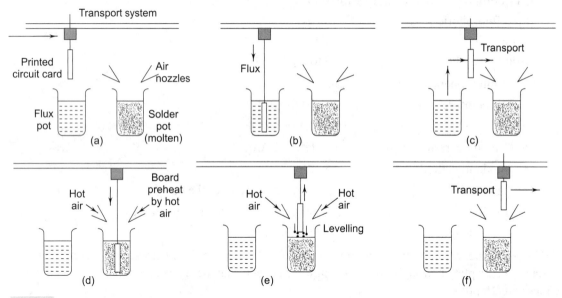

Fig. 8.5 Process flow chart for hot air levelling (a) basic set up (b) printed circuit dipped in flux (c) printed circuit taken out from the flux bath (d) board pre-heated with hot air (e) board dipped in solder with hot air levelling (f) excess solder removed with air knives (redrawn after Haskard, 199)

The quality of solder deposition basically depends upon the cleanliness of the surface. Pre-cleaning is therefore necessary to remove oily deposits, fingerprints and other organic contaminants that may be present on the copper surface on the through-holes. The steps involved in hot air levelling are as shown in Figure 8.6.

Pre-cleaning is generally done by using a mild acid or using persulphate, followed by *rinsing* with clean water. *Fluxing* is done either externally or internally in the HAL machine itself, which has a heated chamber. The flux generally used is water soluble so that it is easy to remove it after solder coating. The board is then transferred to a solder pot in which solder is maintained at a temperature of 250 °C to 270 °C where the soldering action takes place. The solder coating is then *hot air levelled* by hot air knives whose angles can be fixed depending upon the density and geometrical distribution of the circuit pattern on the board. The boards are then allowed to cool followed by cleaning with a slightly warm detergent water which removes the excessive flux, that is left over the board.

In most of the cleaning operations, it will be sufficient to use a mild acid to clean the surface. Hot cleaners should not be used because it will lead to lifting of the edge connector tapes and cause the solder to be deposited on the edge connector gold plating. Persulphates are generally used with other chemicals when the copper surfaces are extremely dirty. The board should then be thoroughly rinsed with clean water.

For hot air levelling, single-unit and high-production conveyorized machines are available. Functionally, the machine consists of the fluxing unit, the solder pot and the hot air knives. The board is initially fixed to the clamps which are operated through a pneumatic foot pedal. The boards are immersed in the fluxing unit for about 15 seconds

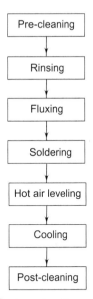

Fig. 8.6 Hot air levelling system

where it is dipped into the hot-solder pot (maintained at about 250 °C) for about 5-8 seconds only. After the solder coating, the boards are withdrawn from the solder pot and passed through a set of hot air knives, whose air pressure and angle are adjusted depending on the geometry of the board. Through the adjustment of the hot air knives, the solder inside PTH as well as on the surface of the board get levelled. HAL machines are generally built with stainless steel. All the pneumatic clamps and heaters are controlled by Programmed Logic Controllers (PLCs).

All multi-layers should be pre-heated before coating, at 120 °C (250 °F) for one to two hours in order to ensure the total removal of entrapped moisture and gases, and to ensure that the PTH wall does not rupture during the HAL operation.

8.3.3.4 Nickel Electroplating

Nickel plating is mostly used as an undercoat for precious metals like gold to act as a barrier layer to prevent copper migration as well as to increase the wear resistance of the coating. Accordingly, it reduces the minimum gold deposit needed for a given application.

Normally for nickel plating, nickel sulphamate baths are used in conjunction with wetting agents. The normal coating thickness of nickel is 4-6 μm. Nickel sulphamate is used both as an undercoat for through-hole plating and on tips.

The composition of a typical nickel plating bath are:

- Nickel sulphamate   350 ml/l to 750 ml/l
- Nickel chloride   15 g/l

- Boric acid   30 g/l
- Additives   As required

The optimum operating conditions are:

- Temperature   45 – 60 °C
- pH   3.5 – 4.5
- Cathode current density $2.5 - 10$ A/dm^2
- Anode   Rolled de-polarized bars or pallets
- Hooks   Titanium
- Bags   Polypropylene
- Length   Rack length minus 5 cm
- Agitation   Solution circulation through filter pump or cathode movement
- Deposition rate   4 micron plating thickness during 10 to 15 min when plating current is 25 A/ft^2

One important parameter related to nickel plating is "stress". This refers to the cause of deposit cracking. While low nickel chloride causes poor anode corrosion (rapid fall in nickel content), high chloride causes excess stress. In order to prevent this, pH, current density and boric acid need to be maintained within specified limits. The operating condition of the bath and the presence of contaminants can be checked by the Hull Cell test.

8.3.3.5 Gold Electroplating

Gold by nature is soft and in order to provide the desired brightness and increased hardness, it is always alloyed with minute quantities of nickel, cobalt or indium. It has good electrical conductivity, good etch resistance and tarnish resistant characteristics. Although early printed circuit board technology used gold extensively, due to soldering problems and high cost, it is now not fully plated on the PCBs. Gold is plated only on the edge connector area and selected areas of PCBs, as it provides low contact resistance and high wear resistance.

Normally, gold plating for the edge connector area depends upon the customer requirement. Mostly 5 micron gold is plated with an undercoat of 4 micron nickel for the edge connecting. Automatic plating machines are used because of the enhanced thickness control, efficient gold usage, productivity and quality. For process control, gold content, pH and density are maintained at optimum values. For control of pH, potassium hydroxide is used to raise it, while for lowering it, acid salts are used. Solution conductivity is controlled by density, while can be adjusted with conductivity cells.

Gold is electrodeposited from alkaline, weak alkaline, neutral or acid bath for contacts and high frequency conductor coatings. A large variety of baths are available to plate both the pure 24 carat metal and gold alloyed with varying amounts of metals such as nickel, cobalt, indium, etc. Consequently, it is possible to tailor the gold deposit to the needs of the PCB designer. Gold plating electrolytes are classified under four categories: (i) acid gold electrolytes, (ii) neutral gold electrolytes, (iii) alkaline gold electrolytes, and (iv) sulphate gold electrolytes.

Acid Gold Electrolytes: They have pH ranging from 3.5 to 4.5 and contain nickel or cobalt. They provide coatings useful for PCB contact tabs, much in the same way as for edge connectors. The thickness of the coating is limited to about 5 microns. However, normal plating thickness on PCB conductors including contacts is 1.25–2.5 microns. Price (1992) explains that a nickel/gold finish on fine-line surface mount PCBs provides the board assembler with significant benefits.

Neutral Gold Electrolytes: Neutral (pH 6.00 to 8.5) gold plating baths facilitate deposition of 99.9 per cent pure coating, when operated at temperatures between 35-45 °C. This type of coating is most suitable for edge connectors.

Alkaline Gold Electrolytes: With pH in the 8.5-13.8 range, these are not very popular in PCB manufacture, because they have a detrimental effect upon copper clad laminates.

Sulphate Gold Electrolytes: They facilitate deposition of gold with a purity of 98.0-99.9 per cent with exceptionally even distribution over the entire panel. This results in saving of gold by as much as 20-40 per cent in sulfite bath as compared to the conventional acid bath. However, their drawback lies in their extreme sensitivity to pH variations: for example, if the pH falls below 8, the gold is quickly reduced.

The acid bath ingredients used are:

- Gold potassium cyanide   6 to 10 gm/l
- Cobalt   0.15 to 0.25 g/l
- Potassium citrate   To be added to electrolyte to maintain the density at 10 to 15Be

The operating conditions for the acid bath are:

- Temperature $30 - 40$ °C
- pH   3.8 to 4.5
- Cathode current   0.3 to 1.5 A/dm^2
- Anode   Platinized Titanium
- Hooks   Platinized Titanium
- Agitation   Oil-free air agitation and cathode movement
- Deposit composition   99.8 or 99.7 per cent gold
       0.2 to 0.3 per cent cobalt
- Deposition rate   5 micron thickness within 13 mts at current density of 1.0A/dm^2

Gold plating is now used only in those applications where no alternative is possible.

The gold plating process in PCB manufacturing has been plagued by excessive loss of goldplating scrap caused by photo-resist breakdown. Anderson (1998) suggests the application of Design of Experiments (DOE) techniques to gold plating line which helps to identify the responsible factors and offer solutions to overcome the problem.

8.4 General Problems in Plating

If the process parameters are not properly maintained, it will result in poor quality plating. The problems that may appear after plating are:

a) Poor throwing power;

b) Poor solderability;

c) Improper anode corrosion;

d) Surface defects [tents, pits, voids, pinholes and inclusions];

e) Bath life; and

f) Dull deposits.

It is therefore, necessary to maintain the process parameters in order to minimize variations in bath composition, temperature and contaminants.

8.5 General Plating Defects

8.5.1 Voids

If the holes are not drilled and cleaned properly during electroless plating, gases or organic chemicals are entrapped or accumulated in the hole walls. When the boards are wave soldered or pre-heated, the entrapped chemicals or gases escape with high pressure and cause discontinuity or cracks on the hole walls. These are called voids. Figure 8.7 shows a PTH vertical cross-section illustrating typical copper voids.

8.5.2 Blow Holes

During plating and copper clad manufacturing, some air can get entrapped between the copper layers or between copper and base material. This air comes out when the board is subjected to high temperature such as in the soldering process. This results in cracks in the plating pattern, which gives rise to blow holes. Small voids also give blow holes.

8.5.3 Outgassing

Pinholes nodules or small 'volcano eruptions' are formed during tin-lead surface fusing. This is shown in Figure 8.8. This is due to co-deposition of organic additives during electroplating of tin-lead. This is corrected by batch carbon treatment

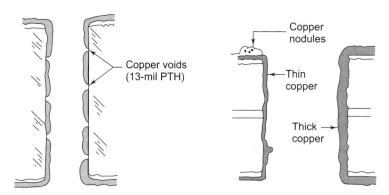

Fig. 8.7 PTH vertical cross-section illustrating copper voids

Fig. 8.8 PTH vertical cross-section illustrating uneven, thick-thin copper plating. Nodules due to particle contamination are also shown

 ## 8.6 Special Plating Techniques

8.6.1 Through-hole Plating

There are various methods for establishing an electroplating-receptive surface on the walls of the hole drilled through the substrate. Known in the industry as holewall-activation, this procedure in the production of commercial printed circuits requires multiple process tanks, each with its own control and maintenance requirement. This procedure is needed as a result of the drilling process. As the bit cuts its way through the copper foil and underlying substrate, the heat generated melts the resin that forms the 'matrix' of the insulating composite in most of the laminates. The melted resin is carried up the hole with the rest of the drilling debris, where it is smeared onto the freshly exposed hole walls in the foil. This effectively spoils the surface for subsequent electroplating. The melting of the resin also leaves behind a heat-glazed hole through the substrate that exhibits poor adhesion to most activation agents. This requires the development of such techniques as de-smearing and etch-back chemistries.

A method that is more suitable to the needs of PCB prototyping involves the use of low viscosity ink that is specially formulated to form a high adhesion, high conductivity coating on the inside wall of each through hole. Instead of multiple chemical treatments, a single application step, followed by a thermal cure, will result in a continuous film on the inside of all the hole walls that can be directly electroplated without further processing. The ink is based on a resin that is essentially a very aggressive adhesive and adheres without difficulty to the most heat-polished holewall, so etch-back is eliminated.

8.6.2 Reel-to-Reel Selective Plating

The leads and pins of the electronic components such as connectors, ICs, transistors and flexible circuits are selectively electroplated to get good contact resistance and corrosion resistance. The plating can be done manually or automatically. Selective plating individually on the pins is very expensive. So, the mass plating is mandatory.

Normally, metal foil rolls of required thickness are punched on both edges, cleaned chemically or mechanically, and are continuously plated (Figure 8.9) selectively with metal like nickel, gold, silver, rhodium, palladium or alloys like tin-nickel, copper-nickel and nickel-chromium.

In this method of selective electroplating, the plated metal foil is first resist-coated on the unwanted portion and the foil is electroplated on the selective portion.

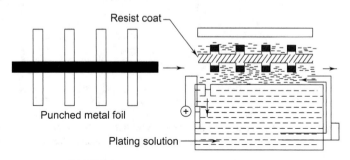

Fig. 8.9　Reel-to-reel selective plating process

8.6.3　Brush Plating

Another type of selective plating is called 'brush plating' as illustrated in Figure 8.10. It is an electro-deposition technique in which all the parts are not immersed in the electrolyte solution.

In this technique, only a limited area is plated, without affecting the rest of the part. Normally, precious metals are plated on the selective parts of the PCBs such as edge connector areas. Brush plating is used more often in electronic assembly shops for repairing rejected boards.

A special anode (inert anode-like graphite), wrapped in an absorbent material (cotton swab), is used to take the plating solution to the point where it is needed.

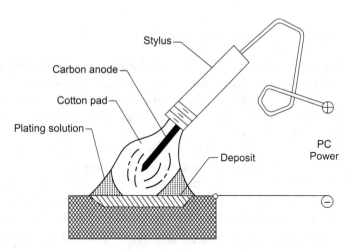

Fig. 8.10　Brush plating procedure

8.6.4　Finger Plating

Precious metals are electroplated on edge connector, tabs, or fingers to provide low contact resistance as well as high wear resistance. This technique is called finger plating or tab plating.

Gold is electroplated with an undercoat of nickel on edge connector tabs. The process is as follows:

a) Stripping;   Removal of tin or tin-lead coating on the tabs.

b) Rinsing;   Water rinsing

c) Scrubbing;   Scrub with abrasive

d) Activation;   Dip in 10 per cent H_2SO_4

e) Nickel electroplating on the tabs;   (4-5 micro thickness)

f) Rinsing;   Demineralised water

g) Gold strike;   Solution treatment

h) Gold electroplating;

i) Rinsing; and

j) Drying.

The fingers or tabs are electroplated (Figure 8.11) by manual or automatic techniques. Gold electroplating is now replaced by rhodium, platinum and palladium electroplating on contact tabs or fingers.

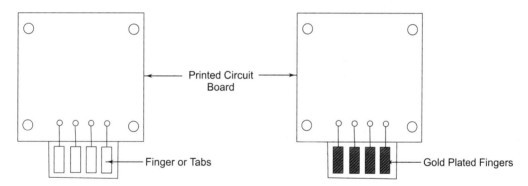

Fig. 8.11 Finger plating technique

8.6.5 Conductor Metal Paste Coating

Conductor metal pastes are generally based on precious metal powder, which are suspended in an organic carrier fluid. Silver, gold, palladium, palladium-silver, palladium-gold or platinum-gold are used for this method.

These types of pastes are used in pattern printing on non-conductors like ceramic substrates, by using the screen printing method. The printed mixture is dried and finally oven-fired at the appropriate temperature.

8.6.6 Reduction Silver Spraying

Reduction silver spraying generally provides the conductive pattern undercoat for subsequent coating by other electroplating techniques on non-conductors. The process is carried out by the following steps:

- The cleaned surface of dielectric material is sensitized with stannous chloride solution;
- The mixed silver solution which is a mixture of silver nitrate, distilled water, ammonia and sodium hydroxide is sprayed through nozzles on the sensitized surfaces
- The reduction solution, which is a mixture of water, d-glucose and nitric acid, is then sprayed.
- This results in a uniform and brilliant silver layer on the surface.

This process is mostly used in flexible additive PCBs.

8.7 Metal Distribution and Plating Thickness

Metal distribution and plating thickness on the PCBs depend upon the process controls of an electrolytic bath. This is done through three methods:

a) Analysis of solution or wet chemical analysis;

b) Physical test on solution; and

c) Testing of electrodeposits.

8.7.1 Analysis of Solution (Wet Chemical Analysis)

The wet chemical analysis is carried out to determine the following in the electrolyte:

- concentration of metal ions;
- concentration of acid;
- concentration of base;
- concentration of chlorides; and
- metallic impurities.

For carrying out traditional wet chemical methods for metals and non-metal plating solution constituents, it is advisable to study the suppliers' literature. These methods mostly make use of pH meters, specific ion electrodes, colorimeters, spectrophotometers, etc. The measurement of organic additives in tin, gold and nickel solutions is done by using various techniques which include liquid chromatography, uv/vis (ultraviolet/visible) spectrophotometer, ion-chromatography, polarography and voltametry. The details of these methods can be found in standard texts on instrumental methods in chemical analysis. The details of these methods are, however, outside the scope of this book.

8.7.2 Physical Tests for Solutions

The following parameters are physically tested in electrolytes:

- Density;
- pH value;
- Surface tension; and
- Hull cell test.

Density: The density of an electrolyte depends upon the salt content in the solution. It can be measured by using an instrument called hydrometer. The density measurement gives an accurate value of the metal content and the solution must be maintained within the recommended value. Density is normally expressed in Twaddell or Baume units. These are derived from the specific gravity. If the density scale is given in Baume gravity, the following formula can be used to calculate the specific gravity:

$$\delta = \frac{145}{145 - °\,Be'}$$

where

δ = specific gravity (kg/dm^3)

$°Be'$ = degree Baume' gravity

Table 8.2 gives representative values of specific gravity, Baume and Twaddel.

Table 8.2 Conversion Table of Specific Gravity, Baume and Twaddel

Specific Gravity	Baume	Twaddel
1.000	0	0
1.100	13.0	20
1.200	24.0	40.0
1.300	33.3	60.0
1.400	41.2	80.0
1.500	48.1	100.0

pH Value: The pH value determines the acidity or the alkanity of an electrolyte. For measuring pH in electroplating solutions, normally an electrometric method using pH meter is used. For high accuracy measurements, a digital pH meter is preferred.

Surface Tension: In order to eliminate gas pitting and to improve deposition, surface active agents (surfactants) are added to the electroplating solution, for example, in baths for acid copper electroplating, non-foaming surfactants are added to minimize the foaming during air agitation. It is necessary to measure the surface tension to the solution for copper maintenance of solutions in process control. Usually, torsion balance is used to measure the surface tension of the solution. This instrument can be used to measure the force required to detach a glass plate of standard area from the surface of the solution under test. The force is indicated by a calibrated dial.

Hull Cell Test: A simple technique to infer when the additive level needs attention and to determine just how much material needs to be added uses a miniature plating cell commonly known as the Hull Cell. The arrangement for Hull Cell test is shown in Figure 8.12. The cell is used to test plate a series of sample boards to determine when the bath needs adjustment and to determine how much of it to add.

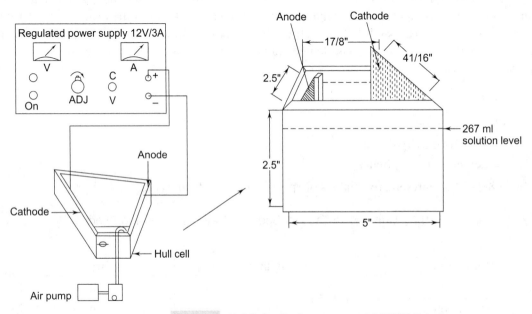

Fig. 8.12   Hull Cell — testing arrangement (NTTF Notes)

The Hull Cell is intended to act as a quick check on the health of the tin/lead plating bath. Using the cell in conjunction with the chemical analyses, it is possible to qualitatively and quantitatively analyse all the major constituents of the bath.

When filled to the line marked on the side of the cell, the volume of the test sample is 267 ml. If we denote:

- V = Volume of main plating tank (litres)
- H = Amount of addition agent added to Hull Cell to produce acceptable test plate (millilitres)
- C = Amount of addition agent needed by main tank (millilitres)

The multiplication factor that relates that you add to the Cell and what you will need to add to your plating bath is given by:

$$C = (V/0.267) \times H$$

When run at specified operating conditions in terms of circuit, time and temperature, the Hull test gives indications of pH control, contaminants and the overall bath conditions. It is a valuable aid in solution and process adjustment. It is particularly useful to determine the catalyst ratio (additive ratio) and impurity level in electroplating solution. Therefore, Hull Cell has been the most popular and is the most important tool in the control, and testing of electroplating and related solutions. The apparatus required is simple and inexpensive and the test takes only a few minutes to determine the quality of the electrolyte. However, the main disadvantage of Hull Cell testing is that defects in copper plating such as dull plating, roughness or pitting, etc. are not shown by this test.

8.7.3 Testing of Electrodeposits

If the quality of plating on circuit boards is not satisfactory, it can give rise to problems like corrosion, current-carrying constraints, adhesion, contact resistance and wear resistance. The quality of plating is usually determined by testing the following parameters:

- Metal thickness test [surface and through-holes];
- Porosity test;
- Adhesion test; and
- Solderability test.

8.7.3.1 Metal Thickness Test

The plating thickness on the circuit pattern and in the through-hole is measured by two methods:

  a) Non-destructive method — Beta back scattering method; and
  b) Destructive methods — micro-section or cross-section method.

Non-destructive Method: The most commonly used non-destructive method for testing PCB finishes is the β-ray backscatter method. This method is particularly useful for measuring the plating thickness on the pattern and plated through-hole during or after the plating and before the etching operation. The method makes use of a radio isotope emitting β-rays, which is mounted on a suitable probe. The probe comprises both the beta-ray source and a receiver. The probe is placed on the surface whose thickness is to be measured. A part of the β-rays which impinge on the surface get reflected. The amount of reflected rays decrease in number as the thickness of the coating increases. With suitable calibration with a standard, a β-ray electronic counter directly gives the thickness of the coating in microns. The instrument gives the average thickness. It is mainly used to measure gold, tin, tin-lead on copper and copper on epoxy and photo-resist on copper. Several β-emitting particles and interchangeable probe operations are necessary to cover a wide range of thicknesses. The technique is a quick, accurate and non-destructive method of measuring thickness. Therefore, it is very popular in the field of quality control of PCB manufacture, as the test can be conducted even by an unskilled operator.

Destructive Methods: The most popular destructive technique which gives a direct measurement of the thickness of the deposit is micro-sectioning, which involves the preparation of a metallographic specimen and examining it under the microscope.

The plated PCB is cut vertically or horizontally and seen under a microscope to observe the visible layer structure and is photographed. The variation of plating thickness on the pattern and the hole walls can thus be determined. Normally, the boards are cross-cut horizontally by using a diamond circular saw and polished with diamond paste or emery paper with rotating wheel polisher. The dried samples are put under a microscope at 30 to 1000 times magnification, examined and photographed at three locations on each plated through-hole wall. The results are generally reported as an average.

The micro-sectioning technique, when applied to thin coatings, may introduce a large error. For example, for a 1-micron deposit, the error may be as high as 50 per cent. However, for thick coatings, the error is considerably less. The error is 2 per cent for a 5-micron deposit.

This technique has an advantage in that it can be applied to any geometry, including hole walls. Additional information on the coatings such as number and nature, uniformity, presence of voids and undercut of conductors, etc., can also be easily obtained. However, the method is time-consuming and requires a skilled operator.

8.7.3.2 Porosity Test

The porosity test is used to detect discontinuities such as pores and cracks in deposits in the plated surface. The test is important especially on contact tabs, as corrosion of the base metal through the pores may have a detrimental effect on any electrical contact. Therefore, the pores and cracks are mainly tested on precious metal deposits.

Porosity on the deposits can be tested by various techniques. However, the electrographic test and some gaseous reagent porosity tests are more common. The *gas test* entails exposing the plated PCB to a twenty-four hours exposure to sulphur dioxide. The specimen is kept in a closed vessel of 10 litre capacity to which 0.5 cm^3 of water and 100 cm^3 gaseous sulphur dioxide are added. After twenty-four hours., the vessel is opened, 100 cm^3 of hydrogen sulphide are injected into it and the vessel is re-sealed. Porosity is indicated by corrosion of the base metal which may be observed either with the naked eye or with a microscope. The gaseous porosity test is mostly used for coating of gold, platinum metals and tin-nickel alloys on substrates of copper and its alloys.

8.7.3.3 Electrographic Test

This test entails using cadmium sulphide paper by interposing it wet between the circuit copper as an anode and an aluminium plate as cathode under pressure, and passing a certain current through the sandwich. The current flows from copper through pores in the plated metal through cadmium sulphide to the aluminium. The black spots formed on the cadmium sulphide paper indicate the presence of pores in plating, whereas pore-free plating does not show up such black spots. Nickel-plated PCBs are mostly tested with this method. Micro-cracks on the plated surface can also be found out by these methods.

8.7.3.4 Adhesion test

Adhesion represents bonding between the metals. When the adhesion of coating is poor, it could result in cracking or peeling which determine the quality of adhesion.

In order to carry out this test, a piece of adhesive tape is placed on the coated surface and pressed uniformly (without air bubble). The tape is then pulled from the edge with force. The metal surface is examined after the pull.

8.7.3.5 Solderability Test

Solderability is a measure of the ability of a surface to be completely wetted by molten solder. It is usually tested for finished PCBs. The test is performed by subjecting the fluxed specimen of usually one inch square to a contact with dross or oxide free solder at the recommended temperature (240 °C/s). The cleaned boards are observed through a 10-times magnifier for the following:

- *Wetting*: formation of uniform solder coating;
- *Dewetting*: irregular solder cover; and
- *Non-wetting*: irregular solder covered with exposed base metal.

The specimen must exhibit 95 per cent complete wetting.

 ## 8.8 Considerations for Shop Floor

It has been established from experience that most electroplating accidents are traceable to carelessness, inexperience, failure to follow instructions, poor engineering of equipment, inadequate layout of plating tanks and ventilation systems, bad housekeeping, improper shop flooring and drainage arrangement, confusion regarding chemicals and poor labelling of tanks and chemicals. Such neglect may result in explosion, fires and the release of toxic gases which are harmful to health.

The common types of hazardous chemicals used in an electroplating shop are:

- Hydrogen;
- Cyanides;
- Chlorinated solvents;
- Sulphuric acid; and
- Chromic acid;

Hydrogen: A hydrogen explosion occurs with electro-cleaners that produce foam. The foam blanket created by wetting agents sometimes traps the hydrogen and oxygen generated at electrical contact points. It can be prevented by keeping electrical contact points away from foam levels and proper exhaust arrangements.

Cyanides: When cyanide baths are used for electroplating, a separate drainage arrangement and exhaust system are recommended. Safety masks for the operators are necessary.

Chlorinated Solvents: These solvents are used in vapour/aqueous de-greasing units. These solvents can affect the brain when the operator or maintenance people inhale this vapour for long periods.

Alkali (caustic soda) vapours exposed to chlorinated materials can cause an explosion. So the degreasing section should have a proper exhaust and ventilation system and should be separated from the other process. A separate cleaning and rinsing tank is needed when chlorinated solvents are being used. Improper design and maintenance of de-greasing units is hazardous to life.

Acids: In electroplating shops, acids like sulphuric acids, hydrochloric acid and nitric acid are used for preparing electroplating solutions. When mixing these materials, protective clothing, aprons, gloves and chemical goggles should be used for safety. Otherwise any accident can harm the face and clothes.

Always add acid to the water slowly, stirring the solution as the acid is being poured. Water should not be added to the acid. This causes an immediate violent reaction which can splash up and result in severe burning of the exposed skin of operators.

The spilled acids and acid vapour also affect the operator and lead to a polluted atmosphere. Proper exhaust, ventilation systems, proper layout and acid-resist flooring are necessary in the plating shop.

8.8.1 Plating Shop Layout

The layout requirement for an electroplating shop depends upon the process, which is of three types:

- Automatic plating process;
- Semi-automatic plating process; and
- Manual plating process.

The above plating systems have the following sequences:

- Racking or wiring system;
- Pre-treatment including rinsing system;
- Electroplating system;
- Post-treatment including final rinsing and drying system; and
- Unloading system.

Regardless of whether the process is automatic, semi-automatic or manual, the arrangement of the above five can be in one unit or in different units. The overall system is connected with a central exhaust system.

The plating room floor should be water-proof and with acid/alkali resist coating (either an epoxy layer or a bitumen layer). During processing, spilling of some drag-out of plating solutions or cleaning solutions on the floor is unavoidable. If the floor is made of cement, it will be rapidly affected and needs to be protected with a protective coating.

Water used for plating and rinsing should be pure water like distilled water or de-mineralized water. All the exposed metal fittings in the room should be protected from the acid and alkali corrosion. For this purpose, the points used should be bitumastic paints "chlorinated rubber paints" vinyl paints or epoxy resin type paints. In the plating shop, the information chart on safety and first aid treatment kits should be displayed prominently.

The used water should be discharged directly from the floor or tanks and connected to drainage channels. The floor should be laid such that there is a slight natural slope towards drainage channels. The collected waste water should be treated in different treatment plants and the treated water must be neutralized upto 7-8 pH. The drainage must be connected to the open evaporator system as shown in Figure 8.13.

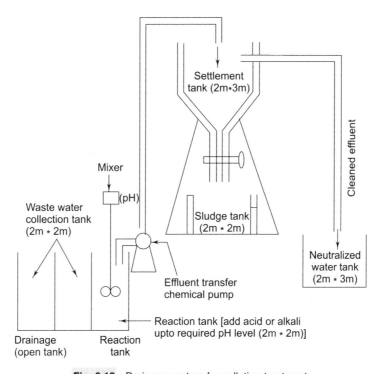

Fig. 8.13 Drainage system for pollution treatment

8.8.2 Equipment

Plating solutions are highly corrosive. Precautions should be taken to protect the walls and bottom floors of the tanks and filters. Solutions of plating tanks, heating elements, filters and filter aids depend upon the type of metal plating solution. The materials used for construction for electroplating equipment are given in Table 8.3.

Table 8.3   Materials Used for Construction of Electroplating Equipment

Bath	Tank Lining	Heaters	Filters	Filter Aids
Copper pyrophosphate	PVC, rubber polypropylene	Stainless steel	Stainless steel, epoxy	Cellulose fibre, asbestos fibre
Acid copper sulphate	PVC, rubber, polypropylene	Carbon	Hard rubber	Non-silicated filter aid
Acid gold	PVC, pyrex, quartz, Teflon	Quartz	PVC lined	Asbestos diatomaceous earth
Alkaline gold	PVC, pyrex polypropylene, Teflon	Titanium	PVC lined	Asbestos fibre
Tin lead	PVC, rubber	Not required	PVC rubber lined	Non-silicated filter aid
Nickel sulphomate	PVC, polypropylene	Not required	PVC rubber lined	Asbestos fibre
Bright acid tin	PVC, polypropylene	Not required	316 stainless steel, hard rubber	Asbestos fibre
Bright cyanide silver	Hard rubber	Not required	316 stainless steel, polypropylene	Asbestos fibre

 ## 8.9   Additive Processing

Additive processing was introduced in the 1960s in the printed circuit board fabrication industry as an economical method to manufacture boards of moderate complexity. The cost savings come primarily from the reduction of process steps and the elimination of waste. Elimination of waste is the very basis of the additive process, which wastes nothing, adding materials to the board only where they are needed.

The 'additive' process basically starts from an unclad laminate, i.e. there is no copper on the base laminate. The copper is deposited selectively on the base laminate wherever it is required as conductors as per the design of the circuit. The process does not involve etching at any stage and therefore, considerable amount of copper is saved. As there is no base copper connecting the conductors, electroless copper is used to build up the thickness of the conductors.

Special types of laminates are required for additive process. The most common laminates are:

- Paper phenolic laminate with a normal thickness of resin on their surface;

- Epoxy laminate with a resin of thickness 30 to 50 microns;
- Glass epoxy with or without catalytic resin bonded with anodized aluminium foil. This laminate is used by the manufacturers for thick or thin film technology.

Epoxy laminates are used directly whereas phenolic laminates need a larger coating to improve the bonding between the base and the copper foil. After drilling, the anodized foil is removed by etching or peeling, and leaves the underlying resin surface in the rough state so that it is good for adhesion.

The additive methods for the manufacture of printed circuit boards are classified into three basic processes:

- Fully additive;
- Semi-additive; and
- Partially additive.

The common step in all these processes is that all of them involve electroless copper plating to form conductors or plated through-holes at one point in their production process.

8.9.1 Fully Additive Process

Fully additive process does not require any electroplating except when precious metal plating is required on edge connectors or contract tabs. Figure 8.14 shows the manufacturing process of a fully additive method. The basic steps followed in this process are discussed below.

The process starts with an epoxy laminate with no copper, but with sufficient resin on the surface. Once the holes have been made by drilling or punching, mechanical abrasion of the adhesive surfaces, for example, with pumice scrubbing is carried out for better adhesion of the plating resist. The plating resist is screen printed or dry film laminated to the surface. The image is a reverse of the conductor pattern.

After the exposed conductor tracks are treated with a strong oxidizing agent such as sulphuric acid or chromic acid, the panel is immersed in the electroless copper plating bath, until the desired copper thickness is achieved. As the thickness to be plated is much greater up to 25 microns, the immersion time is longer. The baths are required to provide a high plating rate and the ability to deposit as much copper as necessary. Plating takes place selectively on the areas not covered by photo-resist. As there is no etching of copper, the problem of creating undercut due to process defect do not exist at all. After this, several additional steps such as solder resist screening and curing, coating with a solderable coating, and dip coating with tin-lead alloy can be applied to improve the board quality.

The following are two basic variations of the fully additive process:

- The *CC-4 process* generally referred to as copper complexer #4, starts with a pre-catalysed base laminate coated with an adhesive which is also pre-catalysed. While facilitating adhesion of copper tracks to the base material surface formed by additive placing, the adhesive also serves as a dielectric material. After holes are drilled in the laminate, a reverse pattern plating resist is applied to the panel surface and cured.

- The *seeded process* starts with adhesive coated base laminates, but neither the adhesive nor the base laminate are pre-catalysed. After the holes are made, the panel is dipped into a strong oxidizing agent, catalysed and dried. The catalysing is done in a Pd (Palladium-based) activator solution with a pH which is approximately 20 per cent of the usual concentration of the conventional PTH process. The additive surfaces should be free from voids, nicks, pits and dents. Scars deeper than 2-3 micron are unacceptable.

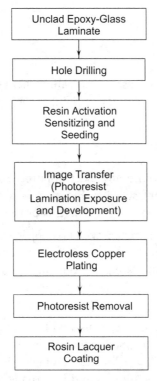

Fig. 8.14 A fully additive process: major steps in the manufacture of a double-sided PTH board from an epoxy laminate with sufficient thickness of resin at the surface

The additive system has several advantages and offers many possibilities for application. For example, a wide range of laminates is available. They offer a cost advantage as they are not supplied with copper foil. The processing is simpler and requires less process steps. There is estimated to be a 10-40 per cent saving. The quality of PCBs provided with additive processing should have a number of improved parameters. The overhang and undercut, which is common in the etching method, is non-existent, thereby permitting finer conductors and patterns, and consequently a higher circuit density. There is a better control of plating thickness over the entire pattern, including inside holes and the thickness uniformity, which offers new possibilities in high frequency applications.

8.9.2 Semi-additive Process

This process (Figure 8.15) starts with the deposition of copper all over the laminate by the electroless process to the required thickness, which is generally 1.5 to 2 microns. The image is then transferred on this board in such a way that only the pattern is visible. This is followed by deposition of more copper on the circuit pattern by the electroplating method. The method has more steps than the fully additive process, and also suffers from the inherent problems of electrolytic plating such as uneven plating thickness and over-plating of the holes on the outer periphery. The advantages of the method are that the stability of the electroless copper is high and less etching solution is consumed, when compared to the fully additive electroless bath.

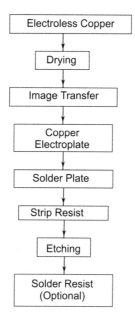

Fig. 8.15   Process flow chart for semi-additive process

8.9.3 Partially Additive Process

This process is a combination of etching and electroless copper plating. The essential process steps are indicated in Figure 8.16. The process starts with a 35 μm copper foil clad laminate. After the panel is drilled and deburred, it undergoes catalysation or activation process. It is followed by application of etching resist by screen printing or photoprinting on both sides of the board. The conductor patterns are then formed by etching and stripping of the resist. The entire surface is screen plated, leaving pads and holes uncovered. The panel is then immersed in an electroless copper plating solution for deposition of about 30 μm thick copper onto the pads and hole walls. The method has several advantages over other additive processes, such as the convenience of using the conventional copper clad laminate, formation of fine line conductors down to 80 μm by the screen printing method, formation of surface bonding pads and uniformly and reliably plated through-holes (Patterson, 1992). PCBs fabricated using this process achieve higher practical densities than conventional boards for two reasons: fine line etching and small hole plating with electroless copper.

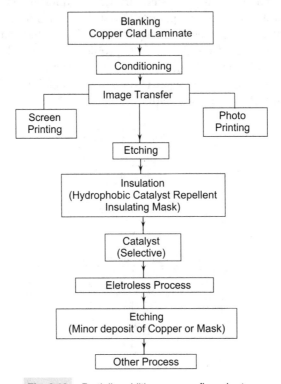

Fig. 8.16   Partially additive process flow chart

Fine line production of outer layers has traditionally consisted of pattern or thick panel plating, followed by etching. Either process suffers from etching control problems due to the uneven distribution of electrolytically plated copper. This is shown in Figure 8.17. More or less etch undercut results in line widths and thicknesses that vary with the thickness of applied copper. Also, etching of external layers below 6 mils is difficult and expensive due to yield loss. Electroless copper is the key to uniform copper thickness, which allows for uniform etching, less line variation and higher yield.

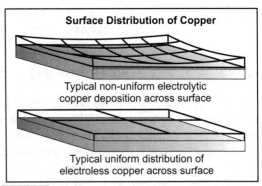

Fig. 8.17   Typical uniform distribution of electroless copper across surface (after Patterson, 1992)

PCBs manufactured by using additive techniques are found in products ranging from cars to computers (*Nargi-Toth, 1994*). The use of additive techniques is particularly attractive for these

PCBs with complex circuit configurations that result in low yields when standard electroplating techniques are employed. The process allows for production of PCBs with less than 5 mil lines and spaces, complex circuit geometry and small (8 mil), higher aspect ratio (8:1) through-holes can be readily demonstrated.

The initial aspect ratio of a bare 14 mil hole, which is the typical hole size drilled in PCBs is 4:1 (*Nakahara, 1991*). Upon completion of plating, one mil of copper in the hole, the aspect ratio will grow to 5:1. To plate this aspect ratio hole galvanically (through electroplating), the current density drops from the normal value of 25-30 amps/sq ft. to about 15 amps/sq. ft.

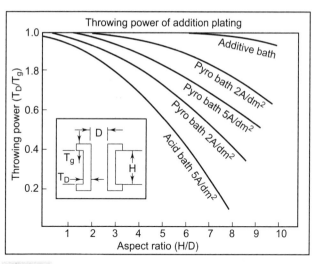

Fig. 8.18   The additive plating process — aspect ratio vs throwing power (after Nakahara, 1991)

Even with this low current density, it is difficult to plate a uniformly thick deposit, and thus maintain a flat surface across an entire panel, particularly with pattern plating. By comparison, the additive process, with its excellent throwing power (Figure 8.18) in small holes and in the absence of any current field, processes a deposition variation of 3-12 per cent. This flatness over an 18×24 in. panel results in excellent etching characteristics. Higher aspect ratio in electroplating results in greater non-uniformity in copper hole wall thickness, as shown in Figure 8.19. This translates into poor reliability. Electroless copper plating deposits uniform copper thickness regardless of hole depth or diameter.

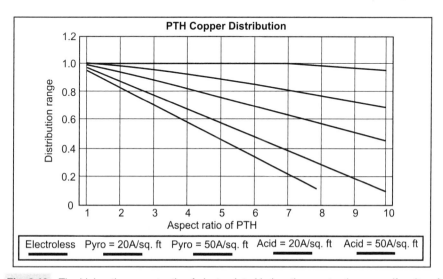

Fig. 8.19   The higher the aspect ratio of electroplated holes, the greater the non-uniformity of copper hole wall thickness (after Patterson, 1992)

8.10 Solder Mask

A solder mask (resist) is a coating used to mask or protect the untinned copper track from chemical and abrasive damage which may take place from the action of an etchant, solder or plating. The solder mask also masks off a printed circuit board surface and prevents those areas from solder shorts during wave re-flow soldering (vapour phase or wave soldering process). Such masks are screen printed and are about 0.1mm thick. In addition, the solder mask provides an environmental protection to the track, acts as an insulating barrier between closely placed cards and prevents the board from getting damaged by dirt, fingerprints, etc. However, the primary function of the solder mask is to restrict the molten solder pick-up or flow to those areas of the printed circuit board, holes, pads and conductor lines that are not covered by the solder resist.

The solder mask in general is expected to perform many more functions than just being a means to restrict the solder pick up (Tennant, 1994). These functions are:

- Reduce solder bridging and electrical shorts;
- Reduce the volume of solder pick-up during wave soldering thereby reducing the weight and cost;
- Provide an environmental protection to the track;
- Act as an insulating barrier between closely placed tracks;
- Prevent the PCB from getting damaged by dirt, fingerprints, etc.; and
- Reduce the solder pot contamination during wave soldering.

8.10.1 Solder Resist Classification

The solder resists are broadly classified into two types:

- Temporary solder resists; and
- Permanent solder resists.

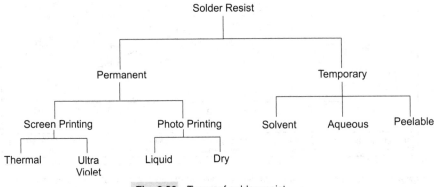

Fig. 8.20 Types of solder resist

Figure 8.20 shows the various types of temporary and permanent solder resists. The permanent resists are applied either by the screen printing method or photo-printing method. The screen printable type can be further classified as thermal curable or ultraviolet curable. The photoprinted method can use either the liquid film solder mask or dry film solder mask. The temporary solder resists are classified on the basis of the chemistry used for developing the resist.

8.10.1.1 Temporary Resists

These resists are used temporarily either to prevent the solder from getting on to certain holes and on to the gold plated edge connectors during wave soldering operations or to prevent certain holes from getting blocked with solder thereby preserving these holes for manual component insertion at a later stage.

8.10.1.2 Permanent Resist

These resists are applied permanently on the PCB surface and form an integral part of the PCB. The chemistry and the demand for this permanent solder masking have undergone a tremendous change with the introduction of surface mounted devices and also due to the fact that the conductor spacing is going down day-by-day. With the increasing complexity of the circuitry, the inspection and the re-work costs are going high. The process of solder masking to a certain extent reduces these costs by generally bringing down the solder bridging and circuit shorts. Also, solder masking increases the environmental protection for the PCB.

8.10.1.3 Solder Mask Application

The solder mask or solder resist are applied either by screen printing or photoprinting method. Figure 8.21 shows the types of solder resist applications.

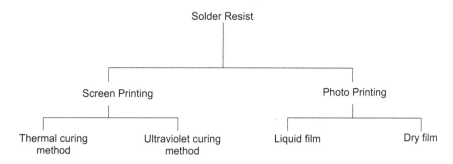

Fig. 8.21 Solder resist application methods

The solder mask for pads or lands used to mount through-hole leaded components is designed to allow a clearance of at least 0.25 mm around the pad, i.e. the edges of the pad are to be covered by the solder mask. This is shown in Figure 8.22.

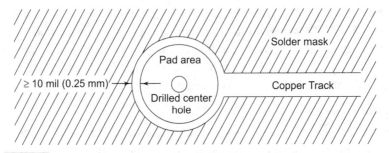

Fig. 8.22 Solder mask minimum dimension overlap for a leaded component land

8.10.2 Liquid Film Solder Mask

The liquid or paste form solder mask is one of the earliest types of solder masks used. This is generally applied by the process of screen printing. The uniformity of the coating in this screen printing process is largely controlled by the maintenance of the consistency of the material during application. For two-component material, the quality of residual solvents and correct mixture of the two constituents are important for achieving and maintaining a defined coating thickness. Only if these parameters are closely controlled, is it possible to achieve consistency in the chemical and physical properties of the solder mask when it is applied to the printed circuit board. These parameters are vital if the required electrical properties of the solder mask are to be achieved in a reproducible way.

If the liquid film has to be applied on a printed circuit board which has been prepared by electroplating a metal resist, the following process steps are followed:

- Strip the metal resist (tin/tin-lead);
- Water rinse and dry;
- Mechanical cleaning;
- Water rinse and dry;
- Micro-etch to increase the adhesion;
- Water rinse and dry;
- Screen print liquid photo-resist; and
- Ultraviolet or thermal cure.

8.10.3 Dry Film Solder Masking

The dry film solder resist, unlike the liquid or paste form, is in the form of a photopolymer film. This film is sandwiched between two protective layers, which prevent the emulsion from getting damaged during the handling process. The steps followed for applying dry film solder mask, are detailed below.

Surface Preparation: In this stage, the tin/tin-lead metal resist, which is placed on the surface, is stripped and the surface is washed with water and dried thoroughly. The copper surface is then subjected to a series of cleaning stages such as:

- De-grease with hot alkali cleaner;
- Water rinse;
- Micro-etch copper for 1 to 2.5 microns; this is done to increase the adhesion of the dry photopolymer film on the surface;
- Inspect for complete removal of tin/tin-lead;
- Water rinse;
- Dip in 10 per cent sulphuric acid dip for 1 to 2 minutes;
- Water rinse;
- Abrasive cleaning [320 grit brush and pumice scrub]; and
- High pressure water rinse and dry.

Controlled copper oxidation with brown oxide or black oxide is used to increase the adhesion of the solder mask film on the surface. The oxide treatment must be controlled to give an oxide thickness of 0.5 μm to 1.0 μm. If the oxidized board is stored for a long time, the surface has to be thoroughly de-greased before lamination.

Pre-lamination Drying: Absorbed water and water residues are some of the most frequent causes of blisters and de-lamination. It is therefore essential that all standing water is removed from the board surfaces and the holes. Water must be physically removed by using high pressure blowers from the surface and the holes. A high volume air knife can also be used for this step.

Absorbed moisture is removed by baking the board for 15 to 20 minutes at a temperature of 110 $\pm$ 10 °C. Conveyorized infra-red ovens drying at 80 to 120 °C for 30 seconds can also be used. Boards of different base materials and different thicknesses require different drying times. Excessive bake time and temperature should be avoided because they increase copper oxidation and may lead to poor adhesion.

Lamination: Vacuum lamination ensures that all conductor lines are totally encapsulated by the photopolymer solder mask and the boards are free from entrapped air pockets.

Selection of Film Widths: Proper widths of the dry film should be selected depending upon the width of the panel. The width of the film should not be more than 10 mm than that of the panel to avoid edge trimming wastages. The lamination could be done either on both the sides at the same time or on one side after another depending upon the type of laminator and the film utilization.

Hold Time after Lamination: The hold time between lamination and exposure should be controlled carefully for optimum results. Immediately after lamination, the required exposure energy is low and increases rapidly with hold time, levelling off with a constant value. Hence the post-lamination hold time should be constantly maintained in order to avoid variation in exposure time from lot to lot.

Exposure (Photo-tools): Generally, the photopolymer solder mask films are negative working and hence they require a positive photo-tool (pad areas opaque) for exposure. Due to the high energy levels required, the photo-tools must have a very high density. Dmax > 4 minimum to avoid solder mask on pads while the clear areas must have a Dmin < 0.15. Since the density of the film can vary with usage, the photo-tools must be periodically checked in order to make sure that Dmax is minimum 4.

Optimum results are generally obtained with 7 mil single-sided emulsion diazo films. In case of registration systems that do not require seeing through the pads, a silver halide film could be used with emulsion in firm contact with the board in order to get proper exposure and edge definition. The phototool also must have a minimum opaque pad areas about 0.1 mm to 0.15 mm larger in diameter than the lands on the printed wiring board.

The exposure time and the consequent degree of polymerization are established and monitored by the usage of stouffer 21 steps tablet. In order to establish correct exposure values, the hold time and development conditions must be consistent. The stouffer step held is defined as the last step with more than 50 per cent resist coverage. A consistent hold time is required to allow the polymerization process to be completed and to stabilize.

Developing: The developing process washes away the unexposed (unpolymerised) resist from the board surface. The developing time for a particular type of photopolymer depends upon the type of chemistry being adopted. The correct developing times are established by controlling the point at which the unexposed film is thoroughly washed away. Thorough removal of the solder resist from within the holes and pads are required to achieve good solderability. The generally used chemistries are: (a) Aqueous, and (b) Solvent developing.

In aqueous developing, generally a dilute solution of sodium carbonate in 99 per cent pure water is used at a temperature of 45 °C. With this solution, the PCB should develop completely within 40 to 60 per cent of the chamber length. In solvent developing, 1-1-1 tricholoroethane is used and the break point should be set at 25 per cent of the total length of the chamber.

Drying: In order to achieve consistent curing results, the boards must be thoroughly dried after development. Moisture absorbed by the resist and the base material during the development and rinsing steps must be completely removed by drying at 90 °C for 15 minutes or by using hot air knives.

Curing: Proper curing of the photopolymer resist is a must to achieve optimum soldering performance and to prevent lifting off of the mask during wave soldering or vapour phase soldering.

Ultraviolet (UV) curing of the photopolymer solder mask could be done in conventional curing machines using mercury vapour lamps. The desired energy levels are achieved by controlling the speed of the conveyor and the lamp intensity. It is preferable to give enough energy to completely cure one side in one single pass. UV lamps should be 200 W/inch high pressure mercury vapour lamps. Since the heat from the ultraviolet lamps plays an integral part in the curing process, 'cool' ultraviolet units designed to minimize the heat reaching the boards are not recommended. Ultraviolet curing with a surface temperature of less than 105 °C often leads to inconsistent results. Care should also be taken not to heat the PCB surface excessively during ultraviolet curing. The typical temperature should be 110 °C to 140 °C. Excessive heating will lead to the photopolymer becoming brittle and blistering from the surface.

In thermal curing, the ovens used to bake boards should be capable of controlling the temperature evenly throughout the chamber and these ovens should not be used for other purposes in order to prevent contamination of the solder mask.

Ovens without a fresh air inlet should not be used since the hot air might condense on to the surface thereby degrading the solderability. If there is no dilution of air, the hot air might reach an explosively dangerous temperature. For the oven to be operated safely and without trouble, it should have a thermostatically controlled timer and temperature recorder, and a circulating fan to provide adequate heating throughout the chamber with an exhaust duct with blower.

The dry film solder mask, which is manufactured under controlled conditions gives a uniform thickness and the application of this film using high pressure laminator, ensures uniformity throughout the panel area. This also allows it to be easily maintained from batch to batch.

The process of screen printing and curtain coating are both single-sided processes. In other words, the solder mask application could be done only on one side at a time and the board has to be baked to remove the tackiness of the ink before it is printed on the other side, which poses the problem of maintaining the second side clean for such a long time. With the use of a vacuum laminator, the dry film can be applied on both the sides simultaneously.

8.10.4   Resolution

Photo-imageable solder mask offers the possibility of taking the solder mask to very close proximity of the solder pad without the problem of bleed-out and smear, which is generally associated with conventional screen printing solder mask inks.

The photo-imageable system is exposed with a positive film master under ultraviolet radiation as in the case of a dry film solder mask. In a dry film system, this can be reliably achieved with the clean removal of the solder mask ink from the solder pad.

Liquid photo-imageable solder mask offers very high resolution since the coating thickness will be of the order of 40 microns only whereas in the case of a dry film solder mask, the coating thickness will be a minimum of 50 microns and even more in the case of certain board designs which call for thicker copper. However, the resolution of the LPISM (liquid photo-imageable solder mask) needs a long exposure time, which necessitates a greater relief area around the solder pad to avoid the problem of mask encroachment on the pad. The same problem is encountered much more in the case of conventional screen printable solder mask wherein the registration between the solder-mask and the solder pad cannot be 100 per cent, because of the inherent sagging associated with the screen mesh, which leads to solder mask encroachment on the solder pad.

8.10.5   Encapsulation

The complete encapsulation of the circuit pattern by the solder mask is desired by all PCB users. This becomes critical when the circuitry becomes more and more dense. Specifications require a minimum thickness of 25 μm solder mask on the conductor pattern at any point on the PCB. The extent of encapsulation also decides the degree of electrical insulation as well as the resistance to environmental attack.

For screen printed solder mask inks, the degree of encapsulation depends directly upon the screen mesh selection, the viscosity of the ink, the temperature and the speed of printing. If there is a variation in the height of the conductor, the degree of encapsulation also varies. The liquid solder mask applied may show a strong directional coating profile. But in the case of a pre-formed dry film, the encapsulation achieved will be uniform throughout the board, provided the conductor height is not more than the thickness of the dry film. A minimum encapsulation of 25 μm can be obtained throughout the PCB surface by using dry film.

Solder mask materials (liquid and dry film) are intended to totally encapsulate underlying circuit elements and to protect them from the corrosive aspects of the operating environment. In order to accomplice this with a dry film solder mask, you must choose a film that is thick enough to flow over and around the pads and traces of your circuit during lamination. A good rule of thumb to use, is to select a film that is at least twice as thick as the copper being covered. In more practical terms, if you are covering 1 ounce (35 micron) copper clad, you should use a film that is at least twice as thick as the 0.0013" thick copper layer. For this application, a 3 mil (0.003") film would be preferred. Failure to follow this simple rule might result in thin capillary gaps along the edges of pads and traces. These gaps are just about the perfect size to wick, and retain the corrosive fluxes used during wave and/or hand soldering, which could lead to the eventual failure of the board.

If the plating thickness is not reliably controlled, then regardless of the type of solder mask used and the method of application, the uniformity in the encapsulation thickness cannot be ensured.

8.10.6   Surface Topography Resist Thickness

There is a considerable difference in the topography obtained between a liquid solder mask and dry film solder mask. In a liquid system, less quantity of the liquid is applied on the board. Due to its fluid nature, the liquid tends to flow downwards to the base material, thus leaving a thin film on the conductor surface. But in the case of dry film solder mask, the thickness of the conductor embeds itself into the solder mask emulsion, thereby offering a virtually flat surface.

8.10.7   Placement Assistance

The presence of solder mask film helps in the location of certain SMD components. The surface tension of the solder can have the beneficial effect of self-alignment of certain SMDs. This positional alignment assistance can be easily achieved with a thicker dry film, whereas it is difficult to achieve with the liquid solder mask ink, which requires multiple coating to get 100 μm deposition on the board. This leads to mask smear on the pad area.

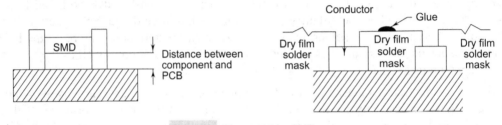

Fig. 8.23   Placement for SMDs

The presence of thick film resist between the mounting pads is an advantage for SMD components, since the clearance between the base of the component and the PCB in some cases should not exceed 100 microns, if the component is to be fixed in place with adhesive. The use of thick dry film solder mask provides this requirement as also a flat surface for the application of the glue as shown in Figure 8.23.

Certain types of flat components are found to exhibit single-side lifting or 'tomb-stoning' during vapour phase soldering operations. This problem could be aggravated by using a thicker dry film. In these cases, it is necessary to use a thinner solder mask coating, which can be obtained by using a 50 μm thick dry, film instead of a 100 μm film.

8.10.8 Reliability of Solder Mask

The reliability of the solder mask is dependent upon the coating integrity and total encapsulation. A minimum 25 μm coverage of conductors provides adequate protection against mechanical damages which may occur during the handling and assembly stages. The long term electrical reliability depends upon the electrical characteristics of the material used for solder mask. The circuits encapsulated by dry film solder mask are protected by a coating with a dielectric strength in excess of 2 KV.

In case of liquid solder mask coated boards, the thickness will be less than 25 μm. Hence, it is prone to have more pin holes and skips, which make the board less insulated when compared to the dry film solder masked board.

8.10.9 Soldering and Cleaning

Different solder mask systems have their own cleaning systems. The degree of cleanliness on and around the pad will affect the flow of flux which, in turn, controls the degree of wetting of the pad area and eventually the solder joint itself.

8.10.10 Tenting of Vias

With SMD technology, the vias tend to become smaller and smaller in size. When liquid mask is used, there is no possibility of reliably tenting the hole. In addition, because of the hole blocking tendency of the many liquid solder masks and the extreme difficulty in dissolving these soft baked coatings during developing, the reliable solder filling of the holes is not assured. Liquid solder mask is therefore incompatible with the process of tenting of vias in SMD boards, whereas by photo-imaging a dry film solder mask in the area where the vias occurs, the vias are reliably tented, as illustrated in Figure 8.24, even with a 50 μm film with thick durable protective coating. These vias stand upto the hot air levelling process also.

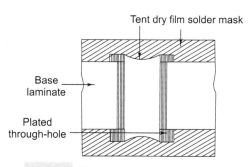

Fig. 8.24 Cross-section of tenting vias

The tenting of vias is desirable from the point of view of providing an environmental and electrical barrier to moisture and flux penetration. Fewer blow holes will be encountered during soldering and there will be less opportunity for solder wicking and bridging. Tenting of vias also provides a significant benefit in bare board testing. Non-protected vias and non-filled vias do not permit adequate vacuum draw down for testing, which is a common technique used to ensure electrical contact in these machines.

8.10.11   Solder Mask over Bare Copper [SMOBC]

This is one of the major solder resist application technologies. SMOBC stands for "Solder Mask Over Bare Copper". One of the major disadvantages of conventional tin-lead electroplating underneath the solder mask is blistering of the solder mask during wave soldering or infra-red re-flow or vapour phase soldering operations.

The flow of the molten tin-lead metal underneath the solder mask leads to the development of a strong hydraulic force, which forces, its way through the solder mask, thereby rupturing the solder mask, which can seriously bring down the reliability of the PCB, by acting as a direct conduit for the flow of liquids and excessive soldering flux on to the base material.

The SMOBC process overcomes these problems by totally eliminating the tin-lead deposition on the conductor underneath the solder mask. An all-copper plated PCB can often be prepared by the "tent and etch" process as shown in Figure 8.25. In this case, the PCB is initially drilled and the holes are metallized by the conventional electroless copper plating process. Then, the copper is built up to the required thickness by panel plating. After this copper build-up, the dry film is laminated and exposed using a negative photo-tool which polymerizes the photo-resist in the exposed areas thereby forming a tent over the hole. The board is then directly taken for etching where the unwanted copper around the hole and the conductor areas are removed, thereby leaving only the holes and the conductor pattern. The photo-resist is stripped and solder masked wherein only the pad areas copper will be exposed and the copper on the tracks will be covered by photo-resist. The copper exposed areas are prevented from getting corroded by applying a thin coating of tin-lead by a process of hot air levelling (HAL).

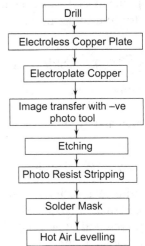

Fig. 8.25   Process sequence for solder mask over bare copper by tenting process

The other alternative process is to process the PCB by the conventional pattern plating process. However, the metal etch resist is stripped and the PCB is solder masked. In this case also, the pads are prevented from getting corroded by the process of HAL, thereby increasing the solderability of the printed circuit board.

8.11 Conformal Coatings

Conformal coatings are employed to enhance the performance and reliability of printed circuit assemblies that get subjected to hostile environments such as marine, aerospace and military applications. Manufacturer of consumer electronics are increasingly using conformal coatings as a cost-effective way to improve product reliability.

When uncoated PCB assemblies are exposed to humid air, the formation of thick films of water molecules on their surfaces reduces their surface insulation resistance (SIR). The lower the SIR, the greater will be the deterioration of electrical signal transmission properties. The typical results are cross-talk, electrical leakage and intermittent transmission, which may lead to permanent termination of the signal, i.e. a short circuit. Moisture films on uncoated PCBs also provides favourable conditions for metallic growth and corrosion, eventually adversely affecting dielectric strength and high frequency signals. Dust, dirt and other environmental pollutants that settle on assembly surface trap moisture magnify these effects, and conductive particles like metal chips can cause electrical bridging.

Conformal coatings are plastic film envelopes that surround printed circuit assemblies. They are applied in film thicknesses of upto 0.005 inches. They seal out dirt and environmental contaminants. They also seal in contaminants that have not been removed by pre-cleaning. It is therefore important to clean the surface before applying a conformal coating.

The functions of the conformal coatings are to:
- Offer protection of circuitry from moisture, fungus, dust and corrosion caused by extreme environments;
- Prevent damage from board handling during construction, installation and use, and reduce mechanical stress on components and protects from thermal shock;
- Resist abrasion in service; and
- Enhance performance and allow greater component density due to increased dielectric strength between conductors.

8.11.1 Materials for Conformal Coatings

A number of materials can be used for conformal coatings. Each one has its own characteristics and areas of application. The main features of different types of coatings are summarized below.

Acrylic Conformal Coatings: They are easy to apply. They cure at room temperature in minutes, and have desirable electrical and physical properties. They are fungus-resistant and do not shrink. These coatings have a long pot life and do not shrink. They also have low or no exotherm during curing, which prevents damage to heat-sensitive components. Their main disadvantage is solvent sensitivity, but this makes them easy to repair.

Epoxy Conformal Coatings: They are usually available as two-component compounds. They provide humidity resistance, and high abrasion and chemical resistance. Epoxy conformal coatings are usually extremely difficult to remove chemically for re-work because any stripper that will attack the coating will also dissolve epoxy coated components and even the epoxy glass PCB itself. Component replacement, therefore, necessitates burning through the epoxy coating with a knife or soldering iron.

Polyurethane Coatings: They are available as either single-component or two-component systems. They provide excellent humidity and chemical resistance as well as outstanding dielectric properties over extended periods. Polyurethane can be burnt through with a soldering iron, thereby simplifying component replacement.

Although cleaning is a vital step prior to the application of any conformal coating, polyurethanes are particularly sensitive to moisture, as it can cause blistering under humid conditions. This can eventually lead to circuit failure. Polyurethane formulations require careful application and close control of coating and curing environments.

Silicone Conformal Coatings: They are particularly useful for high temperature service upto about 200 °C. They provide excellent humidity and corrosion resistance as well as good thermal endurance, making silicone conformal coatings highly desirable for PCBs that contain high heat-dissipating components.

The typical electrical and thermal performance characteristics of conformal coatings are given in Table 8.4. (Waryold and Lawrence, 1991)

Table 8.4 Characteristics of Conformal Coatings

Electrical	Acrylics	Epoxies	Silicones	Polyurethanes
Dielectric strength (short time), 23 °C volts/mil at 0.001"	3,500	2,200	2,000	3,500
Surface resistivity at 23 °C, 50 per cent rh, ohms/cm	10^{14}	10^{13}	10^{13}	10^{14}
Dielectric constant at 23 °C, 1 MHz	2.2–3.2	3.3–4.0	2.0–2.7	4.2–5.2
Thermals				
Resistance to heat, continuous, °C	125	125	200	125
Linear coefficient of thermal expansion, μ/in./°C	50–90	40–80	220–290	100–200
Thermal conductivity 10^4cal/ sec/cm^2/cm/°C	4–5	4–5	3.5–8	4–5

8.11.2 Methods of Applying Conformal Coatings

There are basically four methods of applying conformal coatings. These are discussed below. (Waryold, et al., 1998):

Dipping: In this method, the masked assembly is immersed in a tank of liquid coating material and withdrawn, ensuring uniform coverage and even deposit. Immersion and withdrawal rates are important factors that need to be controlled to allow the viscous liquid material to fill all the voids in the assembly. The typical immersion speeds are 2 to 12 inches per minute to enable the coating to displace the air surrounding components.

When conformal coatings are applied by dipping, the evaporation of the solvents in the liquid plastic may occur so fast that the viscosity of the bath increases rapidly. For this reason, it is important to continuously monitor and maintain proper viscosity.

Spraying: Spraying is the most popular and the fastest method for applying conformal coatings. With the proper combination of solvent dilution, nozzle pressure and patterns, consistent and repeatable results can be obtained. Spraying can be done by both manual or automated means by using computer-controlled systems integrated into existing wave soldering and cleaning lines. A major drawback of the spraying method is that little or no coating is applied underneath components or on components that are shadowed.

Coatings should be sprayed on assemblies using clean, dry gas at the minimum pressure necessary to provide good atomization. The assembly should be sprayed holding the gun at a 45°angle and the assembly should be rotated 90° after each back and forth past.

Brushing: It involves manual operation. It is the least effective method of applying conformal coatings because of the difficulty in achieving uniform and repeatable coverage. The method is practical only for small numbers of printed circuit boards.

Each coating method has its strengths and weaknesses. A combination of dipping and spraying is better than the adoption of a single method.

8.11.3 Standards for Coatings

Some coatings are manufactured strictly for commercial use and are not certified by the manufacturers to meet any particular specifications. The military specification for coatings was Mil-I-46058C, but this is obsolete and the new specification is IPC-CC-830. Coatings are available that also meet UL requirements for components.

 ## 8.12 Useful Standards

- *IPC-4552: Specification for Electroless Nickel/Immersion Gold (ENIG) Plating for Printed Circuit Boards:* Sets the requirements for the use of electroless nickel-immersion gold as a surface finish for printed boards and includes requirements of deposit thicknesses based on performance criteria.

- *IPC-HDBK-830: Guidelines for Design, Selection and Application of Conformal Coatings:* A compilation of the conformal coating industry's practical experience to assist the designers and users of conformal coatings in making informed choices.
- *IPC-SM-840C: Qualification and Performance of Permanent Solder Mask—Includes Amendment 1:* Covers requirements for the qualification and quality conformance of liquid and dry film solder mask.
- *IPC-SM-839: Pre-and Post-solder Mask Application Cleaning Guidelines:* Covers all aspects of cleaning related to solder mask application, including board preparation, in-process control and maintenance of cleanliness during pre-assembly processes.

9

Etching Techniques

Etching is one of the major steps in the chemical processing of the subtractive PCB process. By this process, the final copper pattern is achieved by selective removal of all the unwanted copper to retain the desired circuit patterns. The copper which is not protected by an etch resist is removed by the etching process. The following are the commonly used etching methods:

- Chemical etching or chemical machining;
- Electrochemical etching or chemical milling; and
- Mechanical etching (by milling).

 ## 9.1 Etching Solutions and Chemistry

Several chemicals are used for etching. The most common etchants are:

- Ferric chloride;
- Ammonium persulphate;
- Chromic acid;
- Cupric chloride; and
- Alkaline ammonia.

The following sections give characteristics, chemistry and process details of these etching methods.

9.1.1 Ferric Chloride

Ferric chloride etching solutions are widely used in the 'print' and 'etch' process in the PCB industry. Ferric chloride has a high etch rate and high copper dissolving capacity. It is used with screen inks, photo-resist (wet film and dry film) and gold plated boards. As the ferric chloride etchant attacks tin, this is not suitable for tin or tin-lead plated boards.

This etchant is an aqueous solution of 28 to 42 per cent by weight of ferric chloride. The solution has a specific gravity of 1.353 to 1.402. It operates over a wide range of concentration, but is most widely used at about 35 per cent. The ferric chloride solution has free acid due to the following hydrolysis reaction:

$$FeCl_3 + 3\ H_2O \longrightarrow Fe(OH)_3 + 3\ HCl$$

In order to prevent the formation of insoluble ferric hydroxide, an excess of hydrochloric acid, upto 5 per cent by weight is usually added, which prevents the spontaneous hydrolysis of $FeCl_3$ as per the above equation. Commercial formulations also contain wetting and anti-foam agents.

At the initial stage of the etching process, the concentration of copper dissolved is high due to the high concentration of ferric chloride. As the ferric chloride in the solution gradually gets depleted, the etching time correspondingly increases to the extent that after some time, the solution has to be discarded and replaced by a fresh solution. Better dissolution of copper occurs when the etchant is sprayed perpendicular to the copper surface and the board is moved. The rate of dissolution of copper depends upon the ferric chloride concentration, temperature and agitation rate.

Ferric chloride is the oldest and perhaps the most common etchant. It normally comes in crystal form. The crystals are dissolved in de-ionized water to achieve its desired concentration in the solution. This is typically 500 gm of ferric chloride in one litre of water.

Chemistry

The ferric ion Fe^{3+} oxidizes copper to cuprous chloride (CuCl) as per equation (1). Cuprous chloride is further oxidized to cupric chloride as ($CuCl_2$) as per equation (2)

$$Cu + FeCl_3 \longrightarrow FeCl_2 + CuCl \qquad \text{--- (1) Green colour}$$

$$FeCl_3 + CuCl \longrightarrow FeCl_2 + CuCl_2 \qquad \text{--- (2)}$$

$$CuCl_2^- + Cu \longrightarrow 2\ CuCl \qquad \text{--- (3)}$$

In practice, when a solution contains 8 oz./gal (60 g/l) or more of dissolved copper, the etch time becomes longer than desired. However, the ferric chloride can dissolve copper even upto 120 g/l if the prolonged etching time can be tolerated. In order to increase the copper dissolving capacity and to bring down the etching time slightly, hydrochloric acid (upto 10 per cent of the etchant volume) is added after the copper content has reached 80 g/l. The acid also helps to control excessive sludge formation. Figure 9.1 shows the dependence of etching time versus copper content in the etchant.

For monitoring purposes and to know the exact copper concentration in the etchant, a chemical analysis has to be done. A less accurate but practical solution is to use the colorimetric method of colour comparison with standard solutions of known copper content. On an average, to etch 1kg of copper, 5.1 kg of ferric chloride will be consumed, with etching temperature in the range of 20-45 °C.

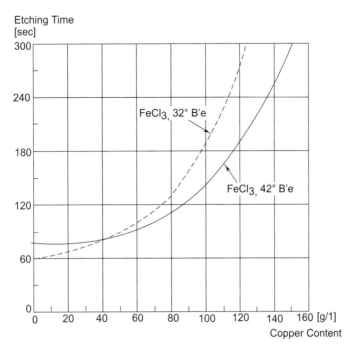

Fig. 9.1 Dependence of etching time with ferric chloride at 43 °C, spray etching, 35 μm cu (after Bosshart, 1983)

Unfortunately, ferric chloride is corrosive in nature and leaves dark stains. Further, it is normally difficult to regenerate for re-use. If solder or tin is used as a mask, it will attack it.

Composition of Ferric Chloride Etchant

- $FeCl_3$ [ferric chloride]   → 450 — 500 g/lt
- Specific gravity   → 1.35 to 1.4
- Copper dissolving capacity   → 120 g/lt
- HCl (hydrochloric acid)   → 10 ml/lit.
- Temperature   → 20-45 °C
- Anti-foaming agent   → 3 ml/l

Process Control and Regeneration: Regeneration of the solution is almost impossible and its disposal is expensive. Regeneration is carried out only due to the need to separate copper chloride from ferric chloride.

Ferric chloride etchant can be regenerated (Figure 9.2) by passing chlorine gas and to recover copper from the ferric etchant by crystallization. The addition of ammonium chloride and cooling to ambient temperature permit crystallization of a chloride double salt ($CuCl_2 \cdot 2NH_4Cl \cdot H_2O$) and cupra-ammonium chloride complex precipitate.

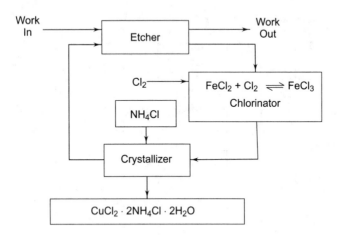

Fig. 9.2   Ferric chloride closed loop regeneration and recovery process using re-crystallization (after Coombs, 1988)

9.1.2   Hydrogen Peroxide — Sulphuric Acid

This system is extensively used for copper surface preparation which is also called micro-etching. It is compatible with organic and metallic resists and provides a steady etch rate with optimum undercut. It is a widely accepted system due to its ease of replenishment, closed loop copper recovery and need for simple waste treatment method.

Chemistry

Hydrogen peroxide is a strong oxidizing agent. It oxidizes and dissolves the metallic copper. The sulphuric acid makes copper soluble and keep the dissolved copper as copper sulphate in the solution, while copper sulphate helps to stabilize etch and recovery rates. The etching reaction is:

$$Cu + H_2O_2 + H_2SO_4 \longrightarrow CuSO_4 + 2H_2O$$

Etchant Composition and Process Conditions

- H_2SO_4 (96 %)   Sulphuric acid   100 ml/lit
- H_2O_2 Hydrogen peroxide   70 ml/lit
- $CuSO_4$ Copper Sulphate   60 g/lit
- Temperature   30 to 40 °C

Process Control and Regeneration

Hydrogen peroxide/sulphuric acid etchant is used in immersion as well as spray etching operations. The process is controlled by composition balance by recovery of copper sulphate as the by-product in closed loop systems. Such systems require continuous re-circulation of the etchant in the etching tank, whereby the etchant replacement is controlled by chemical analysis. The problems commonly encountered include process over-heating, etchant composition balance, etchant contamination and the dangers in handling concentrated peroxide solutions.

9.1.3 Chromic-Sulphuric Acid

Chromic acid mixed with sulphuric acid is used because of its strong oxidizing power and suitability for all kinds of metal resists such as tin, tin-lead, gold, vinyl lacquer and dry or liquid photo-resist.

The etching rate is inconsistent, but it has the advantage of little under-cutting. However, its use is now limited since it is difficult to regenerate and it is highly toxic, polluting and hazardous to health. It is generally not recommended for use.

Chemistry

The reaction between chromic acid and copper is

$$3Cu + 2HCrO_4 + 14H^+ \longrightarrow 3Cu^{2+} + 2Cr^{3+} + 8H_2O$$

In this system, the etching rate for copper is not very high, but it can be increased with an additive like sodium sulphate. The etch rate can also be increased by increases in temperature and agitation.

Composition of Etchant and Process Conditions

- Chromic acid   200 to 240 g/l
- Sulphuric acid   60 to 65 ml/l
- Sodium sulphate   40 to 45 g/l (additive)
- Temperature   20 to 30 °C

Process Control and Regeneration

This etchant is used in immersion etching and not in spray etching. The etch rate is controlled by the colorimetric standard method, while the density is measured by a hydrometer. Regeneration is not in common use because of the corrosive nature of the products and handling hazards. For example, chromic acid is an extremely strong oxidizing agent and attacks clothing, plastics and many metals if safety measures are not adopted to keep fumes out of the air through adequate ventilation. Similarly, spent chromic acid also presents a serious disposal problem, which needs to be handled in such a way that it complies with the state pollution control standards.

9.1.4 Cupric Chloride

Cupric chloride offers an economical solution in the etching of print-and-etch type of PCBs on a larger scale. From the pollution point of view, it offers the advantage of easy regeneratability with the possibility of a relatively easy disposal, high throughput and better material recovery. The dissolved copper capacity, which is upto 150 g/l, is high.

In performance, cupric chloride compares well with ferric chloride, with the advantage that it does not have sludge as ferric chloride does. Basically, it is an aqueous solution having 150-400 grams of cupric chloride. It usually has some hydrochloric acid (HCl) added to it with a large quantity (upto saturation level) of sodium chloride (NaCl) or ammonium chloride (NH$_4$Cl).

The cupric chloride etching system is used in the production of fine line multi-layer (inner layers) and print-and-etch boards. It is compatible with resists like screen inks, dry film, liquid film, photoresist and gold plating, except tin or tin-lead plated resists.

Chemistry

The overall etching reaction wherein the cupric chloride (CuCl$_2$) dissolves copper (Cu) is as follows:

$$Cu + CuCl_2 \longrightarrow 2\ CuCl \text{ --- (i)}$$

The cuprous chloride is then re-oxidized to regenerate etch-active cupric chloride by the oxygen present in the atmosphere.

$$2CuCl + 2HCl + \tfrac{1}{2}\ O_2 \longrightarrow 2\ CuCl_2 + H_2O \text{ --- (ii)}$$

In the process, a small amount of solution is continuously drained off to keep the concentration of copper constant. The level is maintained by the addition of a solution of hydrochloric acid, sodium or ammonium chloride and ammonium hydroxide.

Etchant Composition

- Cupric chloride   200 g/l
- Hydrochloric acid   200 ml/l
- Water   770 m/l
- Initial copper concentration   40 g/l
- Hydrogen peroxide   30 ml/l (optional)
- Temperature   30° to 40 °C

Process Control and Regeneration

The method given in equation (ii) is generally not used for regeneration because the oxygen reaction rate in acids is slow and the solubility of oxygen in hot solution is limited to 4 to 8 ppm. On the other hand, direct chlorination is a preferred technique for the regeneration of cupric etchant because of its low cost, high rate and efficiency in recovery of copper. The chemical reaction of the regeneration process is:

$$Cu + CuCl_2 \longrightarrow Cu_2Cl_2$$

The chlorination process results in:

$$Cu_2Cl_2 + Cl_2 \longrightarrow 2\ CuCl_2$$

The process sequence is shown in Figure 9.3. Chlorine, hydrochloric acid and sodium chloride solutions are automatically fed into the system as required. Various process monitoring equipments such as colorimeter, level sensors, and temperature monitoring and control arrangement, etch rate meter, etc, help to achieve optimum results.

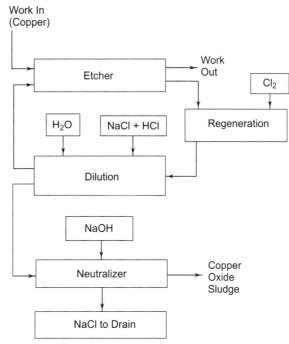

Fig. 9.3 Cupric chloride chlorination regeneration system (after Coombs, 1988)

The above regeneration process is less expensive and has a high rate of copper recovery. Spent etchant can be easily sold for its copper content. There are likely to be several problems with the cupric chloride system if process parameters are not properly controlled. The most significant is the low etch rate which is frequently due to low temperatures, insufficient agitation or lack of solution control. Sludging formation can also take place if the acid concentration is low. Elevated temperatures and the presence of excess acid may lead to the breakdown of photo-resists.

9.1.5 Ammonium Persulphate

Ammonium persulphate (APS) is frequently used as an etchant because it does not have most of the disadvantages of ferric chloride. It is also used in the surface preparation for electroless plating, and electroplating prior to oxide coating in multi-layer PCBs. Unfortunately, while ammonium persulphate is a strong oxidizing agent, it is unstable in solution and decomposes to form hydrogen peroxide, oxygen and peroxydisulphuric acid. The latter is a slow oxidizer at room temperature resulting in a low etch rate, which can however, be accelerated by the addition of a mercuric chloride catalyst when it becomes comparable with chloride etchants. It is compatible to all resists on boards including solder, tin, tin-lead, screen and photo-resist, except gold plated resists.

The formulations of ammonium persulphate containing sulphuric acid and catalysed with mercuric chloride give a good etch factor and easy regeneration in the closed loop system. But this system is costly as compared to other etchants and its use has declined for all practical purposes.

Chemistry

Ammonium (or potassium) persulphate is mixed with sulphuric acid to get a stable salt of persulphuric acid ($H_2S_2O_8$). The persulphate ion ($S_2O_8^{2-}$) is a powerful oxidizing agent and dissolves in water. During copper etching, the persulphate ion reacts with copper metal and the primary reaction is:

$$Cu + (NH_4) S_2O_8 \longrightarrow CuSO_4 + (NH_4)_2SO_4$$

In which a persulphate ion $S_2O_8^{2-}$ generates two sulphate ions, SO_4^{2-} at the same time oxidizing an atom of metallic copper to a cupric ion:

$$S_2O_8^{2-} + Cu \longrightarrow 2\ SO_4^{2-} + Cu^{2+}$$

If tin or tin-lead resist is coated, the printed circuit board is etched in APS. The incomplete etching of copper and darkening due to solder can be eliminated by the addition of 1 per cent phosphoric acid.

Etchant Composition and Process Conditions

- Ammonium per sulphate
  - $(NH_4)_2 S_2O_8$ $\rightarrow$ 200 g/lt
  - Sulphuric acid H_2SO_4 $\rightarrow$ 10 ml/lt
  - Mercuric chloride H_2Cl_2 $\rightarrow$ 0.5 ml/lt or 5 ppm
  - Temperature $\rightarrow$ 30° to 40 °C

Process Control and Regeneration

Ammonium persulphate systems are used in both spray etching and immersion etching operations. The system is controlled by colorimetric measurement. Cooling coils are used to maintain the etchant temperature constant and prevent runaway. The preferred temperature is 45 °C.

The useful capacity of the etchant is about 7 oz./gal copper at 38 to 55 °C. Above 5 oz./gal copper, the temperature of the solution should be maintained at 55 °C to prevent salt crystallization. The etch rate of a solution containing 7 oz./gal of dissolved copper is 0.00027 in./min at 48 °C.

The exhausted etchant consists mainly of ammonium or sodium and copper sulphate with a pH of about 2. Direct discharge of this solution to the sewer is not allowed. Therefore, the dissolved copper must be removed, and subsequently the remaining solution must be diluted, neutralized and discarded.

In general, ammonium, sodium and potassium persulphates, with the addition of certain catalysts, can be used for the etching of copper in PCB manufacturing. However, continuous regenerative systems based on ammonium persulphate have become more common. Persulphate is preferred because it has minimal disposal problems and somewhat higher copper capacity and etch rates.

9.1.6 Alkaline Ammoniacal/Ammonium Chloride

Alkaline ammoniacal etching system is used in both the batch and conveyor spray etching systems and is compatible with metallic and organic resists. The advantages of this etchant are its minimum undercut, high copper dissolving capacity and fast etch rates. Alkaline etchants provide continuous etching rates of 30-60 µm cu/min at a dissolved copper content of 150 g/l in the etchant.

Chemistry

Alkaline etching solutions dissolve exposed copper on PCBs by a chemical process involving oxidation, solubilizing and complexing. Ammonium hydroxide and ammonium salts combine with copper ions to form cupric ammonium complex ions [Cu $(NH_3)_4^{2+}$], which hold the etched and dissolved copper in solution at 18 to 30 oz./gal.

The oxidation reactions for closed loop systems showing the reaction of cupric ion on copper and air (O_2) oxidation of the cuprous complex ion are:

$$Cu + Cu(NH_3)_4^{2+} \longrightarrow 2\ Cu(NH_3)_4^{+}$$

$$4\ Cu(NH_3)_2^{+} + 8NH_3 + O_2 + 2H_2O \longrightarrow 4\ Cu(NH_4)_4^{2+} + 4OH^{-}$$

This process can continue as long as the copper-holding capacity is not exceeded. Due to high copper dissolving capacity, the system (closed loop) is increasingly used in the PCB industry. The functions of different ingredients of this etchant are detailed below.

- Cu^{2+} (copper ions): act as oxidizing agents and dissolve metallic copper;
- NH_4OH (ammonium hydroxide): complexing agent and also holds copper in solution;
- NH_4Cl (ammonium chloride): improves the etch rate and copper holding capacity and solution stability;
- NH_4HCO_3 (ammonium bicarbonate): acts as a buffer to preserve the solder metal surface;
- NH_4NO_3 (ammonium nitrate): increases the etching speed and preserves the solder metal surface;
- $(NH_4)_3PO_4$ (ammonium phosphate): retains clean solder and plated through holes;
- $NaClO_2$ (sodium chlorite): also an oxidizing agent that reacts and dissolves metallic copper.

Composition and Operating Conditions:

- Ammonium bicarbonate   75 g/l
- Ammonium nitrate   80 g/l
- Cupric chloride   200 g/l

- Ammonium chloride   100 to 110 g/l
- pH   7.8 to 8.2
- Temperature   45 to 55 °C
- Specific gravity   1.2 at 20 °C
- Copper content   150 to 160 g/l

Process Control and Regeneration

Etching solutions are operated between 50-55 °C and are well suited to spray etching. It is necessary to have an efficient exhaust system to drive away ammonia fumes which are released during operation.

A constant etch rate can be maintained by using automatic feeding controlled by specific gravity or density. In this method, as the printed boards are etched, copper is dissolved in the solution, resulting in increased density of the etching solution. When the quantity of copper in the solution reaches its upper limit, a pump is activated to replenish the etching solution and to simultaneously remove the etchant until the desired density is achieved.

The system is regenerated by removal of the spent etching solution or by chemical restoration of the spent etchant. This process involves mixing spent etchant with hydroxyl oximes (organic solvent), which is capable of extracting copper. This mixture containing copper is mixed with sulphuric acid to get copper sulphate and the copper-free etchant is reprocessed. Alternately, the processed etchant is cooled (chilled) and the precipitate salt is filtered which is followed by re-processing of the solution by adjustment of operating conditions. Regeneration by these methods is, however expensive and is viable only for large printed circuit facilities.

 ## 9.2   Etching Arrangements

9.2.1   Simple Batch Production Etching

Batch etching is applied in laboratories and small industries wherein small series of PCBs have to be etched occasionally. This simple arrangement for etching of PCBs involves the use of the etchant until saturation or until the etching speed becomes too slow. The etchant is then disposed and fresh solution is filled into the etching machine.

When the etching of a new batch of PCBs is started, the optimum etching time must be initially determined. The typical etchant used in this method is ferric chloride.

9.2.2   Continuous Feed Etching

In this method, a small steady stream of fresh etchant either continuously or periodically flows into the etchant sump while an equal quantity of partially saturated etchant is simultaneously removed.

In order to utilize the full copper solving capacity of the etchant, it is necessary to have etching with a very slow etching speed (conveyor speed). This, however, would result in a low productivity rate. The system is therefore operated in a mode with only partially saturated etchant which gives, more or less, constant results at a reasonable etching speed. However, the copper-solving capacity of the etchant is not optimally utilized. The typical etchant used in the continuous feed etching system is generally ferric chloride.

The etchant economy in continuous feed etching systems can be improved by cascading several etching modules. In such a system, the first module contains almost saturated etchant, whereas the following interconnected modules have decreasing copper content in the etchant and the last module operates with nearly fresh etchant. Usually, three or four modules are cascaded and the etchant flow between the modules goes via the sump overflows. This arrangement gives reasonable etching speed obtained in combination with a practically full utilization of the etchant copper-dissolving capacity. A typical problem with cascaded etching systems is the need to maintain the copper content within certain limits in each one of the modules.

9.2.3 Open Loop Regeneration

The methods described in the previous section depend upon utilization of the addition of pre-mixed full strength etchant, whereas the open loop regeneration systems employ the addition of chemical additives like replenishers and regeneration chemicals in order to maintain the etching performance at a constant level.

In such systems, which are mostly automatic, the composition of the etchant is monitored via a sensor for the pH value, the oxidation-reduction potential (redox potential), specific gravity or colour. The most typical etchant used in open loop regeneration is cupric chloride which permits the dissolution of typically 130 g cu per litre of etchant spent. Open loop regeneration can also be carried out manually, especially for smaller production volumes, but it needs careful and continuous monitoring of the etchant composition.

9.2.4 Closed Loop Regeneration

The closed loop regeneration involves the removal of copper containing by-product from the etchant mainstream while the copper-purified etchant is returned to the etchant sump. Although the investment in equipment is high, it offers efficiency and economy in respect of etching chemicals, constant etching performance and environmental pollution.

Among the etchants economically suited for closed loop regeneration, we find that cupric chloride, ammonium persulphate and alkali etchants are quite suitable. For each one of these etchants, the reactor/separator system is completely different in terms of how it suits the particular chemistry. The conditioning chemicals promote the formation of copper salts in the reactor which is, in certain cases, further supported by chilling of the etchant. The copper salts are then filtered out in the separator and stored as a by-product in a special tank. This by-product with its high copper content can usually be sold without causing any problems to chemical industries in terms of its further use.

The high complexity of the system needs meticulous maintenance for achieving a constant etching quality in the presence of widely variable chemical compositions.

 ## 9.3 Etching Parameters

Etching Rate: The rate of etching is determined by the amount of material removed per unit time.

Etch Factor: The etch factor is the ratio between the depth of etching (d) and the under-etching (b) (under-cut)

$$\text{Etch factor} = \frac{d}{b}$$

 ## 9.4 Equipment and Techniques

The etchants may be applied to boards in one of the following ways:

- Immersion etching;
- Bubble etching;
- Splash etching; and
- Spray etching.

Spray etching is the most commonly used technique due to its high productivity and fine line definitions.

9.4.1 Immersion Etching

Immersion etching is the semi-plast technique which requires only a tank containing etching solution into which the boards are immersed as shown in Figure 9.4. The boards are kept immersed until the etching is complete. This requires a long process time and the etch rate is thereby low. The solution can be heated to speed up the etching process. This method is suitable for small boards or prototyping. Normally ammonium persulphate or hydrogen peroxide with sulphuric acid etching medium is used for immersion etching.

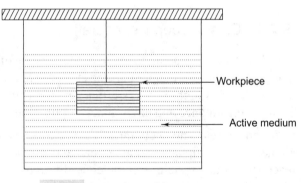

Fig. 9.4 Immersion etching system

9.4.2 Bubble Etching

This technique is a modified form of immersion etching with the difference that air is bubbled through the etching solution. Air, passing through the solution, has two functions:

- To ensure fresh etchant contact at the surface and to rinse away dissolved metal; and
- To enhance the oxidation power and to regenerate the etchant.

Figure 9.5 shows the schematic arrangement of the bubble etching technique.

The rate of etching, to a certain extent, depends upon the air pressure (normally upto 2 psi) to obtain good quality of etching. Chromic sulphuric acid and ammonium persulphate etchants are used in this technique. The primary disadvantage of bubble etching, when used with hydrogen peroxide-sulphuric acid etchant, is that it generates a significant quantity of corrosive aerosol. Effective fume collecting with active scrubbing must be implemented if a bubbler is used.

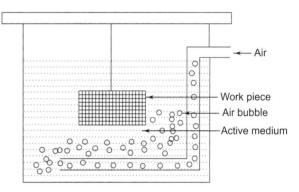

Fig. 9.5 Bubble etching system

9.4.3 Splash Etching

The principle of splash etching involves a paddle or cup attached to a motor-driven shaft. When the motor rotates, the etchant is thrown by centrifugal force towards the boards being etched. The contact of the solution with the boards depends upon the shaft rotation and paddle design as illustrated in Figure 9.6. Splash etching or paddle etching is better than bubble etching with regard to even etching and minimum undercut. But, only a limited number of boards can be etched at a time. Ferric chloride and chromic/sulphuric acid solutions are commonly used in this type of technique. A large

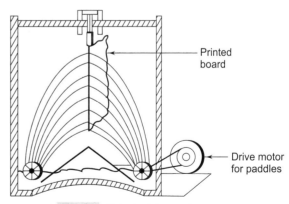

Fig. 9.6 Splash etcher system

volume reservoir is provided at the bottom of the tank to minimize solution replacement. The technique has become obsolete because of the low etch rates as compared to automatic spray etch machines.

9.4.4 Spray Etching

In its simplest form, a spray etching machine consists of a box type chamber having a sump below. The etching solution is pumped under pressure from the sump through a pipe network to the nozzles and splashed onto the board surface. This allows the fresh solution to be sprayed, giving a high etching rate. The factors which determine the evenness of etch are:

- Uniformity of spray pattern, force, drainage and pattern configuration;
- Etchant chemistry, the pump pressure, and nozzle configuration and placement, which determine the rate of etching; and
- The spray, which is done on both sides of the PCB in case of double-sided boards.

The boards are etched continuously in this closed loop system. The etch rate is high in this system with minimum under cut and fine-line definition. Ammonium chloride etchant is commonly used in this technique for double-sided PTH boards. The fabricated equipment should be made of acid and alkali resist material like PVC. However, equipment for sulphuric acid/hydrogen peroxide etchant system requires stainless steel, poly-carbonate or polypropylene material.

There are two types of spray etching techniques, which are:

- *Horizontal Spraying:* In this technique, etching is done from independently controlled spray nozzle banks at the top and bottom. Double-sided horizontal etches are generally preferred in PC manufacturing as a majority of the PCBs are double-sided. The design of a horizontal spray etcher is shown in Figure 9.7.

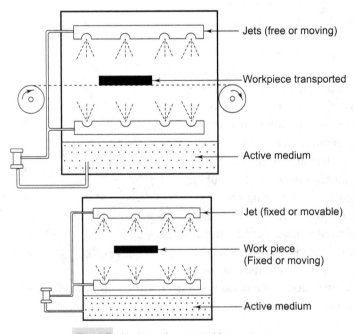

Fig. 9.7 Horizontal spray etching system

- *Vertical Spraying:* Figure 9.8 shows the design of a vertical spray etcher. In this technique, the etching is carried out by placing panels in a rack, which is lowered into the spray box area. A combination of nozzle movement and nozzle oscillation up and down or sideways, with a large number of nozzles, provides optimum results.

Spray etching machines are available in automatic and semi-automatic modes for vertical or horizontal type of spray. The automation contains pressure control, heating, specific gravity indicator and automatic solution regeneration. Automatic machines are designed for high production rates. The boards are loaded on a rack, which is passed through the etch chamber, where it is sprayed on one or both sides by an oscillating bank of spray nozzles. The rack is then used to spray rinse with water and neutralizing chambers. The pressure to each bank of spray nozzle can be easily controlled.

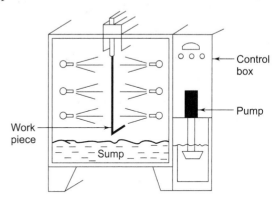

Fig. 9.8 Vertical spray etching system

For the manufacture of prototype and small batch quantities of PCBs, rotary etching machines prove useful. In this machine, the etching solution is contained in a tank at the base of the machine. It is heated by a quartz heater, which provides a short warm-up time and constant temperature that is controlled electronically. Machines have the capacity to handle 300×500 mm boards. In the PCBs, the board holder is rotated $180°$ before the second etching cycle. A provision is made to have a wash tank for the rinsing of the etched boards. This machine facilitates line resolution of better than 0.1 mm and an etching speed of only 90 seconds with fresh ferric chloride.

 ## 9.5 Etching Equipment Selection

The following factors are generally taken into consideration while selecting equipment for etching:

- *Maximum Board Size*: The maximum board size determines the size of the tanks for etching, rinsing, cleaning and neutralizing, as well as the size of the holding rack or conveyor.
- *Quantities of Board*: This determines the type of etching equipment required to meet the production needs. The equipment should be able to complete the normal daily production target while allowing for solution adjusting time, machine maintenance and actual hours of operation.
- *Space Available for Etching*: This is needed to allow generous amounts of walkaround space with extra room for easy loading and unloading. Storage racks, bins, tables and shelves should be provided in each area where boards are stacked, which are regained from plating, screening, photo-resist coating and etching.

- *Types of Etching*: This includes fine line, wide line or print-and-etch only.
- *Types of Boards*: These are metal resist plated, print-and-etch, etc.

9.6 Optimizing Etchant Economy

In PCB etching on an industrial scale, the following are the various desirable features which, in practice, can only be partially fulfilled:

- High etching speed;
- High copper dissolving capacity;
- No attack on the resists used;
- Constant etching speed;
- Easy disposal of spent etchant or by-products;
- Little toxicity and fumes;
- Easy regeneration;
- Low costs for chemicals; and
- Little post-etch cleaning requirements, etc.

The response to this challenge in terms of equipment selection, and the availability of conveyorized spray etching machines is now considered as a standard approach.

9.7 Problems in Etching

The etching process is one of the most important steps in PCB fabrication. It looks simple, but, in practice, several problems are encountered during this stage, which affect the quality of the final boards, especially in the production of fine line and high precision PCBs. The two commonly encountered problems are under-etching and overhang.

9.7.1 Under-etching or Under-cut

During the etching process, it is expected that the etching would progress vertically. However, corrosive action of the etchant works in all directions, and in practice, there is usually an etching action sideways which attacks the pattern below the etch resist. Side corrosion is promoted by the movement of the liquid and the dissolution of copper takes place gradually. The final wall of the conductor becomes an inclined line instead of a vertical one at one end. This is shown in Figure 9.9. This can lead to a considerable reduction of conductor line widths.

The simplest approach to minimize under-etching is by keeping the etching time as short as possible. This is achieved by using fast working etchant and exercising exact control on the etching time.

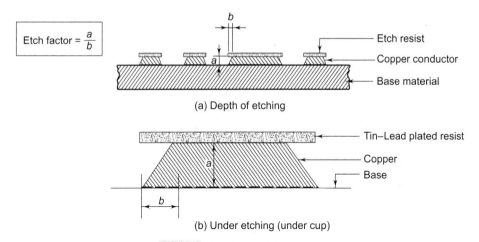

(a) Depth of etching

(b) Under etching (under cup)

Fig. 9.9 Etch factor definition

A common term used to express under-etching is the etch factor, defined as the ratio of etching depth (copper foil thickness) to the width of the side attack, i.e.

Etch factor = a/b

where a = thickness of copper foil

b = width of the side attack

Fine line etching with a minimum of undercut is best achieved with a copper foil of ½ oz or less and is carried out by removing the board from the etching machine exactly at the time of completion.

9.7.2 Overhang

When metal etch resists are used, i.e. in the pattern plating process, the metal plating built-up may show growth sideways in the form of a projection termed as overhang (Figure 9.10). It represents a potential source of trouble because excessive overhang may break or, fall loose in the form of long narrow metallic strips, which can cause electrical short-circuiting between adjacent conductors. Therefore, it is advisable that after etching, the removal of overhang is done by soft brass brushing, ultrasonic agitation and rinsing, or fusion in the case of solder-plated resists. In many cases, only brushing is adequate. The difficulties with overhang are also considerably

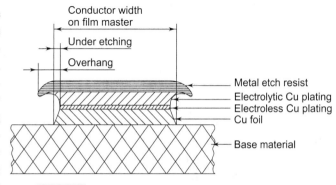

Fig. 9.10 Overhang and undercut in pattern-plated PCBs

reduced with the use of dry film resists. Dry film resists are available in thicknesses of as much as 70 μm and can therefore, act as an effective barrier against sideways growth of plating layers. The figure also shows under cut and overhang for a plated metal resist pattern. In critical cases, the influences of overhang and under-etching are compensated for by modifying the conductor widths on the film master.

9.8   Facilities for Etching Area

The following facilities are required in the etching area:

- Adequate electric power, exhaust systems, storage areas and safety provision are required in the etching facility.
- The floor construction should be chemical-resistant tiling and mortar or acid-resistant epoxy-coated.
- Tanks and pipes must be kept off the floor to allow inspection and to keep them as dry as possible.
- Tanks should be supported well above the base floor and have proper inspection ports.
- Spent solution should never be put into the sewerage system. It should be treated according to the regulatory pollution control practices. Rinses should be waste-treated and discharged as acceptable effluents.

9.9   Electrochemical Etching

In electrochemical etching, the material to be etched is made the anode and the cathode is generally made of a similar metal. The electrolyte used depends upon the metal to be etched. The etching rate is controlled by varying the current.

9.10   Mechanical Etching

In the mechanical etching process, the surface metal is selectively removed by a multi-axis milling machine and a special milling cutter to remove a narrow strip of copper from the boundary of each pad and trace. A number of configurations are currently available for these special mechanical etch bits, but most users report that bits with spiral flutes (vs. a flat 'spade' geometry) are the most effective at removing copper debris and tend to stay sharp longer at higher cutting rates. Tip angles of 60° and 90° are the most common, with 90° seeming to offer the best combination of minimal substrate penetration and longer cutter life. If the circuit design also requires that some (or all) of the non-circuit copper be removed (clear milling), conventional carbide end-mills can be used to accelerate the copper milling process. The typical diameters range from 0.010" (2.5 mm) to 0.050" (1.27 mm). Figure 9.11 shows the typical arrangement for mechanical etching. The carbide etching removes a strip of copper around the boundary of each circuit element. The electrical isolation depends upon the total removal of all copper debris from the milled trough.

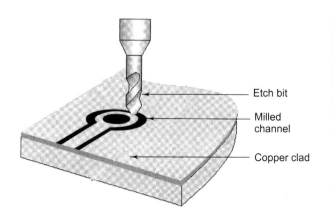

Fig. 9.11   Mechanical etching technique

Fig. 9.12   Rapid prototyping machine (courtesy LPKF, Germany)

The process is only suitable for the design and development of prototype cards. The special machines used for mechanical etching are called 'Rapid prototyping machines'. In these machines, the design information is fed in the Gerber Format which is then loaded into the proprietary prototyping software (Crum, 1995). Once the prototyping process is complete, the software can be used to convert the information back into a Gerber file. The outputted Gerber file can then be used to procure the prototype in production volumes o f the board. The standard LPKF rapid prototyping machine (Figure 9.12) is capable of producing 4 mil lines and spaces; even finer lines and spaces can be produced by the company's laser equipment.

Mechanical Operations

10.1 Need for Mechanical Operations

The PCB fabrication process involves a number of mechanical operations to prepare the circuit boards for the chemical processes of image transfer, plating, and etching. The process starts with the acquisition of laminate sheets which may be as big as 2 m × 2 m or bigger. Therefore, the mechanical process such as cutting to size, drilling holes and shaping play an important role in the final quality of the printed circuit board.

Unlike other PCB processing steps, most of the mechanical operations require considerable manual handling of the board. Such operations do not form part of the transfer-line equipment, and would therefore, have a major impact on the costing of the final product.

Each mechanical operation has its own set of requirements in terms of tolerance and accuracy. In general, tolerances should be made as narrow as they are functionally really needed. Proper choice of tools and their sharpness are very important in each of the mechanical operations for obtaining an acceptable machining finish. Blunt and dull tools result in chipping because of resin brittleness of the laminate. Proper application of machining forces must be kept low. Excessive machining forces may cause partial de-lamination because of the inherent laminate structure, thereby weakening the interlaminar bond strength. A good knowledge of the base material composition, equipment and tools complemented with good operator habits can facilitate good mechanical operations resulting in higher quality PCB yield.

10.2 Cutting Methods

10.2.1 Shearing

Shearing is perhaps the first mechanical operation carried out on PCBs to give them proper shape or contour. It is basically a cutting method applicable to all kinds of base materials, generally of less than 2 mm thickness. When cutting boards have more than 2 mm thickness, shearing results in the edge finish which is coarse and unclean, and therefore, the method becomes unacceptable.

Laminate cutting by shearing can be done either by manually operated or motor-driven machines. Both types however, have common constructional features. A shearing machine normally has an adjustable set of shear blades as shown in Figure 10.1. The blades are of rectangular shape. The lower blade has a free angle of about 7°. The cutting length capability is available upto 1000 mm. The lengthwise angle between the blades is generally preferred between 1–1.5°, though for glass epoxy materials, upto 4° can be used. The clearance between the cutting edges of the two blades should be limited to less than 0.25 mm.

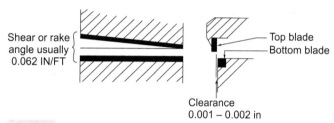

Fig. 10.1 Typical adjustable shear blades for copper clad laminates

The angle between the shear blades depends upon the thickness of the material to be cut. The thicker the material, the greater will be the angle. If the shear angle is too high or the gap between the plates is too wide, feathered cracks appear while cutting paper base materials. However, in case of glass epoxy laminates, even though cracks do not develop due to flexural strength of the material, the material does show deformation if the shearing angle is too wide or the blade gap is too large. For obtaining a clean edge finish in paper base materials by shearing, heating the material in the range of 30–100 °C is helpful.

In order to obtain a clean cut, the board must be firmly pressed down with a spring-loaded hold-down device to prevent the otherwise unavoidable shifting of the board during shearing. Also, parallax errors which may result in errors upto 0.3–0.5 mm, should be minimized and precision stoppers should be used for alignment of the corner marks.

Shearing machines are available to handle jobs of various sizes and to offer accurate dimensional reproducibility. There are larger machines which can cut several hundred kilograms of base material per hour.

10.2.2 Sawing

Sawing offers another method of cutting the laminates. The method is preferred as it gives a smoother edge finish and clean cut, though the dimensional tolerances are similar (0.3-0.5 mm) to that of shearing.

In the PCB industry, mostly circular sawing machines of the moving table type are preferred. The saw blade speed is adjustable between 2000-6000 rpm. But once set, the cutting speed should not vary. This is achieved by using heavy pulleys with more than one V-belt.

High speed steel blades with a diameter of approximately 3000 mm are used at a speed of 2000-3000 rpm for paper phenolic materials. They are about 1.2-1.5 tooth per cm circumference. For glass epoxy laminates, tungsten carbide tipped blades are used. For still better performance, diamond wheels are preferred. Although they may require higher initial investment, they help in future savings due to their long life and improved edge finish.

The following precautions will prove useful during operation of the sawing machine:

- The precision of the bearings has a direct impact on the edge finish. Check the bearings for tightness. No play should be perceptible when inspected by hand.
- For safety, the blade should always be covered by a suitable guard device.
- The alignment of the arbor and the motor mounting should be correct.
- The clearance between the saw blade and bench should be minimum to provide a good support to the board for the cutting edge.
- The circular saw should be so adjusted that the free height of the blade above the boards should be in the range of 10-15 mm.
- A blunt or badly sharpened blade and too coarse teething can result in a bad edge finish. Proper care should be taken to avoid them.
- An incorrect feed rate can lead to a bad edge finish. Adjust it properly. Thick materials need a lower feed rate while thin materials can be cut faster.
- Manufacturers recommendations pertaining to speed should be followed.
- If the saw has a thin blade, a stiffening collar is used to reduce vibration.

10.2.3 Blanking of PCBs

When PCBs are designed to have shapes other than rectangular or have an odd contour, the use of a blanking die is a faster and more economical method. Blanking basically consists of a clean cutting operation done with a punching tool rather than with a saw or a shearing machine. In some cases, even hole punching and blanking are done with the tool in the same operation. However, when a superior edge finish or a tight dimensional tolerance is required, blanking may not fully serve the purpose. In PCBs, blanking is very well adopted on paper-based laminates but rarely on glass epoxy laminates. Blanking helps to achieve PCB dimensions within a tolerance of $\pm (0.1 - 0.2$ mm).

Blanking a Paper-based Laminate

Since paper-based laminate material is soft as compared to glass epoxy laminate, it is a more suitable laminate for blanking. While designing a blanking tool for a paper-based laminate, the resilience or yield of the material is taken into consideration. The blanked part will be, in general, slightly larger than the die which produced it because the paper-based laminate tends to spring back. Hence, the die is made slightly under-sized compared to the print size to compensate for the over-sizing, depending upon the tolerance and the thickness of the base material. It may be noted that when a hole punching is done, the die is over-sized, whereas when blanking is done, the die is under-sized.

In the case of a complex configuration, a progressive tool is preferred i.e. the strip of material progresses from one stage to the next with each stroke of the die. In this way, holes are pierced in the first one or two stages and in the final stages, the completed part is blanked. The finish of the PCB can be improved by doing the piercing and blanking operation in hot conditions i.e. after heating the strip to about $50°–70$ °C. However, care must be taken not to over-heat, which may cause the PCB to shrink out of tolerance when cooled. Also, the thermal expansion characteristics of the base material should be kept in view as paper phenolic material shows differential expansion in x and y directions.

Blanking a Glass Epoxy Laminate

Even though not always recommended, glass epoxy PCBs in special cases are punched out, when it is not feasible to produce the required shapes by shearing or sawing. However, this is resorted to when the finish or the dimensions required are not really stringent. Therefore, the edge finish may not look very clean though it is functionally acceptable. The tool developed for blanking out a glass epoxy laminate will have a closer fit between the die and the punch as the glass epoxy laminate is found to exhibit less resilience as compared to paper-based laminates. Glass blanking is always done at room temperature.

Since the glass epoxy laminate is harder and tough to blank, the punch life will be limited. It will wear out very fast. A better result can be obtained by having carbide tipped punches.

10.2.4 Milling

Milling is a commonly used operation which can be applied for the clean cutting of PCBs and for obtaining good edge finish and overall dimensions with a high degree of accuracy. The generally used cutting speeds are in the range of 1000-3000 rpm. They usually employ straight or spiral tooth HSS (high speed steel) milling cutters. However, in the case of glass epoxy laminates, the use of tungsten carbide tools is preferred due to their long life. In order to avoid de-lamination, the PCB must be given strong support with backing plates during milling. For details of milling machines, tools and other operational aspects, standard texts on workshop or machine shop equipment may be consulted.

10.2.5 Routing of PCBs

For obtaining superior edge finish and higher dimensional accuracy than that obtainable from shearing or sawing, especially for PCBs with odd contours, routing becomes a method of choice. The dimensional tolerances within $\pm$ (0.1-0.2) mm can be achieved with a much lower cost than blanking. Therefore, in some cases, blanking with slight over-size is applied followed by routing to get a smooth surface finish.

The use of present day multi-spindle machines especially ensures that the routing is much faster, and that both the labour cost and total cost much less than that entailed while making a blanking die.

When the board has traces close to the board edge, routing is perhaps the only blanking method that is capable of providing an acceptable quality of boards.

Routing is basically a machining process similar to milling but done at a much higher cutter speed and feed rate. In this method, the boards are moved past a vertical side mill with the aid of a routing jig. The routing jig is guided in relation to the mill by holding it against a bushing which is concentric with the mill. The positioning of the PCB on the routing jig is determined by the material registration holes.

There are three basic routing systems available. They are:

a) Pin routing;

b) Tracer or stylus routing; and

c) NC routing.

Pin Routing

Pin routing is best suited for low volume production requiring very fine edge finish with high dimensional accuracy. For pin routing, an accurately machined template, of either steel or aluminium, made exactly as per the outline of the PCB is required. The template is also provided with pins for registration for positioning the boards. Three or four boards are usually stacked against the pilot pin that protrudes from the table. The cutter used is of the same diameter as that of the pilot pin. The stack is passed in the direction against the direction of rotation of the router bit. Usually, about two or three passes are required to assure proper tracking since the router tends to force the work away from the pilot pin.

Even though this process is labour-intensive and depends upon the skill of the operator, pin routing is found to be the most suitable for low volumes requiring irregular shapes, but high accuracy and finer finish.

Tracer Routing

Tracer router also makes use of a template similar to the one used in pin routing. Here, a stylus traces the board outline on the template. The stylus may control the movement of the spindles over a fixed table, or it may control the movement of the table in case of fixed spindles. The latter approach is commonly used with multiple spindle machines.

The template is machined to the board outline, with the stylus tracing the external edge. The first cut is made with the stylus tracing the external edge. During the second pass, it traces the internal edge, which will relieve most of the load on the router and give a better control of the dimensions. In stylus routing, the accuracy achieved is much higher than that in case of pin routing. With moderate operator skill, tolerance of the order of ± 0.010 inch (0.25 mm) for high volume production can be achieved. With multiple spindle machines, upto 20 boards can be routed at one time.

NC Routing [Numerical Control Routing]

The introduction of CNC (Computer Numerical Control) with multiple spindles is the preferred method of routing in the present-day PCB manufacturing industry. When the production volume is very high and PCBs have difficult contours, NC routing is the method of choice. In these machines, the movement of the table, spindle and cutters are controlled by a computer, and therefore, the work of the operator is limited to just loading and unloading. Complex shapes can be cut with very close tolerance levels, particularly in very high volume productions.

In case of NC routing, the program (a set of commands) to control the movement of the spindle in the x, y and z axes can be written very easily, which makes the machine follow a certain path while routing. Commands for the speed and feed rate are also included in the software program. Design changes can be conveniently made by modifying the software program. Information on the contour is fed to the computer directly from the program.

NC machines are generally operated at an rpm of 12,000 to 24,000 with a carbide router. The motor should have enough power to drive the router without much of a loss in its rpm.

Tooling or registration holes are generally provided inside the periphery of the circuit board. Although routing enables square external corners to be achieved, the internal corners will have a radius equal to the radius of the router bit in the first cut. This could be removed in the second operation by giving a cut at $45°$, thus also giving squared inside corners.

In NC machines, the variables which determine the cutter speed and feed rate are basically the laminate type and its thickness. A cutter rotation of 24,000 rpm and feed rate upto 150 in/min. can be effectively used on many of the laminates. But a softer material like Teflon-glass and similar materials, the laminate binder of which flows at low temperatures, will require a lower rpm of 12,000 and high feed of 200 in/min. in order to minimize the generation of heat.

Generally used cutters are of the solid tungsten carbide type. With the precise control of table movement in NC machines, cutter bits are not subject to shock. Therefore, cutters of smaller diameter can be used successfully.

In case of NC routing, the geometry of the cutter tooth plays an important role. Since faster feed rates are employed, a cutter with an open tooth form is preferred so that the chips could be released faster and more easily. Generally, a diamond toothed cutter will have a life of 15,000 linear inches before it starts eroding. If very smooth edges are required, a fluted cutter may be used.

For expediting the loading and unloading operations, an effective hold-down and chip removal system are provided in the machine itself. There are various methods by which the boards could be mounted on the machine table, while properly registering them to facilitate the routing outline. The most popular method is to have shuttle tables available so that loading and unloading may be accomplished while the machine is cutting.

Laser Routing

Lasers are also being used for routing purposes. The freely programmable and flexible mode of operation makes UV lasers particularly suitable for the precision cutting of HDI applications. The cutting velocities which can be reached, are material-dependent and fall typically within a range of 50 mm to 500 mm per second. The edges are clean and don't need any post-processing, as would usually be required with mechanical routing or punching or when cutting with a CO_2 laser (Meier and Schmidt, 2002).

10.3 Hole Punching

Punching of holes into PCBs is also one of the machining operations like hole drilling. However, the hole diametrical accuracy, hole wall finish and pad de-lamination from the base material are not as good in case of hole punching as compared to drilling. In general, large holes are easier to punch than small holes. For example, failure in holes made by punching below 0.9 mm for paper-reinforced laminates and 1.2 mm for glass cloth-reinforced laminates are quite common. So, punching is used in the high volume production of consumer type PCBs made of paper phenolic and paper epoxy laminates. The other disadvantages of punching are the pad de-lamination and laminate cracks on adjacent holes. In addition, punching results in holes of conical shape and rather rough surface. They are thus not compatible with the professional PCB plated through-hole process where high surface smoothness is desirable.

Along with the above laminations, hole punching also has some advantages. They are:
- Low operation cost because a large number of holes can be punched simultaneously;
- Very high production rate; and
- High accuracy and repeatability of the hole position.

Actual punching involves the use of a 'press' with a capacity of 10 to 40 tons and 100–200 strokes/mm. In single-sided boards, the punching is done on the copper foil side. When paper base laminates are to be punched, it must be noted that the materials are resilient and that their tendency to spring back will result in a hole slightly smaller than the punch which produces the hole. The differences are due to the thickness of materials. So, while making dies, some positive tolerance is required to obtain the correct hole size. The diameter of the punches, in general, should be 0.1 to 0.12 mm bigger than the desired hole diameter in boards of 1.6 mm thickness. The typical diameter tolerance of punched holes is ± 0.1 mm.

For precision work, there must be a close tolerance between punch and die. Generally, the die should be only 0.002 to 0.004 in. larger than the punch for paper-based materials whereas for glass-based laminates, it should be one half of that tolerances. Figure 10.2 shows an example of the required tolerance of a punch and a die.

Punching loads depend upon the types of the laminates and their shear strengths. The shear

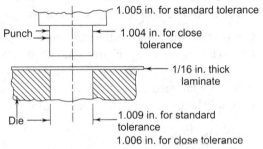

Fig. 10.2 Proper tolerance of a punch and die

strength of the paper-based laminate is 12,000 psi (max) and for glass epoxy laminate, a maximum of it is 20,000 psi. So, the press for paper-based laminate should be for 16 tonnes capacity. In order to provide a good safety factor, a 32 ton press is generally used. Glass-based laminate material, which has a shear strength that is 70 per cent more than that of paper-phenolic laminate, requires press with much higher capacity even for simple boards.

Paper phenolic laminates (XXX and similar) are pre-heated to a temperature from 50 to 70 °C before punching to avoid chipping. Laminates such as XXXPC and FR2 are punched at room temperature provided it is above 20 °C.

Non-woven glass-reinforced laminates (both epoxy and polyester type) have good punchability character. With a die clearance of 50–100 microns, holes with smooth surfaces fit for plating can be obtained.

Many punching problems such as pad de-lamination can be reduced by using the following techniques:

   a) The punch should always cut from the copper side;
   b) Holes are punched prior to etching.
   c) Pads must be sufficiently large for punching.

Punching is economically attractive only if a PCB quantity is substantial, say at least 2000 numbers. Therefore, it is usually preferred to punch non-PTH high volume boards made from paper-reinforced laminates and drill all others.

In the punching process, the occurrence of excessive breakages of small punches could be due to:

   • *Poor Alignment*: This can be easily detected by close examination of the tool.
   • *Poor Design*: It usually means that the punch is too small to do the job required.

When carrying out punching operations, always pierce with the copper side up. It is not advisable to use designs with circuitry on both sides of the board because that could result in the lifting of pads. Also, if the distance between the holes is too small, there is a likelihood of cracks appearing between holes. In such a case, the process should be planned such that the piercing is done before any copper is etched away. The presence of the copper foil has a reinforcing effect and helps in eliminating cracks.

10.4 Drilling

Drilling operation is one of the important mechanical processes in the manufacture of printed circuit boards. Its purpose is two fold: (i) To provide component lead mounting precisely and with structural integrity, and (ii) To establish an electrical interconnection between the top, bottom and sometimes intermediate conductor pathways.

After the drilling process, the drilled circuit board undergoes various processes like plating, imaging, etching and solder plating. Therefore, care is needed to obtain a good surface on the drilled hole and hence its quality assumes great significance. The quality of a drilled hole depends upon various factors such as the quality of the laminate and drills, processes including machine accuracy, drilling techniques and operator skill in control and evaluation of hole quality and drill bits. The important steps in the drilling process are shown in Figure 10.3.

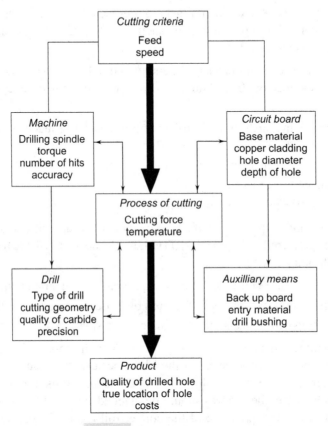

Fig. 10.3 Typical steps in drilling

When all the processes are properly implemented, high quality drilled holes are obtained. Those boards can be plated directly, thereby eliminating other processes such as deburring, de-smearing and etch-back. This results in process simplification, higher yields and lower costs.

Various studies have established that the root cause for as many as 85 per cent of all circuit board failures can be traced back to drilling. On examination of the entire board manufacturing process, it proves that many post-drilling operations are corrective measures designed to overcome shortcomings in the drilling process. For example, the use of mechanical scrubbing to remove burrs, of chemicals to remove resin smear and bonded debris, of etch-back to expose glass fibres, and of acid or alkaline cleaners to remove contaminants are all methods for addressing problems that result from the drilling process. It is, therefore, imperative that greater attention is given towards addressing the source of the problem rather than compensating for drilling problems (Vandervelde, 2001) at a later stage.

Holes are drilled using single head manually controlled machines. The operator centres the pads through an eye piece or by means of a pantograph. Mass production facilities usually utilize numerically controlled drilling machines with several heads.

10.4.1   Drill Bit Geometry and its Importance

Drills used for making holes in PCBs are usually made of high speed steel [HSS] and tungsten carbide. They are available in two shapes: common shank and straight shank. Figure 10.4 illustrates the geometry of the two types of drills.

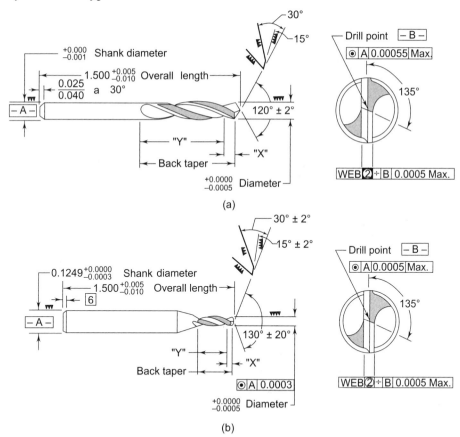

(a)

(b)

Fig. 10.4   (a) straight shank drill (b) common shank drill (redrawn after Coombs, 1988)

The function of the drill bit is to cut and remove the base material and copper. So, the design of a drill is as important as the materials used. The design and its wear and tear during use affect its ability to provide smoothness of the hole wall, while entry and exit burns remove chips and affect drilling temperature. It is therefore important to understand the function and the geometry of each part of the drill. Most bits for PCB drilling are of the common shank design. This allows a drilling machine to use many bit diameters with only one collet. Figure 10.5 shows a typical drill bit geometry. The point angle determines the ability of the tool to cut the laminate material and it usually varies between 90 to 130°. For paper-based material (FR3), it is between 90 to 110° while for glass-based material (FR4, G10, FR5), it is between 115 to 130°. The most commonly used point angle in drilling is 130°.

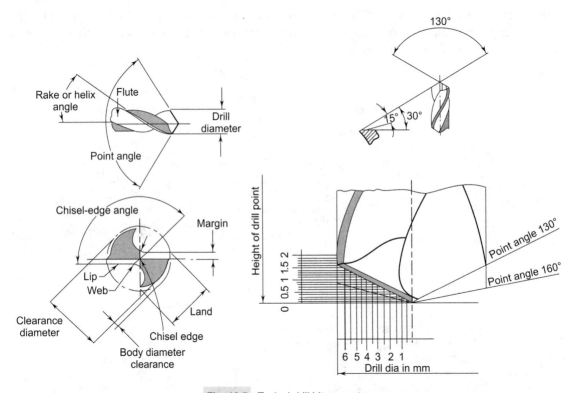

Fig. 10.5 Typical drill bit geometry

The drill point has two primary edges, which are parallel to each other and are separated by the web. The cutting edge is called the chisel edge. The cutting edge must be straight for uniform cutting and should not have nicks or grinding marks. The plate angles (Figure 10.6) must be equal; otherwise the drill will not register at the centre.

The plate angle or helix angle varies from 20 to 50° and it determines the ability of the drill to remove chips from the hole. A large helix angle (50°) is needed to assure good chip removal and prevent breakage of drill. The flute of the drill should be polished to reduce frictional heat and to improve the drill bit life.

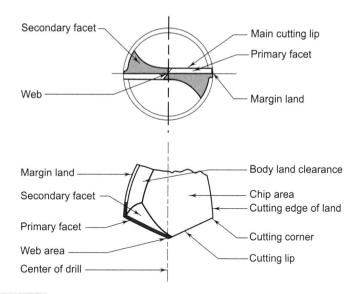

Fig. 10.6   Drill bit geometry showing the position of web and cutting edge

10.4.2   Types of Drill Bits

Normally, most PCBs are drilled with carbide bits rather than high-speed steel (HSS). Carbide bits have good resistance to heat and high hardness. The quality of cutting edges or drill points of the drill bit are very important. Drill bits should not be pressed against metallic or hard surfaces to avoid damage to the bit geometry. The drill bit surface should be cleaned by 1 per cent trisodium phosphate in water for 20 to 30 seconds. This removes the oils and debris from their surface.

Carbides are non-ferrous metals, which are hard in their natural state. In other words, they are not the same as steel. Their hardness is determined by two factors, namely the structure of the carbides and that of the bonding metal (cobalt). The carbides basically determine the wear resistance. The bonding metal determines the strength in accordance with its proportion. The hardness is 9.8 to 9.9 on the mohs scale of hardness at temperature between 1300 to 1600 °C. The fine grained tungsten carbide cobalt alloys produced today are 94 per cent tungsten carbide and 6 per cent cobalt.

The standard PCB drill for holes of 0.024" (0.6 mm) and larger is composed of wear-resistant cemented tungsten carbide crystals. Their composition, i.e. 94 per cent tungsten carbide (WC) and 6 per cent cobalt (co), has provided maximum drilling speed and tool life for years. For holes with diameters of 0.018" (0.45 mm) or smaller, several PCB drilling problems are encountered. These include a higher frequency of drill breakage upon retract, an increase of hole location scrap, and a decrease in output due to a reduction in the PCB stack height. Johnson and Sparkman (1996) showed that diamond-like carbon-coated drills (DLC) improve point life, operate 25 per cent faster than conventional drills, provide hole location improvement, and enhance capability on difficult products such as unused lands and micro-sized holes.

10.4.3 Drill Bit Inspection

It has been observed that the quality of the drilled hole walls decreases rapidly after a drill has drilled 6,000 – 10,000 holes through a 1.59 mm thick laminate of G-10 or FR-4. Drill wear is almost the same for other grades of laminates such as FR-2. However, it is twice as great for G-11 and FR-5 hard drills can give almost twice the life of ordinary drill bits. Therefore drill bit checking is very important in PCB production, because the drilling alone accounts for approximately 25 per cent of the total circuit board cost. So, the entire tool geometry including the point angle, helix angle, clearance angle, web thickness, width margin, body land clearance, faults in the flute section, overall length, flute length and drill diameter should be checked by using the workshop microscope. Mechanical measuring equipment should not be used for drill bit inspection since it may damage the drill geometry (Figure 10.7).

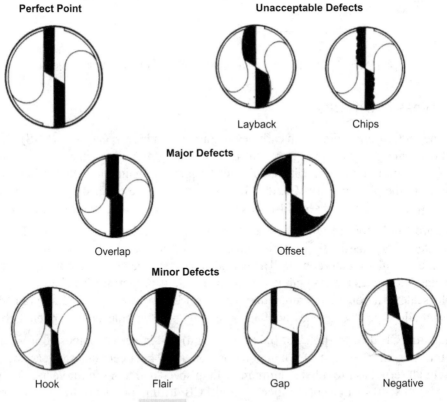

Fig. 10.7 Common drill conditions

The useful life of the drills must be established on the basis of the manufacturing process and the quality required for the PCB so that a constant quality of the holes can be obtained. A practical way to do this is to periodically take a sample of the drilled panel, plate it with electroless copper, micro-section it and examine it through a microscope.

10.4.4 Drill Bit Sizes

Table 10.1 Drill Sizes—Decimal and Metric Equivalents

Size	Decimal	mm	Size	Decimal	mm
85	0.011	0.2794	36	0.1065	2.7050
80	0.0135	0.3429	35	0.110	2.7939
75	0.021	0.5334	34	0.111	2.8193
70	0.028	0.7112	33	0.113	2.8701
65	0.035	0.8890	32	0.116	2.9463
60	0.040	1.0160	31	0.120	3.0480
59	0.041	1.0414	30	0.1285	3.2638
58	0.042	1.0668	29	0.136	3.4543
57	0.043	1.0922	28	0.1405	3.5686
56	0.0465	1.1811	27	0.144	3.6576
55	0.052	1.3208	26	0.147	3.7337
54	0.055	1.3970	25	0.1495	3.7972
53	0.0595	1.5113	24	0.152	3.8607
52	0.0635	1.6129	23	0.154	3.9115
51	0.067	1.7018	22	0.157	3.9877
50	0.070	1.7780	21	0.159	4.0385
49	0.073	1.8542	20	0.161	4.0893
48	0.076	1.9304	19	0.166	4.2163
47	0.0785	1.9939	18	0.1695	4.3052
46	0.081	2.0574	17	0.173	4.3941
45	0.082	2.0828	16	0.177	4.4957
44	0.086	2.1844	15	0.180	4.5719
43	0.089	2.2606	14	0.182	4.6227
42	0.0935	2.3749	13	0.185	4.6989
41	0.096	2.4384	12	0.189	4.8005
40	0.098	2.4892	11	0.191	4.8513
39	0.0995	2.5273	10	0.1935	4.9148
38	0.1015	2.5780	9	0.196	4.9783
37	0.104	2.6415	8	0.199	5.0545

10.4.5   Tool Life and Re-grinding (Re-pointing)

The *tool life* is measured in terms of the number of holes drilled until the point is re-ground. The end of the tool life is identified by the following tool life characteristics:

- Type of base material being drilled;
- Hole quantity; and
- Cutting conditions applied.

While using high precision multi-layer boards, the drill bits are changed after 500 holes at the most. For drilling double-sided circuit boards, the drills are changed after 2000 to 3000 holes in a stack that is three boards high. The end of the tool life is determined by constantly increasing burr formation on the copper layer of the uppermost circuit board. The dull drill will tend to break.

In order to keep drill bits sharp and to avoid breakage, they are generally used for 750 to 1,500 hits on multi-layer circuit boards and for 2000 to 3000 hits on double-sided boards. Hit counts greater than 3000 can be realized on single-sided boards.

Carbide drills for drilling PCBs can be reground 2-5 times on special purpose machines. During re-grinding, the entire length of the worn margin is ground away, otherwise the re-ground drill would jam in the hole. Some drill grinding machines are available with built-in microscopes having magnification upto 20 times. PCBs between 0.2 to 0.4 mm can be re-ground with maximum precision and grinding quality by using such machines. The smaller the bit, the fewer times it can be re-pointed since smaller diameter holes are more critical and require superior drilled hole quality. Drill bit replacement and re-pointing represent substantial expenditures for circuit board manufacturers. Therefore, proper storage, handling and inspection are critical to contain costs and to ensure maximum life span and optimum performance.

10.4.6   Requirements in Drilling

A good drilling technique must satisfy the following conditions:

- Consistent high quality;
- Perfect through-hole plating;
- Smaller diameter, shorter distance between the holes;
- Greater production; and
- Lower cost and simple storekeeping.

All these factors are easily achieved in CNC drilling, which is explained at a later stage.

10.4.7   Drill Speed, Feed and Withdrawal Rates

The speed and feed are the cutting conditions under which a tool operates.

Drill Speed: The drill speed is the speed of the drill spindle expressed in revolution per minute (rpm). For PCB work, 15,000 rpm is considered as the minimum speed. However, the preferred range is 20,000-60,000 rpm.

Feed Rate: The feed indicates the depth to which the drill penetrates into the material in one revolution. The feed is specified in mm/revolution and it is determined as follows:

$$Feed = \frac{\text{Feed in mts/min} \times 1000}{\text{Speed in rpm}}$$

Best results are obtained with feed rates adjustable without steps in a range of 0.01-0.05 mm per drill bit revolution. The drill machines are usually equipped with control of the drill speed and feed rate. With the drill speed in the range of 15,000-60,000 rpm, the feed is adjusted according to the speed so that the drill advances as follows:

- For smooth holes   :   10-30 microns/revolution
- For PTH hole work   :   30-60 microns/revolution
- For low accuracy holes   :   60-100 microns/revolution

Cutting Speed: The cutting speed is the path covered by one cutting corner at the corresponding speed within a given unit time. The cutting speed is stated in m/min and is determined as follows:

$$Cutting\ Speed = \frac{Drill\ dia \times \pi \times Speed\ in\ rpm}{1000}$$

For tungsten-carbide drill bits, the recommended cutting speed is in the range of 70-200 m/min.

Withdrawal Rate: The time taken by the drill bit to come out after the drilling operation is called the withdrawal rate.

10.4.8 Function of Clean Holes

The feed and cutting speed determine the hole quality, and affect the quality and life of the tool. The heat build-up and temperature produced during drilling are directly related to the time spent by the drill in the drilled hole.

A high feed reduces the drilling time and low cutting speed reduces the friction of the drill on the hole wall. This, in turn, reduces heat build-up. The lower levels of heat build-up prevent epoxy resin smearing in the drilled hole and reduce drill wear.

If the feed is too fast, it will result in rough holes or drill breakage. If it is too slow, the drill will turn excessively in the hole. The result will be heat build-up and excessive resin smear along the walls of the holes.

For example, a feed of less than 0.02 mm per revolution should not be used since the cutting tips would no longer cut the material. The result is heat build-up and drill wear. Drill bits for paper- and glass-based material usually drill at 0.0001 to 0.003 inch (0.0025 to 0.075 mm) per revolution of the drill. The smaller drill bits may even drill at 0.0009 inch (0.0225 mm) per revolution. Long twist drill bit with a 100° of point angle is suitable for phenolic laminate drilling. Short twist drill bit with 120° of point angle is suitable for glass epoxy laminate drilling.

In order to compensate for the laminate resilience, the drill bit diameter is chosen which is slightly bigger than the hole diameter expected. Therefore, the diameter of the drilled holes would be usually smaller than the diameter of the drill used. The difference could be 50 microns for hole diameters upto 2 mm and 100 microns for larger holes. For PTH boards, allowance must also be provided for plating thickness. Therefore, a plated hole of 1mm diameter will be made with a 1.1mm diameter drill.

The following hole diameter tolerances are generally accepted, unless otherwise specified:

$$D \leq 0.8 \text{ mm} \quad \rightarrow \quad \pm 0.05 \text{ mm}$$
$$D > 0.8 \text{ mm} \quad \rightarrow \quad \pm 0.1 \text{ mm}$$

where D is the hole diameter.

The general practice is to stack 3-4 panels, each 1.59 mm thick, so that each head can drill the holes simultaneously. Each stack is provided with a back-up panel, usually a laminate without copper foil, to allow the drills to pass right through all panels.

In order to facilitate accurate drilling, the vias and pads normally have copper etched from the centre to help centre the drill. Some manual machines drill from the underside and use a light spot on top to indicate the drilling point.

10.4.9 Drill Entry and Exit (Back-up) Materials

The entry plate and back-up plates are used during drilling to prevent the burr formation at the entry and exit of the drill. The cleaning and cooling of the drill during drilling are also done by the entry/back-up plates. So, proper selection of the back-up and entry plates, and their material is very important.

Drill Entry Material

The main purpose of the entry material is to prevent drill breakage by centring the drill bit. In addition, the entry material helps to avoid damage to the top copper laminate surface, to prevent copper burrs, to reduce contamination in the hole and on the drill bit, and to prevent pressure foot marks from the drilling machine.

Many types of entry materials are available. Aluminium composite, solid aluminium, melamine products and aluminium clad phenolics are the most common. Aluminium composites, in addition, leave no hole contamination. On the other hand, solid aluminium provides good burr suppression and no contamination, but increases the risk of drill breakage for small diameter bits. Phenolic materials are less expensive, but often warp and can contaminate the hole wall which may result in problems in the subsequent process.

The entry materials, which are flat, thin sheets placed on the drill entry side of the laminate, are shown in Figure 10.8.

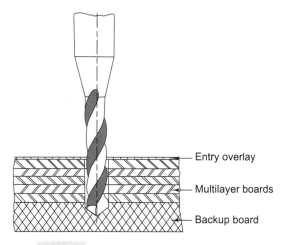

Fig. 10.8   Use of entry and back-up sheets

Back-up Material

Back-up material is placed on the underside of the drilled stack and its purpose is to prevent exit copper burrs and to provide adequate space for drill stroke termination. A good back-up material should not contaminate the hole but help cool the drill bit, thereby improving the hole quality.

A large variety of back-up materials are available for PCBs. However, only few of them are engineered specifically for circuit board drilling. Typical materials include aluminium clad wood core composites, melamine clad wood core composites, solid phenolics and paper resin hard board.

The material generally used is made of specially pressed wood/pulp mixture as core material with sheets of hardened aluminium laminated on both sides. The wood filler is free of resin additives, has an exceptionally high heat absorption capability and acts as a dry coolant for the drill. The chips from the hardened aluminium keep the cutting edge of the drill free and do not allow epoxy resin residue to build up on the spiral flutes.

A back-up plate can be used from both sides, if the correct drilling depth is selected and holes are not drilled beyond the centre of the board.

The back-up material must have a tight thickness and flatness tolerances. It should contain no abrasives that would increase drill wear or contaminants that could be evacuated through the drilled hole. The surface should be smooth and hard to properly suppress exit burrs.

10.4.10   Use of Drill Bush/Collar

The drill bush plays an important role in drilling circuit boards. The sole purpose of the drill bush is to precisely fix the position of the drill hole. Bushes are made of tungsten-carbide materials and are required where the ultimate position accuracy is required. Therefore the bush diameter must match the drill diameter. Excessive dia results in poor centring and drill misalignment, and inadequate play resulting in greater wear on the margins and cutting edges of the drill. The bush also suppresses burr formation at the drill bit entry point. Bushes (Figure 10.9) are high precision tools and are required to be manufactured with tolerances in the micron range.

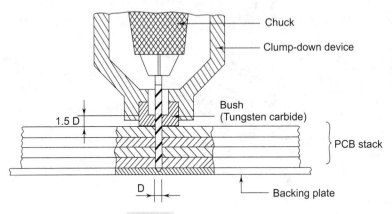

Fig. 10.9   Use of drill bush

10.4.11   Drilling and Types of Laminates

Laminates are manufactured according to standard specifications for required electrical and physical properties. The drilling quality depends upon the quality of resin and supporting fibre of the laminate. Some commonly available copper clad laminates are G-10, epoxy glass/FR4, polyamide glass, phenolic glass, phenolic paper, epoxy paper, Teflon glass and polyamide quartz.

It is essential to attain a good surface finish on the drilled hole walls. The surface finish depends upon the selection of drill, rpm of the drill and feed rate. If they are not selected properly, problems such as smearing of charred resin and friction heat will occur. The chips of reinforced materials will remain on the wall of the holes and provide a very poor base for the metallization process. This will cause high porosity of the plated metal and high absorption of plated liquid and moisture, which in turn, will create voids in the charred resin. Blow holes [outgassing] during soldering are also formed because of the above problems.

Dimensional stability, warp, bow and twist characteristics of the laminates are important for the drilling operations. Laminates which are not flat usually end up with burred holes. The type of weave and fibre thickness of the laminate affect the drill wander. Excessive drill wander results in poor quality holes. There is less drill wander if the laminate fibres are fine.

A proper drill must be selected for each type of laminate to obtain high quality drill holes. Paper phenolic type laminates should be drilled by a long twist with a 100° angle point, while epoxy glass laminates are better machined by a short twist drill with a point angle of about 120°. However, in both cases, the drills used are of the tungsten-carbide type.

10.4.12 Drilling Problems

Defective holes are formed due to improper drill bit geometry, drill speed, feed rates and improper cured base materials. The following are the main defects noticed in the drilled hole:

- Resin smear;
- Nail heating; and
- Roughness.

Resin Smear: The smear is caused when the resin is heated beyond its melting point (115 °C).

Nail Heating: This implies excessive burning on the hole wall when the drill speed and feed are not balanced with the drill geometry.

Roughness: A higher member of chipped drill holes and rough hole walls are due to the poor quality of the drill bits.

The swarf generated during the drilling process must flow out along the drill upto the surface from where it can be removed. This requires proper selection of drill, rpm and the feed rate. If the swarf does not flow out, it will be heated due to friction and cause it to char resulting in smearing of the reinforcing material of the laminate. This smear will remain on the walls of the holes, thus forming a very poor base for subsequent plating. This can cause high porosity of the plated surface and high absorption of plating liquid and moisture, etc. with the consequent problems during soldering. The smearing problem is more severe in the case of multi-layer boards as some resin is smeared on the drilled copper of the internal conductor layers. The plated copper therefore, cannot contact the copper exposed by drilling and the board qualifies for rejection.

However, it is not necessary that the holes to be plated must have perfectly smooth walls. A controlled degree of roughness is acceptable as it helps to increase the anchorage of the cylinder plated copper to the laminate.

10.4.13 Drilling Machines

Drilling machines are of two types:

- Manually controlled machines; and
- Numerical controlled machines.

Small quantities of PCBs can be drilled using single head manually controlled machines while large volumes of PCBs usually require numerical controlled drilling machines.

The selection of drilling machines depends upon their speed, capacity, accuracy and cost. Manual machines require an operator to position the work piece and initiate the drilling cycle. In Computer Numerical Control (CNC) machines, both positioning and drilling cycles are controlled by inputs from the computers.

Even though PCB drilling machines are available in a wide range of designs, they all have a common feature, which is high-speed operation that offers an efficient and economic drilling capability for the various base materials. The speed mostly used is 20,000-50,000 rpm. However, highly sophisticated drilling machines operate upto 1,00,000 rpm. Hudson (2003) describes the development of multi-head drilling system with vision which can operate at 275,000 rpm. The machine allows fully independent control of all axes of each spindle. The spindle uses standard 0.125" (3.20 mm) shank diameter ringless drills while providing two to four times longer useful drill life. The vision system provides each spindle with its own integral camera. Since each spindle is independently positioned in the X and Y axes, it is possible to make corrections to the drill program for each work piece independently of the other stations. Vision registration can be re-calculated at every tool change, if desired, to provide real time process control throughout the drilling cycle.

10.4.13.1 Manually Controlled Machines

The PCB or work piece is positioned for drilling in a manual machine by one of the following two methods: a) Direct sight method; and b) Optical method.

Both these methods require the hole location to be defined either by an artwork overlap or reproduction of the circuit image on the board. Photo-resist is often used when the developed image is sufficient to locate the hole position and the boards may be stacked or pinned for registration to the template, which is usually made of acrylic material. The stack can then be moved under a stylus controlled by the operator.

Spindles or high-speed shaft rotation systems are made up of bearings (radial and axial), motor and stator, tool mounting system, tracking system and cooling system. The spindles are divided into two main groups:

- Ball-bearing spindles; and
- Air-bearing spindles.

The ball-bearing spindle is used primarily in applications that do not require high accuracy or high-speed rotation (above 15,000 rpm). In contrast, the air-bearing spindle allows higher speed rotation and higher accuracy, with lower vibration of the rotating shaft.

10.4.13.2 Direct Sight Bench Mount Drilling Machine

This machine consists of a solid base frame and a column which supports the motor. For drilling applications, the vertical movement of the drilling head is controlled by a lever, which is located on the right side of the drill head. The movement can be locked by means of a small additional lever. The spindle rotation speed is continuously adjustable from 0-45000 rpm. Dust is collected by an exhaust vacuum system.

The operation of the machine is very simple as the board is positioned manually under the point of the drill bit. However, it gives a limited accuracy, usually not better than ± 0.25 mm.

10.4.13.3 Optical Sight High Speed Drilling Machine

This machine is an improvement over a direct sight drilling machine. The drilling spindle whose speed ranges from 15,000 to 60,000 rpm is mounted underneath the work table and fed from the bottom through a jig plate. The feed from the bottom and the clamping from the top are controlled and sequenced by a pneumatic time delay circuit. The complete sequence is affected by a foot pedal. The feed, time delay and rpm can be varied by accessing the respective controls.

For centring the pads to be drilled, an optical magnifier (Figure 10.10) is mounted above the work table in line with the drill axis. The magnification of the optical system is ten times. The illumination is provided by a halogen lamp unit whose light intensity can be adjusted for convenient observation on the viewing screen. Sight marks consisting of concentric circles and a cross-hair on the

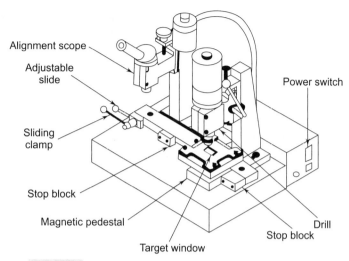

Fig. 10.10   High-speed PCB drilling machine with optical head

screen helps in centring the pad to an accuracy of ±0.1 mm. This machine will produce accurate holes at desired co-ordinates with superior hole finish. The dust collecting vacuum system is also available for producing a dust-free environment.

10.4.13.4   Numerically Controlled Machines

In numerically controlled machines, the machine control logic is obtained from software. The computer is programmed to control the machine. The advantages of NC machine are flexibility, versatility, repeatability and speed.

The drilling of PCB production quantities with NC machines is normally done with a multi-head drilling machine. A typical set-up would include machines with two or more heads, each simultaneously drilling a stack of cards. Dummy cards are placed at the top and bottom of each stack to ensure constancy of drilling. The numerically controlled drill automatically changes the drill adjustment, drill speed and feed rate, and brings the card stack to the correct X-Y co-ordinates for drilling. A record relating to the usage of the drills is maintained so that they can be withdrawn from service before they show signs of wear, deteriorating cutting edge and the possibility of causing burrs. Usually, smaller and inexpensive drills are discarded while the larger and expensive ones are re-sharpened.

In the computer-aided design, post-processing programs are used to produce tools and to run hole-producing equipment. The tools are most commonly in the form of drill tapes prepared directly from the CAD software, or in the case of manual artwork, by using a digitizer. Database information can generate hole size and location drawings for use in fabrication and inspection of the finished PCB. Hole sizes are coded by differently shaped symbols and hole sizes are defined in a hole chart in the field of the drawing. Modern CAD/CAM systems, however, offer a paperless format wherein the computer data can be directly accessed by the PCB fabrication operation and the information transferred by on line or other modern data transfer media.

 ## 10.5 Microvias

New electronic products are becoming smaller, faster, lighter and cheaper in order to be able to compete in today's market. Achieving these requirements, fine pitch area array packaging, fine pitch ball grid array and flip chip-on-board assembly technologies are being implemented. The rate at which these packaging technologies can be adopted is largely being dictated by the availability of higher density PCB technologies with significant reduction in conductor line width and via size at a relatively lower cost.

The real benefit of HDI (high density interconnection) lies in the small holes, identified as **'microvias'**. These holes are very small, defined by the Institute for Interconnecting and Packaging Electronic Circuits (IPC), as equal to or less than 150 micrometer (μm) in diameter. The small size of the hole allows for additional room for conductor routing.

Microvias normally connect only two layers. The term 'capture land' is used to define the start of the microvia and 'target land' is used to describe the bottom of the microvia. HDI layers are normally constructed as the outer layers on a standard double-sided or multi-layer board as a core, using a thin, non-reinforced resin as the dielectric with a thickness of 50 to 80 microns. Vias are mostly created by photo-imaging, laser ablation, plasma etching or by filling a conductive ink into the via.

Although mechanical drilling still accounts for more than 90 per cent of all holes drilled in PCBs, it has never been a serious contender as a method of microvia formation. For hole diameters below 100 μm, three technologies present a solution: plasma etch, photo-imaging, and laser drilling (Keeping, 2000). Each has its benefits and drawbacks but market reports indicate that laser drilling is currently leading in microvia production.

10.5.1 Photo-formed Vias

The photo-sensitive material is applied on both sides to the patterned board in the same way that an ordinary solder mask is applied by using various techniques like curtain coating, screen printing or even dry film coating. The microvia in the photo-sensitive layer (the dielectric layer) is then imaged and developed, and the layer is fully cured. In the following process step, a conductive copper layer is deposited by means of a fully additive process, and the pattern of the copper layer is imaged and etched.

The most difficult aspect of the photovia process is the shrinkage of dielectric materials after hole formation, during the curing process. In the case of multiple build-up layers, manufactures commonly prepare at least 2-3 artworks for the same pattern, each of slightly different dimensions. Photovia has been adopted by only a few manufacturers in the world, with a concentration in Japan.

10.5.2 Plasma Etching

Plasma etching is done by injecting a mixture of gases into the chamber and exciting them into a plasma state with an RF power source. As this occurs, atoms and molecules from the gases split to form ions that react with organic materials in the targeted dielectric. A residual foil acts as a metal mask. The product is replaced into the standard product flow of through-hole drilling, plating, outer-layer image, etc. After removing copper from the microvia locations in a standard image and etch process, a vacuum-chambered plasma driller is used to etch the microvias in the exposed dielectric layer.

Plasma etching can only be performed on exotic materials such as adhesiveless flexible polyimide, pure resin and aramid-reinforced laminate, not on epoxy-glass/fibre materials like FR-4. Resin-coated copper foil (RCC) is laminated on both sides to a double-sided or multi-layer core.

Another PCB fabrication technique described by Buckley (1992) employs plasma etching to produce very small vias which, in turn, mean increased interconnection densities. Called DYCOstrate, this process substitutes a thick Cu-clad polyimide foil for the conventional PCB substrate material and replaces the conventional mechanical drilling operations with a plasma dry-etching process capable of simultaneously producing thousands of small diameter (< 80 µm) holes without dust generation or smearing. These small diameter holes can only be produced in thin laminates, and not in conventional PCB substrates, so the DYCOstrate process employs 25µm thick polyimide which has been clad with half ounce (17 µm) copper. Because the laminate is so thin, the aspect ratio of a typical 75 µm hole is such that plating presents no problems and can be carried out on a normal PCB plating line. With this process, the via holes of 100 µm or less diameter require pad diameters of just 200-300 µm. The small pads combined with track widths below 100 µm facilitate the high interconnection densities. Brist, *et al.*, (1997) detail out the microvia creation process by using the plasma etching technique.

10.5.3 Laser-formed Vias

In recent years, laser technology has become the method of choice for forming microvias in high density interconnects, as well as chip packaging devices. This has been necessitated in view of today's trend in portable electronics to make them smaller, lighter, thinner and faster with added functionalities. Versatile laser technology that is capable of working with both rigid and flexible printed circuit boards for making microvias, is also available.

The term 'laser' has been coined by taking the first letters of the expression "light amplification by stimulated emission of radiation". Although an amplifier, as suggested by the abbreviation, the laser is invariably used as a generator of light. But its light is quite unlike the output of a conventional source of light. The laser beam has spatial and temporal coherence, and is monochromatic (pure wavelength). The beam is highly directional and exhibits high density energy, which can be finely focused.

The main applications of lasers for mechanical operations is in the field of material machining, where operations like cutting, welding, drilling and trimming are performed by different laser systems on a variety of materials such as metals, ceramics, plastic, etc. The advantages of using lasers instead of other more conventional methods for this purpose are: (i) a higher processing speed, (ii) absence of wear and tear of mechanical parts, and (iii) the ability to machine features with very small dimensions. The laser beam is not absorbed by air and can, with the help of fixed or moving mirrors, be directed to the point where it has to work. Laser beams with the wavelength of 1 μm or less can be directed even with the help of optical fibres.

Carbon dioxide (CO_2), Yttrium-Aluminum-Garnet (YAG), excimer lasers as well diode lasers are mainly used today. CO_2 and YAG lasers offer the maximum power outputs and can vapourize or melt materials with the finely focused laser beam. The excimer laser emits UV light (instead of the IR light of the CO_2 laser) and consequently opens new avenues, particularly in the micro-sector. The diode laser has been adopted increasingly, both as a pumping light source for YAG lasers as well as for direct use.

Figure 10.11 shows the wavelength of different laser systems whereas Figure 10.12 represents the absorption characteristics of various materials at different wavelengths.

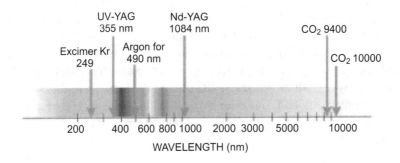

Fig. 10.11   Wavelength of different lasers

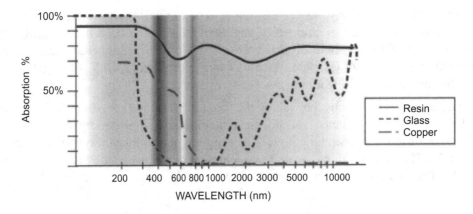

Fig. 10.12   Wavelength vs absorption of laser light

10.6   Use of UV Laser for Drilling PCB

Four types of lasers are currently in use for making PCB microvias: CO_2, YAG, excimer and Cu-vapour. *CO_2 lasers* typically produce holes of around 75 µm, but the beam reflects back off copper and is therefore only suited for removing dielectric. CO_2 lasers are very stable, inexpensive and maintenance-free. *Excimer lasers* are the best for producing high quality, small features. Diameters below 10 µm are typical. The best application for these types is in drilling densely packed arrays in polyimide for use in microBGA devices. *Cu-vapour lasers* are still in their infancy, yet offer some advantages when high production rates are needed. Cu-vapour types can remove dielectric and copper, yet suffer from severe drawbacks that make their current use in a production environment-prohibitive.

The most popular laser system used in the PCB industry is the Q-switched Nd:YAG laser with a wavelength of 355 nm which is in the ultraviolet (UV) range. At this wavelength, most of the metals (Cu, Ni, Au, Ag) that are to be ablated in printed circuit applications, show absorption rates of more than 50 per cent (Meier and Schmidt, 2002). Organic materials can also be accurately ablated. The high photon energy of UV lasers at 3.5–7eV cracks the chemical bonding as the ablation process in the UV spectrum is partly photo-chemical and partly photo-thermal. These capabilities make a UV laser system the first choice for applications in the printed circuit board industry.

The system based on a single laser source, provides energy density (fluence) of more than 4J/cm^2 that is needed for opening the copper surface when drilling microvia holes. The ablation process of organic materials such as epoxy resins and polyimide requires an energy density of only around 100 mJ/cm^2. In order to address this wide spectrum, the laser would need very precise and sophisticated energy control. The drilling of microvias requires a two-step process. The first step opens the Cu with a high fluence and the second step removes the dielectric with low fluence.

The laser typically gives a spot size of approximately 20 µm at a wavelength of 355 nm. The frequency of the laser pulses is between 10 and 50 kHz at a pulse length of less than 140 ns., which produces a no heat-affected-zone in the material.

Figure 10.13 shows the basic principle of such a system. The laser beam is positioned with a computer-controlled scanner/mirror system and focused through a telecentric lens that allows the beam to maintain a right angle to the drilled material. This scanning process allows the software to generate a vector pattern and compensates for both material and layout deviation. The scanning area measures 55 × 55 mm. The system is compatible with CAM software and supports all common data formats.

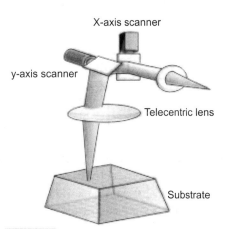

Fig. 10.13   Basic principle of laser scanning system (redrawn after Meler and Schmidt, 2002)

In the laser system produced by M/s LPKF, Germany, the mechanical design is based on a rigid granite construction precisely polished to a surface accuracy of less than 3 μm. The table rests on air bearings and is driven by linear motors. The positioning accuracy is controlled with glass scales that guarantee a repeatability of ±1 μm. An optical sensor integrated in the table itself compensates for the optical distortion and long-term drift based on an accurate alignment of the laser position at various mirror locations. The software creates an array of correction data based on the alignment that is overlaid on the entire scanning area. The calibration for the drift compensation takes about one minute and can be done while a work process is executed. Any variation in the substrate itself, such as inaccuracies in the positioning caused by deviation of the fiducials, is detected by a high-resolution CCD camera and compensated for by the control software.

Such a system is well suited for prototyping, since it can do both drilling and structuring. Its application range extends from flex to rigid PCBs, including polymer materials such as solder masks, cover coats, galvano resists, etc. Raman, *et al.*, (www.esi.com) discuss the latest advancements in solid state UV laser systems and their applications for forming microvias in high density interconnects.

Lange and Vollrath (www.lpkfusa.com) illustrate the versatile applications of a UV laser system (MicroLine drill 600 system) for drilling, structuring and cutting. The system permits drilling of holes and microvias, with diameters down to 30 μm through copper layers and a range of substrate materials in a single-step operation. The system is also able to produce conductors down to a width of only 20 μm in the outer copper layers of a board, which is well beyond the capabilities of photo-chemical processing. The system achieves processing speeds upto 250 drill operations and accepts all standard input formats such as Gerber or HPGL. Its working area is 640 mm × 560 mm (25.2" × 22") with a maximum material height of 50 mm (2") which can accommodate most of the usual substrate formats. The machine table bed as well as its guides are made out of natural granite blocks, precision-ground to an accuracy of ± 3 μm. The machine table is powered by linear drives and is supported on air bearings; thermally compensated glass scales control its position with an accuracy of ± 1 μm. A vacuum unit accomplishes the mounting of the substrate on the processing table.

10.7   Hybrid Laser Drilling Process

Two kinds of laser technologies are commercially available as laser drilling systems; CO_2 laser with wavelength in the far infra-red region of the spectrum, and UV laser with wavelength in the ultraviolet region of the spectrum. The CO_2 lasers are widely used for microvia formation in the PCB industry wherein the microvia design calls for larger vias, 100 μm in diameter (Raman, 2001). The CO_2 lasers have high productivity at these large diameter vias. The high productivity is due to the fact that the CO_2 lasers can 'punch' large vias with very small drill times. The UV laser is widely used when the microvia design calls for < 100 μm via diameters, with the roadmap shrinking to even smaller vias of < 50 μm diameter. The UV laser technology delivers very high productivity at < 80 μm vias. Therefore, given the everincreasing demand to improve productivity of microvia formation, many manufacturers have started introducing dual head laser drilling systems. The following are the three major types of dual head laser drills available in the market today:

- Dual head UV laser system;
- Dual head CO_2 laser system; and
- Hybrid laser system (UV and CO_2).

All the major types of drill systems have their own inherent advantages and disadvantages. The laser drills can be simplified into two categories: dual head single wavelength system and dual head dual wavelength system. Irrespective of the categories, each laser drill has two main components that affect the productivity of the drilling system:

- Laser power/pulse energy; and
- Beam positioning system.

The laser pulse energy and the efficiency of the beam delivery optics determine the drilling time for a via. The drilling time is defined as the time it takes the laser drill to drill one microvia. The beam positioning system determines how fast you can move between the vias. The combined effect of these factors determines how fast a laser drill can produce microvias for a given application.

The *dual UV wavelength* laser systems are best suited for integrated circuit packaging applications for drilling < 90 µm blind vias as well as high aspect ratio through vias.

The *dual CO_2* laser system makes use of Q-switched radio frequency excited CO_2 laser. The main advantage of this system is the high repetition rate (upto 100 kHz), very short drill time, and wide process window. It only requires a few shots to drill a blind via which may result in a poor via quality.

The most popular dual head laser drilling system is the *hybrid laser system* which consists of an ultraviolet laser (UV) laser and a CO_2 laser. The integrated approach of the hybrid laser drilling solution allows the copper and dielectric to be processed in parallel. This means that while the UV laser removes copper, creating the via size and shape desired, the CO_2 laser follows behind, removing the dielectric that is uncovered. The drilling routine is carried out in 2" × 2" blocks called fields.

CO_2 lasers efficiently remove dielectrics, even non-homogeneous, glass-reinforced dielectrics. However, the CO_2 laser alone cannot create small vias (say below 75 µm) and cannot remove copper, apart from the limited success it has achieved in removing pre-treated thin foils 5 µm and below (Justino, 2002). The UV laser can be used to create very small vias and remove all common copper foils (from 3 µm upto 36 µm, 1 oz., and even plated foils). UV lasers alone can also remove dielectrics but the material removal rate is slow. Moreover, the results are generally poor for non-homogeneous materials such as glass-reinforced FR4s because the glass can only be removed if the energy density is increased to levels that can damage the inner layer stop pad. Since hybrid systems include both a UV laser and a CO_2 laser, these systems offer the best of both worlds. All copper foils and small vias can be achieved with the UV laser and all common dielectrics can be drilled fast with the CO_2 laser.

Figure 10.14 shows the architecture of a dual head laser drill with programmable spacing between the heads. The pitch between the heads is automatically set depending upon the layout of the part. This ensures that the laser drill is performing at its maximum throughput.

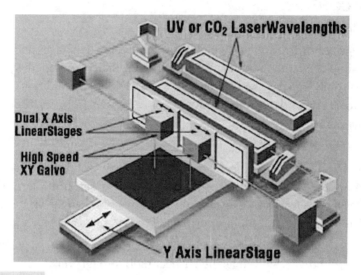

Fig. 10.14 Arrangement of a dual head laser drill (redrawn after Raman, 2001)

Most of today's dual head laser drilling systems have fixed spacing between the heads, together with a step-and-repeat beam positioning technology. The inherent advantage of the step-and-repeat beam positioner allows bigger field sizes (up to 50×50 μm). Its inherent disadvantage is that the beam positioner has to move in fixed field steps as well, with a fixed spacing between the heads. The typical dual head beam positioner has a fixed amount of spacing between the two heads (about 150 μm). For varying panel sizes, fixed head spacing cannot perform at its optimum efficiency as compared to the programmable spacing of head.

Hybrid laser drilling systems are today available with a variety of standard options and features that cater to small PCB shops as well as high volume manufacturing houses.

Ceramic alumina is used in printed circuit fabrication because of its high dielectric constant. However, due to its brittleness, manufacturing processes such as hole drilling needed to attach wiring and trimming, become difficult with standard tools. This then becomes a good case for laser processing, since mechanical stresses have to be reduced to a minimum. Rangel, *et al.* (1997) demonstrated the drilling of perforations in alumina substrates and in gold and chromium-covered alumina substrates, by using laser ablation with a Q-switched Nd: YAG laser. Using a short pulse, low energy, high peak power laser helps to avoid the induction of mechanical stresses that can break up the sample, and to make fine structure perforations of 100 μm diameter or less. The technique was successfully applied in the production of a low noise microwave amplifier in the 8-18 GHz frequency range (Betancourt, *et al.*, 1996).

10.8 Useful Standards

- *IPC-Dr-572: Drilling Guidelines for Printed Boards:* Provides guidelines for drilling quality holes in a wide range of printed board materials.
- *IT-95080: Improvements/Alternatives to Mechanical Drilling of PCB Vias:* Defines and characterizes alternatives to mechanical drilling of small holes and discusses advances in mechanical drilling technology.
- *IPC-NC-349: Computer Numerical Control Formatting for Drillers and Routers:* Defines a machine-readable input format for computer numerical control drilling and routing machine tools used by the printed circuit board industry. The format may be used to transfer drilling and routing information among printed board designers, manufacturers and users.

Multi-layer Boards

 ## 11.1 What are Multi-layers?

Multi-layers, or multi-layer PCBs are circuit boards made up of more than two electrical layers (copper layers) superimposed on each other. The copper layers are bonded together by resin layers (prepreg). Multi-layer boards represent the most complex type of printed circuit boards. Their cost is relatively high, owing to the complexity of the manufacturing process, lower production yields and difficulty of re-working on them.

The need for multi-layer boards has been necessitated by the increasing packaging density of integrated circuits, which give rise to high concentration of interconnecting lines. The printed circuit layout results in unpredictable design problems like noise, stray capacitance, cross-talk etc. The PCB design, therefore, must aim at minimizing the length of the signal lines and avoiding parallel routing etc. Obviously, such type of requirements could not be met satisfactory in single-sided and even double-sided printed circuit boards due to limited cross-over which could be realized. Thus, to achieve satisfactorily performance from the circuit in the presence of a very large number of interconnections and cross-over, the PCB must be extended beyond two-plane approach. This gives rise to the concept of multi-layer circuit boards. Hence, the primary intent of fabricating a multi-layer printed circuit board is to provide one more degree of freedom in the selection of suitable routing paths for complex and/or noise-sensitive electronic circuits.

Multi-layer boards have at least three layers of conductors, in which two layers are on the outside surface while the remaining one is incorporated into the insulating boards. The electrical connector is commonly completed through plated through-holes, which are transverse to the boards. Unless otherwise specified, multi-layer PCBs are assumed to be PTH as with double-sided boards.

Multi-layer boards are fabricated by stacking two or more circuits on top of each other and establishing a reliable set of pre-determined interconnections between them. The technique begins with a departure from conventional processing in that all the layers are drilled and plated before they are laminated together. The two innermost layers will comprise conventional two-sided PCB while the various outer layers will be fabricated as separate single-sided PCBs. Prior to lamination, the inner layer boards will be drilled, plated through, imaged, developed and etched. The drilled outer layers, which are signal layers, are plated through in such a way that uniform donuts of copper are formed on the underside rims of the through-holes. This is followed by lamination of the various layers into a composite multi-layer with wave-solderable interconnections.

The lamination may be performed in a hydraulic press or in an over-pressure chamber (autoclave). In the case of hydraulic press, the prepared material (press stack) is placed in the cold or pre-heated press (170 to 180 °C for material with a high glass transition point). The glass transition temperature is the temperature at which the amorphous polymers (resins) or the amorphous regions of a partially crystalline polymer change from a hard and relatively brittle state to a viscous, rubbery state.

Multi-layer boards find applications in professional electronics (computers, military equipment), particularly whenever weight and volume are the over-riding considerations. However, there has to be a trade-off which is simply the cost for space and weight versus the board's costs. They are also very useful in high speed circuitry because more than two planes are available to the PCB designer for running conductors and providing for large ground and supply areas.

 ## 11.2 Interconnection Techniques

11.2.1 Conventional Plated Through-hole

The most common and least expensive layer-to-layer interconnection technique is the conventional plated through-hole. Figure 11.1 shows an example of a six-layer through-hole board.

In this technique, all holes are drilled through the panel, irrespective of whether they are used as component holes or as via holes. The main disadvantage of this technique is that the through-hole via takes up valuable space on all layers irrespective of the number of layers the hole is connecting.

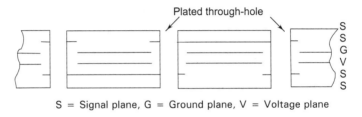

S = Signal plane, G = Ground plane, V = Voltage plane

Fig. 11.1 Conventional multi-layer board: signal plane, ground plane and voltage plane

11.2.2   Buried Via

A buried via is a plated through-hole connecting two or more layers of a multi-layer board, buried inside the board structure, but not appearing on the external surface of the boards. This type of multi-layer board is shown in Figure 11.2.

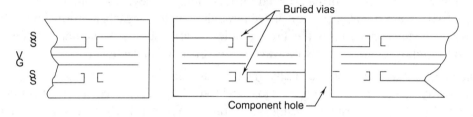

Fig. 11.2   Buried via multi-layer board

As there is considerable saving in this area as compared to conventional plated through-hole construction, the buried via technique is used when signal trace routing is very dense, which requires more via sites connecting signal layers and more channels for signal traces. However, the routing density advantage will result in more costly boards because the technique requires added process steps.

11.2.3   Blind Vias

The blind via hole is a plated through-hole connecting the surface layer to one or more layers of a multi-layer board which does not go through the entire board thickness. Figure 11.3 shows a typical example of the blind via technology. In this arrangement, the hole can be used on both sides of a multi-layer board and can be used in conjunction with the via and component holes which go through the board.

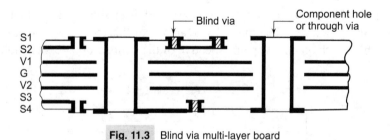

Fig. 11.3   Blind via multi-layer board

Blind vias can be stacked on top of each other and can be made smaller, providing more space or signal lines.

The technology is particularly useful with surface mounted devices and connectors as they do not require large component holes. Only small via holes are needed to connect the external surface to internal layers, which allow the designer to fully utilize the advantage of size and weight reductions offered by surface mount technology for very dense and thick multi-layer boards.

 ## 11.3   Materials for Multi-layer Boards

The three basic sets of raw materials used to manufacture conventional rigid multi-layer boards are: (i) a resin system, (ii) the reinforcing fabric; and (iii) metal foil. The boards are mostly made by using glass cloth, coated or impregnated with a resin as the dielectric material. The glass cloth imparts mechanical strength to the board and its basic function is to carry the resin. The controlled thickness of the glass cloth enables the manufacturer to build multi-layer boards with controlled total thickness and tolerance.

11.3.1   Resin System

The following three resin types are widely used in multi-layer boards:
- Conventional flame-resistant epoxy;
- Modified high performance, high temperature epoxies; and
- Polyimides.

Conventional Epoxy: This is the most common resin used and meets the demands of most commercial and military applications. It has excellent adhesion to copper, exhibits low shrinkage during cure and has good chemical and moisture resistance. However, it has a high per cent expansion when heated to solder temperature. Its glass transition (T_g) is in the range of 120 to 130 °C.

Modified Epoxies: By virtue of their modified molecular structure, these epoxies raises T_g and improve chemical and thermal stress resistance. Their disadvantage lies in their increased brittleness and drill wear, with higher material cost.

Polyimides: Polyimides exhibit excellent thermal stability up to 200 °C, an important property that helps during the repair of large, expensive boards. They can withstand repeated soldering and de-soldering operations, have high copper-to-resin bond strength and a low degree of drill smear. However, their disadvantages are higher moisture absorption, lower flammability rating and higher cost.

11.3.2   Reinforcement Materials

The most widely used reinforcement is the E (electrical) glass: the woven glass cloth sews as a support vehicle for epoxy resins. The material has good resistance to water, fair resistance to alkali and poor resistance to acid, and a high dimensional stability. Glass cloth is available in a wide variety of wear styles. The glass cloth styles are designated in 3-4 digit numbers which determine its nominal weight, thickness and thread count.

11.3.3 Prepreg

Semi-cured glass cloth-reinforced epoxy resin is referred to as **Prepreg** or **Bstage**. At this stage, the epoxy resin is not in a fully polymerized state. The application of heat and pressure makes it a convenient bonding material for multi-layer boards. Generally, thinner fabrics are used in fabricating these boards because they carry a higher resin-to-glass ratio, which helps provides a void-free laminated bond, with the availability of more resin to fill circuit patterns where copper has been etched.

11.3.4 Copper Foil

Copper foil that is mostly used to manufacture multi-layer boards is made by the electrodepositing process. With increased circuit densities which lead to finer circuit and thicker boards, the copper foils must exhibit improved elongation properties at elevated temperatures and special bounding treatments for epoxies and polyimides. Recent developments in copper foils include ultra thin (less than 12 μm) foil for high resolution and fine line circuits.

The substrates for the various layers are selected on the basis of the impedance and signal isolation requirement of the design and the physical limits placed on the total thickness of the laminated panel (www.thinktink.com). Taking into consideration the FR-4, the most common material used in PCB substrates, the key considerations would be as follows:

- For a four-layer design with a finished thickness of 0.063"(1.6 mm), copper clad with 0.0007" (0.017 mm) copper foil on both sides (so called half ounce/half ounce or "half over half" copper clad) and a substrate thickness of 0.025" (0.64 mm) is a good selection for the innermost PCB.
- The outer layers will be constructed from copper clad with half ounce foil on one side and a substrate thickness of 0.017" (0.43 mm). The substrates used are of glass fabric which are held together with a partially cured resin that will re-melt during multi-layer lamination. Re-melting causes the resin to become very tacky and adhere to the inner layers. This is the bonding mechanism that lies at the heart of multi-layer fabrication.

 ## 11.4 Design Features of Multi-layer Boards

The design features of multi-layer boards are mostly similar to those used for single layer or double layer boards, expect that care has to be taken to avoid cramming of too much circuitry into too little space, thus giving unrealistic tolerances, high inter-layer capacitances and possibly a compromised quality. Accordingly, performance specifications should allow complete evaluation of thermal shock, insulation resistance, solder resistance, etc. of inter-layer connections. The important design considerations of multi-layer boards are discussed below.

11.4.1 Mechanical Design Considerations

The mechanical design includes selecting proper board size, board thickness, board lay-up, inner layer copper aspect ratio, etc.

11.4.1.1 Board Size

The board size is optimized on the basis of the application, size of the system cabinet, and limitations and capabilities of the board manufacturers. Large boards have many advantages such as smaller backplane, shorter circuit path between many components thereby allowing for higher operating speed and higher input-output connection count per board, and are therefore preferred in many applications such as personal computers where we come across large mother boards. However, designing large boards is comparatively difficult proposition with regard to routing of signal lines on a board, thus requiring more signal levels or inner lines or spaces and difficult thermal management. Therefore, the designer must consider various factors such as standard panel sizes, fabrication equipment sizes and limitations along with processing limitations. Some of these aspects are covered in IPC-D-322 which provides guidelines for selecting printing circuit/boards sizes using standard panel sizes.

11.4.1.2 Boards' Thickness

The thickness of the multi-layer boards is determined by various factors such as the number of signal layers, number and thickness of power planes, aspect ratio of hole diameter to thickness for quality drilling and plating, component lead length requirement for automatic insertion and the type of connection to be used. The total board thickness will comprise two gold layers on either side of the boards (electrical layers), copper layers, laminate thickness and thickness of the prepreg.

It is difficult to attain tight thickness tolerances on a complex multi-layer board. Tolerance levels of about 10 per cent are considered reasonable.

11.4.1.3 Board Lay-up

In order to minimize the chances of warping of the board and to obtain a flat finished board, the layering of the multi-layer boards should be kept symmetrical. This is achieved by having an even number of copper layers and ensuring the symmetry of copper thickness and the density of the copper pattern on the layers.

In general the warp direction of the fabric material used for the laminate (e.g. fibre-glass fabric) should run parallel to the side of the laminate because the warp direction is subject to definite shrinkage as after bonding. This distorts the layout and is also characterized as variable or low dimensional stability.

However, warping and torsion of the multi-layer can be minimized by improving the design. Torsion and warping are reduced by even distribution of copper over the entire layer and by ensuring symmetrical construction of the multi-layer; i.e. the same order and thickness of prepreg; copper and laminate layers should be present from the centre of the multi-layer layers to both outer layers. The prescribed minimum distance (dielectric thickness) between two copper layers is 0.089 mm.

The rule of thumb for calculating the minimum distance states that the minimum thickness of the prepregs after bonding must be at least twice the thickness of the copper being embedded. In other words, where you have two adjacent copper layers, each of which is 30 μm thick, a minimum prepreg thickness of 2 (2 × 30 μm) =120 μm is required, which can be achieved by using two prepregs (1080 is the type of fibre-glass fabric).

11.4.1.4 Inner Layer Copper

The most commonly used copper is 1 oz. (one ounce of coper foil per square foot area of surface area). However, for dense boards where board thickness is crucial and which require tight impedance control, 0.5 oz copper is used. Heavier copper, of 2 oz or above is preferred for voltage and ground planes. However, etching heavier copper results in reduced control of the desired pattern with regard to line width and spacing tolerances. Special processing techniques are thus required.

11.4.1.5 Holes

The plated through-hole diameter is generally kept between 0.028" and 0.010" from the nominal component lead diameter or diagonal to ensure sufficient volume for good soldering.

11.4.1.6 Aspect Ratio

The 'aspect ratio' is the thickness of the boards as compared with the diameter of the drilled hole. An aspect ratio of 3:1 is generally considered standard, though higher values like 5:1 are not unusual. The aspect ratio is determined by considerations such as drilling, smear removal or etch-back and plating. Via holes are required to be kept as small as possible, while keeping the aspect ratio within a producible range.

11.4.2 Electrical Design Considerations

A multi-layer board is a high performance, high speed system. At higher frequencies, the signal rise times decrease and consequently, signal reflections and line lengths become critical. The multi-layer board is a critical electronic component of the system with controlled impedance characteristics, designed so as to accommodate the above effect. The factors which determine impedance are the dielectric constant of the laminate and prepreg, conductor line width spacing between one layer of the conductor, dielectric thickness between layers, and thickness of the copper conductors. The layering sequence of conductors in the multi-layer board and the sequence in which the signal nets are connected are also critical in high speed applications.

Dielectric Constant: The dielectric constant of the laminate material plays a major role in the determination of impedance, propagation delay and capacitance. The dielectric constant of the epoxy glass used for the laminate and the prepreg can be controlled by varying the percentage of the resin content.

The epoxy resin has a dielectric constant of 3.45 and glass of 6.2. Depending upon the percentage of these materials, the dielectric constant of epoxy glass can be achieved from 4.2 to 5.3. The thickness of the laminate is a good indicator for determining and controlling the dielectric constant.

Prepreg materials with low relative dielectric constants are suitable for use in radio frequency and microwave engineering. The low dielectric constant gives rise to a low signal delay at radio and microwave frequencies. Electrical losses are minimized by low loss factors in the substrates.

Prepreg ROR 4403 is a new material produced by ROGERS CORPORATION (http://www.rogers-corp.com/mwu/index/html). This material is compatible with other substrates (such as RO 4003 or RO 4350, used for microwave boards) used in the construction of standard multi-layers (FR-4 material).

 ## 11.5 Fabrication Process for Multi-layer Boards

11.5.1 General Process

Multi-layer boards are produced by bonding together inner layers and outer layers with prepreg. Prepreg, as explained earlier, is fibre-glass fabric impregnated with partially hardened resin. The individual layers are arranged and bonded by placing them in a pressing tool to prevent misalignment of the layers.

After bonding, the bonded layers are further processed as double-sided through-plated circuit boards. Due to smearing of the hole with epoxy that takes place during drilling, through-hole wall cleaning is required before through-hole plating can take place.

In a multi-layer board, the outer layers may consist of either copper foil and prepreg or of single-sided or double-sided copper clad laminates. The inner layers consist of double-sided copper clad, etched (with structured conductor tracks created) and through-plated board material.

The inner layer etching is done by standard printed circuit techniques. Before bonding, it is important to make a very careful layout for each one of the layers in order to prevent masking of the desired holes. Each layer of B-stage and board substrate requires a different hole arrangement. In order to prevent the flow of resin into aligning pins, the tooling and aligning holes on the prepregs must be 1.25 mm larger in diameter than those of the conductor pads. On the other hand, the holes in the laminates must be 1.25 mm smaller than the pads over which they are to be placed.

In order to illustrate the process, let us take the construction of a four-layer board. The process steps are shown in Figure 11.4. They basically consist of two single-sided laminates and one double-sided laminate with two sheets of prepreg. The process starts by making a sandwich of all the required panels, stacked in the following order from bottom to the top:

- Thermal insulation material (a) to control rate of temperature rise;
- Bottom laminate fixture or caul plate (b);
- Sheet of release material (c) such as Teflon glass cloth;
- Bottom circuit panel (d);

- Prepreg (e);
- Inner circuit Panel (f);
- Prepreg (g);
- Top circuit panel (h as d);
- Sheet of release material (i);
- Top lamination fixture; and
- Thermal insulation material.

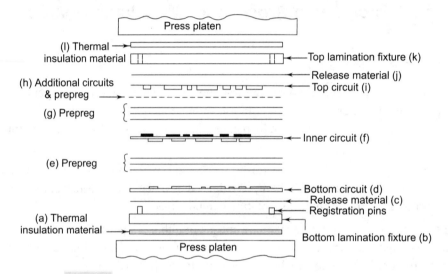

Fig. 11.4 Typical multi-layer board (MLB) lamination process lay-up

11.5.2 Lamination

The bonding is done in a laminating press, which is similar to those used in the manufacture of copper clad laminates. After the different layers are arranged, the sandwich is inserted between the plates of the press. Lamination requires a specific time/temperature/pressure cycle, which depends upon the properties of the prepreg.

During lamination, the resin gets softened due to high temperature when the pressure is applied; it causes it to flow to fill all the voids between the panels. Meanwhile, the material hardens due to a polymerization reaction that takes place, resulting in a single strong panel obtained with the two internal copper layers perfectly embedded in the resin.

The bonding pressure is $150-300 \text{ N/cm}^2$. The curing temperature and the time must be selected according to the type of prepreg used, the number of layers and the thickness of the press stack.

In the case of over-pressure chambers (autoclaves), gas or oil is used to convey the compression force and heat to the press stack. The press stacks are placed on platforms in the tiered stand that has a vacuum connection and vacuum-sealed temperature and pressure-resistant foil. Once the pressing chamber is loaded, it is closed and the inert gas or oil is introduced into the chamber. The isostatic pressure (pressure exerted evenly in all directions) for bonding is 80 to 200 Newton/cm^2. Figure 11.5 shows a typical lamination press for multi-layer prototyping.

In contrast to hydraulic press, different press sizes may be bonding simultaneously in an over-pressure chamber. The advantages of this method of bonding are improved heat transmission and a more favourable thermal time gradient. The all-round application of pressure has a particularly positive effect on the multi-layer stack. It

Fig. 11.5   Typical lamination press for multi-layer boards (Muller, 2000)

prevents resin flow, which is the main cause of stresses in the fibre-glass fabric. Dimensional stability, torsion/warping and thickness tolerance are significantly improved if stresses of this nature are not generated. Furthermore, no resin deficiencies will be found within the board. A lower bonding pressure is required for vacuum bonding (vacuum chamber press, vacuum frame or vacuum autoclave). Fewer stresses are generated in the multi-layer with lower bonding pressure. This gives considerably better dimensional stability of the inner layers, improved thickness tolerance and reduced inner layer misalignment. Since the melting point is lowered in a vacuum, volatile components, including void-free multi-layers can be achieved.

A registration system is required to achieve precise alignment of the several copper layers bearing a layout in a multi-layer during bonding. This registration is done by using locator holes drilled in the production board or in the individual layers. An exception to this is the floating bonding process used for four-layer multi-layers. This involves bonding the inner layer with prepregs and copper foil in the same way as for an outer layer. The locator holes for drilling the multi-layer are then obtained by milling and drilling the targets (registration marks) on the inner layer. Each manufacturing step affects the inner layer registration of multi-layer PCBs. As the number of layers increases and pad sizes decrease, the probability of mis-registration increases dramatically. Hinton (1992) explains the various steps for solving the problems of internal layer registration in multi-layer boards.

The cooling rate for bonding multi-layers must be as slow as possible as too great a temperature gradient within the press stacks gives rise to varying rates of shrinkage between the outermost and innermost layers in the press stack, thus causing distortion in the multi-layers. In extreme circumstances, the press cooling system may be switched off so that the multi-layers take twelve or more hours to cool down.

11.5.3   Post-lamination Process

After removal from the mould, the laminate is inspected for insulation resistance as per the design requirements. The board can also be inspected by radiography. The board is then trimmed of excess material and drilled. The feed and speed of drilling are adjusted so as to minimize burring and epoxy smear.

Before the drilled multi-layers can be through-hole plated, hole wall cleaning must be performed as the action of drilling can heat the resin to above the glass transition temperature, allowing the resin to soften and be smeared over the end face of the inner layer copper by the drill bit. This smear layer must be removed so that copper is present only on the wall faces so that contact between the inner layers is not impeded in any way. This thickness of the smear is generally 2–6 µm; however, it may be as thick as 12 µm if the drilling parameters are not selected properly. Chemical processes or plasma de-smearing may be used to perform hole wall cleaning. Three-stage cleaning with permanganate is the most suitable and widespread among the various chemical processes available.

The use of direct metallization to create the electrical connection between the various conductor track layers in the multi-layer through-holes is an environment-friendly process. Once the holes have been cleaned and coated with carbon particles or palladium (which have no impact on the environment), metallic copper is deposited from a solution of copper salts in sulphuric acid, to which an electrical current is applied. This copper acts as a connector element between the various conductor track layers and as a reinforcement of the external conductor tracks. In the case of some substrate types, through-hole plating for microwave engineering can be performed by using the standard process of direct metallization. Some types of substrate require an additional etching process as part of the standard direct metallization process. Using a similar process, it is possible to fabricate multi-layer boards with as many internal layers as required. However, the production yield and cost considerations become a major limiting factor for higher number of layers.

11.5.4   Multi-layer Drilling

The techniques for drilling copper clad for double-sided and multi-layer PCBs with automated equipment are identical, with the exception that multiple drilling steps will be needed if your multi-layer design includes buried or blind vias.

11.5.5   Schematic Key for Multi-layer Built-Ups

Multi-layer built-ups are designated as per the following (Table 11.1) schematic key: (courtesy, Printed Circuit Boards, GmbH)

Table 11.1   Multi-layer Built-ups

04_188_FR4_L41.35_71.18_p10_20_v1.99_2-3_4-5_6-7_s0

a	b	c	d	e	f	g + h + i
04	188	FR4	55	L41.35_71.18	P10_20	V1.99_2–3_4–5_6–7_s0

(Contd.)

Table 11.1 *(Contd.)*

	Parameter:		**Examples**	**Explanation:**	**Units:**
a	Number of layers:	Core-layer Sequential built-ups	04 (1– 4 –1)	Four-layer MLBs Two-outer layer of sequential built-up and four-core layer	Numeric
b	Total thickness after the built-up and final plating:		188	1880 μ	Per 10 μ
c	Type of material:		FR4	Quality of material	
d	Copper thickness of the outer layer after the built up and final plating:		55	55 μ	Per 1 μ
e	Different kinds of core material and co-pper foils on both sides:		L41.35_ L73.18	L = core material (prefix): core thickness 410 μ + cu foil 2 × 35 μ + core thickness 730 μ + cu foil 2 × 18 μ	Per 10 μ. Per 1 μ
f	Number and thickness of the prepregs:		p10_p20	P = prepregs (prefix)Core thic-kness 410 μ + core thickness 730 μ	Per 10 μ
g +	Buried vias:		v2 – 3	V = buried via (prefix):conne-cts inner layer 2 to inner layer3	Inner layers
h +	Blind vias:		v1.v99	V = blind via (prefix)Connects outer layer 01 to inner layer 2/3 /4 etc and outer layer 99 to inner 7/8 etc	Outer layers: top outer is layer 1 and bottom outer is always layer 99
i	Special code number of the assembly		s0	none	

It may be noted that tolerance on total thickness = ± 5%.

 ## 11.6 Useful Standards

- *IPC-1710: OEM Standard for Printed Board Manufacturers'Qualification Profile (MQP)*: This standard is useful for assessing the PCB manufacturer's is capabilities and allows PCB manufacturers to satisfy customer requirements more easily. The document is aimed at decreasing paper work and enhancing manufacturer effectiveness.
- *IPC-HM-860: Specification for Multi-layer Hybrid Circuits*: This document covers the qualification and performance requirements of multi-layer circuits used in hybrid packaging.

These circuits consist of three or more layers of conductor patterns separated from each other by insulating materials and interconnected by a continuous metallic interlayer connection.

- *IPC-ML-960: Qualification and Performance Specification for Mass Lamination Panels for Multi-layer Printed Boards*: This specification covers the qualification and performance requirements of rigid mass laminated panels for use in multi-layer printed boards. Testing procedures and criteria are also addressed.

- *IPC-TR-481: Results of Multi-layer Tests Program Round Robin*: This report was designed to collect and evaluate data on multi-layer boards and the effects of materials and processing of MLB reliability.

- *IPC-SKILL-201: IPC Skill Standards for Printed Circuit Board Manufacturing*: This document details the industry consensus on PCB skill standards regarding conditions of performance, statement of works, performance criteria, assessment and credentialling approaches for over 40 critical areas of PCB manufacturing.

- *IPC-TMRC-01T: 2001 Technology Trends for Rigid Printed Circuit Boards:* This report details trends in conductor width and spacing, hole processing, electrical and optical testing, metallic finishes and solder mask usage, as also trends in multi-layer production, surface mounting and fine-pitch technology.

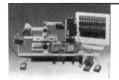

Flexible Printed Circuit Boards

12.1   What are Flexible Printed Circuit Boards?

Flexible printed circuit boards interconnect rigid boards, displays, connectors and various other components in a three-dimensional package. They can be bent-folded or shaped to interconnect multiple planes or conform to specific package sizes. Flex circuits also have the ability to connect moving components, a prime requirement in disk drives, printer heads and other continually moving electronic assemblies. Figure 12.1 shows typical flexible circuits.

Flexible printed circuits constitute a reliable alternative to conventional wiring. They not only improve connection reliability, but also simplify assembly and improve component appearance. By eliminating bulky wires, flex circuits providea cleaner and neater appearance. Generally, flex circuits fit only one way and therefore, cause fewer errors during installation and servicing, thereby reducing re-work and troubleshooting time. Since a flex circuit is more resistant to shock and vibration than a rigid PCB, repair and replacement costs in case of the former are obviously much less. Flex circuits are thin, light in weight and durable. They can be designed to meet a wide range of temperature and environmental extremes. They are excellent for designs with fine line traces and high density circuitry, and are more suited for dynamic applications and vibration conditions than are conventional printed circuit boards. Their high density and light weight make them ideally suitable for redundant circuitry for satellites and avionic instruments, advanced scientific sensors, flexible heating elements, medical equipment, robots and security devices.

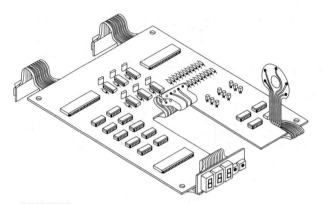

Fig.12.1 (a)Typical examples of flexible circuit layout

Fig. 12.1 (b) Use of a double layer flex circuit in an implantable cardiac pace maker [courtesy Minco Products, USA: minco Application Aid 24 (2000)]

Flexible circuits offer the same advantages of a printed circuit board as are available in rigid circuits, including repeatability, reliability and high density. However, their most important advantage, which has facilitated the adoption of flex circuit technology, is the capability of the flex circuit to assume three-dimensional configurations. These circuits can flex during installation, maintenance and in use. Careful planning can lead to a flex circuit which can save upto 75 per cent of the space and/or weight of conventional wiring.

12.2 Construction of Flexible Printed Circuit Boards

Figure 12.2 shows the constructional parts of flexible printed circuit boards. They are made of a dielectric substrate (film) which is coated with an adhesive over which the copper foil forms the conducting path. The copper foil is protected from corrosive media by a cover layer or special coating.

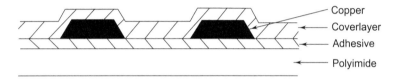

Fig. 12.2 Schematic view of flexible printed circuit board—constructional parts

12.2.1 Films — Types and Their Characteristics

Flexible printed circuits make use of flexible laminate. The properties of the laminate are crucial not only to its manufacturing process but also to the performance of the finished circuit. The flexible laminate consists of a conducting foil and dielectric substrates. The dielectric substances are of two types which are used for flexible printed circuits:

- *Thermosetting Plastics*: These are polyimide, polyacrylate, etc.
- *Thermoplastics*: These include materials which, after curing, will soften by heat input, such as some types of polyester, fluorinated hydrocarbon, polymers, etc.

Copper is the most commonly used foil, while virtually all flexible circuitry is built on polyimide or polyester film. For special purposes, aramid and fluorocarbon films are used.

The selection of a particular film depends upon a number of factors. These are enumerated below.

- High performance flexible circuits, particularly those for military applications, are manufactured with polyimide films because they offer the best overall performance.
- Commercial, cost-sensitive circuits are built on polyester films that provide polyimide performance at a lower cost, but with reduced thermal resistance.
- Aramid non-woven fibre is inexpensive and has excellent mechanical and electrical properties, but exhibits excessive moisture absorption.
- Fluorocarbons, though expensive and difficult to handle, offer superior dielectric properties. They are most suitable for controlled impedance applications.

12.2.1.1 Polyimides

The most common choice of film in flexible circuits is polyimide film. This is because of its favourable electrical, thermal and chemical characteristics. This film can withstand the temperatures encountered in soldering operations. The film is also used in wire insulation and as insulation in transformers and motors.

The polyimide film used in flexible circuit is Kapton, which is a trademark of Du Pont Co., USA. Kapton/modified acrylic has a temperature rating of –65 to 150 °C, though circuits will discolour after a long-term exposure at 150 °C. Kapton type H film is an all purpose film that can be used in applications requiring working temperature ranging from –269 °C to 400 °C. There are some specialized versions of Kapton film which are required for use in applications requiring special properties. One of them is Kapton "XT", which is thermally conductive with twice the heat dissipating capacity of Kapton type "H" film, enabling higher speeds in thermal-transfer printers. Polyimide films are available in standard thicknesses of 0.0005, 0.001, 0.002, 0.003 and 0.005 inches (0.0125, 0.025, 0.050, 0.075 and 0.125 mm)

The main reason for the large usage of polyimide film is its ability to withstand the heat of manual and automatic soldering. Polyimides have excellent thermal resistance and have continuous use ratings approaching 300 °C. At these temperatures, copper foils and solder joints are quickly destroyed through oxidation and inter-metallic growth. The behaviour of the laminate is determined by the combined properties of adhesive and supporting film. It is therefore important to understand the influence of both adhesive and film properties while selecting a laminate.

Polyimides are inherently non-burning and when combined with specially compounded fire-retardant adhesives, produce laminates which can withstand high temperatures. However, many flexible circuit adhesives have much less resistance. Although they can withstand soldering, these adhesives constitute the weak link in a polyimide film laminate. Table 12.1 shows the characteristics of polyimide films.

Table 12.1   Characteristics of Polyimide Films (after Stearns, 1992)

Property, Units	Upilex-S	Kapton-H	Apical
Density, g/cm^3	1.47	1.42	1.42
Tensile strength, psi	56,800	25,000	35,000
Elongation, %	30	75	95
Tensile modulus, psi	1,280,000	430,000	460,000
Flammability	94 VTM-0	94 VTM-0	94 VTM-0
Moisture absorption, %	1.2	3.0	3.0
Oxygen permeability, ml/m^2/mil	0.8	380	~380
Moisture permeability, g/m^2	1.7	84	~84
Dielectric strength, V/mil	6800	7000	7800
Dielectric constant	3.5	3.5	3.4
Dissipation factor	0.0013	0.0025	0.0014
Volume resistivity, MΩ-cm	1×10^{11}	1×10^{12}	3×10^{11}

Note: *Typical values for 1 mil thick (25 μm), at 25 °C

Some polyimide films absorb a great deal of moisture. Prior to exposure to elevated temperatures, such as soldering temperatures, the laminate must be baked dry by keeping it at least one hour at 100 °C or higher for single layer circuits and longer for multi-layer constructions. As the moisture re-uptake is very rapid, the laminate should be stored under dry conditions if the process cannot be completed within an hour's time.

Dimensional Stability: A crucial property of flex circuits is their dimensional stability. Flexible laminates inherently expand and shrink more during exposure to various process conditions than the glass-reinforced rigid system. The stability of a flexible laminate depends upon the film properties, degraded by the properties of the adhesives and process conditions used to form the laminate (Stearns, 1992). Careful laminate manufacture, using low web tensions, vacuum evacuated lamination and thermally stabilized films minimize the chances of shrinkage. After-etch shrinkage of 0.1 per cent is achievable with high performance films having high tensile strength, but shrinkage for laminates made with conventional polyimide film is generally of the order of 0.15 per cent. These shrinkage values may seem trivial, tolerable and predictable if not accompanied by other errors. But many times, they are undesirable and costly to neutralize.

Tear Resistance: Flexible circuits commonly have complex geometries with multiple stress concentration points. That makes tear resistance an important property. For example, a torn circuit cannot be repaired. The adhesive can enhance laminate performance in terms of tear resistance since most flexible adhesives have better tear resistance than polyimide films.

Unfortunately, film characteristics that are essential for better dimensional stability result in lowering tear values because dissipation of tear energy requires a softer film with greater elongation and yield before failures.

In flexible laminates, the primary insulation is provided by the adhesive which has its own insulation resistance and dielectric strength. Thus, the flexible circuit designer must look carefully at the properties of the laminates and not the film, when designing the conductor pattern in a PCB layout.

Polyimide films and adhesives have relatively poor electrical properties for use in controlled-impedance applications because of their high dielectric constant, (3.7 or greater), and dissipation factor (greater than 0.03). This limitation suggests that some other type of laminate should be used in such applications.

12.2.1.2 Polyesters

Polyester dielectric substrate films are mechanically similar to polyimide and electrically superior, and absorb far less moisture. However, they fail to match polyimides in the crucial area of thermal resistance as the maximum temperature upto which they can be used is less than 125 °C for most polyester. Their melting point is below the soldering temperature. Even so, by using special techniques like crimp or pressure, polyester can cut flex circuit cost without lowering circuit performance and quality. Polyester film is most commonly used in automotive and communication circuitry.

As compared with polyimides, polyesters have a lower dielectric constant, higher insulation resistance, greater tear strength and lower cost. The moisture absorption of polyester is well under 1 per cent with excellent dimensional stability. Polyesters have limitations only in the area of thermal resistance, but offer a great cost advantage. Polyester films are highly resistant to solvents and other chemicals. Polyester has a high tensile strength (25,000 psi) and a good dielectric strength (7.5 KV $\times 10^{-3}$ in. for 0.001 inch film).

Polyester film is a polymer. One of the most commonly used polyester films is "Mylar", which is a trade name for the product produced by M/s Du Pont, USA. The temperature range of the polyester film for use is 75 to 150 °C, making it unsuitable for soldering temperatures over 230 °C. This problem can be circumvented by using large solder pads, wide traces and foil thickness of 0.00275 inch (0.07 mm) and using an appropriate mask or jig to keep the heat away from all parts of the circuit except the portion being soldered.

12.2.1.3 Aramids

Commonly used as a motor and generator insulation, non-woven aramid fibre materials are inexpensive and have outstanding dielectric strength and thermal properties. They are rated for continuous use at 220 °C and when quoted with the suitable laminating adhesive, form a very good flexible laminate.

The product has good tensile and tear strength as well as dimensional stability. However, it is very hygroscopic. Like polyimide-based laminates, aramid-based laminate must also be thoroughly dried before solder assembly, and kept dried through the assembly process.

Aramids have an undesirable property of stainability. It develops when the laminate is exposed to the liquid process wherein the process chemistry may wick into the fibre structure, leaving a permanent stain and potential insulation resistance problems. Aramids have many desirable properties and are inexpensive. But their shortcomings make them difficult to use in volume flex circuit applications.

One of the common aramid materials is "Nomex", a trade mark of M/s Du Pont. Nomex is a high temperature paper which can withstand soldering temperatures very well. It has a very low initiation and propagation tear strength.

12.2.1.4 Fluorocarbons

The first flexible circuits were supported by high performance fluorocarbons long before Kapton came on to the scene. Unmatched chemical inertness, extremely high thermal resistance, outstanding dielectric and tough mechanical properties suggested that fluorocarbons would be ideal for flexible circuitry.

Fluorocarbon dielectrics, which, are formed with the fusion process, suffer from dimensional instability. Lamination at the required temperatures (near 300 °C) creates stresses on a semi-molten dielectric that can destroy fine and delicate conductor patterns.

Fluorocarbons have superior characteristics for flexible circuits, specially since their tear values are very good. Because of this property, fluorocarbon patches are sometimes used to reinforce weak corners of polyimide circuit. Fluorocarbon laminates are essentially inert to all common chemistries and inherently incombustible, and do not pose any problem in the production process or in use.

Fluorocarbons are not easily adaptable to the plated through-hole process because they have excellent chemical resistance. Baths used to promote adhesion of electroless copper onto hole walls have little effect, requiring the use of additional process steps.

Today, for ease of circuit and laminate manufacture, fluorocarbons can be assembled with adhesives instead of the use of the fusion process, giving an improved dimensional stability, though not of the level of polyimides. If the adhesive is kept as thin as possible, the circuit will display some of the excellent electrical characteristics of a fusion-made circuit, at a lower cost.

12.2.1.5   Choice of Dielectric

The dielectric substrate of a flexible laminate has a significant effect on the manufacturing cost and performance of the finished circuit. Polyimide films offer the best combination of cost and properties for this use.

Polyester films come a close second, falling short only in thermal resistance. Aramid non-woven fibre has unique properties that suggest use in applications where cost is important and slight imperfections can be overlooked. Fluorocarbons have superior dielectric properties and are suitable for use in demanding controlled impedance applications. Table 12.2 gives the characteristics of polyester, fluorocarbon and aramid films.

Table 12.2   Characteristics of Polyester, Fluorocarbon and Aramid Films (after Stearns, 1992)

Property	Polyester	Fluorocarbon FEP	Aramid Du Pont 410
Tensile strength, K psi	20–40	2.5–3	6–10*
Elongation, %	60–165	300	9
Tear strength, g./mil	50–130	125	550
Propagation	50–300	125	45–80**
Moisture absorption, %	0.25	< 0.01	3–7
Moisture permeability: g-mil/100 sq. in./24 hrs.	1–1.3	0.4	NA
Dielectric strength, V/mi, 1 mil	7500	6500	530
Dielectric constant, 1 kHz-1 GHz	3.2–2.8	2.0–2.05	2.3
Dissipation factor, 1Hz-1 GHz	0.003–0.016	0.0003–0.0015	0.007
Chemical resistance	Good	Excellent	Good

Notes: *Per mil based on 2 mil thickness
**Per mil based on Elmdorf test of 2 mil

12.2.2   Foils

The use of copper foil as a base material in flexible circuits is well known. Knowledge of how it is manufactured, however, is not as common. The production of copper foil requires a number of processing steps to provide the flexible circuits industry with quality foil products.

Two types of copper foils are used for flexible laminates today: (i) rolled annealed (also known as wrought foil), and (ii) electrodeposited foil. The manner in which foils are manufactured, either rolled annealed or electrodeposited, determines their mechanical characteristics. Each type of foil is further categorized into grades, on the basis of its mechanical properties and applications. The copper foil classification is shown in Table 12.3 providing four separate classifications for both electrodeposited and wrought types (Savage, 1992). Generally, electrodeposited foils identified as grades 1-4, are used for rigid printed circuit boards. Flexible circuits use both electrodeposited and wrought copper foils (grades 5-8). Typically, grades 2, 5, 7 and 8 are used in flexible laminates.

Table 12.3   Copper Foil Classification (after Savage, 1992)

Type	IPC Grade	Description	Application
Electrodeposited	1	Standard	Rigid laminates
Electrodeposited	2	High ductility	Automotive flex
Electrodeposited	3	High temp. elongation	Multi-layer board inner layer
Electrodeposited	4	Super high ductility	
Wrought	5	As rolled	Commercial flex
Wrought	6	Special temper	
Wrought	7	Rolled annealed	Military flex
Wrought	8	Low temp. annealable	Commercial flex

12.2.2.1   Rolled Annealed Foils

These are made by first heating copper ingots, then sending them through a series of rollers that reduce them into foils of specified thicknesses. This is shown in Figure 12.3. Rolling creates a grain structure in the foil that looks like overlapping horizontal planes. Both pressure and temperature are used to create stresses between different sizes of copper grains. These produce copper foil properties such as ductility and hardness while also providing a smooth surface. This manufacturing technique yields foils with greater resistance to repeated flexing than that of electrodeposited foils. However, its disadvantage is its higher cost and lack of availability of various thicknesses and widths.

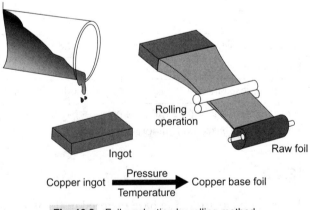

Fig. 12.3 Foil production by rolling method

12.2.2.2   Electrodeposited Foils

They are made by plating copper ions into a cylindrical cathode, from which the foil is continuously stripped. Electrodeposition creates a columnar grain structure. When the foil is flexed, the grains separate. This results in less flexibility and a lower resistance to cracking when folded than in the case of rolled annealed foils. Figure 12.4 shows a schematic diagram of the process for making electrodeposited foil.

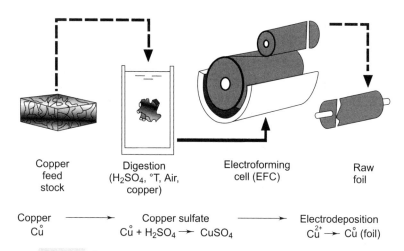

| Copper feed stock | Digestion (H_2SO_4, °T, Air, copper) | Electroforming cell (EFC) | Raw foil |

$$\text{Copper} \longrightarrow \text{Copper sulfate} \longrightarrow \text{Electrodeposition}$$
$$\overset{\circ}{Cu} \qquad\qquad \overset{\circ}{Cu} + H_2SO_4 \rightarrow CuSO_4 \qquad\qquad Cu^{2+} \rightarrow \overset{\circ}{Cu} \text{ (foil)}$$

Fig. 12.4   Process for making electrodeposited foil (Savage, 1992)

The process begins with the dissolution of copper metal in a sulphuric acid solution. Both temperature and agitation are used to control the rate of dissolution. The profile and mechanical properties of the foil can be controlled by using various types of additives.

The copper solution is continuously pumped into an electroforming cell, wherein the application of current between the anode and cathode causes copper ions from the chemical bath to plate on the cathode surface. The cathode is a cylindrical drum that rotates while being partially submerged in the solution. As it enters the solution, copper begins to deposit on the drum surface and continues to plate until it exits. The copper foil is stripped from the cathode as it continues to rotate. The thickness of the foil is determined by the rotation speed of the cathode drum. The electrodeposition process is capable of producing copper foil in many thicknesses and widths.

After raw foil production, both wrought and electrodeposited foils are treated in three treatment stages as shown in Figure 12.5.

Bonding (Anchoring) Treatment: This treatment usually consists of a copper metal/copper oxide treatment, which increases the surface area of the copper surface for better wetting of the adhesive or resin.

Thermal Barrier Treatment: This allows the adhesion of the clad laminate to be maintained in spite of the thermal processing conditions involved in PCB manufacture.

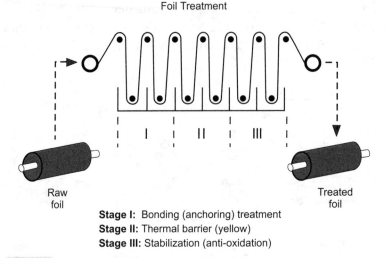

Fig. 12.5 Stages for treatment and stabilization of foils (after Savage, 1992)

Foil Stabilization Treatment: Also called passivation or anti-oxidation, this treatment is applied on both sides of the copper to prevent oxidation and staining. All stabilization treatments are chrome-based. However, some manufacturers use nickel, zinc and other metals in combination with chrome.

After treatment, copper rolls are cut to desired widths and wound on a core after encapsulation in a plastic film to prevent oxidation. The ductility of copper foil is as follows:

- Electrodeposited copper foil : Elongation 4–40 per cent
- Rolled annealed copper foil : Elongation 20–45 per cent

The copper foil is usually covered with a film made of polyimide or liquid polymer solution. The coating of conductors with such type of treatment serves as both a long-term protection against corrosive environments and a solder resist.

12.2.3 Adhesives

The function of adhesives in flexible circuits is to bond copper foil to the dielectric substrate, and in multi-layer flex designs, to bond the inner layers together. A flexible laminate's performance depends upon the combined properties of its adhesive and supporting dielectric film. Bond strength, dimensional stability and flexibility after soldering are the key factors determining an adhesive's suitability for a particular application (Wallig, 1992).

Adhesives such as acrylics, polyimides, epoxies, modified polyesters and butyral phenolics have been used with varying degrees of success to bond the flex circuits. Since polyimide and polyester dielectric films are the two most commonly used substrate materials, the adhesives typically used with these materials are covered in the following sections.

12.2.3.1   Acrylic Adhesives

Acrylic adhesives offer high heat resistance and good electrical properties. They have been used successfully on polyimide film substrates and have made polyimide/acrylic a preferred choice for dynamic flex applications. However, many flex circuits manufacturers are finding acrylic adhesives thicknesses and high Z-axis expansions as limiting factors for more demanding electronic packaging applications. In addition, polyimide/acrylic adhesive laminates are vulnerable to attack by some of the solvents used in the photo-resist process, and to alkaline solutions used in plating and etching. Absorbed solvents are especially difficult to remove prior to multi-layer lamination, resulting in de-lamination or blistering problems, if these volatiles are not removed.

In high density designs, dimensional stability and small hole drilling problems (drill smear) can decrease yields which, in turn, increase unit cost. If not controlled, plated through-hole failures could result.

Most rigid flex systems are made by using acrylic adhesive systems. For these, the most popular etch-back or hole cleaning process in use is the plasma system. The plasma system works with ionized gases, which are generated by a radio frequency source with the ionization of Freon (CF_4) mixed with oxygen. This action de-smears the flexible circuit portion of the assembly but does not harm glass fibres that might be in the holes of the rigid portion. After the plasma treatment, the organic residue left in the holes is removed with an alkaline cleaner at 140 °C for 2-3 minutes in an ultrasonic cleaner.

12.2.3.2   Polyimides and Epoxies

Polyimide substances may also be successfully paired with polyimide adhesives. The chemical resistance and electrical properties of polyimide adhesive are as good or better than those of acrylic adhesives. Additionally, they offer better heat resistance than any of the other adhesives used with flex circuits.

Some polyimide-based flex circuit laminates incorporate epoxies as adhesives. Epoxies generally give good electrical, thermal and mechanical performance. However, they are limited to static flexing applications due to the resin cross-linking that occurs on curing.

The reduced dynamic flexing ability for polyimide and epoxy adhesives is not a serious limitation since the majority of flex circuits produced are used in static flexing applications. The trade-offs for this increased laminate stiffness are better dimensional stability, better processability and lower overall adhesive thickness in multi-layer flex and rigid flex board fabrication.

Epoxy remains in good condition during soldering operations. They exhibit long-term stability at elevated temperatures in environmental conditions upto 120 °C. Epoxy systems include modified epoxies known as phenolic butyrals and nitrile phenolics. They are widely used and are generally lower in cost than acrylics, but higher in cost than polyesters.

12.2.3.3   Polyester and Phenolics

Adhesives commonly used with polyester-based substrates are typically polyester or butyral phenolics adhesives. Polyester adhesives offer excellent electrical properties, excellent flexibility and low resistance to heat. Polyesters are the lowest cost adhesives. They are the only adhesives which can be used properly with polyester films for base laminate and polyester cover film. The low heat resistance may not be a lamination if the application for the circuit does not require soldering as in the case of many automotive and cost-sensitive consumer electronic products.

Butyral phenolic adhesives are more heat-resistant than polyester adhesives, but their electrical properties are not quite as good and they are not as flexible. The flexibilities of phenolics adhesives can, however, be enhanced through the use of additives.

12.2.3.4   Adhesiveless Laminates

Adhesiveless copper clad laminate, a major innovation in new materials, is rapidly gaining the attention of both fabricators and users of flex circuit because it offers improved operating characteristics for single- and double-sided circuits as well as for rigid flex multi-layers (Pollack and Jacques, 1992).

In the new family of laminates, copper is bonded to the polyimide film without adhesives. As compared to adhesive-based laminates, adhesiveless laminates provide a thinner circuit, greater flexibility and better thermal conductivity (Crum, 1994a). Additionally, the thermal stress performance of higher layer count rigid-flex multi-layers is significantly better. Adhesiveless laminates can be manufactured by using any one of the four technologies including: (i) cast to foil, (ii) vapour deposition on film, (iii) sputtered to film, and (iv) plated to film.

The *cast to foil* method involves casting a liquid solution of polyamic acid on to the surface of a metal foil. The entire composition is then heated to a temperature that will evaporate the solvent and imidize the polyamic acid. This process forms a polyimide or amide modified polyimide film. Although the adhesion of the copper to the film is good, this process is usually limited to use with copper thicknesses of 1 oz. or more. Thinner copper, though available, is more difficult to handle and as a result may be too costly. Adhesiveless laminates have been observed to be less repeatable in dimensional changes than adhesive-based laminates.

In the *vapour deposition* method, copper is vaporized in a vacuum chamber and the metal vapour is deposited on a film. A surface treatment on the film enhances copper adhesion. The method is usually limited to a copper thickness of about 0.2 micron. Additional copper thickness can be achieved through electrolytic plating.

The *sputtering to film* method involves placing the film in a large vacuum chamber having a copper cathode. The cathode is bombarded with positive ions, causing small particles of the charged copper to impinge on the film. This results in an ultra thin copper coating, which is followed by the build-up of electrolytic copper to the desired thickness. However, the copper adhesion is not as good as that of the cast or plated method and the dimensional stability does not compare favourably with the adhesive-based materials.

The *plated to film* method of manufacturing adhesiveless laminates is to plate copper onto the polyimide film. The process begins with the surface treatment of a roll of film, followed by an ultra thin coating of a barrier metal to promote good copper adhesion. The copper is then continuously plated onto the barrier metal to the desired thickness. This copper metal deposit can be controlled to provide very thin copper foils, which are increasingly in demand.

A popular adhesiveless polyimide laminate is Pyralux AP from M/s Du Pont. (Du Pont Electronic Materials, Research Triangle Park, NC, USA). The laminates offer continuous thermal stability at temperatures higher than 200 °C, excellent chemical resistance, low moisture absorption, a low coefficient of thermal expansion in the Z-axis and excellent resistance to solder. They are compatible with acrylic, epoxy and polyimide bonding adhesives (Crum, 1994). Pyralux AP is a double-sided copper clad laminate with an adhesiveless composite of polyimide film bonded to copper foil. The all-polyimide dielectric structure of this laminate improves flexible circuitry and is recommended for double-sided multi-layer flex, as well as rigid flex applications requiring advanced material performance and high reliability. The laminate is certified to IPC-FC-241/11 Class 3.

12.2.3.5 Advantages of Adhesiveless Laminates

Significant performance advantages can be obtained when adhesiveless laminates are used to make flexible circuits and rigid flex multi-layers. Many of these advantages are due to the inherent thinner circuit and the elimination of the mismatched characteristics of the adhesives in relation to the film in copper. Adhesiveless clad laminates are thinner due to the absence of the 1 to 2 mils of adhesives currently used in clad materials. This advantage increases in proportion to the layer count of the circuit. It has been found that there is a saving of 4 mils with adhesiveless laminates in the plated through-hole area of comparable double sided circuits.

The acrylic adhesive used in bonding flexible circuits has been recognized as the limiting element of today's material technology. This is due to the high thermal coefficient of expansion between the adhesive and polyimide film, which is incompatible with the ductility of PTH (printed through-hole) copper. Different approaches to circuit construction have been developed to overcome this problem. The common goal of all these approaches is to eliminate as much of the adhesive as possible by selectively reducing the use of adhesive-based material, particularly cover coats and cast adhesives made with modified acrylics, from critical areas such as the PTH portion of the rigid flex circuit.

The thermal conductivity of the thin structure and the absence of thermally resistant adhesives allow adhesiveless circuits to be used in operating environments unsuitable for adhesive-based laminates. The polyimide itself has a continuous operating temperature of 450 °C without degradation of the materials. Adhesiveless circuit bonded to heat sinks can be used in such high performance and high reliability applications such as in automotive electronics.

Another important feature of the adhesiveless circuit is its ability to maintain a uniform thickness. Copper traces will not deform into the base film due to the film's high glass transition temperature. In contrast, conductors are set into adhesives in a relatively uncontrolled manner.

A wide range of tests have been carried out to compare the performance of adhesiveless circuits to adhesive-based circuits. Figure 12.6 compares the results of flex cycling on a ten-layer circuit made from a standard adhesive system with that of the same circuit made from the plated method. The number of cycles until conductor breakage is significantly higher for the adhesiveless circuit.

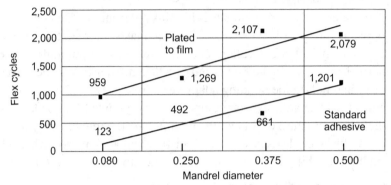

Fig. 12.6 Continuous flex cycling results: nearly twice as many flex cycles are endured by the adhesiveless circuit than by the standard adhesive circuit (after Crum, 1994a)

Figure 12.7 compares the plated through integrity of the plated film material to a standard adhesive on an eighteen-layer circuit. The standard adhesive circuit showed a failure of 175 temperature cycles, between –65 °C and +125 °C, while the adhesiveless circuit was still performing at 500 cycles.

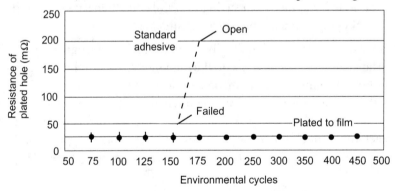

Fig. 12.7 Environmental cycling results: the adhesiveless circuit performs 500 temperature cycles as compared to adhesive circuit giving upto 175 cycles (after Crum, 1994a)

 ## 12.3 Design Considerations in Flexible Circuits

12.3.1 Difference in Design Considerations of Rigid and Flexible Circuits

Most of the design rules for rigid PCBs have to be applied for the design of flexible PCBs. There are, however, a few exceptions plus some new considerations to be taken into account. A few of them are given below.

Current-carrying Capacity of Conductors: Because of less cooling capability by the flexible board itself (when compared to rigid PCBs), sufficient conductor width has to be provided. A guideline for selecting conductor widths for currents of more than one ampere is given in Figure 12.8. Where several conductors with a high current are placed opposite or neighbouring each other, the heat concentration has to be taken care of by giving additional conductor width or extra spacing.

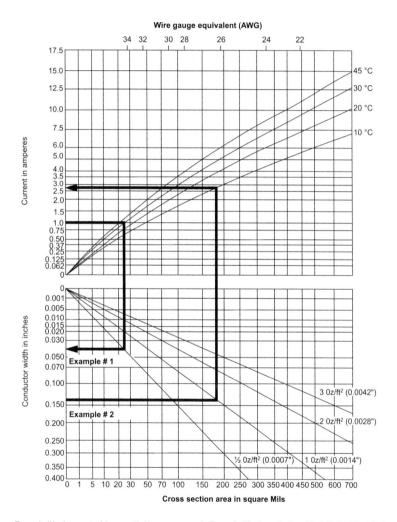

Example #1: A current of 1 amp with ½ oz. copper and 30 °C temperature rise will require a conductor width of 0.040".

Example #2: A conductor with width 0.140", etched from 1 oz. copper (0.0014") will produce a temperature rise of 10 °C at 2.7 amps.

Fig. 12.8   Guidelines for selecting conductor size (www.minco.com)

Contours: Wherever possible, rectangular shapes are preferred because of the better base material economy. There should be sufficient free border space near edges due to the possible dimensional changes with the base materials. Inward looking corners in the contour should be rounded; sharp inward corners could initiate tearing of the board.

Areas of small conductor widths and spaces should be minimized as much as possible. Conductor width should transition from fine lines in tight areas to wider widths where geometry allows. Conductors terminating at PTH vias or component mounting holes should have a smooth fillet transition from the trace into the pad as shown in Figure 12.9. As a general rule, any transition from a straight line to such features as corners or different line widths must be done as smoothly as possible. Sharp corners constitute a natural place for stress to accumulate and for conductor defects to occur.

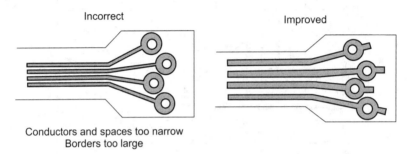

Fig. 12.9   A smooth transition from trace to pad reduces stress and improves reliability. In the top view, the conductors and spaces are too narrow, and the borders too large, the bottom view shows an improved design

Bending: As a general rule, the bending radius should be designed as wide as possible. The possibility of undergoing many cycles is also further improved with thinner laminates (e.g. 50 μm foil instead of 125 μm foil) and larger conductor widths. If subjected to a higher number of bending cycles, single-sided flexible PCBs, in general, show a better performance.

Solder Pads: Around the solder pad, there will be a transition from flexible to rigid material. This zone is highly prone to conductor breakage. Solder pads are therefore, avoided in active bending zones.

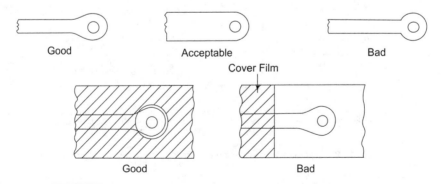

Fig. 12.10   Shape and masking of solder joints (a) shape of solder pads (b) solder joint masking with cover film (after Bosshart, 1983)

The general shape of solder pads should be tear-like (Figure 12.10) and the cover foil must mask the solder joint close by.

Hardboard Stiffeners: The combination of flexible PCBs with adhesively laminated hardboard stiffeners has become extremely popular and cost-effective in the bulk production of small electronic equipment like pocket calculators. The flexible PCB is mounted on one piece of hardboard (e.g. grade G-10) with suitable slots for separating at a later stage. This is illustrated in Figure 12.11. After component assembly and wave soldering, the cutting operation divides the hardboard into different parts, thereby facilitating folding into the planned shape.

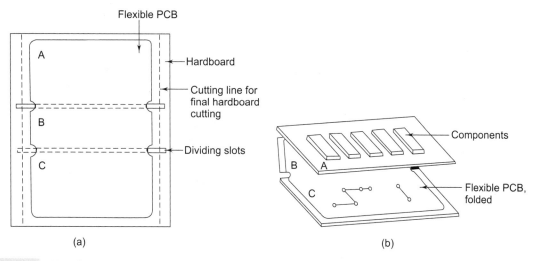

Fig. 12.11 Use of hardboard stiffeners (a) Board as it goes for assembly and wave soldering (b) Board after final hardboard cutting and bending

The above special requirements indicate that designing a flex circuit is only a few steps away from designing a hardboard. However, the important design differences to be kept in mind are:
- Three-dimensionality of a flex circuit is important as creative bending and flexing can save space and layers
- Flex circuits both require and permit looser tolerances than hard-boards
- Because since arms can flex, they are designed slightly longer than required.

The following design tips are useful for minimizing circuit cost:
- Always consider how circuits will be nested on a panel.
- Keep circuits small; consider using a set of smaller circuits instead of one large circuit.
- Follow recommended tolerances whenever possible.
- Design unbonded areas only where they are necessary.
- If circuits have only a few layers, stiffeners can be far less expensive than designing a rigid flex circuit.

- Specify 0.0001" of adhesive on the cover material per 1oz of copper (including plated copper).
- Building circuits with exposed pads and no cover layers is sometimes less expensive.

12.3.2 Step-by-step Approach to Designing of a Flex Circuit

The following steps are guidelines to design a high quality, manufacturable flexible printed circuits (Minco Application Aid 24,):

- It is always good to start with a study of the available literature that is applicable to the intended application. The most useful literature is either the IPC or MIL standards, for example, if the circuit is intended for applications in the military/aerospace field, reference may be made to IPC-6013 and IPC-2223 or MIL-P-50884.
- Define the circuit parameters according to the package that uses the circuit. It is always helpful to cut out a paper template to represent the actual circuit. Experiment with bending and forming the template in order to achieve maximum efficiency. Design a circuit for maximum 'nesting' in order to have as many circuits as possible on a panel.
- Determine the wiring locations and conductor paths. This will determine the number of conductor layers. The circuit cost generally rises with the layer count. For example, two double-layer circuits could potentially be less expensive than one four-multi-layer circuit.
- Calculate the conductor width and spacing according to the current capacity and voltage.
- Decide what materials to use.
- Choose the method of termination and through-hole sizes. Evaluate the bend areas and methods of termination to determine if stiffeners are needed
- Lay down the methods of testing. Avoid over-specification to reduce cost.

12.3.3 Designing for Flexibility and Reliability

The flexible circuits are classified by the type of flexing they will undergo during assembly and use (Corrigan, 1992). There are two types of designs, which are discussed below.

Static Designs: Static designs are those which are flexed or folded only for assembly or on rare occasions during the life of the product. Single- and double-sided as well as multi-layer circuits can be folded successfully for static designs. Generally, the bend radius of the fold should be a minimum of ten times the total circuit thickness for most double-sided and multi-layer designs. Higher layer count multi-layer circuits (eight layers and more) become very rigid and are very difficult to bend without problems. Therefore, they can be designed to have zones with fewer layers for folding. Double-sided circuits requiring tight bend radii are designed to have all the copper traces in the fold area on the same side of the base film. By removing the cover film on the opposite side, an approximation of a single-sided circuit will be achieved in the fold area.

Dynamic Designs: Dynamic circuits are intended to be flexed repeatedly throughout the product lifecycle, such as cables for printers and disc drives. In order to get the highest flex life for a dynamic circuit, the concerned part should be designed as a single-sided circuit with the copper in the neutral axis. The neutral axis is the theoretical plane at the centre of the layers of materials that make up the circuit. By using the same thickness of material on either side of the copper, the base film and cover film, the copper will lie in the exact centre and be exposed to the least amount of stress during bending or flexing.

Designs requiring both high dynamic flex life and high density multiple layer complexity can now be achieved by connecting double-sided or multi-layer circuits to single-sided circuits with anisotropic (Z-axis) adhesive. The flexing takes place only in the area where the assembly is single-sided. Outside of the dynamic flex area, isolated multi-layer zones exist for complex wiring and component needs, without compromising flexibility.

Although flex circuitry is expected to fill applications that require the circuit to bend, flex and conform to fit the specific use, a large percentage of failures in the field are a result of these flexing or bending operations. Using flexible materials in the manufacturing of a printed circuit does not in itself guarantee that the circuit will function reliably when bent or flexed, particularly in dynamic situations. Many factors contribute to the reliability of a printed flex circuit that is formed or repeatedly flexed. All these factors must be taken into account during the design process to ensure that the finished circuit will function reliably. Some tips to increase the flexibility for a reliable operation are:

- A circuit with two or more layers should be selectively plated to improve dynamic flexibility.
- It is advisable to keep the number of bends to a minimum.
- Stagger conductors to avoid the I-beam effect and route conductors perpendicular to a bend as shown in Figure 12.12.
- Do not place pads or through-holes in bend areas.
- Do not place potting, discontinuities in the cover, discontinuities in the plating or other stress concentrating features near any bend location. It should be ensured that there are no twists in the finished assembly. Twisting can cause undue stress along the outer edges of the circuit. Any burr or irregularity from the blanking operation could potentially lead to a tear.

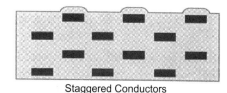

Staggered Conductors

I-beam arrangement

Fig. 12.12 Staggered conductors vs. I-beam effect

- Factory forming should be preferred.
- Conductor thickness and width should remain constant in the bend areas. There should be variations in plating or other coatings and preferably no conductor neck down.
- It is a common practice to provide a slit in a flex circuit to allow different legs to flex in different directions. Although this is a valuable tool to maximize efficiency, the slit represents

a vulnerable point for a tear to start and to propagate. This can be prevented to place a drilled relief hole at the end of the slit as illustrated in Figure 12.13 and to reinforce these areas with hardboard material or a patch of thick flex material or Teflon (Finstad, 2001). Another possibility is to make the slit as wide as possible and to place a full radius at the end of the slit (Figure 12.14). If reinforcement is not possible, the circuit should not be flexed within one-half inch of the end to the slit.

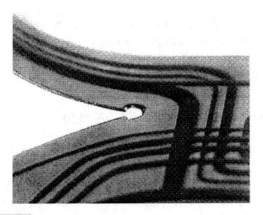

Fig. 12.13 Provision of a drilled relief hole at the end of a slit

Fig. 12.14 The slit is made as wide as possible and placed a full radius at the end of the slit

 ## 12.4 Manufacture of Flexible Circuits

The production processes for flexible printed circuit board are, by and large, similar to those for rigid boards. The flexibility of the laminates requires different equipment for certain operations and a very different way of handling the process. The majority of flexible printed circuit boards are subtractively processed. However, mechanical processes and in-line processing cause some difficulties in the case of flexible laminated panels. One main difference is in the handling of base material. The flexible material is supplied in rolls of different widths. Therefore, during etching, for example, rigid carriers need to be attached for the transport of the flexible laminates.

During the manufacturing process, handling and cleaning of flex circuits is more critical than when processing rigid boards (Lexin, 1993). Improper cleaning or mishandling can lead to defects which can subsequently affect the manufacturing yield. This happens due to the sensitivity of the materials used for flex circuit which plays an important role in the manufacturing process. The substrate is affected by mechanical forces such as baking, lamination and plating. The copper foil is also vulnerable to dings, dents and elongation to ensure maximum flexibility. Mechanical damage to the copper foil or work hardening will decrease the flexibility life of the circuit.

A typical flex single-sided circuit is cleaned a minimum of three times during its manufacturing cycle whereas a multi-layer would require cleaning 3-6 times depending upon its complexity.

Comparatively, a rigid multi-layer board may require the same number of cleaning cycles, but the procedures are different and much more caution is needed while cleaning flexible material. The flexible material is subject to dimensional instability with the slightest amount of stress, which may get introduced during the cleaning process and result in elongating panels in X and/or Y directions, depending upon the stress bias. Flex circuit panels are cleaned chemically through a process that is environmentally safe. The process includes an alkaline bath, a thorough rinse, a micro-etch and a final rinse. The most frequent damage to the thin clad materials occurs with racking of the panels, breaking the surface tension in the cleaning tanks, agitating the panels in the tank, removing racks from the tanks and unracking.

Holes in flexible boards are generally punched, resulting in a high tooling cost. Drilling is also possible by specially adjusting the drilling parameters to obtain smear-free hole walls. After drilling, the holes are de-smeared in an aqueous cleaner with ultrasonic agitation.

It is found that the mass production of flexible boards can be less expensive than that of rigid PCBs. This is because the flexible laminate enables the manufacturer to produce circuits on a continuous basis. The process starts from the laminate roll and produces the finished board directly. Figure 12.15 shows the schematic diagram of a continuous plant for manufacturing print and etch flexible printed circuits. All manufacturing steps are performed in-line by machines that are placed sequentially. Perhaps screen printing is not a part of the conveyor belt-oriented process, which causes an interruption in the on-line process.

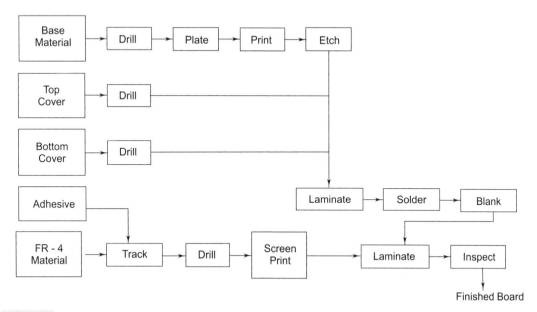

Fig. 12.15   Schematic diagram of a continuous plant for manufacturing print and etch flexible printed circuits (redrawn after www.minco.com)

Soldering is, in general, more critical in flexible circuits due to the limited heat resistance of the base material. Hand soldering needs sufficient experience. Therefore, wherever possible, wave soldering should be employed. The following precautions should be observed while soldering flexible circuits:

- Since polyimide absorbs moisture, circuits must be baked (one hour @ 250 °F) before soldering.
- Pads that are located in large conductor areas such as ground planes, voltage planes, or heat sinks, should be provided with relief areas, as shown in Figure 12.16. This limits heat dissipation and therefore makes for easier soldering.
- When hand-soldering pins in dense clusters, try not to solder adjacent pins one after another. Move around to avoid local over-heating.

Fig. 12.16 Solder relief area around pads

Information on the design and processing of flexible circuits can be obtained from several sources. However, the best sources are always the producers/suppliers of process materials and chemicals. High quality flexible circuits can be produced, by using the suppliers information in combination with the experience of the process scientist.

12.5 Rigid Flex Printed Circuit Boards

Rigid flex printed circuit boards constitute a hybrid system in printed circuit board technology. The manufacturing process uses standard multi-layer technology along with standardized manufacturing methods for flexible printed circuit boards. The various combinations of layers are laminated together. After lamination, the through-holes are drilled and then plated so that they connect the various layers together electrically. The rigid-flex assemblies eliminate the need for jumpers and mother-daughter board combinations, thereby reducing wiring errors while increasing packaging density.

The selection of the *material* for rigid flex assemblies is very important as all the component parts must withstand process in steps without damage. The dimensional changes in the X-Y directions as in the flexible circuits and the vertical stresses in the Z-axis arising due to the stacking of the layers need to be carefully considered. This distortion of materials can cause copper hole barrel cracking when the materials are exposed to final lamination and soldering temperatures.

The *processing* of rigid flex circuits involves four difficult areas: laminations, drilling, removal of adhesive and plating of the holes. In drilling, it is necessary to ensure that the centre of all successive pads in all the layers meets correctly. The adhesive smear removal from the hole may cause hole cracking or misregistration of various layers. These problems can only be overcome with careful processing.

After putting together all the component parts of the rigid flex, the next step is to combine the flexible circuits to the rigid circuits. As most of the component parts are made of similar materials, it is a difficult job. Many companies use the standard platen presses but the modern trend rears towards the use of vacuum *lamination*. Suppliers of flexible and rigid materials usually provide technical assistance on request for carrying out this procedure.

The next step is *drilling* the laminated package. Since many different materials are required to be drilled, the fields and drills should be optimized. After the drilling process, the holes of the stacked materials are cleaned, so that the subsequent copper plating process will yield a hole with no cracks or voids. The *cleaning* can be done by sulphuric acid for epoxy-based system or by plasma etch-back system for acrylic-based materials.

After the holes are drilled and cleaned, they are copper *plated* with electroless copper. Etch-resist placement provides the pattern for the top/bottom services. The final surface and hole electrolytic plating is carried out and the panels are solder plated. The resist is then removed thereby allowing the solder plating to become the new resist. *Etching* then removes the unwanted copper.

Multi-layer rigid flex assemblies often present a difficulty in holding the tolerances. Therefore it is advisable to limit the number of layers as much as possible in rigid flex systems.

12.6 Terminations

There are a variety of possibilities for providing terminations (www.minco.com) for a flex circuit. However, the commonly used terminations are discussed below.

Connectors: Connectors for the flex circuits can be attached (Figure 12.17) by hand soldering, wave soldering or crimping in the case of IDCs (insulation displacement connectors). Connectors can be potted after attachment or conformally coated for protection and insulation with epoxy, polyurethane, etc. Clincher connectors are a good option for many applications.

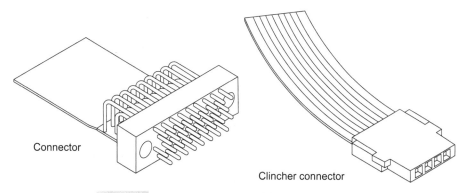

Connector

Clincher connector

Fig. 12.17   Clincher connectors used with flexible circuits

Fingers: Fingers can be supported or unsupported as shown in Figure 12.18.

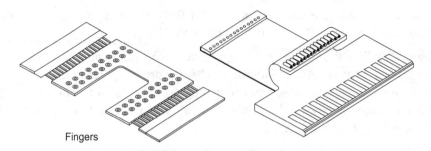

Fingers

Fig. 12.18 Fingers used with flexible circuits

Pins: Socket pins can be pressed in place and then soldered. Pins can also be swaged to the circuit and soldered after the swaging procedure, or pins can be swaged (Figure 12.19) to a FR-4 stiffener and then soldered.

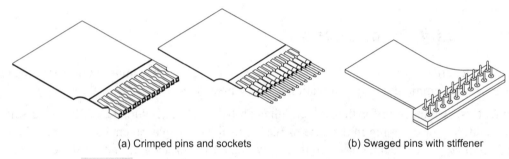

(a) Crimped pins and sockets   (b) Swaged pins with stiffener

Fig. 12.19 (a) Crimped pins and sockets (b) Swaged pins with stiffener

Brazed pins and ribbon, applied with a welding technique (melting point 618 °C), are also available, and can provide a weldable surface for subsequent assembly.

End pins that are in line with conductors can be brazed, soldered, or crimped to conductors. Pins can be sent to form a staggered arrangement as shown in Figure 12.20.

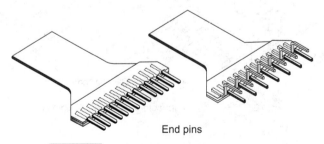

End pins

Fig. 12.20 End pins used with flex circuits

Lap Joints: Flex circuits can interface with hardboards via soldered lap joints, lap joints applied with an anisotropic adhesive (conductive in the Z-axis only), compression dots using a raised metal dot, or 'zebra' strips. Compression methods usually require a stiffener behind the contact area.

 ## 12.7 Advantages of Flexible Circuits

Flexible circuits can solve/minimize design/packaging problems in electronic products. The following are the specific advantages on these accounts:

- Replace multiple hardboards and connectors with a single flex circuit or rigid flex circuit.
- Replace hardboard/ribbon cable assemblies.
- Control EMI with solid or patterned shield layers.
- Control impedance with integral ground planes.
- Provide electrical connections through gaps that are too narrow for round wires.
- Design circuits with exposed conductors for use with conductive elastomer key pads.
- Integrate heaters, temperature sensors or wire-wound antenna coils into a flex circuit.
- Use flex circuits as jumpers between hardboards.
- Specify pressure-sensitive conductive adhesive to attach circuit to cabinets or enclosures.

All these advantages have led to the applications of flexible circuits in some the crucial areas like the military and aerospace field, medical field and commercial applications.

 ## 12.8 Special Applications of Flexible Circuits

There is a growing market for flexible printed circuits which is driven purely by the special applications that they find in some of the most critical areas. Some of these applications are enumerated below.

- Flexible base materials are light and require less space than the rigid laminate. Therefore, they are preferred for use in space, military and mobile applications.
- With the possibilities for three-dimensional interconnections and construction, there is an increasing number of applications in the telecommunication industries and household appliances, which has even led to substantial cost reductions.
- Flexible printed circuits can be shaped into various forms such as tapes, coils or bent shapes. Therefore, they can be used as machine parts where rigid materials are not suitable, for example, printers, aerials, drawing machines; etc.
- The replacement of rigid cable harness by flexible printed circuit boards has made the field servicing of equipment convenient.
- Flexible printed circuits show a higher current-carrying capacity than the wires because of rapid heat dissipation. Thus, they provide a better volume to surface ratio.

12.9 Useful Standards

- *IPC-2223: Sectional Design Standard for Flexible Printed Boards:* Includes the design requirements of singled-sided, double-sided, multi-layer rigid flex or flexible circuits; covers minimum bend radius, differential lengths and board configurations.

- *IPC-6013-K: Qualification and Performance Specification for Flexible Printed Boards:* Covers qualification and performance requirements for flexible printed boards including single-sided, multi-layer or rigid-flex multi-layer.

- *IPC-4202: Flexible Base Dielectrics for Use in Flexible Printed Circuitry:* Covers the requirements for flexible base dielectric materials that are used in the fabrication of flexible printed circuitry and flexible flat cable; includes comprehensive data that will help users to more easily determine both material capability and compatibility.

- *IPC-4203: Adhesive Coated Dielectric Films for Use as Cover Sheets for Flexible Printed Circuitry and Flexible Adhesive Bonding Films:* Establishes the requirements for adhesive coated dielectric film materials used in the cover sheets and flexible adhesive bonding films of fabricated flexible printed circuitry and flexible flat cable; includes comprehensive data that will help users to more easily determine both material capability and compatibility.

- *IPC-4204: Flexible Metal Clad Dielectrics for Use in Fabrication of Flexible Printed Circuitry*: Covers the requirements for metal clad dielectric film materials used in flexible printed circuitry fabrication and flexible flat cable.

- *IPC/JPCA-6202: IPC/JPCA Performance Guide Manual for Single- and Double-sided Flexible Printed Wiring Boards:* Covers the requirements and considerations for single and double-sided flexible printed boards.

- *IPC-FA-251: Guidelines for Assembly of Single- and Double-sided Flex Circuits:* Includes guidelines for the assembly of components and mounting hardware to single- and double-sided flexible printed wiring; covering the type of materials and processes that may be used to accomplish proper electronic assembly.

- *IPC-FC-234: PSA Assembly Guidelines for Single-and Double-sided Flexible Printed Circuits:* Suggests guidelines for the use of pressure sensitive-adhesives in single- or double-sided flexible printed circuits, membrane switches and component attachments.

Soldering, Assembly and Re-working Techniques

13.1 What is Soldering?

Soldering is a process for joining metal parts by making use of any of the various fusible alloys (solder), whose melting temperature is lower than that of the material to be joined, and whereby the surface of the parts create an intermolecular bond, without becoming molten.

Soldering can be classified as soft soldering (soldering), which takes place at temperatures below 450 °C and hard soldering (brazing), which is done at temperatures above 450 °C. Hard soldering is commonly employed on such metals as silver, gold, steel and bronze where in it makes a much stronger joint than soft soldering, the shearing strength being 20 to 30 times higher. However, both thermal joining processes are generally termed as soldering, because in both cases, the molten metal (solder) is drawn into the capillary gap between two closely fitting clean solid metal surfaces.

A soldered connection ensures metal continuity. On the other hand, when two metals are joined to behave like a single solid metal by bolting, or physically attaching to each other, the connection could be discontinuous. Sometimes, if there is an insulating film of oxides on the surfaces of the metals, they may not be even in physical contact. The disadvantages of mechanical joints versus soldering are that oxidation will continually occur on the surface and will increase the electrical resistance. Moreover, vibration and other mechanical shocks may later make the joint loose. A soldered connection does away with both of these problems. There is no movement in the joint and no interfacing surfaces to oxidize, so that a continuous, conductive path can be maintained. Soldering is an alloying process between two metals. In its molten state, solder dissolves some of the metal with which it comes in contact. The metals to be soldered are more often than not covered with a thin film of oxide that the solder cannot dissolve. A flux is used to remove this oxide film from the area to be soldered. The soldering process involves:

- Melting of the flux which, in turn, removes the oxide film on the metal to be soldered;
- Melting of the solder which makes the lighter flux and brings the impurities suspended in it to the surface;
- Partial dissolution of some of the metal in the connection by the solder; and
- Cooling and fusing of the solder with the metal.

Quite often, for locating a problem in the functioning of the circuit, it is necessary to remove a component from the printed circuit board and carry out the requisite tests on it. The process of repair usually involves:

- Disassembly of a particular component;
- Testing of the component;
- Replacement of the component found defective;
- Testing the circuit for performance check.

In this exercise of removal and replacement of electronic components, the process of soldering is employed.

The reliable operation and success of equipment in space, defence, medical electronics, traffic control systems, communication systems or monitoring and control systems all depend on proper soldered connections. Under harsh and hostile environmental conditions like changing of temperature, humidity, vibration etc., even a single incorrect joint can cause the system to either fail completely or partially. As there are thousands of interconnections in equipment, the degree of reliability of these joints should be much higher than that of the equipment itself. Studies have lead to an increased knowledge of materials and their properties and have made many advances in soldering processes possible. Yet soldering is still an evolving technology. As advances in electronics continue to yield more efficient packages and smaller components, soldering techniques are continually being developed to meet the changing demand of the electronics industry as well as of the environmental issues. That is why soldering today has become a very specialized field for scientists and technologists working in the electronic industry.

 ## 13.2 Theory of Soldering

It is essential to understand the theory of soldering so as to visualize as to what happens when a solder joint is formed. Soldering is not a simple physical attachment of one metal to another, but chemically form an intermolecular bond. There are number of important variables in soldering process, which need to be controlled for proper soldering results.

13.2.1 The Wetting Action

When the hot liquid solder dissolves and penetrates the metal surface to be soldered, it is referred to as "wets the metal" or "the metal is wetted". The molecules of solder and say copper blend to form a new alloy, one that is part copper and part solder. It is this solvent action which is called wetting, that forms an intermolecular bond between the parts, an inter-metallic compound. The forming of a proper intermolecular bond is the heart of soldering process that determines the quality and the strength of the solder joint. Wetting can only occur if the surface of the copper is free from contaminations and from any oxide film that forms when the metal is exposed to air. Also, the solder and work surface need to have reached the proper temperature.

13.2.2 Surface Tension

We are familiar with the surface tension of water as a force that retains the cold water in globules on a greasy plate. The adhesive force that tends to spread the liquid on the solid is, in this case, less than the cohesive force. Washing with warm water and using a detergent reduces the surface tension and the water wets the greasy plate and flows out into a thin layer. It happens if the adhesive force is stronger than the cohesive force.

The cohesive force of tin/lead solder is even higher than that of water and also draws the solder into spheres, as it tends to minimize the surface area (a sphere has the smallest surface of any geometric configuration with equal volume in order to satisfy the requirements of the lowest state of energy). Flux acts similarly as the detergent does with the greasy plate. Further, surface tension is highly dependent on any contamination on the surface, as well as, on temperature. Only if the adhesion energy becomes much stronger than the surface energy (cohesive force), ideal wetting will occur.

13.2.3 Creation of an Inter-metallic Compound

The intermolecular bond of copper and tin forms crystalline grains whose shape and size is determined by the duration and intensity of the temperature while soldering. Less thermal application results in fine crystalline structures, which results in excellent solder joints having an optimum strength. Longer reaction times, provided by either longer time or higher temperature or both, result in coarse crystalline structures having less shear strength because of being more gritty and brittle.

With copper as the base metal and tin/lead as solder alloy, lead does not form any inter-metallic compound with copper. However, tin penetrates into copper and the intermolecular bond of tin and copper forms the inter-metallic compounds Cu_3Sn and Cu_6Sn_5 at the interface of solder and the metal being joined. This is shown in Figure 13.1a.

The inter-metallic layer (n-phase plus ε-phase) must be every thin. In laser soldering, the inter-metallic layer has a thickness of the order of 0.1 mm. In wave soldering and manual soldering, the thicknesses of the intermolecular bond of excellent solder joints mostly exceed 0.5 μm. Since the shear strength of the solder joint decreases with an increase in the thickness of the inter-metallic layer, the attempt is always made to keep the thickness less than 1 μm. This can be achieved by keeping the soldering time as short as possible.

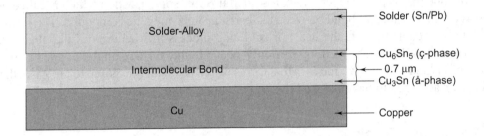

Fig. 13.1(a)   Formation of intermolecular bond

The layer thickness of the inter-metallic compound depends on the temperature and time taken to form the joint. Ideally a soldered connection should be made at approximate 220 °C for two seconds. The chemical diffusion reaction between copper and tin will produce under these conditions the optimal amount, 0.5 μ, of the inter-metallic bonding material Cu_3Sn and Cu_6Sn_5. Insufficient inter-metallic bond, as seen in a cold solder joint or a joint that has not been raised to the proper temperature, can result in a shearing at this interface. In contrary, too much inter-metallic, as seen in a joint that has been overheated or is held too long at temperature, can result in a drastic weakening of the tensile strength of the joint as shown in Figure 13.1b.

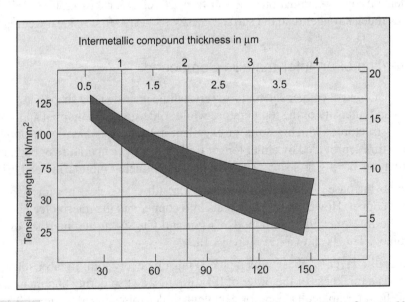

Fig. 13.1(b)   Tensile strength and inter-metallic compound thickness (courtesy Cannon, 2001)

13.2.4   The Wetting Angle

The ability of a surface to be wetted by solder can, to some extent, be assessed by the shape of the meniscus formed when a drop of solder is placed on the hot, fluxed surface, approximately 35 °C above the eutectic point of the solder. If the meniscus has a noticeable undercut edge, like a water drop on a greasy plate or even if it tends to form balls, the metal is not at all solderable. The solderability is good only when the meniscus draws out to a fine angle of less than 30°.

 ## 13.3   Soldering Variables

The important variables of soldering are: temperature, time, tarnish-free surface, right flux and right solder. These variables are important to all the soldering techniques and should be always kept in mind. For achieving good results the golden rule is "Applying the right temperature to the solder as well as to the lands/terminations to be soldered for the correct time on a clean surface by using the right flux and proper solder will provide excellent joint looking bright and shiny".

13.3.1   Temperature and Time Taken for Soldering

The temperature and the time of heat application determine the thickness of the inter-metallic compound. Because of the brittleness of the inter-metallic layer, a too thick layer may cause solder cracking under conditions of thermal or mechanical stress. Beside, the larger thickness of the inter-metallic compound, excess of heat and its longer application may also destroy heat-sensitive components as well as the board. The aim is to keep the temperature low and the time of the highest heat application as short as possible. For surface mounted components, the shear strength is very critical. Due to the different thermal expansion coefficients of the board and components, temperature changes lead to different changes in length and the creation of shear forces especially for larger components.

The temperature for soldering depends on the melting point of solder and its application. For each metal and solder combination, there is a critical temperature below which wetting does not occur or take place to a very small extent. As a rule of thumb, the temperature in re-flow soldering processes should be 30° to 50° and in wave soldering 45° to 60° above the melting point of the solder. A good starting point in manual soldering is a temperature setting of 85° above the melting point of the solder.

The application of heat should be homogeneous since the solder always tends to flow towards the higher temperature.

13.3.2   Tarnish-free Surface

The solder will wet the metal only when the metal to be soldered is free from any tarnish. Although the surfaces to be soldered may look clean, there is always a thin film of oxide covering it. The outer layer of the metal attracts water and various gases get physically bound and below it, they may be chemically bound with oxides, sulphides and carbonates. For a good solder bond, all dirt, grease and surface oxides must be removed before and with the help of flux during the soldering process.

13.3.3 Application of Right Flux and Proper Solder

Fluxes should remove the tarnish from base metals and prevent them from reforming oxides while soldering. The effectiveness of the flux in removing oxide is called "activity" which depends on the activators that are used. A highly active flux will remove oxides. However, the corrosiveness of acids desirable to remove the reaction layer of the tarnish may damage the electronic components. Even mild acids, if not removed, leave a residue that continues to corrode after the soldering process is complete, leading to future failure. The selection of the flux depends on the soldering process chosen, the metal being soldered and on the cleanness of the metal.

The plastic range of a solder varies, depending upon the ratio of the metals forming the solder-alloy, e.g. tin to lead. The 63/37 ratio, known as eutectic solder has practically no plastic range, and melts almost instantly at 183 °C and therefore is usually recommended. Besides, the purity of solder is very important. The presence of more than 0.5 per cent of contaminants within common soft solders may not give the required quality.

 ## 13.4 Soldering Material

The soldering process basically includes an understanding of:
- Soldering material (solder and flux);
- Soldering tools; and
- Soldering procedure.

Following is a description of each of these topics:

13.4.1 Solder

The soldering material or solder usually employed for the purpose of joining together two or more metals at temperatures below their melting point is a fusible alloy consisting essentially of lead (37 %) and tin (63 %). It may sometimes contain varying quantities of antimony, bismuth, silver or cadmium which are added to vary the physical properties of the alloy.

The continuous connection between two metals is secured by soft solder by virtue of a metal solvent or inter-metallic solution action that takes place at a comparatively low temperature. Figure 13.2 is a phase diagram which shows the relationship between temperature and physical state i.e. it shows the tin-lead fusion diagram which explains the alloy or solvent action on molten solder. Pure lead melts at 327 °C while pure tin melts at 232 °C.

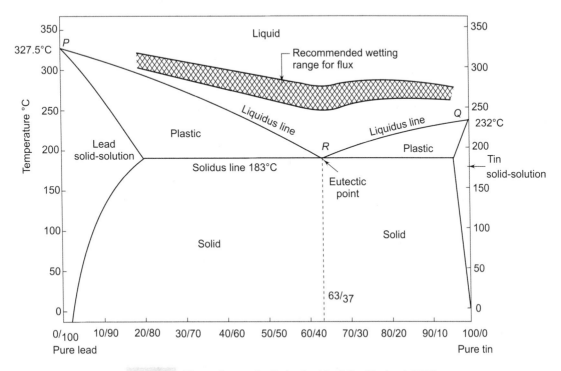

Fig. 13.2   Phase diagram for tin-lead solder (after Haskard, 1997)

When tin is added to lead, the melting point of lead gets lowered and follows the line PR. Similarly, when lead is added to tin, its melting temperature falls along QR. At point R where the two lines PR and QR meet, an alloy of lowest melting point is obtained. The point 'R' represents 63 per cent tin and 37 per cent lead. The alloy at this point is known to have eutectic composition and has a melting point at 183 °C. The eutectic alloy goes directly from the solid line to the liquid line without going through the plastic range. The solder is abbreviated as SN63.

Most common type of solder used in electronics work is an alloy consisting of 60 per cent tin and 40 per cent lead. The alloy is drawn into a hollow wire whose centre is filled with an organic paste like material called rosin. The resulting product is called 60/40 rosin-core solder. Its melting temperature is 375 °F (190 °C) and solidifies as it cools. This alloy is available in wire form in several gauges. Thinner gauges are preferred over thicker ones. Fine solder is easy to position on the joint and requires less heat for the formation of a joint.

The solder alloy wires are commercially available in different diameters from 0.25 mm to 1.25 mm. Usually, 20-22 SWG is 0.91-0.71 mm diameter and is fine for most work. Choose 18

SWG for larger joints requiring more solder. The volume of the flux in the wire is about 25 per cent corresponding to a mass of about 3 per cent.

It is found that with SN63, the stress resistance of the solder joint is the maximum i.e. at the lowest melting point; the alloy has the highest pull strength.

Several other alloys exhibit eutectic behaviour. However, they suffer from certain disadvantages; particularly when they have silver (tin 62.5 per cent, lead 36.1 per cent, Silver 1.4 per cent), as they tend to be more expensive. However, for specialized applications, special wires with high or low melting point and some with 1.5 to 5 per cent silver are required. It is a good practice to buy a solder wire from a reputed manufacturer because both the alloy composition and flux used may vary, which is often detrimental to the product.

The lead present in the solder does not cause any health hazard. However, when handling lead dust, a mask must be used. Smoking during soldering may cause the lead smoke to be inhaled and this must be avoided. Also, thoroughly clean your hands after soldering before eating or smoking.

13.4.1.1 Selection of the Solder Alloy

Even though the alloy Sn60/Pb40 is cheaper and still funds a good market, it is advisable to prefer Sn63/Pb37 for high quality interconnections because of its following advantages:

- The alloy SN60/Pb40 has a 5 °C higher melting point which means that the soldering range is also 5 °C higher, resulting in higher energy consumption.
- The eutectic alloy SN63/Pb37 has, during cooling, a rapid transition direct from liquid to solid. The time for solidification is approximately 40 per cent less than for the 60/40 alloy. A slow solidification always has the danger of unreliable solder joints caused by any vibration during solidification.
- The tensile strength as well as the shear strength of 63Sn/37Pb is higher in comparison to 60Sn/40Pb.
- Only tin forms the intermolecular bond with copper to Cu_3Sn and Cu_6Sn_5. Higher tin content is therefore better.
- The specific gravity of SN63/Pb37 is also lesser than that of SN60/Pb40 that makes the equipment lighter.
- Higher composition of tin increases the electrical as well as thermal conductivity. It also gives brightness to the joint.

13.4.1.2 Solder Wire/Cored Solder

The solder wires for hand soldering are generally combined with flux. The plasticized flux is placed in 1, 3 or 5 channels within the solder itself. The flux to solder ratio is mentioned either by volume or by weight. Cored wires with rosin flux contain mostly 2 per cent to 3.5 per cent flux by weight or 20 per cent to 30 per cent by volume.

Beside the flux-to-solder ratio, the diameter of the solder wire is also important for good soldering. The selection of the correct diameter of the solder wire is necessary. As a general advice, the diameter of the solder wire should be just a little bit less than the half of the diameter of the solder pad. Service Engineers are using mostly two cored solder wires with the first diameter of 0.60 mm or

0.65 mm (or 0.025") and the second wire having a diameter of 0.80 mm or 0.90 mm (or 0.035"). The cored flux in the solder wire is a solid or jelly to prevent outflow of flux while stored.

13.4.1.3   Level of Impurities Permissible in Sn/Pb Solder

The purity of the solder is far more important than the exact tin/lead ratio, because the solder impurities have a substantial influence on the quality of the solder joint. For example: copper and gold form inter-metallic compounds, causing the solder to become gritty and brittle. Their combined presence should not be more than 0.300%.

Zinc is one of the most detrimental solder contaminants. As little as 0.005 per cent zinc will cause grittiness, lack of adhesion, and eventual failure of the joint.

Aluminum, cadmium, zinc and phosphor promote oxidation of the solder surface and have an extremely negative effect on wetting power. Mixed crystal forming metals like bismuth and antimony have a less negative influence and may even improve wetting behaviour. That is why no limit is prescribed for them.

Sources of contamination are first of all the solder alloy that may not be pure. Another source of contamination is the boards and components to be soldered, as foreign metals readily dissolve in solder. Since the soldering process itself is a source of contamination, extreme care has to be taken by continuous observation and analysis of the solder bath.

National standards are available relating to the composition of common soft solders like DIN 1707 in Germany and QQ-571d in USA. A lot of manufacturers of electronic assemblies refer to these specifications. Table 13.1 gives a summary of the contamination limits.

Table 13.1   Contamination Limits in Solder (Courtesy Braun, 2003)

Contaminant	QQ-S-571/JISZ-3282*	New Solder, %	Contamination Limits, %
Aluminum	0.005	0.003	0.006
Antimony	0.500/0.300	0.300	---
Arsenic	0.030	0.020	0.030
Bismuth	0.200/0.050	0.006	---
Cadmium	0.005	0.001	0.005
Copper	0.080/0.050	0.010	0.250
Gold	0.080/0.001	0.001	0.080
Iron	0.020/0.030	0.001	0.020
Silver	0.010/0.001	0.002	0.010
Zinc	0.005	0.001	0.005
Others	0.080/0.010	0.010	0.080

Note. *Limits established by federal specification QQ-S571-E for acceptable contaminant levels for various metals.

  ** Contamination levels, which indicate the solder, should be replaced for any use.

13.4.2 Flux

In order to aid the soldering process, a substance knows as 'flux' is used. Flux is needed to remove the microscopic film of oxides on the surfaces of metals to be soldered and it forms a protective film that prevents re-oxidation while the connection is heated to the point at which the solder melts. Flux is helpful on a stubborn joint that would not accept solder. Most metals tend to form compounds with atmospheric oxygen which leaves a coating of oxide even at room temperature. The oxides are removed by fluxes which remain liquid at soldering temperature, react chemically with the oxides and disperse the reaction products. Fluxes are applied before or during soldering. Thus a good solder flux must simultaneously perform a number of important functions such as promoting thermal transfer to the area of the solder joint, enhancing wetting of the solder on the base metal and preventing oxidation of the metal surface at soldering temperatures (Nasta and Peebles, 1995).

To summarize, the flux actually has three main purposes, which are to:

- Remove the film of tarnish from the metal surface to be soldered;
- Prevent the base metals from being re-exposed to oxygen in the air to avoid oxidation during heating, which means promotion of wetting by preventing from oxidizing while they are being heated to the soldering temperature and soldered.
- Assist in the transfer of heat to the metal being soldered.

The solder used in most electronic work contains this flux as a centre core which has a lower melting point than solder itself. When the molten flux clears the metal, it allows the solder to wet the metal and holds the oxides suspended in solution.

The molten solder can then make contact with the cleaned metal and the solvent action of solder on metal can take place.

In order to perform its function correctly, the flux must have the following desirable properties:

- It must be available in such a state that it can be properly applied on the surface to be soldered (cored in solder wires; liquid state for wave soldering and as paste for reflow).
- It should have low surface tension in order to wet all surfaces to be soldered, i.e. it should penetrate into the very small gap, which sometimes exists between the surfaces to be joined.
- It must be able to destroy the tarnish (dissolution of absorption layer and removal of reaction layer).
- It should start to act at temperatures between 80 °C to 100 °C and not immediately after it has been applied.
- It has to protect the surface till the process of soldering continues.
- There should be no toxicity either by the flux or its volatile products.
- The ideal flux should leave no residues at all on the soldered assembly.

Rubin (1995) describes that no-clean fluxes and solder pastes are rapidly finding an important position in soldering production technology. Their growth has been strongly influenced by the increasing large usage of surface mount components, and accelerated by the need for an alternative to cleaning procedures which incorporate CFCs.

13.4.2.1   Composition of Fluxes

Fluxes have two basic components, solvents and solids as shown in Figure 13.3. The solid portion includes the active components, while the solvent is primarily the carrying medium.

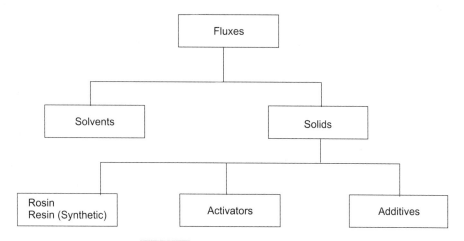

Fig. 13.3   Basic constituents of fluxes

Additves: Additves are wetting agents for reducing the surface tension of the flux in order to promote its wetting ability as well as used as foam regulators or thickeners in solder paste.

Activators: Activators are compounds that decompose at soldering temperatures yielding ammonia or hydrochlorides. They also can be non-halide, e.g. carboxylic acids. The use of strong acids presents a serious problem. The corrosiveness of acids desirable to remove the reaction layer of the tarnish can also damage electronic components and even mild acids leave a residue that continues to corrode after the soldering process is complete, leading to future failure. There are two basic approaches in the electronic industry. Some manufacturers prefer to employ a flux with a strong acid that removes a lot of oxidation and is very corrosive and some prefer using a flux with a mild acid that is not as corrosive, but does not do as good of a job removing the oxidation layer. Because of the difficulty in cleaning flux residues from underneath surface mounted components, flux manufacturers have developed special products that leave very little corrosive or conductive residues behind, if any. There is also halogen and ammonia free fluxes for materials that require active fluxes but are not compatible with halogens or ammonia fumes. Metals, which exhibit good corrosion resistance because of inherently tight oxides, require the strongest or most activated flux.

Rosin/Resin-based Fluxes: Rosin (colophony) based fluxes are made from rosin which is extracted from pine (especially "pinus palustris") sap. It is mostly taken from living trees by tapping them. Its classification and price depend upon the type of tree from which it is obtained and its purity. The most purified product is known as "water white rosin". The active ingredient is an organic acid (85 per cent abietic acid). Equivalents of rosin are also available as synthetic resins. Activated rosin/ resin fluxes are very commonly used in electronics. The flux is inactive at ambient temperature and only becomes active at the elevated soldering procedure. Highly activated fluxes accelerate the soldering process and may permit a reduction in soldering times. However, they go along with rising cost of assembling, cleaning and the growing number of restrictions on the use of chlorinated and fluorinated hydrocarbons. The rosin/activator ratio, controls the activity of the flux, not the solids content. In addition to the rosin activator ratio, the solids content (specific gravity) of the flux can be varied. Higher solids content are used for boards with a high density of connections.

13.4.2.2   Water-soluble Fluxes

The water soluble fluxes are divided into two categories, organic and inorganic, on the basis of composition. Organic fluxes are more active than inorganic, and are the most active of all. Water-soluble fluxes do not mean that the solvent vehicle itself is necessarily water. Although water can be used, it spatters. That is why solvents similar to the systems of resin-based fluxes are often in use. The term "water-soluble flux" refers to the fact that the chemicals used are generally water-soluble. With a few exceptions, all water-soluble fluxes leave residues, which must be carefully cleaned after soldering. They are hygroscopic and contain a considerable amount of ionic substances, which may cause serious corrosion. Water-soluble fluxes can activate most metals used in electronic assemblies.

13.4.2.3   Types of Fluxes

There is a wide range of fluxes available. The different types of fluxes used in soldering electrical connections are described in the standard IPC-J-004. The fluxes are classified into one of the three classes (L for low, M for moderate and H for high). These classes are based on the activity level of the flux, which essentially define their cleaning corrosive nature to the metal. Table 13.2 gives different types of flux materials based on the activity level of the fluxes.

13.4.2.4   Fluxes for Hard Soldering/Brazing

Similar to soft soldering, a flux is required in order to clean the metal surface and to prevent the surface from oxidation during the hard soldering process. The flux is usually applied to the joint in the form of a paste, typically made from borax (boric acid and zinc chloride). The type of flux depends upon the solder being used and the metal being joined. There are more than a dozen different types of hard soldering solders (brazing rods) and fluxes available in the market. Many brazing rods are flux coated or flux cored and do not require additional flux. The molten flux left behind after the job is finished leaves a dark brown residue, which should be removed by immersing the item in a water diluted acid solution generally Sulphuric- or phosphoric acid.

It must be remembered that always acid is added to water, but never water to acid! Always wear rubber gloves and goggles when doing this job!

Table: 13.2   Different Types of Fluxes According to IPC-J-004

Flux Type Symbol	Flux Materials of Composition	Symbol	Flux Activity Levels (% Halide)	Flux Type
A	Rosin	RO	Low (0 %)	L0
B			Low (< 0, 5 %)	L1
C			Moderate (0 %)	M0
D			Moderate (0, 5-2 %)	M1
E			High (0 %)	H0
F			High (> 2 %)	H1
G	Resin	RE	Low (0 %)	L0
H			Low (< 0, 5 %)	L1
I			Moderate (0 %)	M0
J			Moderate (0, 5-2 %)	M1
K			High (0 %)	M0
L			High (> 2 %)	H1
M	Organic	OR	Low (0 %)	L0
N			Low (< 0, 5 %)	L1
P			Moderate (0 %)	M0
Q			Moderate (0, 5-2 %)	M1
R			High (0 %)	H0
S			High (> 2 %)	H1
T	Inorganic	IN	Low (0 %)	L0
U			Low (< 0, 5 %)	L1
V			Moderate (0 %)	M0
W			Moderate (0, 5-2 %)	M1
X			High (0 %)	H0
Y			High (> 2 %)	H1

13.5 Soldering and Brazing

By definition soldering is the process of joining two or more pieces of metal by using a metal alloy whose melting temperature is lower than the metals being joined. It takes place at temperatures below 450 °C. Hard Soldering (Brazing), on the other hand which is commonly used in fine jewellery takes place at temperatures above 450 °C. Hard soldering on such metals as silver, gold, steel and bronze makes a much stronger joint than soft soldering. The shearing strength of in case of hard soldering is 20 to 30 times stronger them softer soldering.

13.5.1 Solders for Hard Soldering/Brazing

All the non-ferrous metals (gold, silver copper, brass or bronze) that have a relatively high melting temperature can be soldered with either gold or silver solder. Both gold and silver solders are available in different melting temperatures and in different shapes and forms.

The melting temperature of gold and silver solder is mainly determined by the content of zinc: the higher the zinc content, the lower the melting temperature. However, zinc is what turns the lower melting temperature silver solders yellowish-grey. Pits in the solder seam are caused when the solder is overheated and the zinc burns out. Again, using a higher temperature solder as well as controlling the heat will help prevent pitting. The hard soldering alloy for use on steel and brass contains silver, copper, zinc and cadmium.

Extra caution must be used in hard soldering. The fire hazard from the torch flame, the hazardous chemical fumes from the heated soldering alloy and flux in poorly ventilated rooms is absolutely unsafe. All equipment and supplier's rules and warnings should be strictly followed.

 ## 13.6 Soldering Tools

Various tools are necessary for facilitating soldering work. The most essential tool in the soldering practice is the soldering iron.

13.6.1 Soldering Iron

A soldering iron is the basic tool for hand soldering. It generates the heat required to heat the surfaces to be soldered and to melt the solder.

It should supply sufficient heat to melt solder by heat transfer when the iron tip is applied to a connection to be soldered. Soldering irons used for soldering electronic components consist of the following three main parts: (i) a handle, (ii) a heating element, and (iii) a bit/tip.

Handle: This is made of a good electrical and thermal insulator having an ergonomic shape so that it is comfortable for the operator.

Heating Element: This must have sufficient thermal capacity so that the set and working temperatures are, as much as possible, the same. The state-of-the-art hand irons are microprocessor-controlled. They compare the tip temperature sensed with a sensor with the set temperature. The heating element responds immediately to heat loss at tip while soldering and returns lost heat to the tip. The heating element must be properly insulated so that there is no electrical leakage appearing on the tip of the bit to cause damage to the components.

Bit/Tip: This is made of copper to provide good heat transfer. It is plated to prevent the solder dissolving them. Iron on the other hand is not attacked by solder, so iron plated copper bits are normally used. Unfortunately, iron is not readily wetted by solder, so the bit is further covered by nickel or chromium in order to provide a hard outer surface that will wet properly by the solder. With the passage of time and usage, the outer plated coat (nickel or chrome) will dissolve away. The bit then must be replaced. Since a bit is the tip of the hand soldering iron, it is often called 'tip'.

Traditional soldering tips which are made of copper, conduct heat well and are inexpensive. However, they have the disadvantage that the tip oxidizes heavily when heated and copper particles are set free into the solder until it has been corroded entirely. New soldering iron tips have been designed which are galvanically plated with an iron coating and is then shielded against oxidation and corrosion by a layer of chrome. The heating element of the soldering iron is protected against over-heating and premature wear due to quick heat transfer. Such tips are available from M/s ERSA GmbH and are called ERSADUR soldering tips.

The selection of a soldering iron is made with regard to its tip size, shape, operating voltage and wattage. Soldering iron temperature is selected and controlled according to the work to be performed. The temperature is normally controlled through the use of a variable power supply and occasionally by tip selection.

Soldering irons are available in various forms. These are delineated below.

Soldering Pencils: Soldering pencils (Figure 13.4) are lightweight soldering tools which can generate as little as 12 watts or as much as 50 watts of heat. A 25 watt unit is well suited for light duty work such as soldering on printed circuit boards. Modular soldering irons use interchangeable heating elements and tips which mate to a main pencil body. Such elements screw into a threaded receptacle at the end of the pencil. A variety of tips (Figure 13.5) are available to handle most soldering tasks. Very fine, almost needle-like tips

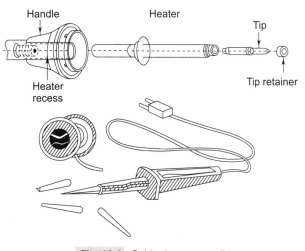

Fig. 13.4   Solder iron — pencil type

are used on printed circuit boards with IC component foil pads which are closely spaced. Larger, chisel and pyramid tips can store and transfer greater amounts of heat for larger, widely spaced connectors. Bent chisel type tips can get into difficult-to-reach areas. Regardless of the type of the tips, it is best to use plated, as opposed to raw copper tips, as these have much longer life.

Iron clad, chrome plated, pre-tinned tips

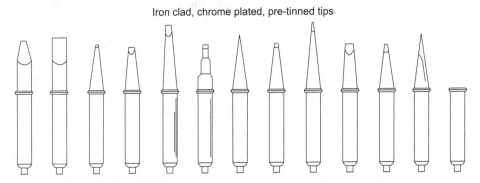

Fig. 13.5   Different types of bits for soldering irons

A pencil type soldering iron takes a few minutes to attain working temperature and it is better to keep it continuously powered even for interrupted type of soldering work. This would need to keep the iron secured in a safe place at working temperature. One method is to keep it in special soldering iron holder which may be a coiled steel form into which the hot soldering iron can be inserted. Most stands of this type also include a sponge which is kept moistened and used periodically to clean the soldering tip.

Soldering Gun: A gun is usually heavier and generates more heat than the average pencil. Soldering of heavy duty conductors or connectors requires the use of a gun because it can generate enough heat to quickly bring a heavy metal joint up to the proper soldering temperature. These soldering tools are called guns simply because they resemble pistols. The gun's trigger (Figure 13.6) is actually a switch that controls application of ac power to the heating element. The working temperature is reached instantaneously. Some guns provide for selection of different heat levels through multi-position trigger switch.

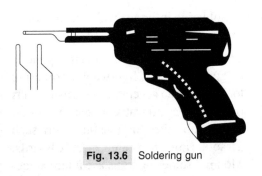

Fig. 13.6   Soldering gun

Soldering Stations: Soldering stations (Figure 13.7) contain an iron and a control console that offers switch selectable temperatures, marked low, medium and high. Obviously, this is more convenient than waiting for a modular pencils' heating element to cool, unscrewing it from the holder and then replacing it with another heater tip combination. The tip temperature is controlled by using a heat sensor and closed-loop feedback control to gate power to heating element. Obviously, soldering stations are expensive compared to basic soldering pencils.

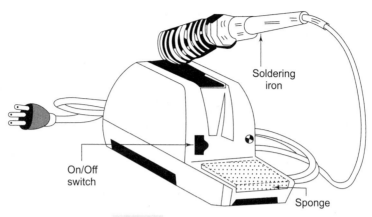

Fig. 13.7 Soldering station

Battery-operated Irons: Sometimes, it is inconvenient to depend on the mains power supply for operating a soldering iron. Battery-operated soldering irons are available which depend upon rechargeable batteries as a power source. Recharging is done automatically when the iron is placed in its charger, which is built on the stand, and is connected to an ac power source. In these soldering irons, the tips attain working temperature in 5-8 seconds and cool off to ambient temperature in about one second. Typically, about 125 connections can be made on one charge. For a standard iron, a typical charging interval of approximately fourteen hours is required to return the cells to full strength. Of course, there are quick change irons also. Sometimes, the soldering irons have built-in light to illuminate the work area whenever battery power is applied to the heating element.

Soldering irons are best used along with a heat-resistant bench-type holder, so that the hot iron can be safely parked in between use. Soldering stations generally have this feature. Otherwise, a separate soldering iron stand, preferably one with a holder for tip-cleaning sponges, is essential.

Electronics catalogues often include a range of well-known brands of soldering iron. The following factors should be kept in mind when selecting a soldering iron for a particular application (Winstanely, 2003).

Voltage: Most soldering irons run from mains supply at 230 V. However, low voltage (12 V or 24 V) type irons are also available and they generally form part of a "soldering station".

Wattage: Typically they may have a power rating of between 15-25 watts which is adequate for most work. A higher wattage does not mean that the iron runs hotter. It simply means that there is more power in reserve for coping with larger joints. Higher wattage irons are required for heavy duty work because it would not cool down so quickly.

Temperature Control: The simplest and cheapest type irons do not have any form of temperature control. Unregulated irons form an ideal general purpose iron for most users as they generally cope well with printed circuit board, soldering and wiring.

A temperature controlled iron has a built-in thermostatic control to ensure that the temperature of the bit is maintained at a fixed level, within preset limits. This is desirable especially during more frequent use, since it helps to ensure that the temperature does not over shoot in between times and also the output remains relatively stable. Some versions have built-in digital temperature readout and a control knob to vary the temperature setting. A K-type thermocouple may be used to measure the temperature of the tip and heating rate is controlled by means of a thyristor. Thus, the temperature can be boosted for soldering larger joints. The necessity of measuring the temperature by means of a sensor as close as possible to the end of the soldering tip is a fundamental principle of a rapid response. The closer to the end of the soldering tip the measurement is taken, the quicker the control system can respond.

Anti-static Protection: For soldering static-sensitive components such as CMOS and MOSFET transistors, special soldering iron stations having static-dissipative materials in their construction are required. These irons ensure that static-charge does not build up on the iron itself. These irons are "ESD safe" (Electrostatic Discharge Proof).

The general purpose irons which may not be necessarily ESD-safe, but can be safely used if the usual anti-static precautions are taken when handling CMOS components. In this case, the tip would need to be well-grounded.

BITS: Bits are available in various shapes and sizes. Keep in mind that the size of a screwdriver bit/tip should be approximately equal to the diameter of the pad. A large tip selection for maximum flexibility should be offered along with the soldering iron to meet the requirements.

The choice of the tip is very important and the shape of its end must be selected so that good heat transfer to the parts to be joined is possible.

The lifetime of the bit/tip depends on the structure and thickness of the material used as well as on proper handling.

It is useful to procure a small selection of manufacturer's bits (soldering iron trips) with different diameters and shapes, along with the soldering iron. They can be changed depending upon the type of work in hand.

Bit/Tip Maintaining: Excess solder and burnt flux should be removed by wiping the bit on water-soaked sponge. The following points should be kept in mind for good maintenance of the tips.

- Never clean the bit/tip with a file or any abrasive tools; not even with brass brushes.
- Never wipe the tip against the surface to be soldered.
- Clean and tin the bit before turning the iron off.
- Turn the iron off when it is not used for more than 15 minutes.
- Clean and wet the bit with some flux cored solder wire after the work is finished.
- Keep all the soldering irons in their respective holders. Don't put more than one soldering iron per holder.
- Never put the iron in any form of a muffle, but keep it in its open spring holder.

Spare Parts: It is preferable to ensure that the spare parts may be available for the iron. So, if the element blows, you don't need to replace the entire iron. This is especially so with expensive irons.

13.6.1.1   Selection of Soldering Iron

The soldering iron should be of the precision type, small but powerful enough to reliably solder components to printed circuit boards. An iron between 25 and 40 watts with a nickel plated tip, or one of the miniature irons capable of a tip temperature of 205 °C are most suitable. Hotter temperatures run a real risk of spoiling the adhesive bond that holds the copper foil to the board. Do not use a higher temperature to make up for an improperly-tinned tip.

The ideal tip is a single flat or chisel tip of about 2.5 mm. The old style unplated copper tips are not very suitable, as they wear away very quickly.

The soldering iron should be examined carefully every time it is to be used. The soldering iron tip should be properly connected or screwed into the holder and it should be free from oxides. The shape of the tip must meet the requirements of the task to be performed. If any one of these items is not as good as it should be the following steps are adopted:

- The oxides from the tip surface are removed by using an abrasive cloth or sand paper.
- The tip generally is made in its proper shape by filing. This is normally done on the unplated copper tip.
- The iron is heated to the minimum point at which the solder melts. Before using the iron to make a joint, the tip is coated or tinned lightly by applying a few millimeters of solder.
- For keeping the tip clean, after it has been prepared, the heated surface of the tip should be wiped with a wet sponge. This is to remove dirt, grease or flux which, if allowed to remain, can become part of the joint and make the joint dry and defective.

If during soldering, excessive heat is generated at the soldering iron tip and the component gets heated beyond its maximum temperature, the component may be permanently damaged, weakened, or affected drastically in value or characteristics. Such effects may not be noticed during assembly or test but may show up later when the equipment is in use.

The tip temperature to be selected must be based on the temperature limitation of the substrate. The circuit boards which have a substrate of fibre-glass epoxy of 280 °C should not be heated for more than 5 minutes. Hotter temperatures reduce the time in inverse relationship; the higher the temperature, the less time the boards will stand it before damage.

Further, heat transmitted along the leads may cause unequal expansion between leads and packages, resulting in cracked hermetic seals. In general, for hand soldering, the recommended soldering iron wattage is 20 watts to 25 watts for fine circuit board work, 25–50 W iron for general soldering of terminals and wires and power circuit boards, 100–200 W soldering gun for chassis and large area circuit planes. With a properly sized iron or gun, the task will be fast and will result in little or no damage to the circuit board plastic switch housings, insulation etc.. For iron temperatures of between 300 °C and 400 °C, the tips of the soldering iron should be in contact with the lead for not more than five seconds. Particularly, the ICs and transistors should be soldered quickly and cleanly.

Temperature: An important step to successful soldering is to ensure that the temperature of all the parts is raised to roughly the same level before applying solder. Heating one part but not the other will produce an unsatisfactory solder joint. The melting point of most solder is in the region of

188 °C (370 °F) and the iron tip temperature is typically 330–350 °C (626°–662 °F). Figure 13.8 shows the temperature range for ideal soldering work. Above this temperature range, there is a risk of thermal damage whereas below this range, cold junctions are likely to develop.

Time: Next, the joint should be heated with the bit for just the right amount of time. Excessive time will damage the component and perhaps the circuit board copper foil. The heating period depends on the temperature of your iron and size of the joint. Larger parts need more heat than smaller ones while some parts (semiconductor devices) are sensitive to heat and should not be heated for more than

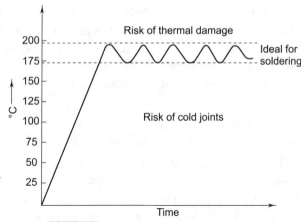

Fig. 13.8   Ideal temperature range for soldering

few seconds. In such cases, thermal shunts or heat sinks are used to protect heat-sensitive components from damage due to heat while soldering. These devices are placed or clamped in place to prevent the heat from reaching the component while its leads are being soldered.

Solder Coverage: In order to achieve a successful solder joint, it is essential to apply only an appropriate amount of solder. Too much solder is an unnecessary waste and may even cause short circuits with at the end joints. Too little solder may not fully form a successful joint or may not support the component properly. How much solder to apply only really comes with practice.

 ## 13.7   Other Hand Soldering Tools

The soldering workstation should be clean, ESD-controlled and organized for an easy access to all the tools. The typical tools may include:

- Solder iron;
- Cleaning material;
- Cored solder;
- Pliers (bent, nose and straight);
- Wire stripper;
- Lead forming tools;
- Toolbox
- Fume absorber;
- Solder iron station including holder;
- Set of soldering iron bits;
- Flux; sponge; soldering braid;
- Tweczers;
- Wire cutter;

- Screwdriver set;
- Board holder; and
- Cleaning solvent and brush.

13.7.1 Cutters

Good cutters are an essential tool for component lead cutting and removing insulation prior to soldering or performing. Ideally, they should be of the side cutter type with insulated grips. They should be slimline and lightweight for precision work. Cutters up to 35 cm long are employed in electronics applications. The cutters should be made of high quality tool steel so that they will make a sharp, clean cut. The tips of the cutters should be tapered to allow the user to reach a particular wire in a crowded area. Cutter jaws should be every well aigned so that cutting edges meet squarely and allow little or no light to pass through when held together. Cutter action should be smooth and clean.

Cutters should be used only for cutting copper wire or leads and not for trimming PCBs or metal parts. The cutting blades are easily blunted if misused. A blunt pair of cutter is worse than useless as it will not crop leads cleanly or strip insulation from wire without snagging and breaking strands. Some cutters have a safety clip incorporated, which traps the cut-off lead and stops it from flying and ending up all over the room.

Sometimes, it is difficult to strip the wire with the cutters. In that case, one can use wire strippers and cutters, which are available with adjustable stops for different wire sizes. The stop ensures that the cutting action is limited to the thickness of the insulation and will prevent nicking the actual wire.

Cutters are specified to cut wires up to a specific diameter and material, usually in electronics, out of copper. One should not try to cut anything larger than the specified sizes in order to avoid damage to the cutting edges. As a rule of thumb, most of the small cutters cut up to 1mm copper wire without any damage. There are different types of cutters available but, in essence, they have variations only in the cutting portions. The cutting portion must be sharp, straight and without any impressions along the edges. After a period of use when the edges of the cutter become blunt, they can be either re-sharpened or replaced. There are two basic variations in the shape of the cutting edge are shown in Figure 13.9:

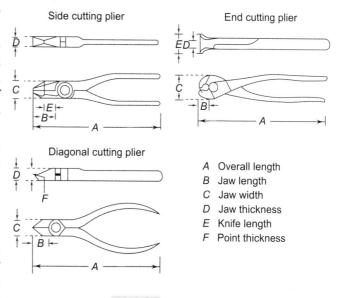

Fig. 13.9 Side cutter

A Overall length
B Jaw length
C Jaw width
D Jaw thickness
E Knife length
F Point thickness

- *Flush Cutter*: Flush cutters are made to cut soft wires like silver, gold, copper etc. and should not be used to cut heavy hardened wires like steel wires. The cutting edge that cut the wire produces a straight face to one side of the wire.
- *Diagonal Cutter*: These cutters are pretty durable and work well on most wires. They make V-Shapes to the wire end.

13.7.2 Pliers

One of the most commonly used tools in electronics shops are pliers. The frequent tasks for which pliers are used include holding wires in place during soldering, acting as a heat sink to protect a delicate component, bending component leads to fit mounting holes on a circuit board and pulling wires through a panel or chassis hole.

Obviously, one single design in pliers cannot meet all the demands. Therefore, there are many types of pliers. The important types are shown in Figure 13.10. Some pliers also have cutting knives. A single pair of long nose pliers is adequate for most jobs, but having several on hand can simplify a task.

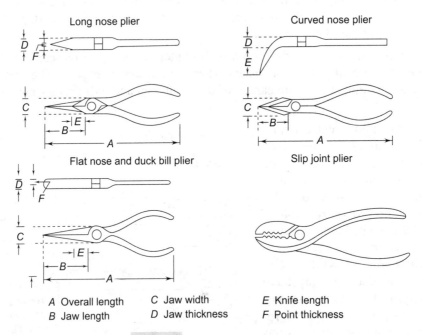

| A Overall length | C Jaw width | E Knife length |
| B Jaw length | D Jaw thickness | F Point thickness |

Fig. 13.10 Various types of pliers

Pliers have different shapes of handles, though the most common are those with curved handles. They are designed for maximum comfort and efficiency. Extended handles are provided on some long nose and Duck Bill pliers for longer reach and increased leverage. While using pliers to cut a wire or metal piece, ensure the protection of eyes with goggles.

A good pair of pliers is invaluable as a precision extension of your fingers when holding and forming components for PCBs. When bending resistor or axial capacitor leads to the correct pitch for your design, the pliers will give you a professional finish and avoid stress to the lead/component joint. Usually, a pair of 'snipe nose' pliers will suit most applications. The fine tip will enable you to use it like a strong pair of tweezers and the serrated jaws will give you a good grip when holding and forming different wires and parts.

The two mostly commonly types are non-serrated round nose pliers and long nose pliers. The pliers having serrated jaws, row of sharp points along the edge can damage the wire and its insulation if not properly used. Round nose pliers are useful for forming wires and component leads providing a smooth curve to any bends. However, one should take care not to squeeze them too hard, otherwise they may cause damage by making indentations in the wire.

Both pliers and cutters are easy to use when they are fitted with springs, which keep them open. This allows one-handed operations.

- *Round Nose Pliers:* Round nose pliers are useful for forming wires and component leads. With the help of this particular tool, one can provide a smooth curve to any bend. However, one has to take care not to squeeze them too hard otherwise they can damage the wires by making indentations on them.

- *Long Nose Pliers:* They are used for accurate assembly work in hard-to-reach areas.

13.7.3 Strippers

Strippers are used to remove insulation from the wires. The most usually employed stripers are of the cutting type (Figure 13.11). These strippers are so designed that they can accommodate various sizes of wire normally used in electronic equipment. To prevent damage to the wire by nicking, it should be ensured that the specific wire size hole is selected in the cutting stripper.

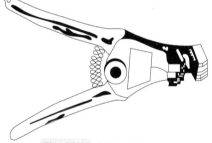

Fig. 13.11 Wire stripper

In the thermal strippers, the wire to be stripped is placed between two electrodes. The electrodes get heated when electric current is passed through them. The resulting heat melts the insulation. When using thermal strippers, toxic fumes emanating from compounds such as polyvinyl chloride or polytetrafluoroethylene must be properly exhausted by using some type of fan ventilation system.

Hot-blade, rotary, and bench wire strippers are generally used in shops where large wire bundles are made. When using any of these automatic wire strippers, the manufacturer's instructions should be followed for adjusting the machine; to avoid nicking, cutting, or otherwise damaging the conductors.

The following procedure is adopted for stripping wire with the hand wire stripper:

- Insert the wire into the centre of the correct cutting slot for the wire size to be stripped. The wire sizes are generally listed on the cutting jaws of the hand wire strippers beneath each slot. After inserting the wire into the proper slot, close the handles together as far as they will go.

- Slowly release the pressure on the handles so as not to allow the cutting blades to make contact with the stripped conductor. On some of the hand wire strippers, the cutting jaws have a safety lock that helps prevent this from happening.
- Continue to release pressure until the gripper jaws release the stripped wire, and then remove it completely.

Be careful not to nick a wire. Even if the wire doesn't break, a cut reduces its diameter and thus its capacity to conduct electricity. This is dangerous and could lead to an electrical failure.

13.7.4 Bending Tools

Bending tools are pliers having smooth bending surfaces so that they do not cause any damage to the component.

13.7.5 Heat Sinks

Some components such as semiconductor devices, meter movements and insulating materials are highly heat-sensitive. They must be protected from damage due to heat while soldering. Devices such as a set of alligator clips, nose pliers (Figure 13.12) commercial clip-on heat sinks, felt-tipped tweezers, and anti-wicking tweezers and similar such devices are usually placed or clamped at the site of soldering so that they prevent the heat from reaching the components.

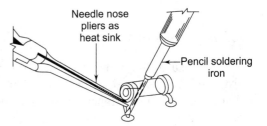

Fig. 13.12   Use of pliers as heat sink in soldering

13.7.6 General Cleaning Tools

Before the soldering process is actually performed, the surface on the printed circuit board or the component leads must be properly cleaned. The tools or devices most commonly used for general cleaning are alcohol dispenser, camel hair brush, small wire brush, synthetic bristle brush, cleaning tissue, pencil erasers, (Figure 13.13) typewriter erasers, braided shielded tool, and sponge with holder, tweezers and single-cut file.

A very useful soldering aid consists of a plastic or wood wand with a pointed metal tip at one end and a

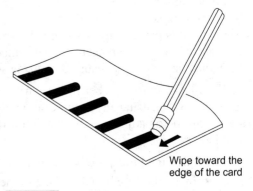

Fig. 13.13   Use of pencil eraser as a cleaning tool

notched metal tip at the other. The blunt end of the aid is used to clear solder from holes in printed circuit boards and from solder lugs. The notched end can be used to make right-angle bends in component leads, to hold leads and wires while the solder joint is made, and to keep leads away from pc boards and lugs during desoldering operations.

 ## 13.8 Hand Soldering

Even though mass soldering techniques have become popular due to economic reasons, hand soldering still has a great relevance. In small and medium scale manufacturing facilities, hand soldering is still practised. Even otherwise, if joints are faulty and require re-work, and repair and touch-up, hand soldering is resorted to. Good soldering skill is essential for operators undertaking manual soldering as well as for field service technicians who undertake repairs at customer sites.

13.8.1 Hand Soldering Requirements

The most important requirement for hand soldering, besides a skilled trained operator working with a proper ESD-controlled, clean workstation, is a good quality and temperature-controlled soldering iron with a suitable, clean and fine bit which is properly earthed.

Making good soldered joints is a skilled task, which requires proper tools, a high level of cleanliness and an ESD-controlled workplace.

Quality solder joints can only be achieved under clean conditions. Boards, component leads, and soldering equipment must be carefully cleaned both before and after the soldering operation. Beside others, two main parameters must be controlled to achieve repeatable quality joints. These are:

- *Temperature:* The metal to be joined and solder need a temperature range from 215 °C to 250 °C. The setting temperature depends upon the equipment used as well as of the shape of the tip and is approximately 50 °C higher.
- *Time of heat application:* There is a minimum and maximum time limit, ranging between 2–4 seconds.

The temperature must be correct for the flux to be activated to remove any remaining oxides and apply a protective coating. The heat source not only melts the solder, but also boils away the resin and flux solvents so that the solder flows over supporting the wetting of the metals joined together.

During the soldering operation, tin and copper forms an intermolecular bond. The thickness of this intermolecular bond should be at least 0.5 μm and should not extend 1μm. The temperature/time applied should not be excessive, otherwise if the thickness of molecular bond is more than 1 μm the mechanical strength of joint deteriorates. Above 300 °C, there is a rapid deterioration in mechanical strength. Furthermore, all components being soldered have a maximum temperature/time profile, and if an excessively high soldering temperature is applied for a lesser temperature and applied for too long a time, the component is damaged, which can affect the printed circuit board resulting in lifted tracks/pads, and blisters in the board. In addition, electronic components may get cracked package seals, increased leakage currents, thermal shock to ceramic components, etc.

13.8.2   Steps in Hand Soldering

The workstation and the area around it should always be kept clean, well-lit and uncluttered. Any dirt, grease, solder splatter, insulation cuttings and other debris should be cleaned away. In soldering of electronic assemblies, flux is generally used to improve wetting. Flux residues left on a board after soldering may become corrosive or can cause electrical leakages. That is why complete removal of the residues is recommended after soldering operations for long-term reliability of the system. The solder on the completed joint should have a smooth, shiny surface. The strength of a joint depends upon the quality of wetting and not on the amount of solder used. Excess solder simply adds weight and increases costs (Daniels, 1991).

The most important part of the hand soldering process is the person who does the job. The commitment to do the job in the right way is the first step.

13.8.2.1   Steps in Hand Soldering for Through-Hole Leaded Components

A good soldering technique must ensure that:

- The solder forms a firm joint.
- The solder should cover all elements of the joint.
- The shape of the elements in the joint is not obscured.
- The solder when solidifies appears as a bright solid and flake-free surface.

To best meet these requirements, proceed as follows:

- Select the right tip/bit (in shape and size) for your particular need (the amount of heat that is transferred to the parts being soldered depends not only on the temperature of the tip, but also on how much of the tip touches the parts being connected). The tip size of a screwdriver shape should be nearly equal to the diameter of the PCB pads. There is no ideal tip temperature, since that will depend on parameters like soldering iron, size of soldering subject and how much the tip touches the parts being connected. However, a setting temperature of 275 °C is a good starting temperature. The use of conical tips transfers heat into a small area. These tips should be used with small lands and components that need very little solder. A screwdriver tip, or chisel tip, will transfer more heat, because more of the tip contacts the parts to be soldered.
- Clean all component terminals and substrates.
- Take the soldering iron and wipe the hot tip two or three times across a wet sponge. This will remove impurities and oxidation but it should not cool the tip too much. Apply some flux cored solder onto the tip. Tinning is especially important on new tips.
- Apply flux to the areas being joined.
- Pre-tin all the areas being joined.

- Prepare the leads of the components by bending and cutting off for the right length. It is always advisable to cut off to the proper length before soldering so that no damage will occur by cutting the leads after assembling.

- Insert the component.

- Select the right solder wire. It is recommended to pick a diameter that is slightly less than half the size of the land that has to be soldered. Using a very thin solder wire for a large connection would take a too long time to melt the needed amount of solder (keep in mind that after three seconds the risk of damaging the component or the board is very high). Otherwise, using an excessively large solder wire for a very small connection any loose the ability to control the amount of solder that ends up in the joint.

- Place the iron at an angle of 45° with the tip touching as many elements of the joint as possible.

- Place the solder near the iron and let it flow. Pass it around the joint till you come back near the iron.

- Remove the iron and let the solder flow into the area from where the iron has been removed. All the elements of the joint should get covered with the solder.

- When the solder has successfully flowed into the lead and track, take the solder away and then remove the iron. So many people make the mistake of removing the iron first and this will nearly always result in a dry joint, due to the solder taking heat from the joint prematurely.

- If there is a particular component that is especially heat-sensitive, a heat sink or thermal shunt should be used to help absorb the heat. This thermal shunt should be attached to the lead near the land in order to prevent excessive heat from damaging the heat-sensitive component.

- Remove the solder wire and then the soldering iron. Three seconds should be the average time to complete a hand solder joint. If it takes more time than that, it may be causing heat damage to the board or to the component.

- There should not be any movement of the component or PCB while the solder is under solidification.

- Inspect the joint. A properly wetted solder typically feathers out to a smooth edge. The solder must not clump like a ball on top of the metal. All the metals being joined are covered by the solder and the outline of the lead remains visible within the solder connection. Re-heating of a joint always needs additional application of flux.

- Wipe the tip on the wet sponge again and apply some flux cored solder onto the tip for tinning the tip. It is a good practice to melt some solder onto the tip, whenever we let the iron unused for a minute or more. This solder coating will keep oxygen from reaching the metal tip in order to protect it from severe oxidation. It is also better to turn off the iron when it will not be used for an extended period.

- Clean the joint. The cleaning operation should always be performed immediately after the hand soldering operation. If it is not done right away, the flux residue will harden and it will become more and more difficult to remove.

13.8.2.2   Steps in Hand Soldering for SMDs

Thermal conduction and thermal convection (hot gas) techniques are mostly used for the hand soldering of SMDs. A common thermal conduction method for hand soldering uses special soldering bits that also permit the exchange of multi-terminal SMDs. The advantage of thermal conduction method is that soldering irons may easily be transported to remote locations for on-site repair. However, it requires a wide range of tooling to accommodate all parts and a high level of operator skill. The steps for soldering are similar to these used for lead-through components with the exception that one has to apply fresh solder onto each pad very evenly and provide correct position of the component with either a clamp or a drop of epoxy glue under the component.

Thermal convection technique using hot gas offers a more controlled process where hot air is blown from nozzles directly onto the SMD solder terminals. The procedure is specially described by the vendors of the respective equipment suppliers. However, the general steps are as follows:

- Select the right size of vacuum nozzle to hold the component under placement.
- Select the right shape and size of nozzle for applying hot air.
- Set the right temperature.
- Apply solder paste over freshly cleaned and tinned land area with a dispenser.
- Pre-heat the board, protecting the component from excess heat.
- Place the PCB in the suitable fixture.
- Hold the component in position by means of vacuum.
- Place the component.
- Check the alignment of component by means of a magnifier or microscope.
- Set the required time for which the heat is to be applied.
- Apply hot air to re-flow (with the shortest needed time) the solder paste.
- Ensure that there is no movement of the component or PCB while the solder is under solidification.
- Clean and inspect the joints.

13.8.3   Soldering Leadless Capacitors

In some equipment leadless capacitors are used. Special techniques are required to successfully solder such capacitors to circuit boards. The following steps will minimize the problems that may be encountered when soldering leadless capacitors (Figure 13.14).

- Tin the capacitor by using a small soldering iron with low heat and holding the capacitor down by weighting the edge of it with a silver coin.

- Tin the area of the circuit board where the capacitor is to be attached.
- Place the capacitor on the board in the desired location.
- Apply heat to the board adjacent to the capacitor without touching the capacitor. Do not attempt to effect a bond by applying heat on top of the capacitor as this will permanently damage the capacitor.
- Press down lightly on the capacitor using a toothpick or other small wooden stick, until it settles down on to the board-indicating that the solder has melted underneath. Remove the heat and allow to cool.

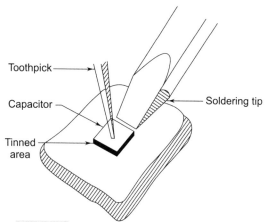

Fig. 13.14 Special techniques for soldering leadless components

 ## 13.9 PCB Assembly Process

The traditional method of the PCB assembly is to attach the leaded components through holes in the board. The bonding of the components with the board is by solder which provides both the electrical and mechanical connections. Subsequently, with the developments in micro-electronics, with the VLSI chip packages having large number of joints on small pin pitches, surface mount technology has evolved. However, not all components are available in surface mount form. Therefore, PC boards continue to have mixed technology components i.e. both leaded as well as surface mount type.

The PCB assembly can be done either manually or by using machines which make the assembly process automatic, fast and reliable. Since the final assembled PCB consists of lot of different types of components with various configurations, an assembly process must be carefully selected in order to make it both economical as well as reliable.

A typical PCB will have a number of dual-in-line packages (DIP), resistors, capacitors, transistors, connectors, etc. mounted on them. All these components have typical hole spacing required for their leads to be inserted. The assembly process, in general, involves the following steps:

- Component collection as per the specifications given in the bill of materials;
- Preparation of the components or component forming (cutting and bending the leads before insertion);
- PCB cleaning: washing with commercial cleaners or rinsing in water etc.;
- Inspection of boards and components for dimensions;
- Placing of components in respective trays in the assembly line;
- Component insertion;
- Soldering of component leads by hand soldering or by using a mass soldering machine;

- Cleaning to remove excess flux;
- Placing of special components which can be either heavy or components that cannot withstand high soldering temperature and cleaning;
- Inspection and testing;
- Solder touch-up;
- Applying conformal coating; and
- Storage.

This flow sequence is illustrative and need not be followed in all the assembly lines. There may be a few changes here and there depending upon the components types, which can be done either manually or by using special machines.

13.9.1   Leaded Through-hole Assembly

13.9.1.1   Manual Assembly

Manual assembly is normally for small volume work and prototyping usually undertaken on a batch basis. Here, most of the process steps indicated above are done manually. The mechanical strength and the electrical connection are achieved with the help of holes in the board which are employed to mount the components. For reliable operation, the size of the hole must be related to the component lead diameter. Similarly, the pad land must also be of correct size, so as to achieve proper mechanical strength to assemble such a board, both axial and radial components are inserted in the holes, their lead lengths are suitably cut and soldered. In this case, all the components have to be pre-formed before putting them on the boards. This performing could be again done totally manually or through die and jigs. In the first case, the performing of components is done by making use of pliers or manual cut/bend machine. The second method is adopted when the number of boards to be assembled is very high. Once all the components have been formed to their required sizes, they are taken for insertion.

A preferred method of manual assembly is to have a rotary table with a number of bins to hold the components. The preformed components are put in individual bins. The operator sitting at one position in front of the rotary table can bring any one of these bins into position, pick the component from the bin, and place it in its position on the board.

In case the component leads have to be clinched, the assembled board is placed in a special fixture consisting of a frame with thick foam supported by a metallic plate. The assembled board is fixed on the frames which could be rotated. By rotating the assembled board, the leads are brought facing the operator and could be clinched easily with a pair of pliers. The lead is grasped at the point where it has to be cut and bent. The work begins from the lower left hand corner of the PCB and proceeds towards the top right hand corner of the PCB.

In order to aid the operator for correct insertion of components, a computer controlled light projection system is used. Different components are placed in the different bins. When a component from a particular bin is required to be inserted, the bin comes before the operator and its trap door opens. Spot-light appears on the board where the component is to be inserted. If the same component is to be used again, the trap door remains open and the spot-light moves to the new position. In case of polarized components, the spot light moves back and forth over the component position and indicate the particular polarity. Well-trained operators can do the assembly job quite fast using both the hands.

After all the long leaded components are inserted, the board is carried in its frame to an electrically operated guillotine which cuts the component leads to the correct size and clinches for clinched joints. The process is followed by insertion of components with short and correct lead lengths such as semiconductor devices, relays, inductances, connectors etc. Once all components are inserted, the board is wave soldered. The printed circuit has a solder mask so that only the pads and tracks where soldering is to be applied are exposed.

Once all the components have been placed in their respective places, bent and clinched as per the requirement, the assembly is subjected to a pre-soldering inspection. The inspection could be done by using a transparent mylar sheet, which has the assembly drawing where the component outlines are marked. By scanning the PCB along with the transparent sheet, the missing or misaligned components can be easily identified and rectified.

Cleaning: Before any component lead, wire or terminal is soldered in a circuit, it is essential to clean it with some braided cleaning tool followed by brushing the cleaned surface with a stiff bristle brush dipped in alcohol. The surface is dried with paper or lint-free rag.

It must be noted that solder will just not take to dirty parts. They should be free from grease, oxidation and other contamination. Old components or copper board can be notoriously difficult to solder because of the layer of oxidation which builds up on the surface of the leads. This repels the molten solder and the solder will form globules, which will go everywhere except where needed. While the leads of old resistors and capacitors should be cleaned with a small hand-held file or a fine emery paper, the copper printed circuit board needs to be cleaned with an abrasive rubber block or eraser. In either case, the fresh metal underneath needs to be revealed.

Component Forming: In order that the components fit properly in the circuit, in which they are to be installed, they must be properly formed (Figure 13.15). Forming of the component has two main functions:

(a) To secure the lead to the circuit; and

(b) To provide proper stress relief. The relief is necessary to prevent rupture of the component lead from the component or in case of a wire, to prevent a stress pull on, the solder joint and rupture of the wire strands.

The following steps are taken to properly bend the leads of the components:

(a) The bend should be attempted no closer than 3-5 mm from the component body.

(b) The radius of the bend should be equal to twice the thickness of the lead wire.

(c) Centre the component between its solder connections.

(d) Bend the protruding lead 45° after insertion into circuit board with the help of a bending tool.

(e) Cut the lead so that no portion when bent exceeds the perimeter of the pad. Press the cut lead firmly against the lead.

(f) On a joint, when the lead is not bent, cut the lead to the thickness of SWG 20.

Forming is not necessary in the case of integrated circuits. They are simply soldered to the board without cutting their leads.

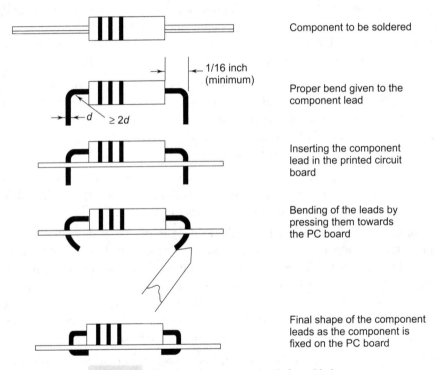

Component to be soldered

1/16 inch (minimum)

Proper bend given to the component lead

$\geq 2d$

Inserting the component lead in the printed circuit board

Bending of the leads by pressing them towards the PC board

Final shape of the component leads as the component is fixed on the PC board

Fig. 13.15   Forming of component leads for soldering

Lead bending is the most important factor in forming. A few sharp back and forth bends in a component lead can easily cause it to break or crack. Bending a lead too close to the component encapsulation may result in excessive stress at the lead entrance, and cause cracks in the encapsulation. Such cracks allow moisture to enter inside of the component and result in gradual degradation of the component resulting in premature failure.

Circuit Board: Although the principle of general cleaning also applies to circuit boards, certain precautions are, however, necessary. This is because circuit boards may contain some components that may be spoiled if a braided brush is used. In case of circuit boards, a sharply pointed typewriter eraser may be used to remove dirt, contaminants or other foreign substances from the pad to be soldered. It is then cleaned with alcohol and a brush, and left to dry.

It may be noted that in the whole process of replacement of components by soldering, the single-most crucial factor is cleanliness and it should be scrupulously followed. With dirty surfaces, there is a tendency to apply more heat in an attempt to force the solder to take. This will often do more harm than good because it may not be possible to burn off any contaminants anyway, and the component may be over-heated. In the case of semiconductors, the temperature is quite critical as they get damaged when excessive heat is applied.

13.9.1.2   Automatic Assembly

The automatic assembly techniques are used for high- volume through-put. Therefore, a flow-line approach is followed for the purpose. Figure 13.16 shows a typical flow diagram for automatic leaded component assembly system . The first versions of the automatic assembly technique was used mainly for the axial lead component insertion into PCBs. Several types of processes are used to bring in the component into the place of insertion such as body taping, lead taping and several magazine feeds. The lead taping process has been standardized and most of the axial leaded components are now available in this lead taped format as shown in Figure 13.17.

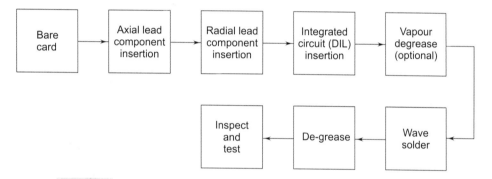

Fig. 13.16   Flow diagram for an automatic leaded component assembly system

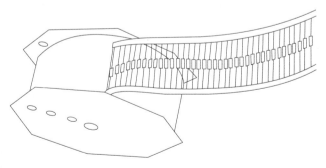

Fig. 13.17   Lead taped components on reel with paper interliner (a layer-to-layer separator)

The earlier versions of automatic assemblers for axial lead components were very simple, bench top units which basically consisted of an insertion head, a clinching end and a reel of taped components with an anvil which could be rotated. The components are cut from the tapes and formed into a staple with an internal or an external former and then pushed downwards with a head through the holes of the PCB as illustrated in Figure 13.18.

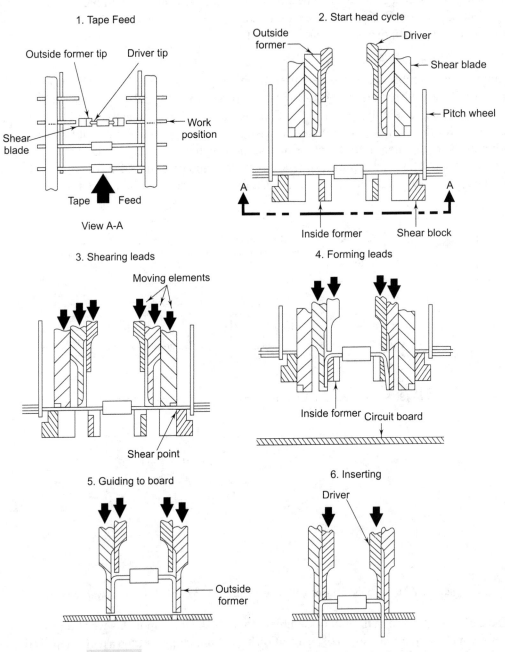

Fig. 13.18 (a) The essential functions of an axial lead insertion head

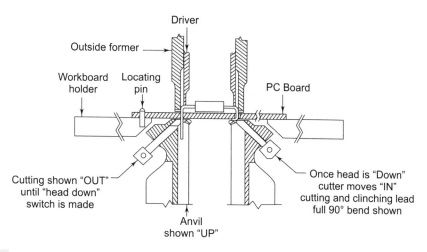

Fig. 13.18 (b) a typical cut and clinch unit, shown in the up position, before cutting and clinching (NTTF, Notes)

In modern machines, the location of each component on the board is checked with a pantograph machine. The manually operated pantograph systems were initially used and by using a stylus and a template, the board was brought exactly in position for component insertion. Later on, the same manual system was upgraded to numerically controlled equipment. With the extended use of computers, computer numerical controlled systems (CNC) were introduced. With such an automatic assembly sequence, the standardization of component packaging became a necessity.

Two types of insertion machines are required, axial and radial for leaded components. Both are of the pantograph type. They allow one or more boards moving simultaneously to enable the heads to insert the components. Components must be fed to the insertion machine in the correct order. This is done by using a sequencing machine which accepts standard axial components in taped form and re-assembles them in the order in which they are to be inserted. Alternatively, some machines use air flow to move components from a selection machine, which cuts them in turn from reels, blows them through a common duct for lead cutting and shaping for insertion. After each component is inserted in the card, it is guillotined and the leads are clinched. This is followed by insertion of transistors, integrated circuits and other electronic hardware components which is then moved over to wave soldering, inspection and testing.

Several factors control the automatic assembly process. For example, the complete process sequence through which a PCB will have to pass through during the assembly sequence and the capabilities of the automatic assembly machines to be used must be studied carefully. Generally, the supplier of the machines would provide necessary application information which will guide the user about designing a PCB suitable for automatic assembly.

Also, while designing a PCB for automatic assembly, one must take care of the clearance between the lead diameter to the hole diameter. Generally, it should be as large as possible to ensure a reliable insertion, but at the same time, it should not be larger than the minimum required for good soldering. Therefore, while designing a PCB, great care has to be taken regarding all the machine parameters.

The automatic insertion machines have certain limitations and drawbacks as compared to hand assembly. For example: a machine cannot insert a component with a lead diameter of 0.6 mm into a hole of 0.625 mm reliably. Manually, components can also be inserted into a PCB at any angle where as, a machine cannot do the same. In hand assembly, components with different shapes and sizes could be easily put in where as in case of a machine assembly, much stricter control over the dimensions and shapes of the component is required. It is, therefore, mandatory to follow a set of rules while designing a PCB for automatic assembly sequence. Generally, in a circuit designed for automatic assembly, the components will be arranged in one particular direction or in two directions perpendicular to each other. Alternatively, the assembly could be achieved by making use of a rotary table.

13.9.2 Surface Mount Assembly

The surface mounted devices which are generally considered for assembly usually will have flat top surface which facilitates its being picked up by a vacuum tipped placement nozzle.

The SMD components can be generally classified as:

- *Chip Devices:* capacitors and resistors
- *Semiconductor Devices Packages:* These are either transistors or ICs in package form

Small Outline Transistors (SOTs):   These are standard three leaded plastic devices which come in several sizes and designated as SOT-23, SOT-89 etc.

Small Outline Integrated Circuits (SOICs):   These are similar to the above devices except that they have more number of leads and their spacing are closer.

Plastic IC Package or Chip Carriers (PLCC):   They have leads which take care of differential thermal expansion which helps in preventing the damage to the device.

Leadless Ceramic Chip Carriers (LCCC):   The use of ceramic base materials with circuits screened on both surfaces and interconnected over the edges. The IC chip is bonded on the top of this carrier and connected to final termination pads on the bottom.

Single In-line Packages (SIPs):   These are special packages used generally for passive networks such as resistors, diodes, capacitors or combinations of them. SIPs are generally available with lead spacing of 0.1 inch or 0.125 inch (2.5 or 3.12 mm) apart. The number of leads varies from 2 to 8 or in some cases more.

Since in this case, all components are surface mounted, it is far easier to automate the placement of these components. There are no holes in the card, except the vias that are used to provide connection between the layers in the card and not for component mounting.

13.9.2.1 Manual Assembly

Figure 13.19 shows a simplified manual surface mount assembly flow diagram. The first step is to apply solder paste to the bare board, which can be done by pneumatic dispensing or screen printing. In manual assembly, the first method is preferred where in a known quantity of the solder paste is applied with a syringe. The tip of the syringe is placed on the pad, the machine is activated with a hand foot operation until each pad receives the requisite amount of solder paste on it.

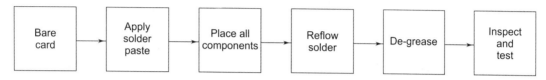

Fig. 13.19 Manual surface mount assembly flow diagram

This is followed by assembly of components, which is manually done by a vacuum operated pick-up pen. The correct component is picked-up by the operator and mounted on the board with correct orientation. However, positioning of integrated circuits with large number of leads is usually quite difficult. For this purpose, a manual pick-and-place machine, using a vacuum system for holding the component can be employed. Components are fed to the machine in several ways. The passive and small semiconductor components are supplied from tapes on reels whereas integrated circuits are fed from stick magazines of varying widths.

After the components have been assembled, they are reflow soldered either in infra-red or vapour phase furnace. The assembly is then degreased to remove all flux and inspected.

13.9.2.2 Automatic Assembly

Automatic assembly of surface mounted components follow the same steps as manual assembly except that all the operations are automated (Alan Roads, 1991). The method chosen for assembling surface mount components on to printed circuit boards depends upon the soldering method i.e. wave or reflow that will be used once the devices are in position. If the wave soldering is selected, the component must be held in place by an adhesive. If reflow is chosen, the solder paste performs this function (Buckley, 1990a). The details on solder paste and adhesives are given in the next two sections.

Once solder paste has been applied to the lands, the circuit board is moved to automated component placement machine which will automatically place the components onto the lands covered with solder paste. A major advantage of SMDs is that the placement process is simplified through the elimination of many functions prior to and during placement; for example, bending, forming and cutting of the component leads is no longer necessary.

The components are delivered to the machine in bulk, in tubes or in a tape reel format. The computer controlled pick and place system removes the component from its packaging and places it on the board in the correct location and orientation. The accuracy of part location should be with in a tolerance of $100\,\mu m$ or less. The majority of components for a pick and place machine are on tape reels of various sizes. Therefore, attention should be paid when changing reels because many components have identical appearances and similar part numbers. These part numbers must be checked carefully against those specified on the documentation. The component feeders are powered either by air pressure, mechanical action or electricity.

Component placement heads can use a vacuum or mechanical pickup tool. For speed and accuracy reasons, the majority of pick-and-place machines use a vacuum tool to pick up components and place them on the PCB. A mechanical jaw also can be used. Some machines use both techniques-the vacuum tool places the device on the board and the mechanical jaw centers it to match the leads to the pads.

Automatic assembly can be broadly divided into two classes: sequential and simultaneous. *Sequential* assembly means that the components are picked up one after another and placed onto the printed circuit board (pick-and-place method). With *simultaneous* assembly, all components are placed on the board in a single operation, which implies very high performance. Sequential machines place one component at a time at speeds from 1,000 to 20,000 ceramic capacitors to large ICs without operator intervention. The standard pick and place machine has a component range of 60 to 120 different components. High performance pick-and-place machines are able to place 120 to 400 component types.

Simultaneous placement will be performed by machines having multiple heads that can place two or more components on a board at a time. As the nozzle places the component, an infra-red beam activates a camera that takes a picture of the component to be placed. That picture is compared to the part description stored in the computer. The computer verifies that the leads are not splayed or bent and that the component is being placed in the correct location on the board. If the component has the wrong size, direction or the leads are defective, the nozzle will drop the component before placing it and the machine will automatically stop. It is important that all components are accurately placed on the board before the assembly goes into the reflow soldering operation.

High performance systems with simultaneous placement have a pick and place rate of 15,000 to 2,00,000 components per hour. These systems are suitable for companies with very high throughput requirements, e.g. for manufacturers of consumer goods.

Both principles, sequential and simultaneous, can also be combined. Although simultaneous assembly machines feature high performance at low cost per component, but many other factors make the sequential method as most suitable. In other works, simultaneous assembly machines are recommended for large lots and sequential machines for small and medium lots.

With the availability of flexible automatic pick and place systems, new possibilities for automation now exist for smaller and medium-sized manufacturing facilities in terms of both component handling as well as capital cost.

Placement performance data about the machine are usually specified in the manufacturers manuals defining the maximum placement rate for a given PCB size and for selected components and feeder types. In practice, the placement performance is sometimes even less than 50 per cent of the theoretical value. The major resources for loss of performance are machine breakdown period, machine set up time, pick up errors and PCB/components related performance reductions.

Placement accuracy is an important quality consideration for automatic placement machines; the key parameter is placement reliability. To achieve high levels of placement reliability, even with today's high pin count components, opto-electronic measurement systems are employed for both PCBs and components.

Error rates of fewer than 20 defects per 1,000,000 (million) or in other words, 20 dpm are usually achievable with sequential pick and place systems. In order to achieve and maintain low defect rates, strict monitoring and control of all the functional aspects of the machines are necessary.

For example, using separately controlled jaw pairs mounted on the placement head, components are centered and aligned while on the vacuum nozzle. During the component X-Y-transfer, components are rotated into the correct orientation for the assembly position to an angular resolution of $0.1°$. When the placement position has been reached, the jaw pairs are opened and the component is lowered on the vacuum nozzle and placed with a programmable set-down force.

The component placement machines play a crucial role in the success of the assembly process. Hodson (1993a) explains the important parameters which must be considered when purchasing a pick-and-place machine. These parameters include: throughput requirements, accuracy required, ease of integration for component type and assembly technique, operational complexity and software support applicable to CAD downloading, Gerber file compatibility, number and style of component feeders, available options (vision system, on-line test, adhesive dispenser, alignment system) etc.; ease of technology upgrade, safety, maintenance and cost. Rhodes (1991) describes the trend towards finer pitch surface mount devices and the parameters to best specify placement machine accuracy and configuration.

For specific production applications, chip-shooters are used for PCB assembly. Chipshooters are chip component placement machines that operate at high speeds to remove components from part feeders, ensure they are the correct parts, and accurately orient them before placing the component on the printed circuit board. The majority of chip components used are leadless ceramic capacitors and resistors. Chipshooters are not used to place larger components because the overall throughput is decreased. The more intricate the part, the more time is required for vision recognition, theta correction, placement and so on (Crum, 1993).

The major difference between pick-and-place machines and chipshooters are the different kinds of components they place and the speeds at which they operate. Most chipshooters place smaller, passive type components (resistors and capacitors), while pick-and-place machines place larger size packages with very small leads.

One of the two jaw pairs is electrically isolated and performs electrical identity testing. The measured values of resistors, capacitors, diodes and coils are compared with the permissible values and rejected in case they do not fall within the specified tolerance range.

Smith (1993) describes equipment for cleaning the surface mount assembly and the post-assembly cleaning systems. The post-solder cleaning of boards with surface mount devices is more effective with ultrasonics as illustrated by Polhamus, 1991).

13.9.3   Combinations of Mixed Technologies

In mixed assembly boards, there are several variations, which are shown in Figure 13.20.

- Single-sided board: mixed components on the same side
- Double-sided boards: leaded components on one side and surface mounted components on the other side
- Double-sided: mixed components on one side and surface mounted on the other

In line with these variations, the assembly procedures are organized accordingly.

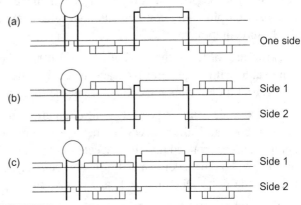

Fig. 13.20   Various combinations of mixed technologies (a) single-sided board with mixed components (b) double-sided board with leaded component (c) mixed components on one side and surface mount on the other. (after Haskard, 1977)

13.9.3.1   Mixed Assembly Surface Mount Components: Single-sided

Flow diagram for assembling the mixed technologies for a single sided board having leaded components on the top side and surface mount component on the bottom side is shown in Figure 13.21.

13.9.3.2   Mixed Asembly

Surface mount components on both sides and leaded components on one side.

The assembly procedure for double-sided surface mounting plus leaded components consists of two steps:

- Screen printing of solder paste on top surface of the board, placement of SMDs and reflow, insertion of leaded compoents
- Application of adhesive to the other side of the board, placement of SMD, curing of adhesive, turning over the baord for a second time, fluxing and wave soldering. The flow chart is given in Figure13.22.

During the assembly process involving a series of mechanical (solder stenciling and component placement for SMT) and metallurgical (solder fusion) operations, it is possible for components to be misaligned or missing upon completion of the process. Therefore, some form of automatic inspection system is essential to improve quality and yield.

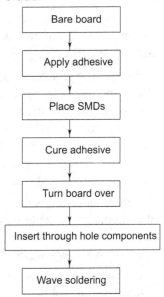

Fig. 13.21   Process steps in a mixed technology assembly with SMD components on one side and through-hole components

Two types of systems are used for performing automated inspection of component placement:

- Camera-based systems; and
- Laser-based systems.

Camera-based Systems: These systems use either gray scale or colour charge-coupled device (CCD) cameras. The cameras capture the images of the circuit board and these images are then analyzed to determine whether there are any defect or abnormality in the area of the board under examination. These systems rely on the brightness of the light reflected from the board and are sensitive to changes in lighting conditions and materials. Therefore, they preferably employ a programmable lighting fixture for creating optimal images of the component or the site. However, as the images become more complex, the image processing takes time and the inspection cycle times get significantly dropped.

Laser-based Inspection Systems: make use of a laser scanner to generate a 3-D image of the circuit board. This 3-D image is based on the height of the board surface and its components. The system is much less sensitive to changes in component colour. The system can also create a 2-D gray scale image which can be used to identify objects where there is little height contrast such as board fiducials and to detect component leads in solder paste. Laser scanning enables accurate position measurement of components and there are lesser errors.

The above referred systems are categorized as AOI (Automated Optical Inspection) systems. These systems, as a minimum requirement, should measure the position of each component along its X, Y, and theta dimension. They should also check that the component polarity is correct. Generally, the actual component positions are compared to computer-aided design data to determine whether each component position is within acceptable tolerance. Component position outside of tolerance are identified and the measurements are used to update statistical process control charts. A good AOI system is one which has minimum number of false calls which include defects identified as good placements and good placements identified as defects.

There is an ever-increasing demand to reduce product generation time-to-market, development costs and production costs. To achieve all this, it is desirable to integrate Design for Manufacture and Assembly (DFM/A) into a product generation framework

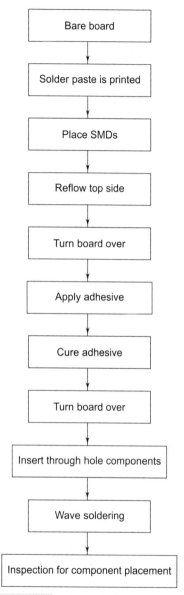

Fig. 13.22 Process steps in a mixed technology assembly. SMD components on both sides and leaded components on one side

(Holden and Kenyon, 1994). DFM/A essentially focuses on several separate domains such as: optimization of printed circuit design and layout, minimization of printed circuit substrate costs, minimization of assembly costs, use of preferred parts and analysis of test coverage. Boothroyd (1990) explains that considering product assembly at the design stage can lead to improved product reliability. He suggests the idea of Design for Assembly so as to look ahead and anticipate manufacturing problems at the design stage itself.

13.10 Solder Pastes for SMDS

Solder pastes, often called solder creams, are used in "reflow-soldering" of surface mount components, where the application of solder and the heat supply are separate steps in the fabrication process. A solder paste basically consists of solder in powder form and a flux with some additives necessary to produce the desired behaviour of the paste during or after its application.

Solder pastes are applied either by stencil or by screen-printing method. Pick-and-place throughput is an important issue in deciding the type of solder paste dispensing system. Erdmann (1991) brings out the stenciling technique including stencil development, stencil cleaning and printing etc. He points out that the demands of stencil are more rigid than that of ordinary SMT screen printing and stenciling, particularly for maintaining near perfect registration.

Large pick-and-place systems require volumes that only screen printers can provide. Pick-and-place throughput in the range of 1500 to 3000 components per hour, however, is ideal for today's dispensing equipment, which can produce 16000 dots per hour in a typical production environment (Cavallaro and Marchitto, 1991).

A rotary positive displacement programmable pump as shown in Figure 13.23(a) can be effectively used for solder paste dispensing involving high speed application of a large number of very small dots. The pump is driven by a DC motor. An electro-magnetic clutch engages and disengages the Archimedian screw. Mounted above the screw is a bellows coupling that aligns the clutch and lead screw and reduces the impact of the Z-axis sensor by more than 60 per cent. The combination of constant motor speed, low air pressure, the software-controlled clutch, and the precise rotation of the Archimedian screw ensures a repeatability that is far superior to pulsed air or piston dispensing systems (Cavallaro, 1994).

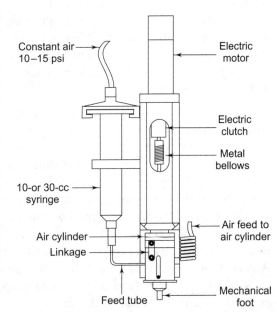

Fig.13.23 (a) The dual-height rotary positive displacement pump, which can perform 25 and 50 mil dispensing within the same program. The pump is appropriate for SMD epoxy or solder paste

The pump can be programmed in one-milliseconds (ms) increments from 10 to 10,000 ms. Most solder paste applications require a shot size of 15 to 20 ms for fine-pitch devices and 50 to 100 ms for 50-mil devices. This short cycle time allows the automated dispensing system to dispense up to 16,000 dots per hour, making it compatible with most pick-and-place machines in the marketplace. The real advantage of dispensing over screen printing is programmability. For example, a 25-mil-pitch application, requires a dot diameter of 0.014 to 0.016 inch (0.35 to 0.40 mm) while a 50-mil-pitch application requires a 0.020 to 0.030 inch (0.50 to 0.75 mm) diameter dot. Dual-height dispensing systems are important to accommodate different dispensing requirements without the need to change either the needle or the programme. Dispensing systems are particularly effective for re-work and mixed technology applications.

13.10.1 Requirements of Solder Pastes

Solder pastes must meet the following requirements:

- The individual powder particles of the solder alloy should have a homogeneous distribution of the metal within the paste as well as a fair equality of shape and surface roughness of the individual powder particles.
- It should develop an adhesive action in order to hold the components in place until the reflow operation has been finished.
- It must not tend to solder balling; if they become detached on the substrate, they may cause short circuits.
- It has to maintain its shape during curing and reflow and must remain on the pads, not leaking to unsolderable parts of the PCB.
- It must have sufficient activity as the solder paste is sometimes in contact with the parts for several hours.

13.10.2 Composition of Solder Pastes

Solder pastes are available with several fluxes and alloy compositions. Their consistency may vary from liquid cream to thick paste. Solder pastes usually contain a certain percentage of silver, most commonly 2 per cent, that gives a solder alloy of 62 % tin (Sn) + 36 % lead (Pb) + 2 % of Silver (Ag). This alloy has a melting temperature of 179 °C. The powder must contain granules of the alloy and not of the individual metals, which make up the solder. This will help to melt the solder at the temperature of the alloy used. Beside homogeneity, an important element in solder powder is the structure and shape of the alloy particles. Therefore, microscopic control of the solder particles is essential. An acceptable solder paste should contain alloy-particles only with the shape of a "sphere", an "ellipsoid" a "tear" or a "dog bone". Chilton and Gaugler (1990) describe the design of solder paste to meet the exacting requirements of fine pitch reflow.

A spherical shape minimizes the surface area and so reduces oxidation at best. However, on melting, the liquid flux flows outwards, also carrying with it the solder particles. The particles with a dog bone shape or a shape of an ellipsoid lock the solder better in place. Solder pastes having particles with extended irregular shapes, shapes of needles or even showing dust are not acceptable. The solder particles are typically 20 μm to 80 μm in diameter. For screen printing, usually a smaller particle size is advisable.

Solder pastes can be manufactured either with rosin based or water soluble fluxes. The paste contains not only alloy powder and flux, but also organic solvents, thickeners and lubricants to determine paste rheology. With screen or stencil printing of solder paste, rheology is critical to ensure excellent print definition.

With the development of lead-free soldering alloys, solder pastes without lead are also available. For low temperature soldering, a composition of 42 per cent tin, 42 per cent lead, 14 per cent bismuth and 2 per cent silver is commonly used.

13.10.3   Solder Paste Application

The reflow soldering process starts mostly with the application of solder paste to the specific areas of the circuit board where the components will be attached. Solder paste printing is commonly applied by stencil or screen-printing. It is wiped across the stencil or screen with a squeegee, which pushes the paste through the openings, depositing it on the lands at the right places. With solder paste printing, the entire amount of solder is deposited on the PCB in one operational step. Solder paste application with dispenser is generally used for laboratory applications because of its flexibility. With a dispenser, the solder paste can be applied not only in form of dots, but also in stripes, which may be more useful and accurate in certain cases.

13.10.4   Handling of Solder Paste

Solder paste ages and changes with temperature, humidity and light. The following precautions may be taken while handling solder paste, keeping in view the recommendations of the paste supplier;

- Store solder paste in a clean, cool, dry and dark location.
- Before opening refrigerated solder paste, keep it for 24 hours at room temperature to avoid water condensation.
- Do not mix old and new pastes since the fluxes and thinners will evaporate at different rates.
- Stir solder paste at least 30 seconds before applying to the stencil/screen
- Use only clean and inert tools (Such as made of stainless steel, Teflon, Polyethylene)
- Apply to the stencil/screen the amount of paste just required for printing.

13.10.5   Stencil Printing of Solder Paste

In stencil printing as well as in screen printing, the entire amount of solder is deposited on the PCB in one operational step. While the squeegee moves over the stencil (or screen), the solder paste is pressed through openings in the stencil (or screen) to the lands on the circuit board. It is important that the stencil (or screen) openings match precisely with the locations of all the land patterns. Therefore, for each circuit, the appropriate stencil or screen must be produced. The important parameters for solder paste printing are homogeneity of speed, squeegee speed, pressure, squeegee angle, snap-off distance and speed of board separation.

The solder paste must be thixotropic. It means, its viscosity should drop during the application process. Thixotropic pastes have an internal structure which breaks down when they are subjected to mechanical action and recover when the shearing force is removed. This property ensures that the paste will flow onto the board properly. The advantages of stencils versus screens are a considerably longer lifetime, higher and controlled paste depth as well as higher accuracy because of less snap-off distance. The snap-off is the distance between the screen/stencil and the surface of the board.

Stencil is usually made out of nickel plated metal or stainless steel. The land patterns in the stencil are mostly performed by means of laser-cutting (but it can also be done by chemically etching from both sides). The stencil is glued with an epoxy onto a sturdy cast aluminum or stainless steel frame which attaches to the screen printer. The stencil has to be accurately adjusted to be in precise alignment with the circuit board. Depending on the stencil thickness and of the hardness of the squeegee, a much higher wet layer thickness of solder paste can be obtained in comparison with screen printing by maintaining an excellent edge definition. The surface tension (adhesion) between the paste and the laminated board ensures that when the squeegee has passed over and the stencil/screen has separated from the board, the paste remains on the board.

Board Fixture: The board fixture which holds the board during printing is mostly provided by means of a vacuum plate and is situated underneath the stencil. The purpose of the vacuum plate is to provide and keep a plane support to the board during the printing operation. If there are parts on the underside of the board, space or standoffs are placed in specified locations on the vacuum plate so that these components will be protected. The board fixture and the stencil will then have a proper alignment by using alignment marks (so called fiducials) on the board and on the stencil.

Squeegee: The squeegee may be made of thin metal. Setting the right and homogeneous pressure is done by putting paper underneath so that one can evaluate proper adjustment done by showing an even contribution of the solder paste. Best approach is to start always with too little pressure rather than with too much, because it may damage the stencil. The solder paste should always roll ahead of the squeegee during application and there should not be a film of solder paste left on the stencil. A film of solder paste left on the stencil indicates too low pressure of the squeegee. The diameter of the roll, rolling ahead of the squeegee, should be approximately 15 mm.

Polyurethane squeegees come as trailing edge or diamond-section (Figure 13.23b) and in several hardnesses. In all cases, the squeegee needs to present a sharp edge, and this is subject to wear and should be redressed periodically.

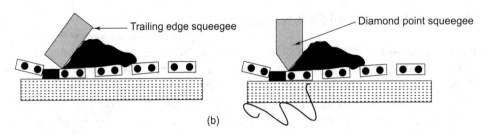

Fig. 13.23 (b) Trailing edge and diamond point squeegees (after Judd and Brindley, 1992)

Various references describe the best contact angles as being anywhere between 45-80°. In general greater print angles give poorer paste transfer through the stencil, while much shallower angles give degraded definition.

As the solder paste is being deposited, the stencil lifts immediately behind the squeegee (snaps off) and returns to its original snap off position, otherwise the stencil would smear the solder paste between the lands.

Fleck (1994) explains that using the laser cutting process to manufacture stencils provides more control over the amount of solder cream deposited on the pads of a printed circuit board. This control becomes more important when attempting to place fine pitch components using no-clean solder cream. There is no cleaning step to remove potential solder balls, making it necessary to control the printing deposit.

In the case of fine pitch printing with its characteristic narrow apertures, metal squeegees are the only practical solution. Their robustness overcomes the wear problem and they do not deform into apertures and scoop paste. The main danger comes from the tendency to ramp up the pressure which can lead to damaged stencils.

13.10.6 Screen Printing of Solder Paste

The advantages of screens over stencils are the lower cost and that they allow printing over a much large area. Disadvantages are limitations in accuracy and coating-thickness. In screen printing process, the solder paste is rolled over a mesh, which has been coated with an emulsion that closes the meshes of the screen where the solder paste is not required on the board. The screen, which may be of stainless steel, metallized polyester, polyester or nylon, is stretched across a strong metal (mostly aluminium) frame. This frame is held in the upper frame of a printing machine..

The action of the squeegee drives the solder paste through the open holes in the mesh, pressing the screen down so that the paste makes contact with the board. After the squeegee has passed, the screen springs back clear off the board, leaving on the board the paste that had been in the openings of the screen. The discrete particles of the solder paste flow together to yield a uniform coverage. Screen Printing needs highly skilled and trained operators in order to get an accuracy of < 0.15 mm shifting for solder paste placement.

Screen Fabrics Classification: Beside the selection of fabric-material, the classification is made according to the mesh counts of the screen that means the number of threads (or openings) it has per linear cm. For example a "55-T" screen printing fabric has 55 threads per linear cm (or 140 threads per linear inch) in both directions. The materials used are:

- *Stainless steel* provides a long life excellent registration definition, good paste flow, prevents the build-up of static charge and needs the lowest snap-off distance.
- *Metallized Polyester* having a high resistance to abrasion, permits a good paste flow and the metallic surface prevents the build-up of static charge.
- *Polyester* has a better elasticity than stainless steel and is cheaper. The higher elasticity can become necessary if the bare board is unevenly soldered.
- *Nylon* having an excellent elasticity

The mesh used should have approximately 30 threads per cm and the opening between the threads should be minimum 3 times the diameter of the largest solder particle within the paste. Table 13.3 shows some selected fabrics for printing of solder paste.

Table 13.3   Some Selected Fabrics for Printing of Solder Paste (Courtesy Braun,2003)

Fabric number cm	inch	Material	Mesh- opening μm	Thread -diameter (μm)	Theoretical Paste thickness (μm)
43	110	Stainless steel	160	71	82
67	170	Stainless steel	100	50	49
24T	60T	Polyester/metal	285	120	105
32T	82T	Polyester/metal	195	105	70
36T	92T	Polyester/metal	165	95	60
43T	110T	Polyester/metal	144	84	55
20HD	51HD	Polyester	300	200	144
24T	60T	Polyester	275	150	111
27T	68T	Polyester	250	120	65
34HD	85HD	Polyester	175	125	73
40 HD	100HD	Polyester	149	100	71
43	110	Nylon	128	88	51

T-normal grade; HD-heavy grade;

Frames: The necessary mesh tension for precision printing can only be obtained when strong metal frames are used. Aluminium is the most used metal for screen printing frames. Steel frames are seldom used, as they rust and are nearly three times heavier than aluminium frames of the same size. However, one has to consider the fact that the thermal linear expansion coefficient of aluminium is twice as high as that of steel.

For example: Steel   :   0.06 mm expansion/meter at 5 °C temperature rise
Aluminium :   0.13 mm expansion/meter at 5 °C temperature rise

The frame size should be so large that the distance between the outer edge of the image and the inner edge of the frame is on all four sides at least 150 mm. Difficulties in adjustment arise if the distance is too small, since the distortion increases with the snap-off distance as well as with decreasing the distance of the squeegee to the frame.

Snap-off and Lift-off: The "snap-off" is the distance between the screen/stencil and the surface of the board. High fabric tension allows a lower snap-off. The additional lifting of the frame at the edge where the squeegee started print is called "lift off". This lifting is provided by advanced printing machines simultaneously with the movement of the squeegee in order to ensure the same snap-off by increased distance. So, the angle between board and screen will be equally maintained by moving the squeegee over the screen. In general, the snap-off should be set as low as possible. With high fabric tension, one can have a low snap-off.

Stretching of Fabrics: Screen printing fabrics are stretched either pneumatically or mechanically. The loss of tension of a newly stretched fabric is usually 10 to 20 per cent within the first hours and is dependent on the type of fabric, the stretching equipment, the stability of the frame and also the rest period before gluing the fabric onto the frame. Therefore, it is recommended that for precision work, the stretched screens are left to rest for 12 hours before stencil making. The tension resistance of a fabric depends on the material used as well as by the thread diameter. The tensile strength of the thread rises by the square of the increase in thread diameter.

De-greasing: It is always advisable to degrease the stencil before every use. That can be done by commercial screen printing degreasing agents or with the help of 20 per cent caustic soda. But after rinsing, it is necessary to neutralize it with 5 per cent acetic acid.

Printing Speed: It is determined by thixotropy of solder paste and typically between 20 to 80 mm per second. It is usually recommended by the manufacturer of the paste. The more fluid the paste is when moved and rolled, the higher print speed can be achieved. The print start has to be at least 90 mm before the aperture pattern is reached so that the paste can roll nearly twice in order to get its thixotropic behaviour.

The mesh count in a screen refers to the number of openings or lines per linear inch (lpi); for printing solder pastes the mesh count is in the range 60-200. Typically maximum particle size should be no larger than one-third the mesh opening to prevent jamming. For example, an 80-mesh screen has openings of about 224 μm so the particle size should not exceed 75 μm. Finer mesh screens tend to be used for thinner deposits. A 180 mesh and a fine powder paste could be used to produce a deposit thickness of 100-150 μm, using an 80 mesh. Hall (1994) brings out the screen printer requirements for low defect process capability, while Noble and Moore (1992) illustrate methods for determining the accuracy of screen printing machines.

13.10.7 Pre-forms of Solder

A pre-form is an appropriately shaped layer of solder (or of solder paste), which contains the amount of solder required to make the joint. It is placed between the parts to be joined and then melted. Pre-forms provide solder in a carefully controlled shape as well as a controlled volume of solder alloy. Typically, pre-forms are punched from a strip of solder alloy. Some pre-forms have flux as an integral part. In the case of flux containing pre-forms, it allows complete control over the placement of the solder alloy, which ensures that the joint is formed in areas specified by the designer and nowhere else. Usually, pre-forms may be applied when the circuit assembly is not plane or because of other reason that does not allow the application of the printing techniques.

13.10.8 No-clean Solder Paste

No-clean soldering processes are becoming the choice of many printed circuit board assemblers (Bauer, 1994). No-clean solder pastes are particularly important in the process. They can be divided into two main categories: Standard no-clean and low-residue no-clean.

Standard No-clean Paste: standard no-clean pastes are typically rosin-based, and have solid content of 35 to 50 per cent in the flux form and 3.5 to 5.0 per cent in the paste form. Most of these pastes do not require a special atmosphere such as nitrogen, because reflow and joint quality is typically good to excellent. Wetting may be an issue with no-clean pastes, but this is often traced to component or board solderability. Excessive residue quantities can interfere with testing.

Low-residue No-clean Paste: low-residue no-clean pastes can be rosin or synthetic-based. Typically, the ultra-low residue pastes use synthetic non-rosin-based ingredients. The main advantage of the low-residue pastes is the reduced residue levels. Low-residue solder pastes usually require an inert atmosphere such as nitrogen.

 ## 13.11 Adhesive for Mixed Technology Assembly

In many of today's circuit board designs, some surface mount components are attached to the bottom side of the board. Through-hole components as well as large surface mount components are attached to the top side of the board. When surface mount and through hole components are combined, the board forms a mixed technology assembly.

With mixed technology assemblies, bottom side surface mount components require adhesive application so that they do not fall off during the subsequent component placement and wave soldering operations.

13.11.1 Requirements of Adhesive

The adhesive should be selected to meet the following basic requirements:

- Hold the component in the given place during the cure process
- Maintain that orientation through the wave soldering operations
- Must have adequate adhesion to different surfaces
- Must not be affected by exposure to the environment of solder flux and wave soldering.
- It must be chemically inert throughout the life of assembly.

Epoxies and Acrylics are commonly used as adhesives for SMT assembly. Adhesives must be stored in a cool, dry and dark location.

Epoxies provide good insulation resistance, high bond strength and low curing temperature. A disadvantage of epoxies is that defective components are difficult to remove during repair. They are typically single-component heat-curing systems and are available in a range of formulations for special requirements.

Acrylic adhesives have fast curing time, high peak profile of dispensed dots and good temperature stability. They are usually cured by applying UV and IR energy. The main disadvantage of these adhesives are lower shear strength and the relatively high temperature required for heat curing.

Acrylic epoxy adhesives represent a considerable improvement on simple epoxy resin or acrylic based adhesives. Acrylated epoxies give a combination of the favourable properties of each type. Their viscosity/temperature stability is excellent enabling them to be stored at room temperature for up to one year. The bond strength is good without being so strong as to make rework difficult. Acrylated epoxy based adhesives can be rapidly cured by heat or UV, or a combination of both. Harris (1991) describes the various adhesive types, paying particular attention to acrylated epoxy based products.

13.11.2 Application of Adhesive

Adhesive can be applied by many different techniques. Most common methods are the syringe dispensing and the stencil printing method.

13.11.2.1 Syringe Dispensing

Adhesive is often dispensed by nozzles on the pick-and-place machine immediately prior to placing the component. For each component being glued, an appropriate adhesive dot is dispensed according to the space available and the size of the component. There are many different types of mechanisms used to force the adhesive through the nozzle and onto the PCB, but mainly all are air-driven. Each system has advantages and disadvantages. For example, one system may be easier to clean and another type may have better repeatability of dot size. Today's ultra high-speed dispensers have a dispensing rate of 100 000 dots per hour. The main advantage of syringe dispensing is the flexibility of operation.

13.11.2.2 Stencil Printing

In this method, a squeegee pushes the adhesive down into the apertures and ensures full aperture fill and contact to the board by stroking across the stencil. Proper alignment is a must to ensure the correct locations for the application of adhesive on the circuit board. The important parameters for stencil printing are homogeneity of speed, squeegee speed, pressure, snap-off distance and speed of board separation. The board must be kept flat during printing. The following points may be noted:

- Squeegee speed strongly depends on the viscosity. The lower the viscosity the higher the speed. For example, an adhesive with a low viscosity, "runs" at about 20 cm per second and a high viscosity adhesive is printed at about 1.3 cm per second.

- Pressure: Adhesives with higher viscosity need higher pressure than one with lower viscosity. A rule of thumb is to have sufficient pressure so that the stencil is wiped clean of adhesive with each printing stroke.

- The adhesive used in printing should be thixotropic, that means its viscosity should drop during the application process. This ensures that it will flow onto the board properly.

Stencils for adhesive printing are usually stainless steel. However, plastic stencils are becoming more common now. The advantage of plastic stencils is their flexibility that reduces the need for periodic cleaning of the stencil like stencils of stainless steel. However, plastic stencils are not as durable as the one of stainless steel.

Metal stencils tend to have a longer useful life than mesh screens, which can lose their resilience and shape after much use, compromising print accuracy. Stencils have the potential for up to 50,000 prints compared with a life of 5000 for the mesh versions. Stencils are virtually the only way to produce so-called fine pitch prints (defined as any thing below 20 mil) and have been proven in applications down to 12 mil.

The adhesive for stencil printing must be designed for exposure at room temperature and ambient humidity. Good adhesive printing results are seen with a hard polyurethane squeegee or with one out of metal.

While stencil printers can be made to deposit adhesives on PCBs, the quality of the depositions is often inconsistent. This is because the reheology of solder paste for stencil printing is totally different from that of SMD epoxy. Solder has a slippery surface because of the properties of its lead and flux. Epoxy adhesive, on the other hand, is sticky and stringy.

The positional accuracy in a dispensing system is data-driven. Each point of deposit is measured from the datum position in the CAD system. Any rotational offset or expansion/contraction of the PCB pattern is compensated, point by point, using fiducial correction. Duck (1996) points out that stencil printing is not based on positional correction but, instead, on a "best-fit" algorithm. A stencil cannot be changed to meet the dimensional variations of the PCB material caused by fluctuations in board fabrication. To overcome this problem, dispensing systems compensate for fluctuation in board thickness and board warp by utilizing support pins and vacuum supports.

The rotary pump (also known as auger screw pump) is not accurate enough for precision dispensing on ever shrinking package designs. A linear pump, shown in Figure 13.23(c), which is a true positive displacement piston pump overcomes these limitations. The pump is not affected by fluid viscosity, simply pressure, needle size or fluid/pump temperature. The pump's servo drive mechanism allows programmable shot sizes and flow rates.

The metal step after a job is completed is to remove and properly dispose off all unconsumed materials. Next clean and remove adhesive from the squeegee, check the squeegee blades for nicks. Then clean immediately the stencil. The adhesive will eventually cure at room temperature, making it more difficult to remove later.

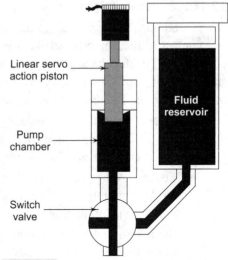

Fig. 13.23 (c) Positive displacement linear pump

13.12 Mass Soldering

The term "Mass Soldering" or "Automatic Soldering" is used to describe all methods for making simultaneously a number of solder joints 'en mass' without manual application of solder and/or heat to the surface to be joined. Thus, mass soldering methods speed up the manufacture of electronics assemblies. Besides, most automated soldering systems have provisions for adequate ventilation and thereby provide proper employee protection from solder fumes.

The driver for automatic soldering is not only the large throughput required for modern high-volume production but the reliability of a product having a huge number of very tiny solder joints, which may not be possible with hand soldering. For high-density interconnection (HDI) boards, assembling is done only by automatic means, as there will be smaller pad sizes and solder lands, thinner materials and finer circuitry features.

13.12.1 Dip Soldering

Dip soldering is the simplest method of mass soldering in which the fluxed board, moving in a horizontal position, is lowered vertically into a tank containing molten solder. The board is immersed in the solder bath to the required depth until the surfaces become wetted by the solder. The arrangement is shown in Figure 13.24. After maintaining the contact for the required dwell time, the board is withdrawn from the tank. In general, the temperature for mass soldering of PCBs is kept around 240-250 °C with the average contact period not exceeding 5 seconds. Usually, 2-3 seconds contact period is good enough for satisfactory soldering. Basic dip soldering process has been considerably modified for speedier and automated operation.

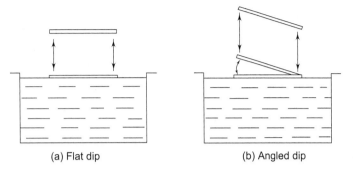

(a) Flat dip   (b) Angled dip

Fig. 13.24   Dip soldering principle (a) flat dip (b) angled dip

13.12.2   Drag Soldering

The earliest form of automated soldering is "dip and drag soldering" in which the assembled and fluxed board in a horizontal position is lowered onto a bath of molten solder and drawn along the surface. The board is dragged along the surface of the solder in a stationary bath for a predetermined distance and then lifted from the bath. Usually, the boards are lowered into the bath at an angle of 15° and as the board moves along gradually, this angle is reduced to zero. Again at the time of withdrawal of the board, the angle is increased from zero to 15°. In this way, icicle formation is avoided. Mostly, the boards were mounted on carriers having a blade at their front edge in order to remove dross from the top of the solder bath.

Even though drag soldering can produce quality joints, but to solder large modern boards, drag soldering is not a method of choice because of the following disadvantages:

- Longer contact time between PCB and solder, thus increasing the heat of the base material and components.
- Larger area of contact does not allow the gases generated to escape. So, the number of blow hole defects are generally large.
- Dross formation starts very fast on the surface of molten solder.

The dip and drag soldering method has been mostly discarded these days and a technique called wave soldering has been mostly employed.

13.12.3   Wave Soldering

The standard method for mass soldering of leaded through hole components onto PCBs is by wave soldering in which the loaded boards pass over a wave of solder. The equipment and the processes are well developed and can be readily utilized for wave soldering of surface mount assemblies. This soldering method has a significant advantage in that it can be used successfully for assembly of 'mixed technology' boards containing both insertion mounted and surface mounted components (Buckley, 1990).

The basic elements of a wave soldering unit are the conveyor system that transports the assembly or populated board either continuously or in a stepwise fashion from a loading position to, in turn, a fluxing station, preheating stage, the solder wave and a cooling station before removal from the conveyor as the unloading station.

After fluxing, the conveyor moves the assembly into the preheating zone, where it is slowly heated to a temperature close below the T_G-Value (glass-transition-temperature of the base-material). Up to that temperature no separation takes place between resin and copper and resin and reinforcement. The pre-heating zone is followed by the solder bath where the molten solder is delivered to the underside of the assembly in the form of a wave. Wave soldering is carried out at a temperature in the range 235-260 °C with a contact time between 1 and 4 seconds. The components which move through the molten solder must be able to withstand this treatment and remain unaffected either by the high temperature or the temperature gradients involved. Also, the solderable surface on the components must not unduly dissolve into the molten solder. Within the cooling section, the assembly cools down and the solder solidifies, providing the desired mechanical and electrical connection.

Wave soldering machines require careful adjustment of flux application, of setting the right pre-heating temperature, solder-wave height, conveyer speed, solder temperature and smooth transportation of assembled boards especially during the cooling down process.

The companies that employ low residue fluxes make use of the soldering machine equipped with a nitrogen blanket. This blanket reduces the oxygen concentration to a very low level. Without oxygen, there will be minimal oxidation or corrosion. Thus, these machines eliminate or at least reduce subsequent flux removal.

In these machines, gas jets in the soldering zone blow in the nitrogen gas that circulates through the tunnel. This displaces the oxygen from the process chamber resulting in high soldering quality, avoiding dross formation and having better heat penetration, because of the higher convection coefficient of nitrogen versus air. Also, it provides homogeneous temperatures throughout the entire working width. The equipment contains a process gas cleaning facility with low nitrogen consumption. The channel is made transparent for assembly observation throughout the whole process. Gothard (1991) explains the developments in inert atmosphere wave soldering including the retrofit kits for use on the existing machines.

13.12.3.1 Fluxing the Board

Fluxing in wave soldering equipment is the process of applying flux to the underside of the assembly in order to remove oxides from the surfaces of the parts to be soldered as well as to protect these surfaces from further oxidation during the pre-heating zone.

As the flux contacts the exposed metals on the heated assembly, it chemically removes the oxides and contaminants, allowing them to be carried away by the molten solder. There is an optimum range for the amount of flux that is retained on the board to ensure satisfactory soldering. This depends upon the:

- Method of application of flux;

- The quantity of liquid applied; and
- The fraction of solvent to flux solute, and hence its viscosity and its evaporation characteristics between application and soldering.

The flux is always applied in a liquid form in order to cover all the solder areas quickly and evenly.

There are different methods of applying flux to the board. Among them, foam fluxing, wave fluxing or spray fluxing are mostly used. The application method should produce a continuous film of flux on the underside of the board. It can only then facilitate capillary rise up into the plated holes. The wet flux usually forms a layer of 5-20 µm thickness. During soldering, this layer helps to remove the oxide film, thus reducing, the occurrence of excess solder drag-out from the wave to form solder bridges.

Foam Fluxing: In this method, the liquid flux is applied from a large tank to the board by means of an aerator to produce a turbulent bubbling surface through which the underside of the populated board passes. The arrangement is shown in Figure 13.25. Compressed clean and dry air, which bubbles through a porous stone (or tube) that is submerged into a flux reservoir, forms a head of foam at the top of a wide chimney that ends just under the conveyor. When the low pressure air is blown through the pores of the tube, it generates fine bubbles, which are guided to the surface by baffle plates. The bursting of the bubbles at the surface assists in the coating

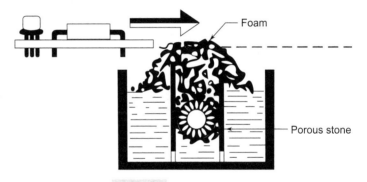

Fig. 13.25   Foam fluxing method

of the walls of the through-board holes. The height of this foam head above the chimney is limited. If greater height is required because of components with long leads to go through the foam fluxing unit, brushes can be added to create a supported foam head. The viscosity of the flux is an important parameter for controlling the height of the foam. If the viscosity is too high, the bubbles will not burst correctly. In that case, the foam may rise uncontrollably high and overflow. If viscosity is too low, there might be difficulty in achieving any foaming action. The advantage of this method is that it is very quick and the amount of flux applied is independent of the conveyor speed.

Wave Fluxing: The application of flux, in this method, consists in passing the board across the crest of a standing wave of the flux. The arrangement is shown in Figure 13.26. A pump forces flux out through a wide chimney where the liquid spills over the top in order to produce a wave, over which the board assembly is passed. Wave fluxing is one of the simplest systems to operate and maintain. However, this method often results in the application of more flux than it is required due to excessive hydrostatic pressure. Thus, it is critical factor to control the impeller of the pump.

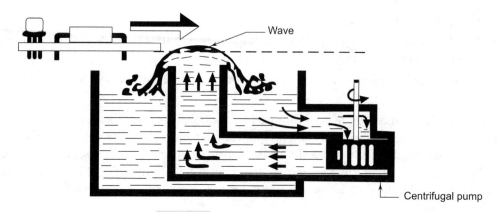

Fig. 13.26 Wave fluxing arrangement

Many wave fluxing units employ an air knife immediately following the flux application in order to remove excess of flux. The air is directed at a slight angle backward with a carefully adjusted low flow that does not drive off all the flux. The air knife helps to spread the flux and pushes it up into the holes. If excess flux is not removed, there is a possibility of dripping of excess flux on the preheater where it may cause fire.

Spray Fluxing: Spray fluxing depends upon the production of a directional spray of liquid flux on to the underside of the populated printed circuit board. The systems employ either reciprocating spray nozzles or fixed spray nozzle technology. Both systems are controlled by a computer programmed to deliver flux fairly accurately to the width and length of the board. In either application, the computer will sense the board speed using fiber optics or proximity switches and from this information calculate the coverage required. One type of spray fluxing arrangement is shown in Figure 13.27. It consists

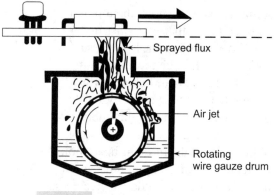

Fig. 13.27 Spray fluxing arrangement

of a drum of closely spaced radial spring leaves which are loaded with flux by rotating through the reservoir. As the drum rotates, the flux is fluxed off each leaf in turn to fall on the underneath of the circuit board.

The amount of flux applied to the board depends on the amount of solid flux dissolved in the solvent career. This parameter is usually monitored and maintenance through the liquid density. It can be either done automatically or with the help of a hydrometer.

Fluxing process is generally accompanied by contamination which depends on the method of flux application being used. The level of contamination needs to be checked regularly when using foam fluxing and wave fluxing application, since the flux will continuously remove some contaminants while fluxing the board. So, the fluxing unit should be cleaned and refilled regularly, depending on the type of flux used and the area already coated. Spray fluxing, by nature, will not have flux contamination problem.

The need to clean circuit boards after wave soldering is very necessary because the rosin fluxes demand it. Synthetic fluxes having no rosin content eliminate the problems associated with rosin residues and offer a true no-clean solution. Taylor (1991a) describes the development of complex synthetic fluxes which overcome most of the problems associated with the rosin based fluxes and details out the wave soldering process control needed to ensure high joint quality when using synthetic fluxes (Taylor, 1991b).

A preheating stage is always incorporated between the fluxing stage and the wave soldering of the board. There are several purpose for which one needs a preheating zone: The important reasons are:

- *Activating the Flux:* Preheating supports fluxes to rise up through the holes in the assembly and by doing so, it assists top side hole fill on the assembly.
- *Shorter Soldering time* is made possible by raising the temperature of the assembly closer to soldering temperature. If there was no pre-heating, all of the required heat would have to come from the solder wave and would need a longer dwell time.
- *Thermal shock* could occur by raising the temperature of the assembly very quickly from room temperature to wave soldering temperature. This extreme temperature impulse may damage some of the heat-sensitive components. It may also cause bow and twist of the board.
- *Evaporate the flux solvents:* otherwise the remaining solvent could cause blow holes by entering the wave.
- *Evaporate moisture* from the board, which also could cause blow holes.

Pre-heating is an important part of the "thermal profile". It is achieved either convection of circulating hot air or by radiation from infra-red lamps or hot plate panels or a combination of both. The radiant heat is applied from above in addition to the underneath or solder side of the board.

The preheating stage of a wave soldering machine raises the temperature of the board in the range of 80-120 °C. The most common flux carrier is isopropyl alcohol whose boiling point is 82.4 °C. Therefore, the evaporation and ultimately volatilization of the flux carrier is very rapid during the period the board is pre-heated.

13.12.3.2 Solder Zone or Solder Wave

The solder wave has two basic functions: (i) it carries and transfers heat to the component leads, lands and plated through holes and (ii) delivers the solder that makes the mechanical and electrical joint.

To carry out these functions, a continuously replenished wave of molten solder is generated by pumping upwards from a sump. The assembled printed circuit board is made to traverse across the crest of the wave. The arrangement is shown in Figure 13.28.

The arrangement uses an electronically controlled pump motor in order to avoid running of the pump during the process of solidification or switching on the pump if the solder is not molten yet. Some solder pots have only one nozzle creating one wave while some are equipped with two nozzles thereby raising two waves.

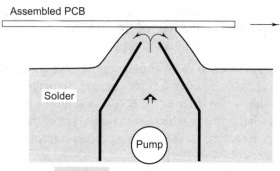

Fig. 13.28 Wave soldering machine

Each solder bath should be equipped with facilities of controlled heat setting. For one wave, the temperature is set between 250 °C to 257 °C and for baths with two waves, the bath-temperature is maintained between 245 °C to 255 °C.

On single-sided boards, the inserted component leads are soldered to lands on the board underside. In case of double-sided plated-through hole and multilayer boards, the solder rises through the hole around the component lead by capillary action and hydrostatic pressure. It thus fills the hole and flow over the solderable land on the top surface of the board. The leads of the component are slightly crimped after insertion in order to avoid their dislodgement due to buoyancy effect which may make the components to float up from the board as they pass over the board.

Various wave soldering shapes have been used by different equipment manufacturers. In the simplest solder-wave arrangement, the wave falls back to the sump on either side of the nozzle. An improvement over this system is shown in Figure 13.29 which is provided with extension plates on either side in order to define better the wave profile. This arrangement helps in drawing the excess solder thereby reducing the possibility of bridge formation. Even excellent wave soldering shapes have problems with shadows, which often occur when surface mount components are on the underside of the board, because the component body prevents the solder from reaching the parts at the SMD on the rear side.

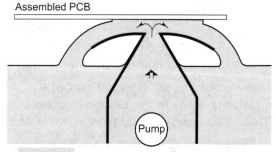

Fig. 13.29 Use of extension plates to control the shape and contact time of the wave

To overcome this problem, *dual wave* soldering machines have been developed. A schematic diagram showing the principle of the machine is shown in Figure 13.30. The machine combines a first wave that is turbulent and a second wave that is smooth. The turbulent wave is generated by a jet mechanism, which enables the molten solder to drive between the components and

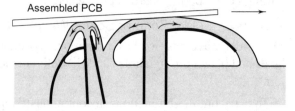

Fig. 13.30 Schematic of a dual wave soldering machine— turbulent first wave and a smooth second wave

achieve full wetting. The second wave is smooth which controls the meniscus of the molten solder at each joint. The board leaves the smooth or calmer wave at near zero relative velocity. Dual wave machines specifically meet the requirements of surface mounting and mixed technology boards.

Some systems may have a hot *air knife* at the exit of the wave in order to blow off any excess solder to find its way back to the solder pot. Immediately after the solder wave, while the solder is still molten, an air stream forcibly removes solder where wetting has not been achieved. Excess solder can cause bridging, especially where joints are closely spaced. Furthermore, non-wetted areas are easy to see, reducing inspection times. However, the angle of impact of the air, air temperature and air pressure are important parameters that need to be properly adjusted for effective action of the air knife.

The use of oil as an additive to the molten solder helps to eliminate solder bridging and heavy solder build-up by controlling the formation of an oxide skin on the surface of the wave, especially in the dual wave machines. The oil layer on the surface of the solder in the sump also inhibit dross formation. The oil used in wave soldering is a mixture of mineral oils and fatty acids known as tinning oil. The oil should be virtually chemically inert at 250 °C and should suffer little decomposition at that temperature.

Dross must be removed periodically from the solder well. Otherwise, dross can be emulsified with the solder and reach the board being soldered to cause bridging, grainy joints etc. Stopping the pump when the wave is not needed for some time and using the minimum possible wave height can also reduce dross formation on wave soldering machines.

13.12.3.3 Conveyor System

The conveyor carries the assembled boards from the loading to the unloading position, over all the other units. It controls the speed at which the boards are passed through the process. Basically, it controls the preheat time, temperature and dwell time in the solder wave. Conveyors are generally of two types:

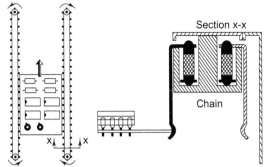

Pallet type: Pallet conveyors have two parallel rails on which pallets are driven. The conveyor can be either horizontal or inclined, with an inclination angle adjustable from 4 to 8°. Pallets consists of rectangular frame with a device for holding boards on a fixed position. They can also be adjusted to accept boards of different dimensions.

Finger type: Here the boards are held in spring fingers. These can be adjusted to accommodate the width of the board. They are more convenient to load and unload. These conveyors also have two parallel rails, which carry a chain with suitably sharpened stainless steel or titanium projections which hold the board on their edge. Figure 13.31 shows a simplified diagram of a finger conveyor.

Fig. 13.31 Simplified diagram of a finger conveyor — each of the two rails contains a closed loop chain which rotates around two sprocket wheels (after Leonida, 1989)

The critical adjustments for the conveyor system are the width and parallelism of the rails/fingers; the speed of the conveyor, the angle of the rails to the solder bath, smooth and vibration free drive, especially while solidification and loading. Segmented conveyors are also available which have independent speed control through the three main stations of fluxing, pre-heating and solder-bath. Some manufacturers of solder wave equipment provide even PTFE (Teflon) coated titanium fingers for the parts having contact with molten solder.

The conveyor angle adjustment for passing the solder wave can be done manually or motorized. The time (t) for which a point on the solder side of a board being soldered stays in contact with molten solder should be more than 2 seconds because this time will be needed for creating an intermolecular bond having at least 0.5 µm thickness. However, the time more than 2 seconds should be such as to keep the thickness of the intermolecular bond at less than 1µm.

Contact Length(l):   can be measured by observing a borosil-glass plate riding on the conveyor over the solder wave. The transparent plate of borosil-glass has a built-in scale, so we are able to recognize contact length(l) while the glass plate is passing the wave instead of the board.

Conveyor Speed(v):   can be checked by measuring the distance moved by a board within a defined time and compared with the setting speed at the potentiometer. Always measure the distance moved in one minute, so you can read directly the conveyor speed in meters per minute.

$\qquad$ T = contact time (measured in seconds)

$\qquad$ L = contact length (measured in mm)

$\qquad$ V = conveyor speed (measured in meters per minute)

$\quad$ T (sec) = L (mm)/V (mm/s)

$\qquad$ = L (mm) /V (1000 mm × 60/1000 × 60s)

$\qquad$ = 0.06 × L (mm) /V (m/min)

13.12.4   Reflow Soldering

Reflow soldering (Filleul, 1990) is a process to make a soldered joint by means of re-melting a previously applied solder deposit without the addition of any more solder during the soldering process. The solder deposit is usually in the form of solder paste, which is applied to the printed circuit board. After attaching the components, the reflow process is performed by the application of heat.

The two most common methods for reflow soldering are infra-red and forced convection. Vapour phase is also used, but is not as common. A combination of convection and infra-red is preferred to accomplish the soldering process. About 60 per cent of the heat transfer comes from convection through hot gas and the remaining 40 per cent of the transfer comes from infra-red radiation or heat panels. The heating that takes place in a reflow soldering oven has to be precisely adjustable. If an assembly is heated too quickly or at too high a temperature, there could be damage to the board or some of the electronic components. On the other hand, if an assembly is not heated enough, proper soldering will not take place. The measurement of temperature in each station of the whole reflow soldering line versus the time is called thermal profile. Each oven as well as each assembly has a unique and specific thermal profile, which is shown as a graph of temperature versus the time that the assembly takes to pass through each stage. Generally, a temperature profile has five phases, taking eutectic point of solder as 179 °C.

- Pre-heating zone
  - Phase 1 heats up the assembly slowly from environmental temperature to approximately 80 °C with not more than 2 °C per second.
  - Phase 2 raises the temperature to the T_G-value of the laminate (135 °C to 145 °C for FR4) with about 3° to 4 °C per second
  - Phase 3 increases the amount of heat slowly with less than 0.5 °C per second and takes upto 155 °C.
- Reflow zone
  - Uniformly raising the temperature to 215 °C to 235 °C.
- Cooling Down
  - Quite fast, however limited to less than 5° per second.

Pre-heating Zone: This zone transfers a low amount of heat to allow for a slow, uniform and gradual temperature rise up to nearly 155 °C. The "pre-heating zone" performs the following functions:

- Activates the flux
- Prevents the board and the components from getting a thermal shock by providing all the required heat at one time. This thermal shock may damage the board and some of the components.
- Vapourizes moisture and volatiles on the board otherwise they may explode out through the solder and cause blow holes.

Reflow Zone: In this zone, reflow of the solder and wetting action take place. The method of heat transfer, the temperature in the reflow area and the speed of the conveyor are important inputs for this zone in order to provide an intermolecular bond.

Several methods are available to measure the temperature rise at given locations on the board. Accordingly, providing the right temperature at the place of bonding is an important issue.

Cooling Zone: The cooling down process solidifies the solder and the assembly slowly approaches to room temperature and the components are finally electrically and mechanically bonded to the board. The cool down phase is properly controlled so that the temperature change is not too quick.

13.12.4.1 Convection System

In convection system, circulating hot gases are blown onto the assembled board to reflow the solder. The inert gas (nitrogen) is superheated by passing over electric heater coils. They are then directed through a nozzle/nozzles to the location where reflow solder is to be carried out. Gas temperatures in the region of 150-170 °C and nozzle diameter less than 2.5 mm are used to control the gas flow. The important parameters which are critical and should be controlled for a good quality process are flow time and gas temperature.

In free convection, the layer of highly energized gas particles just below the heat source transitions into a buffer layer. Beneath that is a laminar or boundary layer. This sub-layer can cause heat transfer to slow down due to less surface contact between the gas and solid and the insulating properties of certain gases.

The transfer rate is accelerated by using an uneven surface conveyor (Figure 13.32); increasing the source temperature or resorting to forced convection. Altering the source temperature may cause some assemblies to burn and others to reflow inadequately. Forced convection, on the other hand, enables a faster heat transfer at a lower temperature heat source, regardless of conveyor condition.

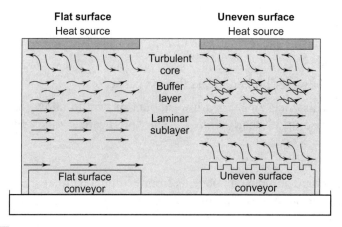

Fig. 13.32 An uneven surface conveyor promotes greater particle flow and thermal energy exchange in a convection system. (after Dytrych, 1993)

13.12.4.2 Radiation System

In radiation systems, rays of radiant energy are used for heating any surface that absorbs them. The absorption coefficient depends on the wavelength of the radiation as well as on the material, colour and surface properties. Several types of infra-red emission sources are available. The commonly used sources are the tungsten tube, the nichrome tube and the panel secondary emitter types. Each type of source radiates in a different portion of the infra-red spectrum. Consequently, they have differing heating effects on the materials of which the furnace is constructed and on the gas that forms the environment. The heating mechanism consists of an integrated effect of direct radiation from the source, radiation from the furnace walls, conduction and convection from the gas.

Furnaces designed for surface mount assembly are generally constructed with alumina/silica back-up insulation around a fire brick inner shell. The whole structure is encased in the outer steel case. The conveyor belt rides on quartz rods as it moves through the tunnel. The conveyor speed must be accurate which is achieved by using a feedback loop controlling the motor. The degree of control of temperature is fairly tight, limited typically to + 3 °C including hysteresis. Figure 13.33 is the schematic diagram of the furnace for radiant heating. Nichrome tube radiator gives a lower emission temperature than the tungsten tube.

The basic tube-emitter infra-red furnace comprises two temperature zones. The first zone runs at about 1200 °C with peak emission at a wavelength of 2 µm. Here, the laminate and the solder is pre-heated, uniformly. The second reflow zone is set around 2100 °C with a peak wavelength 1.2 µm. This shorter wavelength energy is absorbed by the solder but transmitted by the laminate. Thus, the solder temperature rises while maintaining the laminate temperature at a lower level. However, the furnace type which has proved most successful has four heating zones. Radiant heating reflow systems generally have three primary control functions. These are: the radiant power, the exposure time and the temperature of the workplace. Figure 13.34 shows the temperature profile of solder and epoxy-glass fibre laminate as they pass through the four zone radiant emission furnace. If there is a radiation only system, there may be areas which get shadowed, i.e. the areas not in a straight unobstructed line from the energy emitting source will not receive direct heating.

Reithinger (1991) states that the board and components get heated unevenly by the IR reflow process. Therefore, in

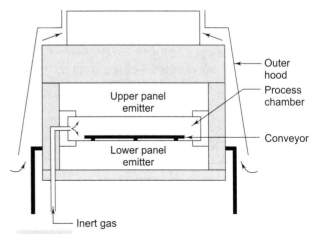

Fig. 13.33 End view of area source furnace tunnel especially suitable for soldering surface mount assemblies.

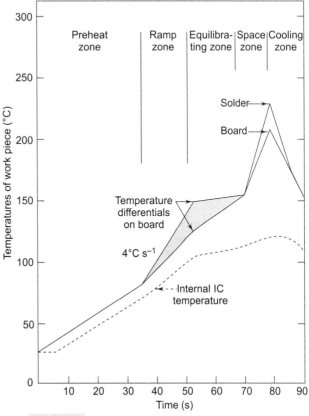

Fig. 13.34 Typical temperature profile of a work piece passing through a four-zone radiant emission furnace

addition to choosing the correct type of emitter, it is also necessary to establish correct temperature-profiles. Trials have shown that a component's heat-up rate depends mainly upon its mass. Figure 13.35 illustrates the maximum temperatures reached by components with different masses, all IR soldering system parameters being constant.

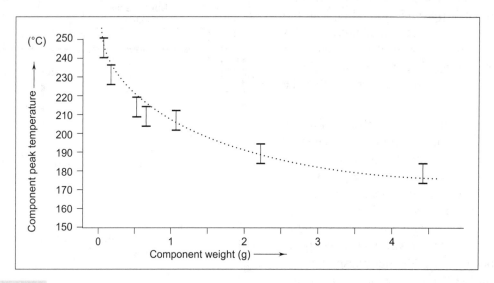

Fig. 13.35   Component peak temperature vs component mass during 1R processing (after Reithinger, 1991)

Laser soldering is a special method of radiation heating. These systems use YAG or CO_2 lasers to heat each solder joint. They have the major advantage of heating only the component leads/pads and do not subject the component to high temperatures. However, because they process — under programmed control only one joint at a time, they are considerably slower than mass soldering systems.

Radiation systems in combination with forced convection are the most common method for soldering SMD-boards. There are numerous versions of infra-red soldering equipment, which differ widely in price and performance. The most important features in a radiation system are the wavelength, the radiators used, use of nitrogen and percentage of convection contribution.

Although most of the heat is provided by convection, a small percentage comes mostly from infra-red radiation. This combination provides uniform (convection) and quick heating (radiation) to the surfaces. In convection, there is a direct contact with a heating media, usually air or nitrogen. The gas is blown over and around the assemblies through many holes above and below the conveyor in such a way that the assembly is gradually and uniformly heated. Both the maximum temperature and the rate at which the temperature is reached can be controlled. Component leads, terminals and board lands will reach soldering temperature at the same time. However, the convection system as such alone has also disadvantages like:

- It needs longer time to heat up the assembly since the gas can not be forced too strong, otherwise it will misplace the components.

- Especially the blown forced hot air may sometimes dry out the surface of the solder paste and form a skin of resin.

Machines vary in size from small benchtop models to large floor standing versions which may incorporate computer control of the reflow profile, and even the conveyor width setting, the former being displayed on aVDU (video display unit) and stored in memory to be called up for future processing of a particular PCB assembly.

13.12.5   Vapour Phase System

In vapour phase soldering, an inert liquid is heated to its boiling point to create a saturated vapour zone above the liquid. Thermal energy is transferred from the heater and absorbed from the heat that is generated through condensation (Dytrych, 1993a). In this process, the entire PC board is uniformly heated until a defined temperature is reached, with no possibility of over heating. The defined temperature i.e. 215 °C in a saturated vapour zone is obtained by heating an inert (neutral) fluid to the boiling point.

The vapour system is a condensation reflow soldering process and relies on condensing vapours of a high boiling liquid. The heat comes from a vapour provided by a liquid, which is heated in a tank and boils at a specific temperature. The assembled board, which rides on a conveyor through the vapour chamber, is entirely immersed either in the vapour or direct in the liquid. In case the assembly is passing above the liquid media, so the vapour condenses on the assembly by transferring it's heat to the assembly and the solder paste. Cooling coils are provided on the upper area as well as at the input-zone and output-zone.

The vapour system has a fixed upper temperature determined by the boiling point of the fluid used. It is the most uniform method of heat application as the vapour/liquid surrounds all components, heating from all sides and providing the targeted temperature depending on solder paste composition. For example for Sn62/Pb36/Ag2, it may be 215 °C. Also, it is a rapid method of heating and quite tolerant of large mass differentials on a given assembly. The heat-transfer liquids are inert and prevent oxidation of the soldering surfaces. The vapour phase systems however, have the following disadvantages:

- There is an immediate temperature rise from pre-heat temperature to solder temperature that may damage some components.
- High capital cost of equipment.
- High operating cost as some liquid may still remain with the assembly.
- Process may be subject to environmental restrictions, since halide-activated hydrocarbons are mostly used as heat-transfer medium. However, the known media in use are non toxic and stable up to a temperature of 25 °C.
- Needs strong cleaning process because the heat-transfer media acts also as solvent for the flux contained in the solder paste and consequently will be enriched with flux residues.
- Some residues may still remain under the components.

The chief drawback to vapour phase pertains to its sharp rise time during heat transfer. Initial heat transfer rates can be as high as 108 °C to 208 °C per second. If preheat is inadequate, it causes components to crack due to mismatches in coefficients of thermal expansion. Another potential problem is solder wicking resulting from thermal mass differentials and makes heating faster than the printed circuit boards. These disadvantages has now been overcome with a availability of online control of various heating zones required for satisfactory soldering. Linman (1990) explains that by adding an infra-red preheat to a vapour phase reflow soldering process provides gradual heating prescribed by the component manufacturers, thereby reducing the temperature differentials at the solder joints and reducing the tendency for wicking. Siemens (1991) provide a comparison of radiation and vapour phase systems for mass soldering

13.12.5.1   Hot Bar Reflow System

The IR reflow system suffers from the so-called "pop corn" effect where packages crack due to the expansion of absorbed moisture as the devices are heated beyond 160 °C. Also, with fine pitch surface mount components with finer pad dimensions and spacing, lead coplanarity and skew may give rise to quality problems and the solder of the plastic leaded chip carrier using IR reflow systems cause several manufacturing problems. All these problems can be avoided using hot bar reflow technique (Smith, 1991).

Hot bar systems consists of a pick and place machine fitted with component specific heating bars which, after the device has been placed, press the leads onto pre-tinned pads (no solder paste is required) and apply a programmed heating cooling cycle. Flux applicators can be incorporated, and processing times can be as low as 10 seconds per component.

The technique uses previously plated or reflowed solder rather than solder paste since the device is held down by a slight pressure during the reflow cycle, the coplanarity concerns inherent with mass reflow are eliminated.

The component can be either placed by hand before bringing down the reflow head-a technique often used for re-work and repair—or the component can be handled automatically using appropriately hot bar soldering machine which incorporates a pick-and-place mechanism. Using the latter, a pick/place/solder cycle rate of 150-200 components per hour is possible.

Present developments indicate that the combination of convection and radiation heating by the use of an inert gas like nitrogen will be the system of choice in the future. Radiation to add the amount of heat quickly and convection to provide uniformity of heating to the surface. The adaptation of nitrogen helps to reduce the amount of oxide formation.

13.12.5.2   Pulsed-heated Reflow System

With the growth of miniaturization, the use of flexible interconnection has become a requirement for many new designs. The examples include flex circuit to PCB, flex-to-flex and multiwire to PCB. These combine modular assemblies with a variety of special purpose devices, such as LCD displays, microphones, speakers, etc. For such types of applications, pulsed reflow systems are now being increasingly used whereas conventional soldering devices maintain a constant preset temperature at the tip (which meets the parts to be joined at temperatures as high as 425 °C), with pulse-heated soldering, the contact is at ambient, after which the parts are briefly heated (typically to 350 °C), then cooled. Pulse-heating thus is performed with minimum risk of heat damage to the parts being joined.

Pulse-heated reflow system makes use of a special soldering tool—the thermode—which presses the parts together, adds heat, melts the solder and then holds the position as the solder cools. Two factors namely pressure which ensures intimate contact of all solder surfaces and heat, produced by electrical current through a resistive element are important for working of the thermode. The details of the system are provided by Boyd (1998).

 ## 13.13 Post-soldering Cleaning

Post-soldering cleaning is an important step in assembling of PCBs. It affects the ultimate reliability of the board and is carried out for the purpose of removing all contaminants such as:

- Flux residues and their derived compounds;
- Plating residues used to manufacture the bare boards; and
- Residues such as dust, oil, grease, etc., which get attached onto the board during handling and storing.

The cleaning of assembled boards is essential to:

- Reduce the corrosion.
- Prevent any reduction in electrical insulation between adjacent conductors;
- Eliminate poor electrical contact on plated or printed connectors/contact tabs;
- Reduce the amount of dirt that boards collect from the atmosphere;
- Remove substances that may be attacked by fungi; and
- Enhance the cosmetic appearance of boards.

13.13.1 Types of Contamination

There are two categories of contaminants:

Polar (ionic) Contaminants that can carry a current in the presence of moisture and it can enter into chemical reactions associated with corrosion. Ionic contamination typically results from plating residues, flux activators and salts associated with finger prints.

Non-polar (non-ionic) Contaminants typically consist of rosin from solder fluxes, creams, oils, dust and grease associated with handling operations and storing. It can form insulating films on contact surfaces.

13.13.2   Solvents and Cleaning Methods

Solvents used in the final cleaning process must be capable of dissolving both kinds of contaminants and should be compatible with the assembly to be cleaned.

Brush cleaning process, which is used for conventional leaded-through hole assemblies are not recommended for surface mount assemblies as flux residues might be swept beneath the component and even remain there. This effect is less likely to occur with spray cleaning, as the spraying action washes away the flux residues together with the solvent.

The ultrasonic cleaning methods can also be used for cleaning operation. The ultrasonic energy is used to separate flux and other residues adhering to the surface. The frequency used for ultrasonic cleaning process is generally greater than 40 kHz. However, intensive ultrasonic cleaning, particularly at resonant frequencies may damage some components.

The solvent for reflow soldering is different from hand soldering or wave soldering. The ideal solvent for mass soldering should have the following features:

- Capability to remove polar and non-polar residues
- Non-flammable
- No tendency to react chemically
- Low surface tension
- Non-affecting nature for the board
- Low toxicity
- Low cost
- Environment-friendly

There are several types of solvents suitable for mass-applications available in the market. However, CFC(chloro-fluoro-carbons)-based products used in the cleaning process are found to deplete the ozone layer, allowing harmful radiations to reach the earth which may lead to skin cancer and other diseases. Some solvents used are:

Chlorinated Hydrocarbons like trichloroethylene (C_2HCl_3), tetra-chlorethylene (C_2Cl_4) or 1, 1, 1,-trichlorethane ($C_2H_3Cl_3$) having a high power to dissolve non-polar contaminants. However, toxicity is also high and their power to dissolve polar contaminants is very low.

Fluorocarbons like Freon 113 ($C_2F_3Cl_3$) are very effective for removing non-polar contaminants. Their low surface tension promotes their penetration into very small crevices. Fluorocarbons are ozone-killer as well and their solvency for ionic residues is low.

Alcohols like ethyl alcohol [C_2H_5OH] or isopropyl alcohol ($2(CH_3)CH–OH$) are often used for removal of resin based fluxes after hand soldering. They are relatively cheap and have a low toxicity. They dissolve polar contaminants as well. The disadvantage is that it is highly flammable and has a high vapour pressure. Isopropyl alcohol is much more toxic than ethyl alcohol.

Water plus detergent Non-CFC (chloro-fluoro-carbons)-based methods of cleaning like aqueous cleaning method are gaining popularity. Since all the CFCs, VOCs (volatile organic compounds) and ODS (ozone depleting substances) are proposed to be phased out by the year 2005, as per the Montreal Protocol, the electronics manufacturing industry is increasingly turning to other cleaning alternatives which are CFC-free and also environment-friendly.

There are five basic non-CFC cleaning processes available to the industry. These are: semi-aqueous, pure-aqueous, solvent-solvent, perfluorinated rinse, and water-miscible.

Aqueous cleaning (Andrus, 1990) is a viable alternative to CFC-based cleaning. It is an environmentally safe, cost-effective and efficient method of cleaning the PCB assemblies. Water is an excellent means to remove polar contamination. De-ionized water (DI-water) removes all contamination from the board. Since ionic polar contamination is water soluble, tap and soft water also remove ions but at less efficiency than Dl-water. The non-ionic residues have to be removed by the detergent, a saponified bath or so-called "rosin soap" (which is one to 10 per cent alkaline concentration) diluted in water for the purpose of cleaning. Undoubtedly, water is a better solvent for flux activators than any organic solvent currently available as it obviates the legal and environmental requirements inherent in solvent cleaning.

However, non-polar residues like rosin will not dissolve in water. The detergent used to react with rosin (called "saponification") depends strictly on time and temperature. If the temperature is too low or time too short, the reaction may not reach completion stage and the dirty board may not get cleaned. The rosin soap as well as the detergent must be thoroughly rinsed away. A final rinse in de-ionized water is essential. High pressure air blowers have to be used after final rinse for quick drying of the board. It is necessary to use saponifiers with de-foaming agents, because the alkaline solutions tend to create lot of foam.

13.14   Quality Control of Solder Joints

Solder joints need to be inspected and checked before being accepted as good and reliable. In machine soldering, it is extremely important to continuously monitor the soldering process in order to provide the feedback necessary for maintaining the solder line under best operating conditions. The present day trend is to employ computer assisted inspection techniques.

The very look of a solder joint would normally provides a good indication of a proper job done. If a surface is not adequately wetted by the solder, a good joint cannot be achieved. A proper solder joint can be achieved by:

- Using the right temperature, approximately 30 °C to 35 °C above the melting point of the solder as well as the metal being joined.
- Clean and deoxidized metal surface.
- Using the proper and non-contaminated solder
- Use of proper flux in order to remove oxides and prevent new oxidation while soldering.
- Keep the contact time between the base metal and the molten solder as short as possible.

In addition to the optical inspection as well as monitoring the process parameters during the soldering process, the complete evaluation of a solder joint involves many other tests and procedures. Some of the commonly used tests are:

Mechanical Tests:

- Pull Test
- Vibration Test
- Micro-section: measuring the intermolecular bond thickness which should be between 0.5 mm and 1 μm.

The structure of the intermolecular bond is made by crystals which tend to grow under the influence of higher temperature and as a function of time. Absence of crystals does not provide sufficient physical strength, where as larger crystals reduce the bond strength. Therefore, it is necessary that proper intermolecular bond is established.

Electrical Functions:

- *Resistance Testing:* It is often not very helpful because even a bad joint may sometimes show a very low resistance. The difference sometimes will not be discernible.
- *Joule Test:* It is carried out by providing a constant current to solder joints in series. They heat up to a different extent according to their different resistances.
- *X-Ray Inspection:* This is quite often integrated in the equipment line, like automated optical inspection (AOI), especially for BGAs.

13.14.1 Good Quality Solder Joints

A good quality solder joint is a shining and smooth surface with an intermolecular bond of less than 1 μm and more than 0.5 μm. The shape of the solder joint should be concave formed by a small wetting angle. The angle should be less than 90°. However, it is preferable to have it less than 40°. Solder surface should be smooth and finely grained and should not show blow holes, voids, inclusions or cracks. The placement of the SMD components should be accurate so that more than 75 per cent of the component termination cover the land or solder pad.

13.14.2 Common Soldering Faults

The standard "IPC-A-610-C, Acceptability of Electronic Assemblies" provides the basis for solder joints that are acceptable or otherwise. The common soldering faults are detailed below.

13.14.2.1 Inaccurate Placements/Misalignment

The guidelines for inaccurate placements for different kinds of SMD components are provided in the standard IPC-A-610-C. For different shapes of component terminations, the following criteria applies (Figure 13.36):

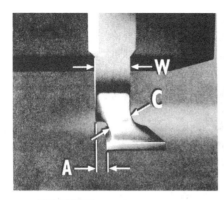

Fig. 13.36 Acceptance criteria for inaccurate placements and misalignments

- It is a defect for class 3 if the side overhang A is larger than 25 per cent of component termination width W or 25 per cent of land width P, whichever is less.
- It is a defect for Class 1 and 2 if the side overhang A is larger than 50 per cent of component termination width W or 50 per cent of land width P, whichever is less.

From the above, it may be concluded that a solder joint is rejected if the solder covers < 75 % of the edges of the SMD terminals resting on the pad for Class 3 and for Classes 1 and 2, it should be < 50 %.

However, these rules for misalignment are only valid for the side overhang A and the end joint width W. They are not valid for the axially situated side joint length. Any axial side joint length is acceptable if all other joint parameter requirements are met, but the terminations of 'Rectangular or Square End Components' must not overhang the land.

13.14.2.2 Non-wetting

Non-wetting or poor wetting occurs when the solder does not wet the PCB completely. Therefore, the board must be inspected in its entirety, and not joint by joint. There can be a number of causes for non-wetting, of which the most typical in wave soldering are:

- Presence of contamination such as oil, grease, etc. on the surface to be soldered, which may prevent the flux from coming in to direct contact with the surface;
- Inadequate solderability of the base metal;
- Unsuitable flux for the surface to be soldered; and
- Improper soldering conditions such as improperly controlled time, and temperature cycle during the soldering process.

In hand soldering, non-wetting occurs due to insufficient heating of the joint, improper lack of flux and lack of solderability of the surfaces.

In fact, non-wetting is a serious defect and calls for stopping the manufacturing process, if a significant number of joints (say 5 per cent or more) are found to be defective.

13.14.2.3 De-wetting

De-wetting is a condition in which the molten solder wets the complete pad/land and because of low adhesion, it forms an irregular film which may be every thin or very thick at places. It implies that the molten solder withdraws from the base metal after initial wetting and forms irregular droplets. De-wetting is generally caused by certain types of contaminants on the surface of the base metal, for example, the contaminant may be embedded in the cleaning abrasives. Similarly, metallic impurities, present in sufficient concentrations in the solder bath can also result in de-wetting. Another cause of de-wetting is the use of wrong flux. De-wetting is acceptable only if at least 75 per cent of the land size meets the solder joint criteria and the angle between the solder and the thin coated area is less than 90° for 75 per cent of the circumference (Figure 13.37). Re-soldering a board with dewetting on wave soldering equipment usually does not improve the situation. The only way to re-work on such a joint is to mechanically remove it from the surface to be soldered with a fine sand paper down to the copper and then resolder it.

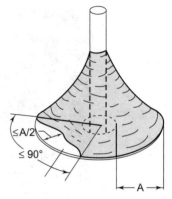

De-wetted area

Fig. 13.37 Acceptance criteria for maximum extent of de-wetting on a land if the de-wetted area is properly tinned (redrwan after Leonida, 1989)

Non-wetting and de-wetting are mostly caused by surface oxides during improper storage or by deposits of other contaminants as well as by large crystallites of the inter-metallic layers which grow on the solder surface and cause the solder to recede. Non-wetting of SMDs can also occur by clogged solder paste screen or wrong squeegee pressure.

13.14.2.4 Bridging

Bridging is a short which occurs when an excess of solder makes an unwanted electrical connection between two adjacent conductors, or two leads or one lead and a conductor as shown in Figure 13.38. Bridging is a major defect and is usually not accepted, except when it shorts two conductive parts which are otherwise electrically connected with one another on the PCB. In wave soldering, the cause of bridging is quite often a too low temperature or insufficient flux. The too low temperature is generally associated with the speed of the conveyer belt, with the contact time as well as with the temperature of the pre-heating zone. Usually, boards with a large area of copper or a high density of terminal areas and terminations tend to act as heat sink which may cause bridging. Other factors, such as the form of the wave and the angle at which the assembled board approaches and leaves the molten solder during wave soldering may also have a strong effect on the tendency for bridging to occur.

In manual soldering, bridging is due to either a lack of skill on the part of the operator or the use of improper equipment with a too large iron bit.

Bridging can also take the form of a web of solder joining the legs and adjacent conductors. In webbing, non-metallic surfaces can even be involved and many conductors may be thus shorted together.

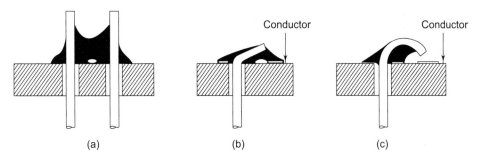

Fig. 13.38 Typical examples of bridging (a) between two leads (b) between lead and a land (c) reduction of clearance between a lead and a conductor due to bad assembly

A special form of bridging when the short is caused by a thin and relatively long mass of solder is called 'whisker'. This defect is usually difficult to detect on visual inspection. As it usually appears where the clearance is small, sometimes it may be required to revise the board design.

13.14.2.5 Disturbed Solder Joints

Disturbed solder joints are often mentioned as "cold joints" because in manual soldering, they are often caused by inadequate heating. The 'disturbed solder joint' results from any movement of the solder during solidification, which may cause an irregular surface that appears, at least partially, rough and wrinkled. The cold joint also sometimes manifests itself as a crack in the joint. If cold joints are a continuous problem, it may be due to the working of vibration of boards on the soldering machine, due to the working of a poor conveyor. This may easily be corrected by providing a smooth transfer of the printed board during solidification (freezing). Besides vibration of the conveyor, any other movement of the solder during freezing like outgassing should be examined. It may however, be remembered that a cold joint causes a serious doubt about the integrity of the joint and the joint should be re-worked.

13.14.2.6 De-lamination/Blistering/Measling

De-lamination, blistering and 'measling' are problems of the laminate caused due to excessive exposure to heat and may also be combined with entrapped moisture or any stress during the temperature above the T_G-value. They are briefly defined below.

Delamination: This implies the separation occurring between adjacent layers of the laminate or between the resin and the copper foil. De-lamination of the adjacent layers starts from the edge of the board or from the holes. De-lamination is generally not acceptable at all.

Blistering: This is a localized swelling and separation occurring only locally between adjacent layers of the laminate or between the base material and the copper foil. It looks like air bubbles inside the laminate and is acceptable if they are far from plated holes and the imperfections are non-conductive and no conductor from inner layers is affected.

Measling: It is an internal condition occurring in laminated base material in which the glass fibres are separated from the resin at the weave intersection. It appears in the form of small white singular spots or crosses internal to the laminate and is due to the separation of the filaments of the glass fibres. It is acceptable if present to a limited extent only and if all the white spots are still covered with resin. Glass fibre must not be exposed to the surface by the laminate.

If the white spots are not singular, and the condition is in the form of connected white spots or crosses below the surface of the base material, this kind of laminate damage is called 'crazing'. It is usually related to mechanically induced stress.

The causes for the above mentioned defects are either excessive exposure to heat during soldering (including curing) or due to handling (stress) when laminate temperature exceeds the glass transition temperature (T_G-Value), which for FR4-Material, is normally 135 °C.

In PTH boards, it must be ensured that the solder has risen in all the plated holes and has fully wetted the walls of the holes. There shall be no non-wetting or exposed base metal on any plated through-hole. If this is not achieved, the joint is defective.

A hole may be considered filled if there is a minimum of 75 per cent vertical fill of the hole. A maximum of 25 per cent depression, including both primary and secondary sides is permitted. Figure 13.39 shows such a condition in a through-hole.

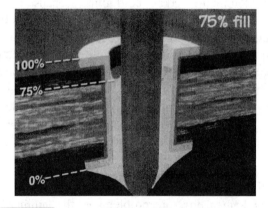

Fig. 13.39 Acceptability criteria for vertical fill of the hole

If there are problems with the vertical fill of the hole, the contact angle relative to the hole wall may be checked. If the contact angle is larger than 90°, the bare PCB is out of standard and must be rejected.

13.14.2.7 Solder Fillet Extends onto the Component Body

The solder fillet height may sometimes exceed the component termination. However, it must be ensured that it does not extend further onto the component body.

13.14.2.8 Outgassing/Blowholes/Pinholes

A blow hole is a small spherical deep cavity in the solder fillet of the joint. It occurs when moisture or flux, which may be trapped in the board, is vaporized by the hot solder and blows out through the joint as the solder is cooling. If blow holes are caused by emission of a gas through the solder fillet, the defect is termed as *outgassing*. The problem can usually be alleviated by baking and/or by increasing the pre-heating parameters. However, the temperature in the pre-heating zone as well as the baking temperature (100-110 °C) is kept less than the T_G-value of the laminates.

Excessive flux or insufficient evaporation of moisture and/or flux solvent before soldering can give rise to blowholes and internal porosity in the joint.

The moisture may be from water vapour absorbed by the boards during storage. Tests have shown that even after 24 hours of baking, the moisture content again gets drastically increased when the boards are exposed to high relative humidity. That is why such boards should be soldered as soon as possible.

Blowholes, which are very small cavities in the solder fillet of the joint are called "*pinholes*".

The major causes of blowholes on PTH boards are:

- Hole being too large as compared to the lead diameter;
- Incorrect insertion of the component;
- Organic residues like inks, photo-resists, solder mask, which vaporize upon heating;
- Moisture or other liquids absorbed by the plated walls of the hole;
- Excessive flux application;
- Thermal profile for preheating zone not optimal; and
- Too quick freezing of the solder fillet.

Figure 13.40 shows blow holes defect in a solder joint.

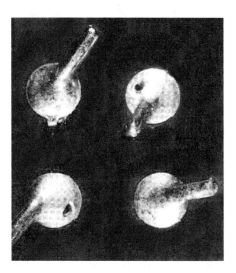

Fig. 13.40 Blowholes: a defect caused by the trapping of liquid or vapour inside the joint as the joint is forming

Blowholes usually occur on the solder side because the solder cools when rising into the hole and starts freezing from the top. The gas escapes through the path of least resistance, which is usually towards the side, where the solder is hotter.

13.14.2.9   Minimum Side Joint Length

The entire fillet for soldering should be properly wetted along the full length of the lead. The joint is not at acceptable, if the side joint length D is less than the lead width. The lead width is measured from the toe to mid-point of the heel bend radius as shown in Figure 13.41. The joint is acceptable if minimum side joint length (D) is at least equal to the lead width or 75 per cent of the lead length, which ever is less.

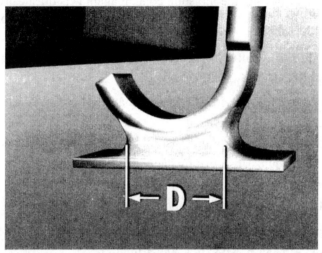

Fig. 13.41   Minimum side joint length — the joint is not acceptable if 'D' is less than the lead width or 0.5 mm, whichever is less

13.14.2.10 Solder Balls/Splashes

Solder balls/splashes are commonly caused by incomplete curing of the solder resist or in PCBs, without solder mask by not fully cured resin of the laminate. The defect often disappears when the board is re-soldered. However, the presence of non-soluble contaminants in the solder bath, sputtering due to use of wrong fluxes, or lack of cleanliness of the working area are likely to cause this kind of defect.

There should be no evidence of solder balls on the printed circuit assembly.

If the splashes are not entrapped or encapsulated and the adhesion is low, the splashes may detach and cause erratic shorts.

13.14.3   Solder Joint Defects and their Common Causes

A wide range of defects is observed in the soldered boards along with their common causes. Table 13.4 shows the most common troubleshooting summary for soldered boards.

Table 13.4   Common Solder Joint Defects and their Causes (Braun, 2003)

Symptom \ Cause	Contaminated surface by improper storage	Contaminated surface by fingerprints	Contaminated surface by other contaminates	Solder temperature too low	Solder temperature too high	Temperature application not homogeneous	Conveyer speed too low	Conveyer speed too high	Preheating temperature too low	Preheating temperature too high	Flux insufficient or contaminated	Flux application too less	Excessive flux application	Vibration of conveyor during solidification	Solder wave uneven	Solder contaminated	Board not seated correct
Bridging/ Icicling				√				√	√		√	√			√	√	
Delamination/ Blisters					√		√			√							
Disturbed Joints/ Cold joints				√				√	√					√		√	
Insufficient Solder flow				√	√	√			√	√	√	√		√	√	√	√
Non-wetting/ De-wetting	√	√	√	√	√		√		√		√	√			√	√	√
Outgassing/ Blowoles *				√	√			√	√					√	√		√
Solder Balls/ Splashes			√	√	√	√	√	√	√		√					√	
Tombstoning/ Lifted Component					√	√								√	√		√
Warpage/Twist					√	√	√			√							√

Samsami (1990a) summarizes the causes of various fine pitch soldering defects. Examples of commonly encountered assembly faults are shown in Figure 13.42(a).

a. Solder bridging

b. Tombstoned component

c. Lifted lead

d. Insufficient solder

e. Misaligned component

f. Unclipped lead

g. Missing component

h. Misaligned solder paste

Fig. 13.42 (a) common types of assembly faults (a) solder bridging (b) tombstoned component (c) lifted lead (d) insufficient solder (e) misaligned component (f) unclipped lead (g) missing component (h) misaligned solder paste (after Samsami, 1990b).

A lot of tools are available for hand soldering and repair work worldwide. Some are excellent and some are not very suitable for modern electronic re-work. A good address for soldering tools is www.ersa.com with its "Soldering Tools and Inspection Division". For assembling of modern electronic items, tools such as de-soldering systems like Microprocessor Controlled Soldering Stations, SMD Soldering and Repair Systems, BGA-Placement and Re-work Systems, Optical Inspection Systems, Quality Assurance and Process Control Software are available from M/s ERSA.

Of course, everybody has to select the most suitable equipment for his shop (on the basis of price, performance and service), that best meets his need.

It is with experience that one learns the difference between a good or bad soldered joint. However, the following points should be kept in mind:

(a) The solder should be uniformly distributed over the elements and base metal. All solder joints, particularly in the high voltage circuit paths, should have smooth surfaces. Any protrusions may cause high voltage arcing at high altitudes.

(b) The quantity of the solder should be only so much that it does not obscure the shape of the element.

(c) No residue such as flux or oxide should be left on the surfaces.

(d) No solder should reach the shield of the wire.

A good solder connection will be quite shiny, not dull gray or granular. If your result is less than perfect, re-heat it and add a bit of new solder with flow to help it re-flow. The examples of bad solders are given in Figure 13.42(b).

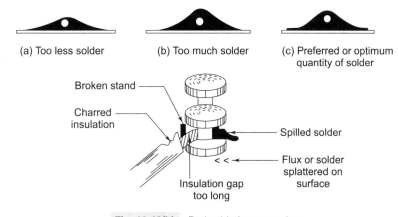

(a) Too less solder   (b) Too much solder   (c) Preferred or optimum quantity of solder

Broken stand

Charred insulation

Spilled solder

Flux or solder splattered on surface

Insulation gap too long

Fig. 13.42(b)   Bad soldering examples

13.15 Health and Safety Aspects

- Don't eat any thing on the soldering worktable. The solder wire you will be using may contain lead, which is poisonous, if taken internally. Some of the lead from the solder can transfer onto the hands and then onto any food that one eats. Wash your hands properly after soldering since lead is poisonous.

- Use a rag or a brush to clean your workstation from any dirt, grease, solder splatter, etc. The common practice to clean up the workstation by wiping with the open hand over the workstation may hurt your hand.
- The fumes given off during the hand soldering process do not contain lead, since lead does not vaporize at the relatively low temperatures involved in hand soldering. However, some people are allergic to the flux fumes that are released during soldering. Removing the fumes with the help of fume extractors is a common practice.
- Molten solder can easily burn the skin and can cause serious eye-damages.
- The soldering bit will burn skin and material. That is why keep the iron in an open holder when not being used.
- Always use the wet sponge to clean the tip.
- Never flick the soldering iron.
- Protective clothing, if provided or needed, should be worn.
- Obviously try to avoid dropping the iron but if you do, don't try to catch it. You always get the hot end.
- The mains power must be earthed to the line earth and frequently checked for damage in order to avoid shock hazards.

If you happen to receive burns during soldering/de-soldering operations, the following first aid steps are necessary:

- The affected area should be immediately cooled with running water or ice for 10-15 minutes.
- Remove any rings before swelling starts.
- Apply a sterile dressing to protect against infection.
- Do not apply lotions or ointments.
- Do not prick any blisters which may form later on.
- Seek professional medical advice where necessary.

It may be noted that the work of soldering/de-soldering should be carried out carefully to avoid any possibility of electric shock or burns. In order to avoid electrical shock, disconnect the equipment under repair from ac power before removing or replacing any component or assembly. It is a good practice to periodically inspect the grounding lead, the state of the handle insulation and cord insulation. The ground terminal must be the last line to be disconnected while pulling the plug.

 ## 13.16 Electrostatic Discharge Control

When two conductive objects with different potential levels or polarity of charge come close together or even in contact with each other, the charge rapidly moves from one object to the other. This sudden transfer or discharge of electricity from one object to another is known as electrostatic-discharge (ESD).

ESD is one of the most serious problems faced by the electronics industry today. Many electronic components are damaged by normal electro-static discharge by humans as well as by objects.

More and more electrostatic-sensitive devices are appearing in the modern electronic market. The amount of voltage, causing the components to fail by ESD is rapidly decreasing. Till a few decades ago, the common man was concerned with ESD associated only with thunderstorm and how we could provide safety to the buildings by proper grounding with lightning conductors. However, today a lot of precautions are required to be observed for handling electrostatic-sensitive devices like CMOS, EPROM, MOSFET, laser diodes, VHSIC (very high speed integrated circuits), hybrids, thin film resistors, precision voltage regulating diodes and detector diodes with frequency response greater than 1GHz, surface acoustic wave (SAW) devices or VMOS. An ESD voltage can cause or may cause 'catastrophic failure', resulting in a metal melt, junction breakdown or oxide failure. The so called 'latent failures' are more difficult to identify, because the component may be partially degraded, yet continue to perform its intended function. However, the operating life of the device may be reduced dramatically. Latent defects are extremely difficult to detect by using current technology, especially after the device is assembled into a finished product. The assembly often passes inspection, gets shipped to a customer and may later on, lead to a dissatisfied customer. On the other hand, proper ESD control will help to improve product quality, reduce costs and ensure better customer satisfaction. Therefore, the modern assembly lines for electronic equipment are housed in air-conditioned rooms with controlled humidity, and are installed with air ionizers, static dissipative floors as well as work surfaces. However, all these methods of ESD control will be successful only if all workers understand the basic concepts of ESD control and practise the same.

13.16.1   Fundamentals of ESD

Every material has the ability to take on an electrostatic charge. The charge of the object is called static electricity and figuratively, it means that there is some charge just sitting on the item and waiting for an opportunity to move. The amount of static electricity created depends upon the material, the area of contact, the speed of separation and the relative humidity. Higher relative humidity creates less charge. If two conductive objects with different potential levels or polarity come close together or in contact with each other, the charge rapidly moves from one object to the other. The terms associated with static electricity are defined as:

Coulombs: Static electricity is measured in Coulombs. The charge of 1 Coulomb on an object is determined by the product of the capacitance of 1 Farad and a potential of 1 Volt on the object.

   1 Coulomb = 1 Farad × 1 Volt;

   1 C = 1 F × 1 V;   1C = 1As   where   A = amperes   S = seconds

   With the introduction of the electron Volt, which is the energy taken by an electron for passing the potential difference of 1 Volt we have:

$$1 \, eV = 1.602 \times 10^{-19} \, C \times 1V$$
$$= 1.602 \times 10^{-19} \, As \times 1V$$
$$= 1.602 \times 10^{-19} \, Ws$$
$$= 1.602 \times 10^{-19} \, J \, (Joule)$$

Furthermore the potential difference between two points in an electric field is defined as the amount of work done in moving a unit positive charge from one point to another point, which means the

Potential difference = Work done divided by quantity of charge transferred

1 Volt = 1 Joule/1 Coulomb = AsV/As = 1 Volt

The potential difference between two points is said to be 1 Volt if 1 Joule of work is done in moving 1 Coulomb of electric charge from one point to another.

13.16.2   Electrostatic Voltages Generated by Various Operations

You must have sometimes experienced crackling and sparks that occur when you take off some clothes or the shock you sometimes feel when you walk across a synthetic carpet and then touch a doorknob. These examples of ESD, which we feel for a fraction of a second can contain anywhere from about 2 000 Volts (the lowest level most people can feel) to more than 25,000 V.

The human body can easily pick up static electricity like other materials. The skin can store relatively large amounts of the electric charge. The voltage generated by various operations is given in Table 13.5. Different relative humidity develops different electrostatic voltages. A higher relative humidity is sometimes useful as lower electrostatic voltages are generated. However, due to other important reasons, it is not recommended to have relative humidity above 65 per cent.

Table 13.5   Electrostatic Voltage Generated by Various Operations Depending on the Relative Humidity

Operation	Voltage within a Relative Humidity of	
	10–20 %	**70–90 %**
Walking on synthetic carpet	35 000 V	1500 V
Pulling tape from a PCB	12 000 V	1500 V
Cleaning a PCB with an eraser	12 000 V	1000 V
Freon circuit spray	15 000 V	5000 V
Poly bag picked up from bench	20 000 V	1200 V
Sitting on a foam cushion	18 000 V	1500 V

13.16.3 Sensitivity of Various Components to ESD Voltages

Damages to an ESD-sensitive item by the ESD event is determined by the device's ability to dissipate the energy of the discharge or withstand the current levels involved. This is defined as device "ESD sensitivity" or "ESD susceptibility".

Integrated circuits appearing in the market are so designed that the equipment are becoming smaller, faster and higher in performance. The conductive paths within the components are drastically reduced. Such tiny circuit paths can be burned up or damaged by a small ESD voltage.

Table 13.6 shows that various component types and the ESD voltage with which they can be damaged. It may be noted that we feel an ESD of minimum 2000 Volts.

Table 13.6 Various Component Types to ESD Sensitivity (Measured in Volt)

Device Type	ESD sensitivity (V)
VMOS	30V–1800V
MOSFET	100V–200V
EPROM	100V–2500V
CMOS	250V–3000V
TTL	300V–2500V
Film Resistor	300V–3000V
SCR	680V–1000V

Different models and test procedures are used to characterize, determine and classify the sensitivity of components to ESD.

ESD-sensitive devices should always be classified in order to alert a potential user of the component to the need for a controlled environment.

In order to find out the classification of a particular component, the first source would be the manufacturer or supplier of the component itself. An additional source is ITT Research Institute/ Reliability Analysis Centre, NY, which publishes ESD susceptibility data for 22,000 devices, including micro-circuits.

13.16.4 Electrostatic Protection

Even if any sensitive component is already soldered onto a printed circuit board, it can be damaged by a discharge that passes through the board's conductive pattern to the ESD-sensitive component. The amount of voltage needed to damage or destroy varies from component to component, but as these components become more complex, their sensitivity to ESD will increase, like we have seen in para 7.2.2. The most important thing we should learn about ESD control is that we ourselves are the most important part of ESD control.

13.16.5 Anti-static Workstation

Anti-static workstations are normally placed in an ESD safe work area. The key ESD control elements comprising most workstations are:

- A static dissipative work surface;
- A means of grounding personnel (usually a wrist strap);
- A common grounding connection;
- An ESD protective mat; and
- Appropriate signage and labelling.

The most important requirement for ESD control is to provide a ground path to bring ESD protective materials and personnel to the same electrical potential. All conductors in the environment, including personnel, must be bonded or electrically connected and attached to a known ground, thereby creating an equipotential balance between all items and personnel.

ESD Association Standard ANSIEOS/ESD 6.1- Grounding recommends the following two-step procedure for grounding ESD protective equipment:

Fig. 13.43 Symbol for ESD common point ground

a) Ground all components of the work area (work surface, people, equipment, etc.) to the same electrical ground 'common point ground' so they have the same electrical potential. This ESD common point ground should be marked and the ESD Association Standard EOS/ESD S8.1-1993 recommends the use of the symbol illustrated in Figure 13.43.

b) Connect the common point ground to the equipment ground (electrical ground connection). This is the preferred ground connection because all electrical equipment at the workstation is already connected to this ground. Connecting the ESD control materials or equipment to the equipment ground brings all components of the workstation to the same electrical potential.

If a soldering iron used to repair an ESD-sensitive device, which is connected to the electrical ground and the surface of the workstation containing the ESD-sensitive components, are connected to an auxiliary ground (water pipe, building frame, ground stake), there could be a difference in electrical potential between the iron and the ESD-sensitive item. This difference in potential could cause damage to the component. Any auxiliary grounds present and used at the workstation must be bonded to the equipment ground to minimize differences in potential between the two grounds.

13.16.6 A Proper Assembly Environment

A proper assembly environment can be ensured by taking the following steps:

- Define ESD-safe areas: Define the specific electrostatic protected areas in which sensitive parts are handled. Typical areas requiring ESD protection are: receiving inspection area, store, assembly line, test/inspection area, packaging (dispatch), R and D/Field Service Repair, and Clean Rooms.

- Create a "dust-free Environment": This should be similar to clean room conditions of 'Class 10,000' where no more than 3,500,000 particles larger than 0.5 μm exist.

- Control Relative Humidity: Keep relative humidity controlled at approximately 60 per cent. Conductive smocks cover static generating clothes and drain any charges to ground. If conductive smocks are not provided, the clothing should be made of cotton rather than wool or synthetic materials. Some anti-static materials become ineffective if the relative humidity drops to 30 per cent or less. The best solution for ESD control is to eliminate the materials that generate and hold static charges.

- Use Air Ionizers: Set up air ionizers blowing ionized air into the work area in order to neutralize any positive or negative charges on non-conductive items that are not connected to ground. Grounding does not remove an electrostatic charge on plastics and other insulators because there is no conductive pathway. The ionization process generates negative and positive ions that are attracted to the surface of a charged object, thereby effectively neutralizing the charge.

- Packaging Susceptible Devices: The use of proper (conducting) materials for packaging may effectively shield the product from charge, as well as reduce the generation of charge caused by any movement of product within the container. All goods supplied from vendors must be delivered in sealed conductive containers which will dispel any charge (Grant, 1990).

- Use Static Dissipative Floors and Static Dissipative Work Surfaces: Static dissipative materials have an electrical resistance between conducting and insulating materials, which means that they have a resistance between 10 kilo Ohm (k) and 100 giga Ohm (g). Thus, there can be electron flow across the dissipative material, but it is controlled by the specific resistance of the dissipative material. Of course, charge can be generated tribo-electrically on a static dissipative material like on insulators or conductors. However, like the conductive material, the static dissipative material will allow the transfer of charge to ground in a controlled manner. The use of anti-static floor material is especially essential in those areas where increased personnel mobility is necessary. In addition, floor materials can minimize charge accumulation on chairs, carts and other times that move across the floor. However, those items require dissipative or conductive castors (small additional wheel that is attached to the floor) or wheels to make electrical contact with the floor. The resistance to ground including the person, footwear and floor must be less than 35 megohms.

- Label ESD-Sensitive Devices: The components, assemblies, and finished products which are ESD-sensitive should be labelled with the ESD susceptibility symbol.

The ESD susceptibility symbol consists of a triangle, a reaching hand and a slash through the reaching hand. The triangle means 'caution' and the slash through the reaching hand means 'do not touch'. The symbol is applied to integrated circuits, boards and assemblies that are ESD-sensitive. It indicates that handling or use of this item may cause damage from ESD if proper precautions are not taken.

- Use ESD-protective Material: Mats, chairs, writs straps, garments, packaging and other items that provide ESD protection should be indicated with the symbol of ESD-protective material

The ESD protective symbol consists of the reaching hand in the triangle. An arc around the triangle replaces the slash and will indicate protection. It may also be used on equipment such as hand tools, convey or belts or automated handlers that are especially designed or modified to provide ESD control.

Neither symbol is applied on ESD test equipment, footwear checkers, wrist strap testers, resistance or resistivity meters or similar items that are used for ESD purpose, but which do not provide actual protection.

In addition,

- Remove all the unnecessary items from the assembly room.
- Eliminate all materials generating and holding static charges.
- Ensure a clean working table.
- Deploy trained manpower wearing cotton clothes, wrist straps as well as heel straps and avoiding activities generating charges.

13.16.7   Component Handling

The following precautions should be observed while handling components:

- Handle components such as integrated circuits by the non-conductive portion of the body, rather than the leads. The leads are the most conductive pathways for ESD susceptibility.
- Touch PCBs only on the edges, never on the solder/component side, because the conductors could be conductively connected with the sensitive component.
- Non-conductive materials should be eliminated from the work area, whenever possible, because the electric field will cause charge separation in the ESD-sensitive device. If the ESD-sensitive device comes into contact with a conductive item while exposed to the field, the device can be damaged.
- If static generating materials are essential for the job, the workstation should be arranged so that the static sensitive boards do not get closer than 30 cm to the static generating materials.
- The use of two shoe grounders for standing activities and a wrist strap for seated operations is most advisable. It is also important that wrist straps as well as shoe grounders should be tested at regular intervals to make sure that they are working properly.

Wrist Strap: Wrist straps used by people seated at their workstations are very effective for ESD control. The wrist strap is worn snugly against bare skin. It should be attached securely to the common point ground with a current limiting resistor of 1M (0.25 Watt with a working voltage rating of 250 Volts) and some form of quick connect/disconnect arrangement. The current limiting resistor provides a slow but controlled drain of any charge to limit the current level and prevent damage when the charge is drained to ground.

In order to test the wrist strap, use the opposite hand to press the test button. Shake your wrist to check for intermittent failures and do not stand on the metal plate while checking the wrist strap.

Shoe Grounders or Heel Straps: They are designed to connect the body through the socks to a conductive floor or mat. Shoe grounders are very effective for stand-up operations and preclude the use of a long cord from the wrist strap. While seating, shoe grounders are not so effective because people may raise their feet when seated and lose contact with the conductive floor.

While testing shoe grounders, place only one foot with the strap on the metal plate and use the opposite hand in order to press the test button. That means you have to test each foot individually. Testing both feet at the same time will not indicate a failure unless both straps fail.

There is no need to remove the socks. The moisture in the socks provides a complete electrical connection to the outside of the socks.

13.16.8 Special Considerations for Handling MOS Devices

MOS (metal oxide semiconductor) devices are highly sensitive devices and get damaged easily by accidental over-voltages, voltage spikes and static-electricity discharges. The human body can build up static charges that range upto 25000 Volts. These build-ups can discharge rapidly into an electrically grounded body or device, and particularly destroy certain electronic devices. The resultant high voltage pulse burns out the inputs of integrated circuit devices. This damage might not appear instantly, but it can build up over time and cause the device to fail.

The most common causes of electrostatic discharge (ESD) are: moving people, low humidity (hot and dry conditions), improper grounding, unshielded cables, poor connections and moving machines. When people move, the clothes they are wearing rub together and can produce large amounts of electrostatic charges in excess of 1000 Volts. Motors in electrical devices, such as vacuum cleaners and refrigerators, generate high levels of ESD. ESD is also most likely to occur during periods of low humidity say below 50 per cent . Any time the charge reaches around 10,000 Volts, it is likely to discharge to grounded metal parts.

An important point to remember is that 10,000 to 25,000 Volts of ESD are not harmful to human beings whereas 230 Volts, 1 amp current produced by the mains power supply is lethal. The reason for this is the difference in current-delivering capabilities created by the voltage. The ESD voltages, though in the kilovolts range, produce currents only in the micro-ampere range, which are not harmful for human beings. However, the level of static electricity on your body is high enough to destroy the inputs off CMOS (complementary metal-oxide semiconductor) device if you touch its pins with your fingers.

Special care is needed while storing, handling and soldering MOS devices. The following precautions must be observed when using such devices:

- While storing and transporting MOS   devices, use may be made of a conductive material or special IC carrier that either short circuits all leads or insulates them from external contact.

- The person handling MOS devices should be connected to ground with grounding strap as shown in Figure 13.44. These anti-static devices can be placed around the wrists or ankle to ground the technician to the system being worked on. These straps release any static charge on the technician's body and pass it harmlessly to ground potential.

- Anti-static straps should never be worn while working on high voltage components, such as monitors and power supply units. Some technicians wrap a copper wire around their wrist or ankle and connect it to the ground side of another. This is not a safe practice because the resistive feature of a true wrist strap is missing.

- The work areas should preferably include anti-static mats (Figure 13.45) made of rubber or other anti-static materials they stand on while working on the equipment. This is particularly helpful in carpeted work areas because carpeting can be a major source of ESD build-up. Some anti-static mats have ground connections that should be connected to the safety ground of an ac power outlet.

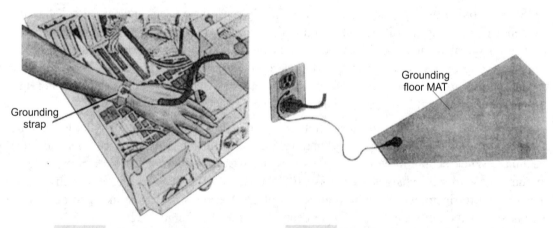

Grounding strap

Grounding floor MAT

Fig. 13.44   Use of anti-static strip   **Fig. 13.45**   Use of anti-static mat in the work area

- Before touching any component inside the system, particularly that containing MOS devices, touch an exposed part of the chassis or the power supply housing with your finger. Grounding yourself in this manner ensures that any static charge on your body is removed. This technique, however, works safely only if the power cord is attached to a grounded power outlet.

- Mount MOS integrated circuits on printed circuit boards after all other components have been mounted.
- When replacing a defective IC, use a soldering iron with a grounded tip to extract the defective IC and while soldering the new IC in place.
- After the MOS circuits have been mounted on the board, proper handling precautions should still be observed. In order to prevent static charges from being transmitted through the board wiring to the device, it is recommended that conductive clips or conductive tape be put on the circuit board terminals.
- In order to prevent permanent damage due to transient voltages, do not insert or remove MOS devices from test sockets with the power on.
- Avoid voltage surges as far as possible. Beware of surges due to relays, and switching electrical equipment on and off.
- Signals should not be applied to the inputs while the device power supply is off.
- All unused input leads should be connected to either the earth or supply voltage.
- Personnel handling MOS devices are advised to wear anti-static clothing; synthetic fibre clothing should especially be avoided.
- Do not insert printed circuit boards into connectors that have voltages applied to them.
- Workstations should have non-conductive table tops, non-conductive trays, grounded soldering irons, etc.

The switching action of some controlled-output soldering implements can generate voltage spikes, which can be transmitted to and adversely affect MOS devices. Care should be taken while selecting soldering irons so that they have low voltage spikes.

13.16.9   Education/Certificate for ESD Control

Every engineer, technician or operator, who is handling ESD-sensitive devices, whether by performing incoming inspection, storing items in the stockroom, kitting, doing assembly operations or testing and troubleshooting, must have a basic knowledge of the safe handling of ESD-sensitive devices.

IPC (www.ipc.org) has designed a videotape 'IPC-VT-54' which provides complete courses on ESD control. The IPC training video is designed to help you create a receptive and efficient learning environment. Developed by industry experts, the tape also has an evaluation test.

The ESD Association, Rome, NY/USA provides awareness, solutions, standards and education in the field of ESD and is available online under www.esda.rog.

 ## 13.17   Re-work and Repair of Printed Circuit Boards

When printed circuit boards are inspected and tested, whether bare or fully loaded (assembled), and defects are found, it is necessary to evaluate the cost-effectiveness of repairing the board, and at the same time provide the user with the same reliability as the original product. In case of simple boards with a few defects, it is usually not economical to re-work on them. However, many boards are highly complex and a fully loaded board could be expensive. It may turn out to be more economical to re-work on the same so that it will pass all the tests.

Bare boards, in general, are often not repaired because of the reliability risks associated with their later utilization in assemblies and because of their comparatively low cost as compared to assembled boards. In view of these factors, repairs or re-work of bare boards are not allowed in high reliability and military applications. Bare board re-work is permissible for boards which are used for commercial applications, as part of in process connections. However, the repaired board must meet the original design requirements and the expected reliability and quality standards.

Further, repair and re-work is also required on the boards received for repairs from the field. In most such cases, there may be a requirement of removing and replacing a component with a new one. The exercise is normally undertaken manually. For plated through-hole boards, the repair work can be done by simple tools such as soldering iron and wicked braid. For surface mount components, on the other hand, special re-work stations are required which depend upon hot air re-flow soldering units. During repair, a number of chemicals are used specially for cleaning, moisture displacement, flux removal, wiper lubricants and freeze sprays to locate thermally sensitive components.

13.17.1   Approaching Components for Tests

Most designers provide test points at convenient locations on the circuit board. These points are defined by specific dc and ac voltages, along with the waveform pattern. Figure 13.46 shows a test point as it would appear on a circuit board. This is usually a vertically mounted pin to which a test prod can be attached.

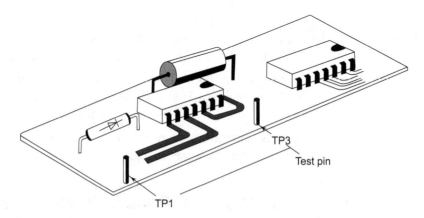

Fig. 13.46   Typical test points indication in a printed circuit board

If specific test points are not provided, measurements can be made at various points on the circuit by approaching various components. In that case, proceed as follows:

(a) For transistors, make test prod connection to the legs under the case

(b) To read the signal on a circuit board trace, locate a component that is connected to the trace as shown in Figure 13.47. Clip your test lead onto the leg of the component that is connected to the trace.

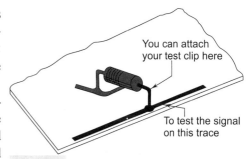

Fig. 13.47 Taking measurements from the circuit trace by connecting a test prod on the component

(c) Connections to the ICs can be made more conveniently by using an IC test clip. This is illustrated in Figure 13.48. Be careful not to touch more than one conductor at a time, otherwise you can easily create a short-circuit. Since digital circuits are usually densely packed on a board, make use of only as narrow a prod as possible.

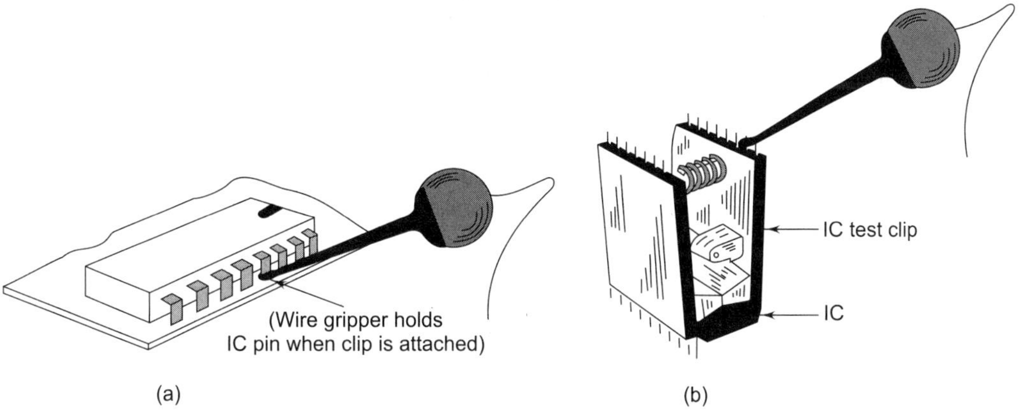

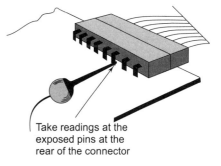

(a)   (b)

Fig. 13.48 Use of test clip for taking measurement on IC pins (a) test clip on the IC directly (b) Test clip on the IC connector

(d) Flexible flat wire with connectors often offer good place to take readings. The connector pins themselves are usually well protected, but you can take readings at the conductors behind the connectors as shown in Figure 13.49.

13.17.2   De-soldering Techniques

Desoldering means removal of solder from a previously soldered joint. The two techniques common in soldering are:

- Wicking, and
- Sniffing.

Fig. 13.49 Taking test readings from a connector

13.17.2.1 Wicking

In the wicking process, a heated wick, well-saturated with rosin, is placed on top of the joint to be de-soldered. The solder will flow rapidly into the rosin area due to capillary action leaving the joint to which it was previously affixed.

A wicking solder remover may consist of a braided shield wire with the core removed or it may be a piece of multi-strand wire. Wicks are available commercially which are suitable for de-soldering work. The de-soldering technique using the wicking process is as follows (Figure 5.15):

- Place the wick on top of the solder joint to be de-soldered.
- Position the iron tip on top of the wick. The heat of the iron will melt the solder. The solder will readily flow into the wick.
- Cut off the wick containing the removed solder. Repeat the process until all the solder is removed from the joint.

Take extreme care to ensure that the solder is not allowed to cool with the braid adhering to the work, otherwise you run the risk of damaging PCB copper tracks when you attempt to pull the braid off the joint. This technique is more effective specially on difficult joints where a desoldering pump, described below, may prove unsatisfactory.

13.17.2.2 Sniffing

In sniffing, a rubber ball (Figure 13.50) is employed as a solder sucker (sniffer). The sniffer uses the forced air pressure to accomplish the sniffing (removal of solder) action.

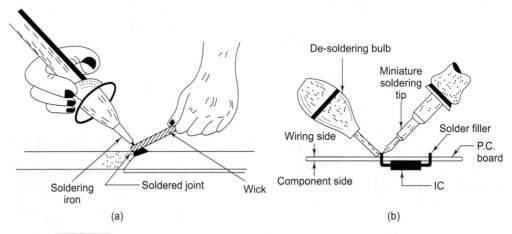

Fig. 13.50 (a) wicking process for de-soldering (b) sniffing technique in de-soldering

Another vacuum type sniffer uses a spring loaded plunger.

The following steps are adopted in sniffing:

- The air is first squeezed out of the rubber ball.

- With the ball depressed, the pointed end of the sniffer tube is placed next to the solder to be removed.
- The joint is heated with the soldering iron. The tip of the iron should be kept in the solder and not on the sniffer.
- The pressure on the sniffer ball is slowly released to allow air to enter the ball through the sniffer tube. As the air enters, it pulls the molten solder into the tube with it.
- After the solder has been completely pulled into the tube with it, the sniffer is removed from the joint. By depressing the ball again, the collected solder can be forced out.

A de-solder pump is another device for solder removal. It uses a spring-loaded mechanism. For using the device (Figure 13.51), the spring is cocked and the tip of the vacuum pump is held against the solder joint. When the solder melts, the trigger is operated which releases the spring, thereby creating a powerful vacuum action. Some of these devices can generate a static charge. Be sure to get a type that is specified as 'anti-static.'

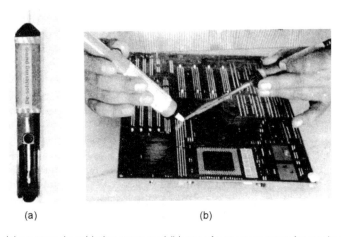

(a)   (b)

Fig. 13.51   (a) vacuum de-soldering pump and (b) use of a vacuum pump for use in de-soldering

For stubborn joints or those connecting to the power planes (surface or multi-layer boards), you may need to add some fresh solder and/or flux and then try again. Generally, if you only get part of the solder off the first time, repeated attempts will fail unless you add some fresh solder.

A very important consideration which must be kept in mind while de-soldering is that the heat required may damage the base materials and adjoining components. The de-soldering should be carried out by using appropriate tools so that a minimum amount of heat is used during the de-soldering process.

During any repair work, it is well worth taking time and care so as not to damage or lift copper back from the printed circuit board, as the printed circuit board is usually a very expensive item.

Do not use a sharp metal object, such as a twist drill for removing solder from component mounting holes. Sharp objects may damage plated through-conductor.

Removing multi-lead components such as integrated circuits presents a special problem. If the component to be removed is still functional, it must be de-soldered quickly lest it be damaged by heat. Alternatively, if the device is defective, it also needs to be removed fairly quickly to avoid lifting of printed circuit foil conductors by excessive heat.

Specialized devices are needed to solve this problem. One such device is a special DIP-shaped soldering iron tip (Figure 13.52) and a spring-loaded IC extractor tool. The tool is placed above the IC to be removed and locked into position. When the tip is hot, it is applied to all the dual-in line IC pins or the foil side of the board. The extractor tool lifts the IC off the board as soon as the solder holding it melts. Special desoldering tools are available for use with other IC and transistor cases.

Circuits coated with silicone conformal coatings may be repaired after removing the coating using solvent-swell or mechanical abrasion techniques, the defective device can be de-soldered and removed. Standard burn-through techniques can also be used.

After removing the old solder, the area should be thoroughly cleaned with a solvent-soaked swab to ensure a good replacement joint. After component installation, re-coating can be accomplished.

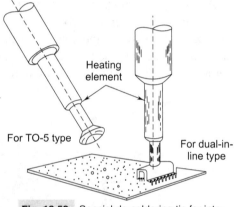

Fig. 13.52 Special de-soldering tip for integrated circuits

13.17.3 Replacement of Components

Printed circuit boards used in modern equipments are generally the plated-through type consisting of metallic conductors bonded to both sides of an insulating material. Before a component replacement is attempted, the following precautions should be observed:

- Avoid unnecessary component substitution. It can result in damage to the circuit board and/ or adjacent components.
- Do not use a high power soldering iron on etched circuit boards. Excessive heat can dislodge a conductor or damage the board.
- Use only a suction device or wooden toothpick to remove solder from component mounting holes. Never use sharp metal object for this purpose as it may damage the plated through-conductor.
- After soldering, remove excess flux from the soldered areas and apply a protective coating to prevent contamination and corrosion.

The following steps are to be followed for replacing a component:

- Read carefully the replacement procedure from the service manual of the instrument.
- Switch-off the power, if applicable.

- Remove any assemblies, plugs, wire that will facilitate repair work.
- Label the component to be removed.
- Observe carefully how the component is placed before removing it. Record information regarding polarity, placement angle, positioning, insulating requirements and adjacent components.
- Be careful to handle the printed circuit board by the edges only. Fingerprints, which though invisible, can cause an accumulation of dirt and dust on the boards, resulting in low impedance bridges in portions of the circuit board, which normally should have a very high impedance. Gloves should preferably be worn to prevent fingerprint problem if the boards must be handled.
- Remove the dry film or the hermetic sealer from the solder joint that is to be worked on. This is done by using a cotton tipped applicator dipped in the recommended chemical. Large quantities of solvents should not be allowed to drip on the board because the impurities will then only be shifted from one place to another on the board. This cleaning is necessary because it can be difficult to burn through a layer of dry film with a soldering iron. In addition, if the dry film is not removed before heating, the appearance of the board will be badly changed.
- Heat the solder fillet on the solder side of the printed circuit board. Using the de-soldering tool (suction device), gently and carefully remove the component. Too much soldering iron heat should not be used otherwise the foil is lifted or plated through-holes get removed.
- In case of multi-lead component, the vacuum desoldering tool must be used to remove almost all the solder from the component leads before the component can be removed from the board. This procedure must be carefully followed because multi-lead components multiply the probability of printed circuit board damage during repairs.
- Some components are difficult to remove from the circuit boards due to a bend placed in each lead during machine insertion of the components. The purpose of the bent leads is to hold the component in position during a flow solder manufacturing process which solders all components at once. In order to make removal of machine inserted components easier, straighten the leads of the components on the back of the circuit board using a small screw driver or pliers while heating the soldered connections.
- After removing the component from the printed circuit board, the area around the removed component must be cleaned up by using the cotton tipped applicator in a solvent. Also, there may be solder in the plated through-holes or other areas of the board that must be removed in order to allow easy insertion of a new component.
- Clean leads of new component or element with a cleaning tool, such as a braided tool. Use abrasives if required. In case of a wire lead, the insulation must be removed. The secret to a good solder joint is to make sure everything is perfectly clean and shiny and not depend on the flux alone to accomplish this. In case of multiple strands, form the strands. Tin to about 3 mm from insulated part.

- Shape the leads of the replacement component to match the mounting hole spacing. Insert the component leads into the mounting holes and position the component as originally positioned. Do not force leads into mounting holes because sharp lead ends may damage the plated through conductor.
- Start with a strong mechanical joint. Don't depend on the solder to hold the connection together. If possible, loop each wire or component lead through the hole in the terminal. If there is no hole, wrap them once around the terminal. Gently anchor them with a pair of needle nose pliers.
- Heat the parts to be soldered, not the solder (Raby, 1994). Touch the end of the solder to the parts, not the soldering iron or gun. Once the terminal, wires or component leads are hot, the solder will flow via capillary action, fill the voids and make a secure mechanical and electrical bond. Apply the soldering iron to the joint and feed solder into it. The solder should be applied to provide a complete seal covering all elements. Be careful with the amount of solder and the amount of heat. Check the component side of the board for good solder flow. Remember SN63 is the best type of solder for soldering electronic components. SN60 is acceptable.
- Remove the soldering iron and allow the solder to cool and solidify. Do not disturb the board for a while, otherwise you will end up with a bad connection, what is called a 'cold solder joint'.
- Clean the area of splattered rosin flux and residue using isopropyl alcohol. Be careful not to leave cotton filaments on the printed circuit board. Allow the circuit board to air dry completely.
- Apply protective coating, if possible, on the repaired area and allow this to air dry.
- It is always advantageous to check the integrity of the joint soldered or repaired. This check can be performed with an ohmmeter (multimeter) by measuring the resistance between the solder and the component lead. Any reading except a short reveals a defective joint. Recognize defective solder joints, that are cracked, pitted and cold stressed, have excessive flux or the impure solder.
- When working with semiconductor devices and microelectronic IC circuit components, a heat sink may always be used, while soldering. Also, when working on equipment having components like these, the specifications of allowable soldering iron sizes, voltage ranges and other factors must be studied. This is essential to understand the damage one can do if a unit is repaired improperly.
- While replacing components, it may be noted that mechanical shocks can seriously damage the components. For example, semi-conductors can get damaged by the high impact shock if dropped in a concrete floor even from a table height. Cutting of leads can also cause shock waves which may damage delicate or brittle components. Therefore, cutting or scratching of surfaces of components by the careless use of tools or sharp test probes should always be avoided.

- It is always wise not to remove or replace any component while the power is on. This may well produce voltage or current surges that could damage the component itself and other sensitive components in the circuit.

 ## 13.18   Repairing Surface Mounted PCBs

In the next few years, it will be hard to find a manufacturer who will build anything that will not contain surface mounted devices. Having the proper tools to do the job and the knowledge to use the tools that you choose will be important to your survival in the electronics repair business. Surface mount technology has been around for a long time, but the tools for removal and replacement of surface mount devices have been slow to be accepted by the service and repair personnel and organizations. Possibly, it was because of the high cost of such tools. The technicians looked for innovative ways to use the tools at hand to do the work, sometimes even ruining the whole board by not having the right tools. The justification for investment on the repair and re-work stations come from the fact that the high priced equipment certainly deserve to be treated in a better way so that you are in a position to perform the job worthy of your knowledge and experience.

As the assembly of electronic components moved away from the use of single-sided printed circuit boards towards double sided boards with plated through-holes, component removal became somewhat easier. Also, the associated damage occurring to PCBs during re-work/repair became less. However, the SMT printed circuit boards are, as far as re-work is concerned, essentially single-sided PCBs. Therefore, with the increasing use of surface mount components, we are seeing an increasing number of damaged pads and tracks due to inappropriate or careless component removal practices. It has been observed that these damages are mostly due to the inadequate training of the operators/repair workers in properly understanding and handling the SMT PCBs.

Re-working faulty SMT assemblies usually requires component removal and replacement. Occasionally, the replacement of damaged PCB pads and tracks also becomes necessary because of poor re-working practice. The methods of removing a faulty multi-lead surface mount component are (Morris, 1990) discussed below.

13.18.1   Cut All Leads

Cutting all leads is the simplest method to remove a faulty component. It is recommended if other methods are not practical. The technique is to carefully cut through each leg in turn and take off the device. Each joint is then melted with a fine tip, temperature controlled soldering iron and remaining IC leg are removed with tweezers. After allowing a cool down period, excess solder can be removed with a de-soldering braid.

The advantage of this method is that it is cheap and can be carried out in the field as it does not require any special tool. The disadvantage is that it damages the component and there is a possibility of damaging the PCB substrate and copper pads. Also, soldering the replacement component in position using a soldering iron requires processing one lead at a time, a difficult if not impossible task with fine pitch multi-lead devices.

13.18.2 Heating Methods

There are two basic heating methods for re-working PCBs whose components include SMDs: conductive and convective. Conductive re-work involves a heated tool that contacts the solder joints to effect re-flow. The convective approach employs heated gas or air to melt the tin lead alloy.

Conductive Method: Soldering tools fitted with tips designed to heat all the component's leads are available. They rely on electrodes coming into contact with the component legs and holding them flat to the copper pads on the PCB. The more sophisticated re-work stations employ a precisely controlled pulse of current which passes through the electrode heating them to solder reflow temperature every quickly (approximate three seconds). This melts the solder on the joints and a built-in vacuum pick-up will lift the component from the surface. The technique enable all the leads to cool down rapidly after the soldering operation and so allows the leads to be held in position while the solder solidifies.

This method has several advantages. It is very fast, repeatable and there is no heating of the component body. It is very good for replacement as the electrodes will hold the legs flat to the pads during solder reflow, while the alignment and positioning is ensured with a microscope. The disadvantage is that it is expensive and machines are dedicated solely to gull wing (QFPs) and TAB components.

Dual Heater with Vacuum Pick-up: This is a special tool for handling larger component removals. Dual heating brings the larger tips upto the required temperature quickly and the built-in vacuum pick permits one-handed removals once re-flow is established. The tool enables to remove all conventional flat packs as well as several BGAs.

The *thermal tweezer*, with dual heaters and a squeezing action, can remove a variety of parts ranging from small chips to large PLCCs and leadless packages. The tweezer action permits the tips to contact the solder joints, thus ensuring high heat delivery, but at the lowest possible temperature.

Convective Method (Hot Gas Soldering):   Most production and re-work stations use a hot gas or hot air as the heat transfer medium. With a single point nozzle, small parts such as chips, transistors, SOICs and flat packs can be removed. The hot gas is swept over the leads until full re-flow is achieved, after which the part is lifted with a tweezer. Although removal timer are longer than with a conductive tool, one tool and nozzle shape handles several applications. With longer components, a component specific nozzle is fitted to the hand-piece and brought around the part to re-move almost any two or few-sided SMD. The provision of vacuum provides component lift-off after re-flow. A re-work station that uses infra-red radiation to reflow the solder joints is also available. Ancillary features frequently include avacuum pick-up mechanism for removing the faulty device and magnification systems, sometimes with video display unit (VDU), to aid observation of the work-in progress.

13.18.3 Removal and Replacement of Surface Mount Devices

The following steps should be taken to remove a component using a hot gas machine :

- Apply a small amount of liquid flux to all joints.
- Choose the correct head to suit the component.
- With the PCB in place, activate the gas flow to re-flow the solder on every joint (use microscope /VDU to check).
- If the component has been bonded with an adhesive, rotate the head to shear the bond.
- Remove the component with the vacuum pick-up and allow the PCB to cool.
- Remove any remaining solder by the use of a fine de-soldering braid.
- Allow a further cool down period.
- Inspect the pads to ensure that they are not damaged.

For replacement of the component, the following procedure is followed;

- The new component should be carefully inspected to ensure that the legs are not bent or distorted. Ideally the legs of the device will slope down from the body by 1-2 degrees. This will allow the legs to flatten on to the pads when the component is placed on to the PCB.
- A thin film of flux is lightly applied to the pads.
- The component is then placed into the head of the hot gas machine and carefully lined up to the PCB. The fingers on the SolderQuick tape will help to align the component.
- Before gas flow is initiated, the component should be lifted away from the board surface until the legs are just clear of the pads.
- The gas glow should then be applied. The gas will heat up the legs and the solder on the pads.
- When the solder flows, the component should be carefully brought down on to the board, ensuring that the legs of the component are sitting between the fingers of tape and hence are central over the pads.
- Allow gas flow to continue for a few seconds to ensure that the solder flows correctly around each leg.
- When the solder flows correctly, switch off the gas and allow the board to cool at least for one minute to avoid disturbing the joints before removing the PCB from the machine.
- After removing the PCB carefully remove the SolderQuick tape and clean all excess flux from the joints.
- Inspect all joints with a X10 magnifier to ensure correct re-flow.
- Clean the PCB with isopropyl alcohol in the aerosol form to ensure penetration of solvent under the component to wash out any flux. The area can then be brushed to remove all traces of flux.

13.18.3.1 Repairing Damaged Pads

The most common damage on surface mount boards is lifted pads on quad flat pack (QFP) layouts. The most probable reason for this is when operators have difficulty in knowing when the solder joints on all four sides of the device package are molten. The following method is suggested to repair such type of damage:

- Remove the damaged pad/track and clean the immediate area on the board.
- Select appropriate replacement track/pad (These are available from a number of suppliers).
- Solder the replacement pad/track to the undamaged track on the board. Figure 13.53 (a) shows the replacement pad and portion of track together with the undamaged track to which it will be joined. The replacement track is cut so that it overlaps the undamaged track and the two parts are soldered together as shown in Figure 13.53(b). Purdie (1991) explains how damaged or incorrectly designed surface mount PCBs and PCB assemblies can be modified or restored to a good as new condition.

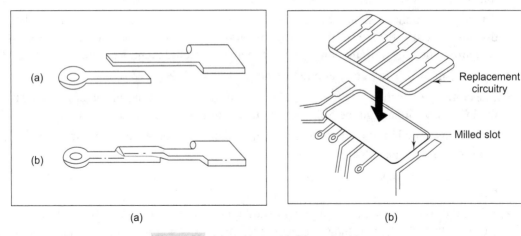

(a)   (b)

Fig. 13.53   (a) working repair (b) serviceable repair

- Using an adhesive capable of withstanding high temperature, glue the new pad/ track to the PCB substrate. Clamp together until the adhesive has cured.
- Solder the replacement component in place.
- Clean off flux residue.
- Re-apply any solder resist that has been removed.

Removing a surface mount component can be compared to 'steaming a stamp off of an envelope'. It is actually done by simultaneously melting the solder around a component's joints and then pick the component off of the PCB. The substrate is then cleaned and a new component is soldered back on to the circuit board. The best way to apply heat to the solder and component leads is a method of choice. There are conductive tools, convective tools, single point, multi-point, tinable and non-tinable tips which can be used for this purpose.

13.18.3.2 Repairing Damaged Plated Through-holes

There are three methods of remaking a damaged through hole connection on a double-sided board (Willis, 1992). These are:

- Through-hole Copper Plating: The process involves forcing a series of plating solutions through the hole under repair, thus simulating the original plating process. Not suitable for single piece repair.

- Fused Eyelets: The use of eyelet has been standard in the industry for a number of years. Designed to be formed flush with a circuit board, their overhang, on modern circuits where the track spacing is limited, may pose a constraint on their use.

- Use of Copper Bails: Copper bails are made by plating approximately 30 microns of copper onto solder wire of different sizes and then over-plating to protect the surface solderability. When positioned in the (reamed out) damaged through-hole, formed and fused, the copper bail does not take up any more room than the original through-hole plating and is undetectable as a repair after component insertion and subsequent soldering. Figure 13.54 shows the use of copper bail to replace a damaged PCB plated through-hole.

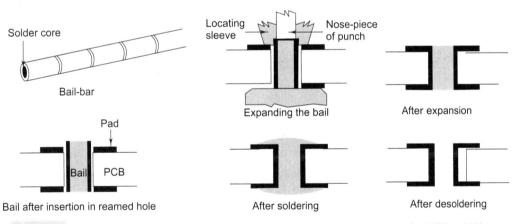

Fig. 13.54 Use of copper bail for repairing damaged plated through-holes (redrawn after Willis, 1992)

On a through-hole assembly, a defective component is de-soldered before the component can be removed and replaced. It is ensured that the solder is removed from the hole, and then cleaned, before a new component is placed on the board. On a surface mount board, it is unsoldered. The difference between the two is that on the through-hole board, molten solder is removed or sucked away from the lead and in plated through-hole by vacuum. With the use of hot air tool or solder pot, all leads on a through-hole can be reflowed simultaneously, allowing the component to be removed. On the surface mount board, all of the device leads must be heated simultaneously, the component must be lifted off the board before it can resolidify. If all leads are not heated concurrently and the device is pulled off before all the solder has been melted, the foot print on the board may be damaged. This can affect the co-planarity of the new component when it is placed on the PCB.

Hunn (1990) explains that for putting the heat where it is required, a range of heads has been designed which carry a series of fine nozzles to direct the air to the joint. Care has to be taken to reduce the amount of heat radiated from the main body of the head by keeping all of the hot metal as far away from the chip as possible and producing a cool zone above the chip. The effort is that the solder joint is the item that directly receives the heat and the chip body remains cool.

As more and more manufacturers include fine-pitch technology in their surface mount designs, the re-work process becomes even more complex. As board pitch becomes finer, boards are more sensitive to component misalignment and PCB heat damage. Re-working fine pitch boards usually requires some sort of a vision system. As the lead count becomes finer and finer, vision system that allows simultaneous viewing of the PCB and component is essential. Therefore, optical devices should be used for placement of fine pitch components to ensure proper alignment. Considering all these factors, an ideal re-work station would include :

- A vision system that can be used when placing and soldering the component;
- A placement tool that will allow for movement that is smaller than the smallest pitch being used on the board;
- A heating method that can control the heating process and can heat the board and the component in a manner that approximates the method used in the original production–it must be able to apply uniform heat without de-laminating the board or damaging the component during removal or replacement; and
- Facilities that are simple to use, both by an operator and engineer, without much training.

13.18.4   Re-work Stations

Today's printed circuit boards, with BGAs, DCAs, CSPs and fine-pitch SMDs require a level of precision and performance that cannot be met with hand-held tools. Adding to the difficulties of re-work are area array components. Since the bumps are on the bottom of the chip, interconnections with the pads are not easily aligned and inspected, and voids, bridges and other defects can go unnoticed until functional testing discovers them. Also, with manual de-soldering, using soldering iron and a wick control is required over several parameters such as tip temperature, dwell time at each pad, applied pressure, affected area, contact area and location. On the other hand, vacuum de-soldering tools require control over vacuum flow, distance from pad, hot air flow (if applicable) temperature source pressure etc. Most of these parameters are directly related to the operator skill may result in over heating and damage to pads, traces and solder masks. Automated workstation which eliminate depending on technician skills offer a practical way to secure consistent quality and cost-effectiveness in re-work operations.

Many different types of re-work equipment are available in the market today (Hodson, 1993b). One typical example of a re-work station is that of Model SD-3000 from M/s Howard Electronic Instruments, USA. It is a microprocessor controlled equipment using single nozzle blowing out hot air which traces along the soldered points of the SMD. The equipment is suitable for any size and shape of QFP, SOP, PLCC, PGA, BGA etc. to remove and/or reflow (solder). It is able to handle all

SMDs without changing the nozzle head. A built-in timer enables to prevent damage of PCB and nearby parts caused by over heating.

The various controls provided on this equipment are shown in Figure 13.55 and are as follows:

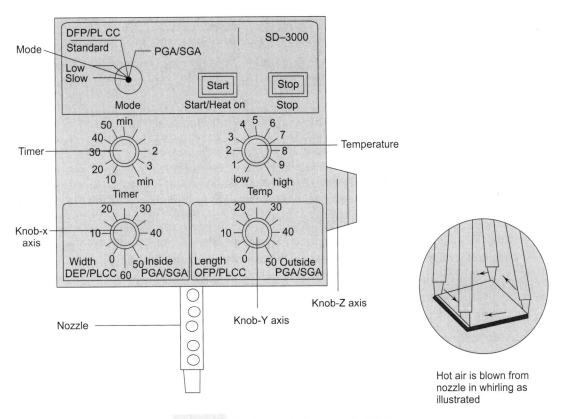

Fig. 13.55   Controls on typical re-work station

- X-Axis: This knob is used to adjust the nozzle width of the component to be re-flowed. It is also used as the inside adjustment for BGA/PGA removal.
- Y-Axis: This knob is used to adjust the nozzle length of the component to be re-flowed. It is also used as the outside adjustment for BGA/PGA removal.
- Z-Axis: The Z-Axis control adjusts the height of the nozzle above the solder points to be re-flowed.
- Nozzle: The nozzle is adjusted by the X, Y and Z-Axis knobs to whirl around the solder points of the component to be removed. Holes in the nozzle allow the operator to visually inspect the temperature of the heater according to the colour of the heater coils.
- Timer: The timer is used to set the time required to reach solder melt temperature after

the start button is pushed. At the end of the time cycle, the unit automatically goes into its cool down cycle and shuts off after reaching its cool down temperature.

- Temperature: This control is used to control the temperature of the heater in all modes, at the discretion of the operator.

- Mode: The mode switch provides facilities for removal of BGA/PGA, QFP/PLCC standard packages. The mode also has high air flow (12 litres/min.) and high temperature.

  In the LOW mode, lower rate of air flow (6 litre/min.) can be obtained. The mode is used for heavily populated boards so that small chips won't accidentally be blown from the board during re-flow.

  In addition, SLOW mode is available which is used for replacing the QFP/PLCC packages after old solder has been removed from pads and new solder paste has been applied to the new component to be re-flowed.

- Start: Pushing the start button the first time starts the nozzle rotating to allow adjustment of the width and length of the nozzle. After adjustment is complete, pushing the start button a second time starts the air flow, heat and timer.

- Stop: Pushing the stop button at any time will stop the heat and raise the nozzle approximately one half inch to allow vacuum picking the component from the PC board.

  The other facilities available on the re-work station are a mechanism for holding the PCB, applying vacuum to pickup the IC to be removed from the board and providing ease of sliding the PCB for alignment of component under the hot air nozzle.

While re-working on QFP or PLCC components, the following will assist in getting faster re-flow times and lower temperatures:

- Keep nozzle height at 1 or 2 mm above the board at all times. This might require a fixture to hold the board and heater head/nozzle assembly.

- Use as high an airflow rate as possible without over heating peripheral solder joints.

- Use flux if desired.

With these steps, the technician should be able to develop his own re-work process using connective tools and to understand the effects on assemblies. Buckley (1990b) details the procedure for cleaning, inspection, re-work and testing of surface mount assemblies.

Re-work stations are also available which make use of medium wavelength infra red radiation, emitting radiation in the range of 2 to 8 μm. However, it is desirable that the re-work system should completely protect heat-sensitive components. This is possible using IR technology as the radiation can be shielded by the use of heat resistant tape or aluminium foil, thereby keeping the solder joint temperature of an adjacent chip well below its melting point, even at a distance of 0.5 mm. Such a system is available from M/s ERSA GmbH and is shown in Figure 13.56.

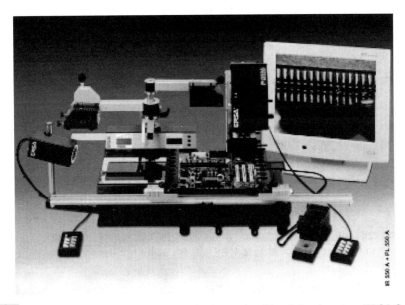

Fig. 13.56   Re-work station based on re-flow technology using IR radiation (courtesy ERSA GmbH)

 ## 13.19 Useful Standards

- *IPC-ESD-20 20: Association Standard for the Development of an ESD Control Program:* Covers the requirements necessary to design, establish, implement and maintain an Electrostatic Discharge (ESD) control program; offers guidance to protect and handle ESD sensitive times, based on the historical experience of both military and commercial organizations.

- *IPC-SA-61A: Post-solder Semi-aqueous Cleaning Handbook:* Covers aspects of semi-aqueous cleaning such as chemical, manufacturing residues, equipment and processes, process control, environmental considerations and safety.

- *IPC-AC-62A: Aqueous Post-solder Cleaning Handbook*: Describes manufacturing residues, types and properties of aqueous cleaning agents, aqueous cleaning processes and equipment, process and quality control, environmental controls and considerations, worker safety, cleanliness determination, measurement and cost.

- *IPC-DRM-40E: Through-hole Solder Joint Evaluation Desk Reference Manual*: Includes computer-generated 3-D graphics, as well as close-up photographic illustrations of component, barrel and solder-side coverage as per requirements in the standards; covers solder fillet, contact angle, wetting, vertical fill, land coverage and numerous defect conditions for solder joints.

- *IPC-TA-722: Technology Assessment Handbook on Soldering*: Contains 45 articles on all aspects of soldering covering general soldering, soldering materials, manual soldering, mass soldering, wave soldering, re-flow, and vapour phase and infra-red soldering.
- *IPC-7525: Stencil Design Guidelines*: Provides guidelines for the design and fabrication of stencils for solder paste and surface mount adhesive; also discusses stencil design for surface-mount technology, as well as mixed technology with through hole or flip chip components including overprint, two-print and step stencil designs.
- *IPC/EIA J-STD-004: Requirements for Soldering Fluxes-Includes Amendment 1*: Covers requirements for qualification and classification of rosin, resin, organic and inorganic fluxes according to the activity level and halide content of the fluxes; also addresses solder fluxes, flux-containing materials and low residue fluxes for no-clean process.
- *IPC/EIA J -STD -005: Requirements for Soldering Pastes–Includes Amendment 1*: Lists requirements for qualification and characterization of solder paste; also includes test methods and criteria for metal content, viscosity, slump, solder ball, tack and wetting of solder pastes.
- *IPC/EIA J-STD-006A: Requirements for Electronic Grade Solder Alloys and Fluxed and Non-fluxed Solid Solders*: Provides the nomenclature, requirements and test methods for electronic grade solder alloys; for fluxed and non-fluxed bar, ribbon, and powder solders, for electronic soldering applications; and for "special electronic grade solders.
- *IPC-Ca-821: General Requirements for Thermally Conductive Adhesives*: Includes requirements and test methods for thermally conductive dielectric adhesives used to bond components in place.
- *IPC-3406: Guidelines for Electrically Conductive Surface Mount Adhesives*: Covers guidelines for selecting electrically conductive adhesives for use in electronics manufacturing as solder alternatives.
- *IPC-AJ-820: Assembly and Joining Handbook*: Contains descriptions of proven techniques for assembly and soldering; includes terms and definitions; specification references and synopsis, design, printed circuit boards, component/lead types, joining materials, component mounting; solderability; joining techniques and packaging; cleaning and coating; and quality assurance and testing.
- *IPC-7530: Guidelines for Temperature Profiling for Mass Soldering (Re-flow and Wave) Processes*: Provides guidelines for the construction of appropriate profiling test vehicles and various techniques and methodologies for temperature profiling.
- *IPC-TR-460A: Trouble shooting Checklist for Wave Soldering Printed Wiring Boards*: Provides a checklist of causes/recommended corrective action for wave soldering.
- *IPC/EIA/JEDEC J-STD-003A*: Solderability Tests for Printed Boards.
- *J-STD-013: Implementation of Ball Grid Array and Other High Density Technology*: Establishes the requirements and interactions necessary for printed board assembly process for interconnecting high performance/high pin count IC packages and covers information on design principles, material selection, and board fabrication, assembly technology, testing strategy and reliability expectations based on end-user environments.

- *IPC-7095: Design and Assembly Process Implementation for BGAs*: Delivers useful and practical information to anyone who is currently using BGAs or is considering a conversion to area array packaging formats; provides guidelines for BGA inspection and repair and addresses reliability issues associated with BGAs.

- *IPC-M-108: Cleaning Guides and Handbooks Manual*: Includes the latest editions of IPC cleaning guides and handbooks and assists manufacturing engineers when making decisions on cleaning products and processes and provides guidance for troubleshooting.

- *IPC-CH-65-A: Guidelines for Cleaning of Printed Boards and Assemblies*: A roadmap for current and emerging cleaning issues in the electronics industry includes description and discussions of various cleaning methods; explains the relationship between materials, processes and contaminants in fabrication and assembly operations.

- *IPC-SC-60A: Post-solder Solvent Cleaning Handbook*: Addresses the use of solvent cleaning technology in automated and manual soldering operations; discusses properties of solvents, residues, considerations on process control and environmental issues.

- *IPC-9201: Surface Insulation Resistance Handbook*: Covers the terminology, theories, test procedures and test vehicles of surface insulation resistance (SIR) or temperature-humidity (TH) testing; includes failure modes and troubleshooting.

- *IPC-DRM-53: Introduction to Electronics Assembly Desk Reference Manual*: Includes photographs and graphic drawings to illustrate assembly technologies, for both through-hole and surface mount assembly.

- *IPC-M-103: Standards for Surface Mount Assemblies Manual*: The set includes all the 21 IPC documents for surface mounting.

- *IPC-M-104: Standards for Printed Board Assembly Manual*: Contains ten of the most widely used documents on printed board assembly.

- *IPC-CC-830B: Qualification and Performance of Electrical Insulating Compound for Printed Wiring Assemblies*: An industry standard for qualification and quality conformance of conformal coating.

- *IPC-S-816: SMT Process Guideline and Checklist*: This troubleshooting guide lists all types of processing problems and solutions for surface mount assembly; covers bridging, skips, misalignment, placement, etc.

- *IPC-CM-770D: Component Mounting Guidelines for Printed Boards*: Provides effective guidelines in the preparation of components for printed circuit board assembly and reviews pertinent design criteria, impacts and issues; contains techniques for assembly (both manual and machines including SMT and flip chip) and consideration of, and impact upon, subsequent soldering, cleaning, and coating processes.

- *IPC-7129: Calculation of DPMO and Manufacturing Indices for Printed Board Assemblies*: Industries consensus documents on calculating benchmark indices for defects and quality; provides consistent methodologies for calculating benchmark indices for DPMO (defects permillion opportunities) index.

- *IPC-9261: IN-Process DPMO and Estimated Yield for PWAs*: Defines consistent methodologies for computation of in-process defects per million opportunities (DPMO) metrics for any evaluation stage in the assembly process.

- *IPC-D-279: Design Guidelines for Reliable Surface Mount Technology Printed Board Assemblies*: Includes design concepts, guidelines and procedures for reliable printed circuit assemblies covering SMT or mixed technology boards.

- *IPC-2546: Sectional Requirements for Shop Floor Equipment Communication Messages (CAMX) for Printed Circuit Board Assembly*: Describes material movement systems like conveyors and buffers, manual placement, automated screen printing, automated adhesive dispensing, automated surface mount placement, automated plated through-hole placement forced convection and infra-red reflow ovens and wave soldering.

- *IPC-PE-740A: Troubleshooting for Printed Board Manufacture and Assembly*: Contains case histories of problems and corrective action in the design, manufacture, and assembly and testing of printed circuit products.

- *IPC-6010: Qualification and Performance Series*: Includes IPC's qualification and performance specification standards for all printed circuit boards.

- *IPC-6018A: Microwave End-product Board Inspection and Test*: Covers requirements for qualification and performance of high frequency (microwave) printed circuit boards.

- *IPC-D-317A: Design Guidelines for Electronic Packaging Utilizing High Speed Techniques*: Provides guidelines for the design of high-speed circuitry including mechanical and electrical considerations and performance testing.

Quality, Reliability and Acceptability Aspects

14.1   What is Quality Assurance?

Quality Assurance (QA) is the activity consisting of all those planned or systemic actions necessary to provide adequate requirements for ensuring quality. QA basically provides the evidence needed to establish confidence among all concerned, that quality-related activities are being performed effectively. Quality assurance is analogous to the concept of the financial audit which provides assurance of financial integrity through an independent audit. In today's world, as independent certified public accountants have become an influential force in the field of finance, quality assurance experts are finding an increasingly important role in manufacturing and service sectors.

All manufacturing houses, including PCB manufacturers, have QA departments, which are concerned with all quality-related activities such as quality planning, quality control, quality improvement, quality audit and reliability issues.

In quality assurance, the assurance comes from evidence or a set of facts. For simple products, the evidence is usually some form of inspections or a set of facts and test data. However, for complex products, the evidence includes not only inspection and test data, but also review of plans and audits of the execution of plans. Thus, a family of assurance techniques is available to cover varied needs of products.

Quality, which is defined as fitness for purpose, will be achieved only if integrated into the design and manufacturing process. Therefore, the amount of testing involved in the manufacture of the printed circuits should be carefully planned so that all the faults or defects are picked up, rectified and brought within acceptable limits.

14.1.1 Classification of Defects

Any aspect or parameter of the printed board that does not conform to specified requirements is known as a defect. Defects are normally classified according to the degree of their seriousness or their eventual effect on the performance of the board. Defects are of three types:

- *Critical Defect*: A defect, that judgment and experience indicate, is likely to result in hazardous or unsafe conditions for individuals using, maintaining or depending upon the product, is called a critical defect. This type of defect may also lead to an unsatisfactory performance of a space vehicle, aircraft or ship, etc, which *could be catastrophic.*

- *Major Defect:* A major defect is a defect that is likely to result in failure or to reduce substantially the usability of the unit or board for its intended purpose.

- *Minor Defect*: This is a defect which is not likely to materially reduce the usability of the unit or product or board for its intended purpose.

The number of defects per hundred units of any given quantity of units of product is 100 times the number of defects contained therein, divided by the total number of the units of product:

$$\text{Defects per hundred units} = \frac{\text{Number of Defects}}{\text{Number of units inspected}} \times 100$$

14.1.2 Defectives

A defective is a product, unit or board which contains one or more defects. Defectives are usually classified as critical, major and minor depending upon the nature of the defect.

The defectives per hundred unit is defined as:

$$\text{Per cent defectives} = \frac{\text{Number of defectives}}{\text{Number of units inspected}} \times 100$$

14.1.3 Acceptability Quality Level (AQL)

The 'Acceptability Quality Level' is the maximum per cent defective or the maximum number of defects per hundred units that, for purposes of sampling inspection, can be considered satisfactory as a process average. This implies that AQL is the designated value of the per cent defective (or defects per hundred units) that the consumer would accept.

14.1.4 Quality Control Programme

A quality assurance programme in the PCB industry requires controls of various steps, starting from the incoming material to the finished product in the form of a fully tested and assembled board. The steps for implementing a quality assurance programme are detailed below.

- *Incoming Material Control*: All incoming raw materials are sampled and inspected for compliance with drawings, specifications and other conditions of procurement contract. The inspection may involve the use of specialized equipment, chemical analysis and tests within the organization or in an outside test laboratory.

- *In Process Control:* This involves the identification of critical inspection points in the production flow and the material inspected by random sampling. This enables one to take remedial action in case the process goes out of specifications or control. It also helps to isolate non-salvageable rejects and they are removed from further processing. Examples of such inspections are the measurement of plating thickness and conducting the plating adhesion test. Similarly, each board is tested for its electrical functions and for shorts, circuit continuity and circuit verifications.

  For proper process control, production and inspection, equipment and tools also need to be checked periodically to determine wear and tear, mechanical alignment and electrical calibration, and maintenance of proper documentation on the parameters.

- *Complete Product Control:* The final check is made on at least one board from each 300 completely packaged boards, to check completeness of packaging, identification and overall quality, including the functional electrical test.

14.1.5 Statistical Process Control and Sampling Plan

Statistical process control is the application of statistical methods to the measurement and analysis of variation in process. A process is said to be in a state of control if the variation is such as would occur in random sampling from stable population. The inspection of materials in the semi-finished or complete state is called 'acceptance sampling'. The objective of acceptance sampling is to evaluate a certain lot of material about whose quality a decision must be made. This is done by inspecting a sample of material, by using statistical standards, to infer from the lot if it is acceptable (the term 'lot or batch' means a collection of units of a product from which a sample is to be drawn and inspected to determine conformance with the acceptability criteria).

The sampling plan is set according to what is required of the product rather than the inherent capability of the process. The sampling plan in case of printed boards is decided on the basis of the following factors:

- The complexity of the operations to be controlled;
- The defect occurrence levels of the operations;

- Required frequency of examination per lot quantity;
- Allowable delay time for in-process testing; and
- The size of the lot to be sampled.

The sampling plan chosen in each control category and the method used to perform the quality control are determined from the standards on the subject. For example, the sampling plan for incoming materials requires that 125 pieces be drawn at random from 3000 for inspection and evaluation. If three or more major defects are found, the lot is rejected. Sampling plan for in-process control is determined by the quality control supervisor on the basis of the number of out-of-control processes.

14.2 Testing for Quality Control

Testing for quality control is conducted in the following stages:

- *Pre-production Testing:* It includes visual and dimensional examination, micro-sectioning, tests for plating adhesion, terminal pull, warp and twist, water absorption, solderability, plated through-hole structure, thermal shock, thermal stress, interconnection stress, moisture and insulation resistance, dielectric strength and current breakdown.

- *Production Testing:* Testing is done to a minimum level during the day-to-day production phase provided the quality is maintained. Under the normal conditions, the following tests are performed:
  - Visual and dimensional inspection;
  - Micro-sectioning in one plane;
  - Plating adhesion, moisture and insulation resistance tests on the test coupons; and
  - Circuitry tests on the boards.

For effective sampling tests, destructive tests are done every week or after every five thousand board, whichever occurs first. The following tests are done:

- Visual and dimensional inspection;
- Micro-sectioning in the principal planes; and
- Tests for terminal pull, water absorption, solderability, plated through-hole structure, thermal shock, thermal stress, interconnection resistance, dielectric strength, current-carrying capacity etc.

If failures occur in any one or more parameter(s), normal production is resumed only after the defect is detected and corrective action taken.

- *Final Testing:* Final testing is very crucial for the testing of multi-layer boards. The two most critical requirements which must be tested are:
  - Micro-sectioning of one test coupon from each production panel; and
  - Electrical circuitry testing of the finished boards.

In multi-layer boards, the mechanical integrity of the plated through-holes is of the utmost importance.

14.2.1 Characteristics for Testing of Quality Assurance

As explained in the previous section, a number of characteristics are required to be tested to ensure proper quality assurance. Table 14.1 shows all such requirements. In this Table 14.1, the following define the various levels:

- LR = Commercial boards, with limited requirements of tests;
- HR = High reliability requirements;
- MIL = Military specifications; and
- 1,2,3 = Three levels of quality designated by IPC.

Table 14.1 Characteristics for Testing Various Levels of Quality (after Coombs, 1988)

Requirements	LR	1	2	3	MIL	HR
Circuitry electrical test (100 %)					√	√
Current-carrying capacity	√		√	√		√
Dielectric strength		√	√	√	√	√
Etch-back						√
Flammability			√	√	√	√
Insulation resistance		√	√	√	√	√
Internal shorts	√	√	√	√		√
Mechanical shock						√
Micro-sectioning					√	√
Moisture resistance				√	√	√
Outgassing						√
Plating adhesion		√	√	√	√	√
PTH structure					√	√
Solderability	√			√		√
Terminal pull		√	√	√	√	√
Thermal shock					√	√
Thermal stress					√	√
Traceability						√
Vibration						√

(Contd.)

Table 14.1 (*Contd.*)

Requirements	LR	1	2	3	MIL	HR
Visual and dimensional characteristics	√	√	√	√	√	√
Warp and twist				√	√	√	√
Water absorption						√

The quality of the final printed circuit board (PCB) depends upon the cumulative quality of many successive fabrication steps. The entire process, beginning with raw material and tooling through the final inspection and test, may incorporate as many as forty or more process steps, each with its own opportunity for defects. The cumulative impact of an uncontrolled operation can be devastating on both the quality and reliability of the products (Watts, 1993).

For this reason, a growing number of companies are electing to invest in a balanced quality assurance (QA) programme, which typically combines incoming inspections, in-process product checks and process controls with quality conformance and reliability assurance tests.

The first three serve as preventive measures for minimizing the value added and probability of defects. Quality conformance tests act as screens, filtering out escapes from the preventive measures in place. The reliability evaluations expose latent or hidden defects that cannot be detected by any other means.

The PCB manufacturer should include all types of tests in an overall quality programme. These tests should be tailored to match resources, process capability and customer requirements.

14.2.2 Designing a QA Programme

Given the number of process steps in a typical PCB fabrication line, a step-by-step evaluation programme for monitoring product quality at every significant step is both impractical and impossible. The competitive company will find the right optimum of process control to supplement product inspection (Table 14.1) during the fabrication process.

An optimum quality assurance programme is achieved best by prioritizing the value of inspection at each process step. The final choice of evaluation steps can be based on any number of criteria but should take into consideration the impact of quality as it relates to the following:

- Circuit functionality;
- Process capability;
- Circuit long-term reliability;
- Containment of defects;
- Re-work capability;
- Customer requirements;
- Subsequent processing (assembly); and
- Product cost.

From the standpoint of process control, not enough can be said about the importance of a well-equipped and well-staffed control lab. Circuit fabrication relies upon chemical consistency in 75 per cent of its processing steps. If a manufacturer is really serious about preventing defects through process control, investing in a control lab is as fundamental as the PCB fabrication process itself. Although most chemical analyses can be performed by using a basic titration test, more control is achieved with cyclic voltametric stripping (CVS), atomic absorption, UV visible spectroscopes and gas chromatographic analysis.

14.2.3   Incoming QA

Despite the numerous raw materials that feed a PCB fabrication process, very few are actually inspected on-site by the manufacturer. Again, it is both impractical and unnecessary to do so. Decades of development work has resulted in very uniform chemistries, resists, laminates and drills. Stiff competition within the supplier sector has also resulted in products that are continually upgraded to keep pace with ever-changing customer demand.

Manufacturers with available laboratory facilities can perform relatively quick tests on laminates and prepregs to monitor the consistency in flow, cure and resin content of materials entering their processes. A visual audit of copper clad laminates and prepregs is sufficient to sort out the lots with gross defects.

14.2.4   Traceability

For all boards used in high reliability applications/products, it is necessary to establish and maintain a traceability programme. This is required because of the possibility of latent defects which may appear at a later stage and affect the reliability of the product. The traceability programme starts at the lamination panel level and continues throughout the process of flow upto the serial number of the equipment in which it is used. Similarly, a record of the rejected boards is also useful to identify the weak links in quality control in the process.

 ## 14.3   Quality Control Methods

One major aspect in any manufacturing is to maintain the quality. In PCB production reliable electrical and mechanical performance of the PCB must be achieved by using the proper QC method. The control of the quality in the manufacture of a PCB starts right from the procurement of the basic raw materials and the design stage.

The basic quality control methods in the manufacturing stage of a PCB can be classified into the following three categories:

   a) Physical inspection;
   b) Optical inspection; and
   c) Electrical inspection.

In the physical examination, the PCB is subjected to various dimensional measurements and the neatness of the pattern. In optical inspection, the PCB is subjected to various observations under a microscope and other inspection equipment which reveal a lot of information about the manufacturing process and the product. One of the major and the most reliable methods of inspection is called micro-sectioning, which reveals a lot of information about the quality of the board by giving an insight into the quality of the plating process, thickness of plating, etc.

14.3.1 Micro-sectioning

One of the primary ways to determine PCB quality is micro-section evaluation. A micro-section is a tiny portion of a PCB, mounted in a hard moulding material to reveal a sample of the semi-cylindrical plated through-hole (PTH).

Microscopic analysis of this section provides an inside view of the processes and quality controls that were used in the manufacture of the board. The micro-section can be used to assess operations such as copper plating, etch-back, drilling and lamination.

Most PCB suppliers have the equipment and skills needed to fabricate and evaluate micro-section samples. Many can supply customers with evaluation data and cross-sections.

However, producing micro-sections in-house can also be beneficial. Cross-sectioning a sample board can be used to evaluate a new board supplier or as a standard incoming procedure. This technique also helps to discover the causes of process-related failures, such as barrel cracking, blowhole formation and mechanical stress effects.

Equipment and Cost: A micro-sectioning system includes a shop microscope, mounting compound and other miscellaneous materials which must have a minimum magnification power of 100X and a 0.010-mm-resolution reticle (Falco, 1991).

Micro-sectioning is a destructive method. It is generally carried out on a test piece which is processed along with the actual PCB having the necessary circuit pattern with the same parameters. Micro-sectioning is done by carefully removing a small portion from the bulk material and then encapsulating it in suitable plastic to preserve the metallic layers during moulding and grinding operations.

After the mould is cured, it is subjected to a series of abrasive operations to obtain a highly polished surface that is suitable for microscopic examination. A lot of information including the layer thickness and geometric irregularities in plating becomes available through this examination. Verification of plating thickness is usually done either individually or as an average as illustrated in Figure 14.1.

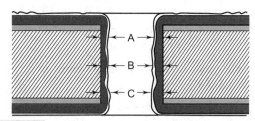

Fig. 14.1 Verification of plating thickness in holes by use of vertical cross-sections (after Coombs, 1988)

Steps Involved in Micro-sectioning: Micro-sectioning involves the following basic steps involved in performing a successful sectioning of PCB through-hole:

- Bulk cutting;
- Precision cutting;
- Mounting or encapsulation;
- Fine grinding;
- Rough polish;
- Final polish; and
- Observation under microscope;

Bulk Cutting: It involves removing a small portion which includes the area to be observed from the parent board. The bulk cut is done away from the area under observation, to avoid damage to the adjacent metallic layers.

Precision Cutting: Precision cutting is a critical procedure to produce a damage-free cut at or nearer to the plane for examination. A low speed saw is used for this purpose. The bulk cut board is fed into the thin rotating diamond blade, which is best suited to make a precision cut on the printed circuit board.

A micrometer, fixed on the feed arm gives a measure of the cut which enables the operator to precisely perform the cut to obtain the exact plane to be inspected.

Mounting or Encapsulation: Mounting or encapsulation provides the necessary protection needed to prevent the delicate outer layers of metal from getting damaged during subsequent grinding operations. It also makes it comfortable to hold the specimen during observations under microscope. The specimen is generally moulded in a room curing epoxy resin which has very low shrinkage, good hardness and is transparent. The common epoxies take about 6-8 hours for complete curing.

Fine Grinding: Fine grinding is done through a series of abrasive steps with 240, 320, 400 and 600 grid silicon carbide papers to reduce the damage to the specimen. The grinding can be done either manually or on a rotating grinding wheel. After the grinding process, the specimen is thoroughly cleaned with water prior to cloth polishing in order to remove the abrasive particles.

Rough Polishing: Rough polishing is undertaken to remove the remnant fine grinding scratches and to produce a flat surface for accurate results. This is done by using diamond paste and holding the specimen against a rotating grinding wheel. The diamond paste should be evenly distributed on the surface of the specimen using a cotton ball. The specimen is thoroughly cleaned before final polishing.

Final Polishing: Final polishing is done to produce a lustrous, scratch-free and flat surface to obtain accurate results. This is generally done with an aqueous slurry of alumina powder and then by holding the sample against a rotating wheel.

Microscopic Observations: The microscope used for observing the section should be of metallurgical type having an integral vertical illumination system. The other pre-requisites of a suitable microscope are:

- Objective lenses with a magnification range from 5x to 50x with flat field lenses;
- Eye piece with cross-hairs — to make the thickness measurement of the plated layers.

- Illumination — Tungsten bulbs are adequate for simple observations. Xenon or halogen lamp is preferred for photography.
- Camera — to take photographs of the inner layers.

Applications of Micro-sectioning in Printed Circuits

Testing Stage/Parameter	Functions Checked
Incoming laminate	• To evaluate fabric build • Detect substrate and foil defect
Electroless copper	• Plating quality, thickness • Drilled hole quality
Solder fusion	• Electrolytic copper and tin lead quality and thickness • Plating uniformity and thickness
Gold tab	• Plating uniformity and thickness

Since micro-sectioning is a destructive test, a series of extra holes are provided on the PCB outside the actual circuitry and processed under the same conditions as the actual circuit. An analysis is then carried out on these holes thereby getting a very good insight into the process and quality of the PCB without destroying the actual circuit pattern.

14.4   Testing of Printed Circuit Boards

The manufacture of printed circuit boards comprises a large number of process steps. In order to produce a quality product, it is necessary that proper inspection and testing are carried out at every step in the process. In spite of conducting in-process quality control measures, final testing is vital to ensure the reliability of the product.

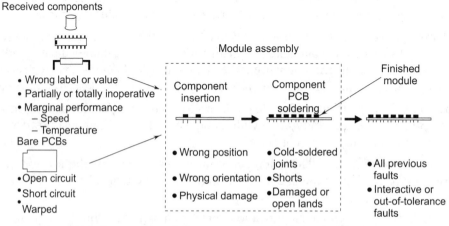

Fig. 14.2   Spectrum of module faults: approximately 75 per cent of module faults occur in assembly, 20 per cent are component faults, and 5 per cent PCB faults (Adopted from Coombs, 1988)

From experience and statistics, it has been estimated that approximately 75 per cent of faults in an assembled PCB occur in assembly, 20 per cent are component faults and 5 per cent are the board faults. Figure 14.2 shows some common faults in assembled printed circuit boards.

14.4.1 Automatic Board Testing

With the introduction of surface mount technology, the board packing density has increased manifold. So, even for boards of modest density and moderate board quantities, automatic board testing is not only essential but economical also. Two approaches have been common in the testing of complex boards: bed of nails method and two probe or flying probe method.

14.4.1.1 Bed of Nails Method

This method consists of a spring loaded pin brought down on each test point on the board. The spring action enables one to achieve a pressure of 100-200 grams to ensure a good contact on each test point. An array of such test pins is called 'bed-of nails'. Under the control of test software, test points and the signals to be applied to them can be programmed: Figure 14.3 shows a typical bed-of-nails fixture set-up. The tester is provided with all the information about the test points and only those pins which correspond to the required points are actually set-up. While designing the board, it is advisable to keep all the test points on the solder side of the board, though it is possible to conduct tests simultaneously on both sides of the board with the bed-of-nails approach. The equipment is expensive and difficult to maintain. Nails come in different pin head arrangements depending upon their applications.

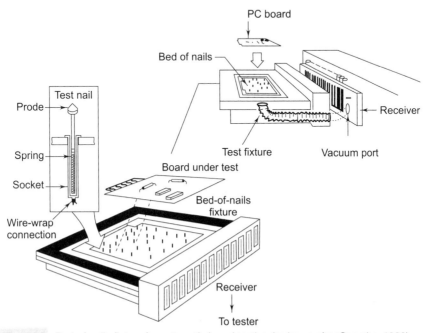

Fig. 14.3 Bed-of-nails fixture for automatic board testing (redrawn after Coombs, 1988)

A basic universal grid handler consists of a drilled plate populated with pins spaced on 100, 75 or 50 mil centres. The pins act as probes and make direct mechanical contact with electrical connections or nodes on the circuit boards. If the pads on the boards match the test grid, a mylar sheet drilled to specification is placed between the grid and the circuit board to enable design specific probing. Continuity is tested by accessing the end points of a net, defined as the X-Y co-ordinates of a pad. Since every net on the board is tested for continuity, an isolation test is thus performed. However, the effectiveness of a bed-of-nails fixtures is limited by pin proximity.

14.4.1.2 Two Probe or Flying Probe Method

The flying probe tester does not rely upon a pattern of pins built onto a fixture or holder. Depending upon the system, two or more probes mounted onto tiny heads move freely in an X-Y plane to test points as directed by CAD/Gerber data. The two probes can come within 4 mil of each other. The probes move independently and there are no real limitations in terms of how close they can get to each other.

A tester with two roving arms can be based on the measurement of capacitance. The board forms one plate of a capacitor by being pressed onto a dielectric layer over a plate. In case of a short-circuit between the tracks, the capacitance will be larger than it should be at a particular point. If there is an open circuit, the capacitance will be lesser.

The speed of testing is an important criterion for the selection of the tester. While a bed-of-nails fixture can test literally thousands of test points at a time, the flying probe is only able to test two or four points at a time. Also, while single-sided testing on a bed-of-nails may take only 20 to 30 seconds, depending upon the complexity of the board, a flying probe requires as much as an hour or more to perform the same evaluation. Shipley (1991) explains that even though the moving probe technology is considered slow by high volume PCB manufacturers, the method does offer benefits to the lower volume producer of complex boards.

For testing of bare boards, special types of fixtures are designed (Lea, 1990). A cost-effective method is to have a universal fixture. Although this type of fixture is initially more expensive than a dedicated fixture, its initial high cost is offset by reducing the individual set-up costs. In a universal grid, the normal grid for leaded component boards and for surface mount devices is 2.5 mm. Here, the test pads should be greater than or equal to 1.3 mm. For grids of 1mm, test pads are designed to be greater than 0.7 mm. In case of smaller grids, the test pins are small and fragile and easily get damaged. So, it is preferable to have the grid greater than 2.5 mm.

Crum (1994b) illustrates that using a combination of universal (standard grid testers) and flying probe testers enables one to provide accuracy and economy in testing high density boards. Another approach suggested by him is the use of conductive rubber fixtures. This technique makes it possible to test off-grid points. However, the varying heights of solder pads created by hot-air solder levelling prevents all the test points from being contacted.

Testing is usually carried out at the following three levels:

- Bare Board Testing;
- In-circuit Testing; and
- Functional Testing.

The testers can be universal type that may be useful for a range of board styles and types or they may be dedicated for a specific application.

14.4.2 Bare Board Testing (BBT)

The increasing track density and increasing number of via holes make it important to test the PCB before the assembly operation begins. An earlier PCB on an average had about 400 through-holes out of which about 25 per cent were via holes and an average of 200 networks. As the density has increased, a typical PCB is likely to have about 2000 through-holes, 40 per cent of which are via holes and a network connection of about 600. These highly populated PCBs have increased the failure rate, which may at times be even upto 20 per cent.

Some extensive failure mechanisms have been identified, which can lead to a total circuit failure at the advanced level. These failures may prove to be extremely expensive in the case of high density and multi-player PCBs, which have a track separation of less than 0.01 inch (0.25 mm).

It is also observed that under extreme humid conditions, there will be an electrochemical growth on the copper which can lead to some shorts. Improper applications of dry film solder mask leads to voids appearing on the board, especially between closely spaced conductors. The voids can entrap moisture and dust particles which can lead to shorts. These types of failures occur because of shorts, opens, cuts, leakage and contamination. Hence, the boards must be tested before PCB assembly.

With the increasing track density and number of through-holes, it has become necessary to test the printed circuit board before assembly. It has been observed that the failure rate in highly populated printed circuits may be as high as 20 per cent. If the boards are not tested at the pre-assembly stage, the failures at a later stage may prove to be extremely expensive in the case of high density and multi-layer boards. Before populating a board with expensive devices such as application-specific ICs and microprocessors, it is cost-effective to first check whether the bare board meets expected quality standards. Bare board testing is thus becoming mandatory for the PCB manufacturers.

14.4.2.1 Causes of Boards' Failure

Normally the reliable functioning of a PCB is taken for granted unless a serious field failure occurs. A PCB is expected to perform perfectly well with respect to the application for which it is designed. It should be able to withstand the environmental effects depending upon the mechanical, electrical and chemical nature of the product. Generally, the performance and reliability of the PCB to a large extent, depends upon the electrical, mechanical and chemical characteristics of the copper clad laminate used in the manufacture of the PCB.

The presence of dust or any other form of contamination on the board can result in electrical leakage, corrosion and failure of electronic components which can lead to the failure of the total equipment. These contamination tests can be performed on a bare board level and some of the faults that are generally found on a single-sided and double-sided PCB, such as opens, shorts, silvers, etc. can be easily corrected whereas the same faults occurring on the inner layers of multi-layer boards are difficult to eliminate.

14.4.2.2   Testing Techniques

As the PCBs are becoming more and more dense, the testing of a bare board is becoming mandatory for the PCB manufacturers. There are fully automatic equipments available which can test anywhere between 10,000 to 50,000 nodes and networks. The electrical tests must be performed at very high voltages to check for contaminations.

The commonly used methodologies for bare board testing are automated optical inspection (AOI) and electrical test. AOI is an in-process tool to check on the inner layers but is limited in its ability to verify, for example, the electrical continuity of plated through-holes. Similarly, AOI cannot identify poor multi-layer interconnections after lamination (Dytrych, 1993b). The electrical tests, on the other hand, help to identify potential defects like ionic contamination and hairline cracks, which are difficult to detect through AOI. In case of small geometries, most fabricators are doing a combination of electrical tests and automated optical inspection on final product. Straw (1992) details the solutions to some of the fine pitch bare board electrical test challenges.

Bare board testing generally checks for short-circuits between tracks and continuity of tracks. These tests can be performed by fully automatic machines which can test upto 50,000 nodes and networks.

The cuts and shorts in case of single-sided and double-sided PCBs can also be detected by the age-old method of visual inspection by optical means. A more sophisticated method of inspection of multi-layers is by the use of X-ray scanning, which gives an in-depth inside view of the board structure, thereby enabling one to detect the faults quickly.

14.4.2.3   Electrical Tests

Isolation Resistance: For performing this test, initially the test voltages are accurately determined in order to ascertain the resistance. Thus, the test is performed from one network to all other networks. Once the leakage is established, the tests can be repeated from network to network in order to determine the exact leakage path and the same could be re-worked. Generally, the isolation resistance is performed by applying 100 to 250 V and the failure threshold will be typically in the range of 10 to 100 Mohm.

Breakdown: The voltage breakdown tests are dependent upon time and the voltage applied. Generally these tests are performed for 10 ms during which a consistent result can be achieved. However, during this time, the energy discharged should be controlled in order to avoid damage to the PCB under test. This breakdown test may not be necessary in case of boards where their operating voltage is low.

Continuity: This is one of the commonest and most widely used tests performed on any PCB. These tests are generally conducted between two nodes on the same network in order to ensure accurate results. Continuity measurements are usually made in a 10 ohm to 500 ohm range.

14.4.2.4   Fixtures Used for Bare Board Testing

Even though most of the bed-of-nails fixtures used for BBT look alike, several factors need to be considered in order to maximize the use of the fixture and to make them cost-effective. For this reason, most of the grids are wired upto 50,000 test points and used as a general universal fixture even though individual test fixtures could be designed or purchased.

The Standard Grid Pattern: The basic idea that has to be borne in mind while selecting any grid fixtures is to minimize the nodal test points but at the same time to be able to test maximum area. One of the simplest ways of doing this is to etch a number of smaller circuits on one bigger panel which saves lot of handling time and consequently the cost involved in checking. The smaller boards can thus be cut out of the bigger panel even after the assembly time. Testing these small boards will be more time-consuming and on the long run will work out to be more expensive. But the testing of these small boards as part of a bigger panel will not only reduce the time and cost of testing but eventually result in a very high output.

Another important feature that determines the purchase of a grid system is its cost. The purchase of a dedicated grid system helps to reduce the initial set-up cost. But universal test grid, though it may initially work out to be more expensive, will offset its initial cost by reducing the individual set-up costs. Also, dedicated fixtures require lot of storage area as compared to a universal grid pattern. But one of the major problems faced with universal grid fixtures is their inability to accommodate off grid test probes.

Even with CAD designs, it is often found that some of the test points will not be exactly on the grid points but will be between two such grid locations. With dedicated grids, these points could be easily accommodated. It needs to be mentioned here, that with the increasing track density of the boards, the spacing between the tracks becomes much less than the diameter of the nail. Hence there is a question mark in the industry about the bed-of-nails method for high density PCBs.

The latest generation of PCB technology comprises quad flat packs (QFPs) with centres of 0.020 in. and less, ball grid arrays (BGAs) with centres of 0.050 in. and less, and other devices with small pad geometries. These components have dictated that new fixture designs be developed to accommodate the new technology. Hallee (1996) suggests that in certain fixture designs, longer pins with greater deflection can virtually eliminate the need for expensive double-density grid electronics. It is also pointed out that if reliable, repeatable mechanical contact with the PCB fails to occur, the testing process is compromised, regardless of the quality of the electronic measurement system involved.

14.4.2.5   Hold-down Systems in BBT Equipments

There are generally two types of hold-down systems available for bare board testing. It could be either a vacuum hold-down or a pneumatic hold-down. The dedicated fixtures are mostly designed to be used with a vacuum hold-down system. Vacuum hold-down is much less expensive than pneumatic systems. In this system, the board to be tested is placed over the test probe and a rubber blanket is placed on the board. Air is drawn out of the fixture by using a pump, thereby giving a contact between the probes and the test points. In the case of pneumatic hold-down system, air cylinders are used to compress and hold the board firmly against the bed of nails. Many universal grid fixtures cannot operate properly with a vacuum hold-down system because the pressure applied on the probes will not be sufficient to establish a perfect control.

Some universal grid fixtures cannot operate satisfactorily with a vacuum hold-down system because the pressure applied on the probes will not be sufficient to establish a perfect control. For example, a universal pattern with about 20,000 test points and each test point requiring a 100 gram pressure, would need 2000 kg or 2 ton of air pressure to keep all the pins in good contact with test pins on the board. If the pressure is not adequate, the test consistency is not satisfactory. On the other hand, with pneumatic systems, the pressure from air cylinders could be much more to give a high throughput. Cronin (1995) explains the various methods used to generate bare board test programmes which provide different levels of test coverage.

Probe Assembly: In a universal grid fixture, when the centre-to-centre spacing is 0.075 inch (1.875 mm) or more, the probes (Figure 14.4) are interchangeable. Hence they are mounted on individual sleeves or sockets. The plunger head should be so designed as to make perfect contact with the test point. In order to assure positive electrical contact, the probe should exert a pressure of 4 to 8 oz of force during normal engagement. In the universal grid system, the probe should have a minimum traverse of 0.2 inch (5 mm) to accommodate the use of the through-hole mask. Internal resistance of the probe should be less than 10 Mohm for proper testing.

The testing of the boards could be done with the already existing programme, which is generated during the design stage itself, if the PCB design is done by using a computer. Most of the bare board testing machines have the self-learn mode. In such cases, the already established "good" board is placed as a reference to obtain the necessary test parameters. With these parameters as the base, the other boards could be tested.

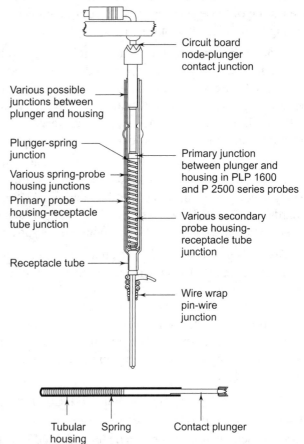

Fig. 14.4 Spring contact system for probe assembly

M/s MicroCraft have introduced a bare board tester which is a moving probe fixtureless tester and performs continuity and isolation tests simultaneous on both sides of a PCB (www.vikingtest.com). The design includes four moving probes, high speed closed loop servo motors, firmware and Windows interface. The four moving probes (two front, two rear) move in X, Y and Z directions and test all necessary pad locations for continuity and isolation. The continuity test involves high precision resistance measurement to detect open traces, splits between trace and pad, buried, blind and micro vias. The isolation test is done with menu-driven adjustable voltages up to 500 volts.

14.4.2.6 Optical Inspection

With the density of the PCBs increasing, the age-old method of visual inspection during the manufacturing stages is proving to be less effective for double-sided and multi-layer boards. Although the cuts and shorts could be easily detected by visual or by electrical continuity check, in case of single-sided and double-sided PCBs, it is almost impossible to detect such faults in case of the inner layers of a multi-layered PCBs. Hence a more sophisticated way of inspection of multi-layers through X-ray scanning has been developed, which can give an in depth inside view of the board structure thereby making it easy to detect the faults. However, optical inspection cannot give information on the leakage, contamination or isolation resistance tests. Hence, the high level electrical inspection along with optical scanning will lead to the production of zero defect PCBs.

Optical methods basically make use of magnifiers. They are of several types. The most commonly used magnifiers are delineated below.

Single Lens Magnifiers: Hand or bench magnifying lenses work well for low-power magnification applications, such as identifying wrong, missing or misaligned parts, but are limited to about 2½X magnification. The curvature required for higher magnification can only be obtained with smaller diameter lenses which provide an inadequate field of view. Another limitation is that as magnification increases, the focal length and working distance decrease.

A problem associated with single lens magnifiers is eye strain. A single lens will progressively distort an image as the distance from the focal point increases radially. Although this distortion may not be immediately perceptible, several minutes of viewing can cause severe eye strain and/or headaches. A planatic magnifier designs use two convex lenses to reduce this distortion.

Screen Projection Systems: There are several products which project a magnified image onto a vertically oriented screen, one of which claims to provide a three-dimensional image regardless of whether it is viewed with one or both eyes. These devices eliminate the fatigue associated with binocular devices and the problem of losing an image by tilting the head. Binocular devices also require adjustment for the distance between an operator's eyes and separate focusing of each eyepiece.

Another screen projection device is the optical comparator. Comparators work on the basis of showing "what should be" vs. "what is." In one method, mirror images of a master and a production circuit are shown on a split screen. This method is easily understood method in which differences readily stand out.

Stereo Microscopes: Stereo microscopes are perhaps the most commonly used inspection tool. They provide an upright, three-dimensional image and, in some cases, the ability to zoom in on a feature of interest. Broad goals for improvement are to increase accuracy and throughput capability, and to decrease operator fatigue. Evolution of the microscope has brought about the incorporation of angled eyepieces, screen projection, oblique angle viewing, camera and video interfaces, 360 degree mirror rotation, and movable and programmable stages.

Microscope Types: There are basically two types of stereo microscopes: the Greenough and the Galilei or telescope. The more traditional Greenough scopes have two separate optical paths, each with its own objective. The more common Galilei scopes use one main objective to create the two paths of the stereo image (Samsami, 1990b).

Greenough scopes provide especially good contrast and colour correction because each optical path can be independently corrected. These microscopes also provide the highest resolution because the optical axis runs through the centre of each objective. However, the interchangeability of objectives and tubes on these microscopes is limited. Also, flat objects are only in focus within a narrow zone in the centre. Greenough microscopes that are not ruggedly constructed, will eventually be prone to skewing the stereo images, leading to operator fatigue and even rendering the scope useless.

Galilei microscopes, on the other hand, provide interchangeability of the main objective for selection of the most convenient working distance or magnification. These scopes can also incorporate a magnification changer or zoom. Intermediate tubes, such as dual observation, drawing, TV or camera tubes, may also be added. Finally, the depth of field planes for both optical paths will always coincide. On the other hand, Galilei microscopes can distort flat objects, creating the illusion of a convex surface.

Usually, quantitative measurements of optical quality, such as line pairs/mm, are not published because they do not give an accurate performance picture. For example, lighting, in addition to optics, must be carefully considered to achieve optimum scope performance.

Generally, the lighting level should be high in order to cause the iris of the eye to contract, giving greater field depth and resolution. Military specifications call for shadowless lighting, but this reduces contrast and makes edge detection and surface quality assessment more difficult.

The two basic types of lighting are bright field and dark field. Bright field or coaxial light travels along the optical axis of the scope and through the objective. It is the only technique in which the numerical aperture of the condenser is equal to that of the objective lens. Bright field images provide excellent colour contrast and undistorted shapes. These images also allow accurate measurement, because the relief edges are not obscured, as in dark field. Bright field is considered an excellent way to view polished surfaces and is commonly used in semiconductor applications.

There are many features however, such as surface defects, that are more easily seen with dark field lighting. Dark field images are achieved by directing light from outside the objective toward the sample. Only light that is reflected or diffracted by features on the sample enters the objective. Dark field illumination increases the visibility of details which would otherwise be washed out by bright field illumination. While the numerical aperture of dark field condensers does not approach that of coaxial lighting systems, their cost is significantly less.

A typical dark field illuminator is the fibre optic spotlight. Because of its specular nature, it shines light over a narrow arc. A 2 degree arc corresponds to a maximum condenser numerical aperture of 0.0175, contributing very little to the resolution of a stereo microscope system. One problem sometimes associated with this type of lighting is that it can cause shiny solder surfaces to be completely obscured.

The other parts of the illumination puzzle are brightness, contrast and depth of field. Increased resolution contributes to a brighter image. In fact, brightness or intensity is directly proportional to numerical aperture (NA). However, brightness also contributes to operator fatigue and must be carefully controlled. Increased resolution also decreases contrast and depth of field.

Working distance is important when the board is manipulated by hand and especially when rework is performed under the viewing device. Although greater working distance will decrease image brightness in a microscope, it will also decrease the resolution of the microscope system.

Field of view (Figure 14.5) board which is magnified, is critical in minimizing the time needed to view a board. It is also important for photographs or video images taken from a stereo microscope. The area shown must be sufficiently large to allow fault location later or else more than one photograph must be taken.

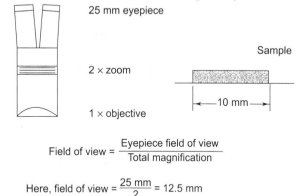

$$\text{Field of view} = \frac{\text{Eyepiece field of view}}{\text{Total magnification}}$$

$$\text{Here, field of view} = \frac{25 \text{ mm}}{2} = 12.5 \text{ mm}$$

Fig. 14.5 The microscope's field of view allows the entire sample to be viewed

14.4.3 Testing of Assembled Boards

For the majority of boards tested after assembly, the fault spectrum is broadly as per Figure 14.6. It may be seen that solder opens and missing, misaligned or wrong parts are the number one and number two causes of failure. Although it is a typical pattern of faults, it varies with products and processes. In fact, every trace, every solder joint, every component and every connector pin presents an opportunity for process failure. Also, board size, routing space, package types, component types, handling etc, all contribute in a good or bad way to the probability of failure in assembled boards. The testing procedures therefore, have to be so designed that they look at these possibilities to ensure the quality of the product.

14.4.3.1 In-circuit Testing

The purpose of in-circuit testing is to locate defects and isolate incorrectly placed or soldered components. Basically, in-circuit testing is used to check the correct assembly of the board. It makes use of a matrix of probes

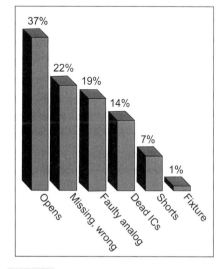

Fig. 14.6 PCB fault spectrum (redrawn after Oresjo, 1990)

which contact circuit nodes on the board. The tester applies a signal to stimulate an individual node and compares the measured response to the expected one. In-circuit testing follows three basic approaches:

Impedance Testing: The impedance between two points is measured and compared with the impedance of a known good board.

Component Testing (Analog and digital): The basic approach is to apply a signal and monitor the output. Usually, the integrated circuits, analog or digital, are not isolated and therefore, the output for a test may not match without data given in the manufacturer's data books. By composing the output at specified points and comparing it with the good board, correctness of the integrated circuit and the correct polarity of the components can be ensured.

14.4.3.2 Functional Testing

In-circuit functional testing is an effective method of locating faulty components in the circuit under real life working conditions while the device is still in-circuit or wired on a PCB. In principle, it compares the function of the device under test with the corresponding ideal device in the fault locator device library, which comprises a database of device models. Each device model in the library includes a sequence of test patterns that will initialize the device, drive the input pins of the device and check for appropriate responses on the device output pins.

Functional testing is most useful in locating faulty digital devices. For example, it can easily check that a simple logic gate operates in accordance with its truth table or that a counter or shift register correctly responds to a known number of clocks. During testing, the logical operation of the device in respect of the relationship between inputs, clocks, control signals and outputs is displayed by the fault locator in diagrammatic form and compared with the "ideal" device in the database (Polar Instruments, 2002b). With the modern functional tester, you can check that programmable devices such as ROMs or EPROMs have been properly programmed or that programmable interface devices respond correctly to control commands.

14.4.3.3 Boundary Scan Technique

With the increasing use of surface mount devices, many boards are assembled with components on both sides and without through-holes that enable test probes to reach every node from one side of the boards. Therefore, the targets that the probes must hit are becoming substantially smaller. The use of 50 mil probes that are smaller, more fragile and more expensive than the standard 100 mil probes used on less complex boards becomes necessary. The node counts are rising and the entire systems now fit on to single boards. In-circuit testing, probably the most popular board test technique, does not adequately handle all the testing of complex boards. Functional testing in such cases, may be applied but developing a functional test programme is time-consuming.

An alternative test method is the boundary scan, which complements these test methods and has been adopted as the Institutes of Electrical and Electronics Engineers (IEEE) standard 1149.1. The method is based on the statistical data which demonstrates that most loaded PCBs fail due to external manufacturing errors including opens, missing or wrong components, faulty analog components, dead ICs, shorts and fixture problems (Oresjo, 1990).

The boundary scan implementation involves assigning all ICs a Boundary Register composed of boundary cells associated with device inputs, outputs and by directional signals as well as embedded device control signals. Boundary scan cells are located along with perimeter of IC, surrounding the internal logic. The details of the test are given by Evans (1991).

14.5 Reliability Testing

The purpose of reliability testing is to identify latent defects that may arise in the PCB after extended or excessive operation of the circuit in service. These types of defects are not visually apparent as the product is being built or following simulated conditions as repeated cycling from extreme low to high temperature (typically–65° to +125 °C). Bond strength and re-work simulation assess the structural integrity of plated holes after repeated solder and de-solder cycles. All these reliability tests are outlined in the IPC standard IPC-TM-650 (Sections 2.6.3, 2.6.7.2, 2.4.21.1 and 2.4.36).

Figure 14.7a shows printed circuit board testing hierarchy in relation to the requirements of the users, capability of the tests and the percentage of products which should be tested to ensure the reliability of electronic products.

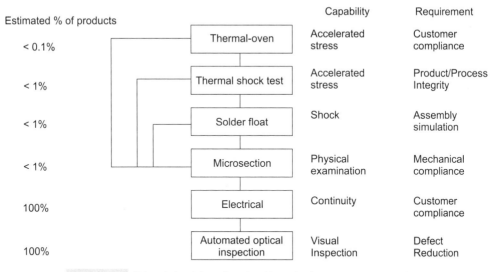

Fig. 14.7(a)   Printed circuit board testing hierarchy (www.smartgroup.org)

14.5.1   Reliability of Printed Circuit Boards

Two major sources of failure of printed circuit boards affect their reliability. These are:

Plated Through-holes: Their failure is due to fracture, since they consist of thermally incompatible materials and also incorporate multi-layer interconnection.

Loss of Electrical Insulation: Degradation of impedance levels (open or short circuits), affects the performance of electrical circuits.

14.5.1.1   Plated Through-Hole Reliability

The main factor affecting plated through-hole reliability is due to the thermal expansion mismatch between copper and the FR-4 epoxy-glass composite, encountered during thermal stressing of the circuit board. The thermal stresses develop during wave soldering, re-work or thermal cycling during assembly, use or storage. Thermal excursion generally leads to fractures such as corner cracks, inner-layer cracks or barrel cracks. Engelmaier (1991) explains the reliability aspects of plated through-holes.

Plated through-hole reliability is normally tested under simulated conditions of thermal excursions of a PTH throughout its life. The military specifications which are applied to testing the thermal shock are MIL-P-55110 (Military) specification for printed circuit boards. Table 14.2 gives the test conditions for thermal shock testing.

Table 14.2   Test Conditions for Thermal Shock Testing

Low temp. °C	Time minutes	High Temp.	Time minutes	Laminate type
− 65	15	105	15	PX
− 65	15	125	15	GE
− 65	15	150	15	GB, GH, GP, GT, GX, GR, FEP
− 65	15	204	15	GI

The samples are tested for electrical continuity or shorts. The laminate types, as per military designations are:

- PX = Paper base, epoxy resin, flame-retardant
- GE = Glass (woven-fabric) base, epoxy resin, flame-retardant
- GB = Glass (woven-fabric) base, epoxy resin, heat-resistant
- GH = Glass (woven-fabric) base, epoxy resin, heat-resistant and flame-retardant
- GP = Glass (non-woven fibre) base, polytetrafluoroethylene resin
- GT = Glass (woven fabric) base, polytetraethylene resin
- GX = Glass (woven fabric) base, polytetraethylene resin, for microwave applications.
- GR = Glass (non-woven fibre) base, polytetraethylene resin for microwave applications
- FEP= Fluorocarbon unfilled
- GI = Glass (woven fabric) base, polyimide resin, general purpose

Thermal Stress Test: The PCB specimen is baked between 120°–150 °C, fluxed and floated in a solder bath at 287° ± 5 °C for 10s.

Following the thermal stress, the samples are micro-sectioned and examined for defects.

PTH Reliability Improvement: The studies carried out to relate the PTH fabrication process with reliability have indicated that the following steps lead to improvement in PTH reliability:

- Use 2 oz copper on inner layers.
- Increase plating thickness.
- Use ductile copper.
- Use non-functional pads, wherever possible.
- Eliminate the use of small holes.
- Bake laminates prior to processing.
- Bake after drill and/or etch-back.
- Minimize etch-back.

14.5.1.2 Insulation Resistance Reliability

Reliability problems due to changes in insulation are of two types: insulation resistance failure and low impedance failure.

Insulation Resistance Failure: This indicates a change from high impedance to low impedance. It generally occurs slowly over a large period of time. The reduction in impedance causes serious problems in analog measurement circuits. The problem is particularly of great concern in the case of medical instruments, wherein the sensors are attached to the patient and any deterioration in the insulation resistance may have a potential electric shock hazard. This condition leads to the development of unwanted current paths (leakage current) on the surface or through the bulk of the board.

As the component density in the boards increases and the spacing between its conductors decreases, any degradation of the insulator material between the circuit traces has a pronounced effect on the board reliability. While the initial level of current leakage may not be serious enough to cause a failure during electrical testing, this leakage will increase over time, ultimately leading to a loss of board reliability. Also, in the case of PCBs having high sensitivity analog circuitry, if the surface of the PCBs and its components are not free of the contamination, circuit performance is likely to suffer. For such cases, surface insulation resistance (SIR) testing is one of the most important evaluation tools for this purpose (Kamat, *et al*, 1995).

SIR is a measure of an insulating material's surface resistance to the flow of current between conductors. This is a parameter of great interest to PCB manufacturers because residues left on a board's surface during manufacturing or contamination due to improper cleaning or careless handling, can reduce the insulator's ability to resist current flow. With perfectly cleaned PCBs, the current flow is negligible, but any conductive contaminants show as increased current (or a reduced SIR reading). If left unchecked, this current leakage can eventually cause the board to fail in use.

To make the SIR measurement, a constant test voltage (100 to 500 V forward or reverse polarity, depending upon the specific test requirements) is applied for a pre-defined period, usually 60 seconds, and the resulting current is measured. Resistance is calculated using Ohm's Law: $R = V/I$. This resistance is usually at the gigaohm level or greater. So instruments capable of measuring extremely low currents accurately, such as electrometers and picoameters, are needed. The measurement process is repeated after the boards are exposed to various temperature and humidity levels up to 85 °C and 90 per cent relative humidity.

Low Impedance Failure: In case of PTH, the low impedance of the circuit must be ensured. The usual failure due to thermal shock results in a fracture which usually produces an abrupt change from low to high impedance.

14.5.1.3 Insulation Resistance Reliability Testing

A few precautions taken during the manufacturing process of PCB can lead to tremendous improvement in reliability and reduce the possibility of insulation resistance failures in PCB. Some of these are:

- Process cleanliness: Eliminate any ionic or hygroscopic residues on the board during processing.
- Provide protective coatings over exposed metal.
- Use metals that tend to migrate carefully.
- Provide separate ground and voltage planes and connection pins.
- Use the final product in an environmentally controlled area, which should be free from corrosive pollutants, dust, high temperature and humidity.
- Increase the component lead spacing, if any.

As first order of approximation, the reliability of an assembled PCB is inversely proportional to the number of component interconnection joints. Thus the weakest link in a product is the *connection*. Care must be taken to employ such techniques which give the highest joint reliability. Ideally, the joint must provide adequate mechanical strength and low electrical resistance.

14.5.1.4 Bath Tub Curve

For mechanical parts such as relays, plugs, sockets and motors, their wear out occurs throughout the entire life of the component. Therefore, they show higher failure rate generally defined by the traditional bath tub curve as shown in Figure 14.7(b).

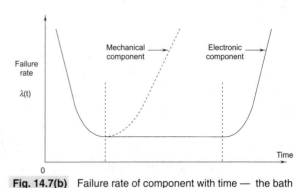

Fig. 14.7(b) Failure rate of component with time — the bath tub curve

14.6 Acceptability of PCBs

Today's printed circuit board test engineers face significantly more test challenges than just a decade age. Board complexity is dramatically increasing with more components, more joints, higher demonstrates and new package technologies (Oresjo, 2003). The higher component and joint counts create more defect opportunities, which may lead to lower yields for a given defect level. Along with complexity, the manufacturing process also influences the number of defects. Through several studies, it has been found that defects are added at every process step.

The complexity of the PCB assembly directly impacts the number of defects it has, which, in turn, affects the test strategy selection. The higher the complexity of the circuit, the more difficult it is to achieve high yields without tightening the test and inspection procedures.

14.6.1 Acceptance Criteria

The institute for Interconnecting and Packaging Electronic Circuits (IPC) has established certain criteria and guidelines for acceptance of PCBs based on the applications and functionality of the equipment. The equipment has been classified into the following classes:

Class 1: General Electronic Products: This includes consumer electronic products such as TV sets, toys, entertainment electronics, some computers and computer peripherals suitable for applications wherein cosmetic imperfections are not important and the major requirement is the function of the completed electronic assembly.

Class 2: General Industrial or Dedicated Service Electronic Products: This includes telecommunication equipment, computers and sophisticated business machines, and industrial instruments. This class of equipment is expected to show high performance, long life and uninterrupted service, but it is not critical. Certain cosmetic imperfections are tolerable and accepted.

Class 3: High Performance Electronic Products: This includes equipment and products where continued performance or performance on demand is critical. Examples are life support systems in the medical field or the flight control system. In this class, equipment downtime cannot be tolerated as the end-use environment may be uncommonly harsh.

The acceptability criteria for each group have been separated and represent three levels of quality for each specific characteristic. They are: target condition, acceptable and non-conforming.

Target Condition: This is a condition that is close to perfect. It is a desirable condition and is not always achievable, and it may not even be necessary to ensure reliability of the board in its service environment. In the past, it was labelled as 'Preferred'.

Acceptable: It indicates that the condition, while not necessarily perfect, will maintain the integrity and reliability of the board in its service environment. However, the acceptable condition may be applicable for at least one or more classes, but may not be acceptable for all classes.

Non-conforming: This implies that the condition depicted may be insufficient to ensure the reliability of the board in its service environment. However, the non-conforming condition is considered unacceptable for at least one or more classes of product, but may be acceptable for other classes, when it satisfies the acceptance criteria.

The user has the ultimate responsibility of identifying the class in which the product is evaluated. It is a general practice to prepare specifications based on the requirements that apply to each single type of board which must be strictly related to its use and the customer's policy on quality and reliability. Any requirement eventually causes an increase in the cost of the PCBs, either directly or indirectly. However, the accept and/or reject decisions must be based on applicable contract conditions, drawings, specifications, standards and reference documents.

14.6.2 Inspection of Assembled PCBs

The requirement for printed circuit board inspection has spawned a variety of equipment for performing this task. Automated optical inspection (AOI) systems routinely inspect inner layers before lamination; X-ray systems monitor registration accuracy and pinpoint defects after lamination; and scanning laser systems provide a means of solder paste inspection before re-flow. These systems, together with production line visual inspection techniques and parts inspection integral to automated component placement, help ensure a reliable final assembled and soldered board.

However, even though these efforts minimize defects, the final inspection of the assembled PCB remains an additional requirement, and perhaps the most important since it represents the last chance for evaluating the product and the overall process.

The final inspection of an assembled PCB can be performed by manual methods or by an automated system, and often by both. 'Manual' refers to an operator visually inspecting the board via optics and making appropriate judgments as to what constitutes a defect. An automated system employs computer-aided image analysis for determining defects. Many also consider automated systems to encompass any method other than manual optical inspection.

X-ray techniques offer a method of assessing solder thickness, distribution, internal voids, cracks, bond separation and the presence of solder balls (Markstein, 1993). Ultrasonics will detect voids, cracks and unbonded interfaces. AOI evaluates external features such as bridging, solder volume and shape. Laser inspection can provide three-dimensional images of external features. Infra-red inspection detects the presence of internal solder joint faults by comparing the joint's thermal signature to a known good joint.

Interestingly, it has been found that limiting assembled PCB inspection to these automated techniques does not uncover all defects. Therefore, manual visual inspection must be used in conjunction with automated methods, especially for small volume applications. X-ray inspection combined with manual optical inspection is the best defect detection method for assembled PCBs.

An assembled and soldered PCB is prone to the following defects

- Missing parts;
- Faulty parts;
- Part misalignment;
- Tombstoning;
- Poor wetting;
- Bridging;
- Insufficient solder;

- Solder balls;
- Solder pin holes;
- Contaminants;
- Wrong parts;
- Improper polarity;
- Lifted leads;
- Excessive lead protrusion;
- Cold solder joints;
- Excessive solder;
- Solder voids;
- Blow holes; and
- Poor fillet formation.

14.6.3 Inspection Techniques

14.6.3.1 Visual Inspection

Visual inspection involves a study/inspection of the characteristics of the printed circuit board with the unaided eye. It takes about $1/500^{th}$ of a second for a human to recognize an object, even if it is incomplete or obscured. Human vision is quite different from machine vision. Machine vision sees lines and spaces without any interpretation, whereas human vision sees objects and can readily spot variations in an object. In addition, humans can readily adapt to new situations, communicate findings, instruct others and correct problems. Humans also have the capability to improve equipment and processes through the application of imagination. The challenge therefore, is to make the most of these human capabilities.

However, visual examination for external attributes is preferably conducted at 1.75X. Magnification, upto 40X is used for defects, if not readily apparent with unaided eye. Dimensional requirements such as spacing or conductor width measurements sometimes require magnification devices. Plated through-holes are internally examined for foil and plating integrity at a magnification of 100X. For viewing of micro-sections for defects such as resin smear, magnification of 50 to 500X are required, depending upon the characteristics. In general, visual inspection should be 1.75X minimum and 10X maximum. Defects which can be easily detected by visual inspection fall into various categories: These are discussed below.

14.6.3.2 Dimensional Inspection

Dimensional inspection involves the measurement of the printed circuit board parameters to determine compliance of the dimensional values with functional requirements. While visual inspection is carried out by the naked eye or magnifying tools, dimensional inspection invariably makes use of gauges, measuring microscopes, co-ordinate measuring equipment, micro ohm meter and beta back scatter gauges.

In order to carry out dimensional inspection, a sampling plan is prepared and data collected and statistically analysed. Usually, the sampling plan is based on the criteria of AQL (Acceptable Quality Level) which specifies the maximum percentage of defects which can be considered statistically satisfactory for a given product.

14.6.3.3 Machine Vision

PCB assemblies are becoming increasingly complex with large quantities of miniaturized surface mount and through-hole components. Pitches on multi-leaded surface mount and through-hole devices have decreased dramatically. Furthermore, we deal with a large variety of part numbers requiring short introduction lead times.

This complexity presents a challenge to the entire manufacturing process. This is particularly true with the quality and inspection processes due to the significant increase in the opportunity for error and the inherent human limitations of manual inspection (Qazi and Calla, 1997).

Machine vision is one solution for these challenges that has alleviated the adverse effects of complexity in the inspection process. This solution has been in place since 1987 and is facilitated by an experienced cross-functional vision team with members from engineering and manufacturing.

These vision systems have defect detection capabilities that can be divided into two categories: placement defects and solder defects. Most placement and solder defects are identified with 99 per cent reliability, which is referred to as a detection rate.

Placement defects refer to the positioning of the component or leads on the pads. Placement defects that can be detected by vision systems include missing, skewed, misregistered, bill-boarded, tombstoned, tilted and bent or lifted leads. Solder defects refer to the characteristics and quality of the solder joint. Solder defects that can be detected by vision systems include bridging, opens, insufficient, excessive, de-wetting and cold joints. This provides a multitude of information in all key areas for continuous process improvement, while ensuring that defects are detected effectively.

Finally, vision systems can be implemented as in-line or off-line systems. In-line systems enhance the feedback process and facilitate repairs before solder, directly on the line. Off-line or stand-alone systems offer flexibility and strategic deployment to new process developments since the main focus is not defect "screening" but using the system as a process improvement tool. Machine vision is a technology that acquires images optically, using CCD (charge coupled device) cameras and specialized high-intensity lighting that is optimized to highlight relevant features of the card, such as joint and component features. Images are captured and stored in a digitized format and processed at a high speed using vision computers. The images are analysed by complex algorithms to determine if a defect exists. Since image acquisition is optical and the machine relies on gray scale analysis, the machine must be able to recognize the signature of a defect, i.e., some visual characteristic that allows that algorithm to distinguish the joint as defective. Figure 14.8 shows the image of a missing discrete component, while Figure 14.9 illustrates a bent lead defect on a fine-pitch component.

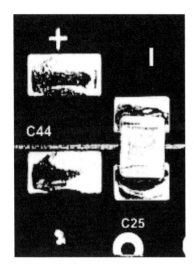

Fig. 14.8 A missing discrete component
(adopted from Qazi and Calla, 1997)

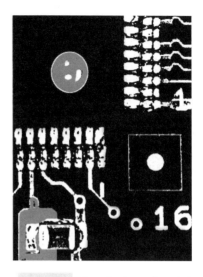

Fig. 14.9 A bent lead on a fine-pitch component

A small field-of-view (FOV) is required for resolution of fine-pitch components. The systems have FOVs as low as 10.16 mm to 12.7 mm, resulting in high resolution. A high-accuracy X, Y stage is used to position the cameras to an accuracy of ± 0.0254 mm.

The vision system relies upon some form of database, such as a CAD file, for the placement of inspection windows that direct the system on the location for image capture and analysis. At least one inspection window is usually defined per joint. A typical PCB can have 5000 to 6000 joints per side, which results in at least as many inspection windows to be generated by the vision systems programmer.

Automated X-ray and laser inspection systems are also used to provide benefits similar to those provided by vision systems. They are also deployed to complement test and inspection capabilities.

Vernon (2003) reviews the current status and future direction in the machine vision in the electronics and PCB inspection industry. Visual inspection systems are used for sample inspection, to support the production set-up or for special investigations in regard of failure analysis. However, the evaluation of the images is always a matter of experience and available specifications. Software packages are now available which can be used to support operators with adequate knowledge to allow quick and objective decisions (Neues, 2002).

14.6.3.4   X-ray Methods

X-ray inspection systems make use of the X-ray transmitting principle and image processing technology. The method relies on the X-ray blocking characteristics of the lead content in the solder to assure the integrity of solder joints. They also screen for solder splash and solder flow shorts as well as open joints, even beneath the ball grid array packages. They are better suited to process or laboratory applications including diagnosis for functional or field failures. Such systems are particularly good at detecting corner cases and rare types of errors such as open on BGAs. Automatic X-ray method is the fastest growing screening technique and is the only method that can quantitatively as well as qualitatively assess solder joints integrity.

While some fabricators rely on film-based X-ray inspection systems that produce radiographic images, the majority use fluoroscopic or real-time systems, which provide immediate feedback on a video monitor. Either type of system projects an X-ray beam through the board, producing a shadowgraph that is subject to the same geometric factors that influence the sharpness and magnification of a shadow created by natural or artificial light falling on an object. The sharpness of the shadow depends upon the size of the light source and its distance from the object as illustrated in Figure 14.10.

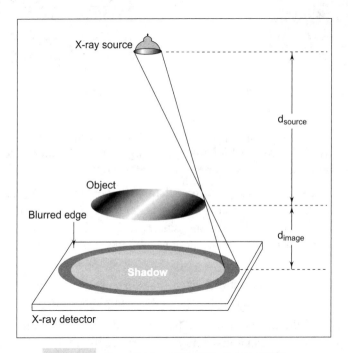

Fig. 14.10   X-ray shadowgraph (after Zweig, 1996)

The shadowgraph is a two-dimensional image, revealing the internal, three-dimensional anatomy of the board. Just as radiologists must understand the body's three-dimensional anatomy in order to interpret two-dimensional medical X-rays properly, users of industrial X-ray systems must understand the anatomy of a multi-layer board in order to make a correct interpretation of the different shades of grey displayed in a shadowgraph (Zweig, 1996).

However, X-ray inspection systems are generally not suitable for production line use because the system throughput is lower than typical production lines — typically 30 to 50 solder joints per second. If the line rate is 20s or 30s for the entire board, the X-ray system will not be able to inspect 100 per cent of the connections on the board with the required throughput (Tong, 2003).

X-ray microlaminography (Figure 14.11) uses X-ray imaging to visualize the internal details of thin multi-layer structures such as ball grid arrays and printed circuit boards. It allows layer by layer information to be measured in slices as small as 10 microns. Questionable solder joints, disbands and voids are easily detected. This technique works in exactly the same way as the X-ray tomography systems (CAT-scans) used in medicine but with much finer resolution. A compact X-ray scanner with high resolution three-dimensional visualization is available from M/s Micro Photonics Inc., USA

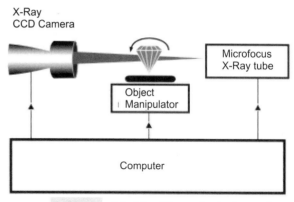

Fig. 14.11 Principle of X-ray microlaminography

(www.microphotonics.com). The computer separates 20 layers simultaneously during 20 seconds of scanning and 80 layers for 90 seconds of scanning. Application includes non-destructive testing of defects for BGA, flip chips, PCB-inspection, and inspection of other micro-devices.

The three common process faults that are difficult to catch with electrical in-circuit test are: (i) missing/misloaded by-pass capacitors; (ii) solder opens on pins that bring power to an IC; and (iii) solder opens on pins which provide ground return. These faults have significant consequences specially on high speed microprocessor assemblies. Three-dimensional X-ray laminography has the ability to screen for these and other similar manufacturing defects on boards with components mounted on both sides, including multi-chip modules.

The instrumentation comprises a finely focused sealed X-ray tube, a precision sample manipulator, an X-ray CCD camera and a dual Pentium workstation with tomography software running under Windows 2000/XP. The X-ray source is 80 KV, 0-100 μA with less than 8 μm spot size. The X-ray detector is scintillator with a 28–65 mm field of view integrated to 768×512 pixels, 8-bit CCD in on-chip integration mode.

14.6.3.5   Automatic Optical Inspection (AOI)

Automatic optical inspection machines are being increasingly integrated into the PCB production flow and consequently linked to CIM (computer-integrated manufacturing) systems. Figure 14.12 indicates where AOI fits into the production flow (Doyle, 1991) and how comparison of data from the design database with the image of the panel allows problems at any stage of manufacture to be rapidly detected. For this purpose, it is desirable to obtain the reference data from as early as possible in the overall process. For instance, if AOI is carried out by point-to-point comparison with the 'Golden Board', any defects occurring upstream of the production of that board will remain undetected. A better approach is to use the photo-tooling as a reference, but this will not detect errors introduced by the photoplotter or its associated software. Better still is to use the Gerber data input to the photoplotter and so detect any error introduced into the process between the original CAD data and the actual panel. The ease with which this can be done depends upon the processing algorithms used. Point-to-point comparison requires vector-to-raster conversion, which can require very large amounts of processing time and data storage. Netlist comparison requires much less data and processing times can be much shorter. The technique depends upon extracting the netlist from the image so that only netlists need be compared.

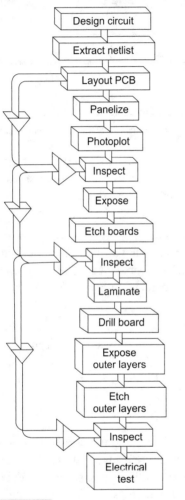

Fig. 14.12   Place of automatic optical inspection in the production flow

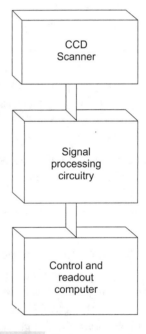

Fig. 14.13   Basic components of automatic optical inspection (AOI)

A block diagram of a typical present day AOI system, shown in Figure 14.13 employs either CCD scanners or laser scanners. In a CCD scanner, a light-sensitive silicon chip detects light shone onto the surface of the board by a suitable illumination system. In a laser scanner, a laser spot is mechanically scanned across the surface of the board, causing the substrate material to fluoresce. This fluorescent light is detected and the signal from the detector provides the image of the board. The principal advantage of this technique is that it overcomes the requirement to illuminate the board in such a way as to provide a usable contrast between the copper and the substrate. It has, however, a number of disadvantages. Since the substrate is detected rather than the copper, it will not detect a dish-down that may extend almost all the way through the copper. Non-fluorescing substrates such as PTFE, ceramics and glass, which are likely to become more common in the future, cannot be detected by a laser scanner. Also, expensive and relatively fragile components such as lasers and rotating prisms are used by this technique.

These problems limit the use of laser scanners, whereas CCD scanners are witnessing advances in CCD technology arising from their widespread use in fax machines and video cameras. Devices now becoming available have speeds ten times greater and sensitivities a hundred times greater than the devices of a few years ago. This greatly simplifies the problems of illumination, and will allow the possibility of multiple devices providing colour discrimination for the inspection of even the most difficult material combinations.

The data processor is the next unit in to be considered. The volume of data produced by scanning a single panel is so great that a conventional software-based processor is too slow by several orders of magnitude and, despite the rate of improvement of such processors, this will remain the case in the foreseeable future.

Attempts have been made to solve this problem by using arrays of processors, with each processor handling a small portion of the image, but such solutions have proved to be expensive and slow. A far more effective solution is a pipelined data processor in which the data flows through a set of circuits, each one of which performs some small operation in an assembly-line fashion. If each circuit is instructed as to what operation it should perform by the controlling processor (which is a conventional computer), it is possible to use a relatively small number of circuit types and retain a high degree of flexibility to deal with different circuit patterns.

The last item in the figure is the control and human interface computer. This carries out some of the final stages of data reduction in software, and provides numerous facilities for helping with the set-up and test. This part of the system will benefit from the improving price/performance ratio that is characteristic of standard computer equipment.

The use of proper image processing technology in automated optical inspection of assembled printed circuit boards is extremely important. Frame grabbers provide the capability of handling and processing the image captured by the camera. Image processors either process colour or gray scale images. Obviously, the colour image processor is preferred because it has the ability to process true colour images without reduction in detail (Fishburn, 2002).

Some automatic optical inspection systems scan and identify components that appear different from the reference "golden board". Only the questionable components are brought to the operators' attention and displayed at a high resolution 10x magnification. The operator reviews only these items using a set of software tools. The productivity is increased 4–5 times. This technique is known as "inspection by exception" in which the attention is focused on the differences and not on the similarities. The system can detect faults such as missing components, reversed components, polarity marks, solder bridges, placement defects, tombstoning, billboarding, skew and, no loads (www.testronics.com), etc.

Highly sophisticated optical inspection systems are available which make use of a fibre-optic light source and are equipped with extremely accurate X-Y table where the assembled PCB is held (Zarrow, 2000). The instrument called ERSAscope 3000; contains over 30 internal lenses and gives upto 35x magnification. It contains a prism to offer a 90-degree view angle. By scanning along the component, the solder joints can be examined row by row. By adjusting the field of view, internal interconnections can be seen and evaluated. All this is projected on a flat screen monitor via a high-resolution CCD colour camera incorporated into the scope.

Automated Optical Inspection (AOI) and Automated X-ray Inspection (AXI) are the most widely used technologies for inspecting the placement and soldering defects in today's PCB assemblies. It is therefore important to understand the type of faults covered by each technology, which are given in Table 14.3. (www.agilent.com/go/manufacturing).

Table 14.3 Comparison of AOI and AXI

Automated Optical Inspection	Automated X-ray Inspection
Excellent for finding placement defects — missing, skew, incorrect polarity, part marking, etc.	Excellent for finding solder joint defects that often cause latent field failures
Excellent for measuring accuracy of component placement (pre-reflow)	Excellent for measuring solder thickness
Capable of finding many visible solder joint defects	Capable of finding solder joint defects not visible to AOI or human inspectors
Capable of specific process inspection and feed-back (post-paste, post-placement, post-solder)	Capable of inspecting double-sided boards top-and-bottom in a single inspection cycle
Programming time typically less than one day	Programming time typically two to three days
Lower capital investment than AXI	Higher capital investment than AOI

In order to meet the requirements of testing, today's complex and high node count printed circuit boards, manufacturers are adopting multiple types of test systems to accurately detect, diagnose and repair the manufacturing faults. The multiple test style environment combines two or more different types of PC Board testers. One such approach is to combine automatic X-ray inspection and traditional in-circuit and functional testing techniques (www.techonline.com). This approach capitalizes on the strength of each test technique to maximize fault detection and minimize redundancy of test. For example:

- *X-ray inspection* identifies manufacturing solder defects such as shorts, opens insufficient or excess solder.
- In-circuit test verifies device type, orientation and operation; reduces overall required nodal access and number of fixture probes, which minimizes fixture problems and re-test occurrences.

- *Functional test* concentrates on verifying the printed circuit board's overall operation, with a lower incidence of board failures and higher yield resulting from the previous test step.

14.6.3.6 Infra-red Thermal Imaging System

Despite the availability of a variety of test and inspection systems, the components and assembly faults which may result in early life failure, may not be detected. However, infra-red imaging of populated circuit boards has the capability of mitigating these early life failures, potentially replacing/reducing the current electronic tests and alternative traditional test methodologies. The technique, known as thermography is the process of converting the heat emitted from an object (like a PCB) into a visible, dynamic TV-like picture. The picture is obtained by means of an infra-red mechanical scanning system or by the use of a phased array of detector elements that can be electronically scanned. By creating a two-dimensional temperature pattern (thermogram) of the surface under test, information on temperature is obtained from several thousand points in the field of view and displayed on a CRT.

The infra-red portion of the spectrum starts from 0.7 μm to 1.3 μm (near infra-red), from 1.3 μm to 3 μm (mid-infra-red) and beyond 3 μm to microwave region (thermal infra-red). The thermal infra-red is directly related to the sensation of heat whereas near and mid infra-red energy is not. The infra-red imaging cameras (with optical-electronic sensors) capture the invisible infra-red energy naturally emitted from all objects in proportion to their temperature and material characteristics.

The thermographic system makes use of an infra-red camera which can detect temperature differentials as low as 0.2 °C at ambient temperature. A typical thermal imaging system for PCB defect analysis is shown in Figure 14.14. Usually, a subtraction method is used for locating defects, which involves scanning the unit under test (UUT) and then subtracting the image from that of a standard golden board and displaying the differences. Utilizing appropriate analysis software, the subtraction method can be automated to highlight common failure types and to generate failure reports.

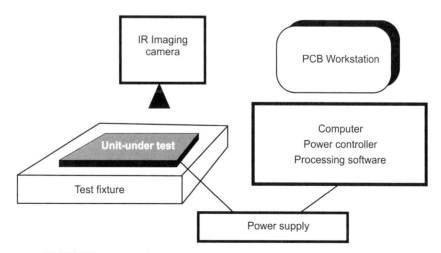

Fig. 14.14   Infra-red data acquisition system (www.infrared.com/images)

14.6.4 Acceptability Criteria

14.6.4.1 Base Material

Visual inspection may be used to check several major characteristics of the base materials. These are delineated below.

Type and Manufacture: One must verify that the laminate is of the type specified and is supplied by the approved manufacture.

Surface Defects: Visual inspection may be carried out to look for the following surface defects shown in Figure 14.15:

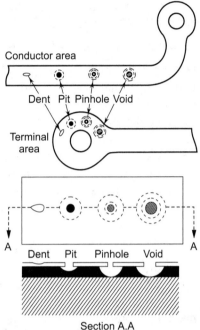

Fig. 14.15 Some common defects (i) dent (ii) pit (iii) pinhole (iv) void (redrawn after Coombs, 1988)

- Void (absence of material in a localized area);
- Dent (depression in the conductive foil, which normally does not significantly decrease foil thickness);
- Pit (depression in the conductive material, that does not go up to the total depth);
- Cracks (discontinuity in the conductive material);
- Surface smoothness (surface having bumps, projections, etc. and which is not smooth); and
- Pin hole (a small hole which covers the entire depth of the material).

Surface defects, which are minor, usually have little or no effect on functionality. However, their presence can be detrimental in some areas, such as edge board contact area. The acceptance/rejection of the PCB with any or more of the above surface defects then depends solely upon the judgment of the inspector with reference to the standards on the subject.

Blistering and De-lamination: A localized separation or swelling between any of the layers of a laminated base material or between base material and copper foil is called blistering. Blistering generally gives the appearance of air bubbles inside the laminate and is acceptable if the bubbles are small and far from plated holes.

De-lamination is the separation between plies within the base material or between the base material and copper foil or both. It is usually unacceptable.

In general, blistering and de-lamination are considered as major defects. There is a great possibility that the de-lamination or blister area will increase to the complete board separation, normally during assembly process.

Measling: Measling is a typical defect in glass cloth reinforced laminates in which the glass fibres are separated from the resin at the weave intersection. It usually appears in the form of discrete white spots or crosses internal to the laminate. Generally, the effect of measling over functional characteristics of finished PCBs is insignificant, and is therefore, acceptable if present to a limited extent.

Haloing: Haloing is mechanically induced fracturing or de-lamination on or below the surface of the base maternal. It appears as a light ring around holes or other machined areas or both.

Haloing is not acceptable if strength specifications are followed. However, in many situations it is accepted upto about 1.0–1.5 mm from the machined edges.

Weave Exposure: This is a surface condition typical of glass cloth reinforced laminates, wherein the fibres of woven glass cloth are not completely covered by resin. Weave exposure is considered as a major defect because the exposed bundles of glass fibres may allow wicking of moisture and entrapment of processing chemicals. The condition may be accepted to a limited extent if the peel strength of the material is still satisfactory.

A more serious condition is fibre exposure in which semi-forcing fibres within the base material are broken and exposed. It is a major defect and the boards are not usually accepted with this condition.

Contour Dimensions: Verification of board contour dimensions is important to ensure that the outside border dimensions are within the drawing requirements. Both undersized and oversized printed boards can affect functionality with respect to fit requirements. Contour dimensions can be measured/verified by using simple ruler, caliper or digital distance measuring instruments.

Edge Finish: The edges of boards should be properly sheared, with no burrs, fragments, projections, free fibres, etc., except to a very small extent.

Bow and Twist: The deviation from the flatness of a board, which manifests itself in the form of cylindrical or spherical curvature, is known as *Bow*. In case of a rectangular board, its four corners are in the same plane. This is shown in Figure 14.16.

Twist on the other hand, represents deformation parallel to a diagonal of a rectangular sheet such that one of the corners is not in the plane containing the other three corners. Referring to Figure 14.17, points A, B and C are touching the base whereas 'D' is lifted up. This is called twist. Two methods are generally used for measuring bow and/or twist:

- Height gauge method
- Feeler gauge method.

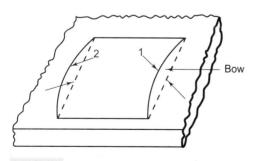

Fig. 14.16 Definition of bow

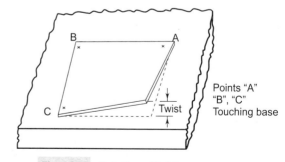

Fig. 14.17 Definition of twist

Referring to Figure 14.18, the percent bow can be calculated as

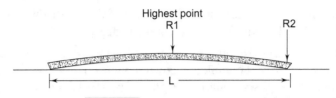

Fig. 14.18   Bow measurement

$$\% \text{ Bow} = \frac{(R_1 - R_2) \times 100}{L}$$

Where R_1 = Maximum vertical displacement measured with a micrometer

R_2 = Thickness of the sample, measured with a micrometer

L = diagonal of the sample (rectangular board), for non-rectangular boards, L is the diagonal between corners exhibiting displacement

The permissible percentage of bow and twist depends upon the thickness of the board. Thickness codes are given in Table 14.4.

Table 14.4   Board Thickness Codes

Thickness Code	Nominal Board Thickness
T1	0.2 mm, 0.5 mm, 0.7 mm, 0.8 mm, 1.0 mm
T2	1.2 mm, 1.5 mm, 1.6 mm
T3	2.0 mm, 2.4 mm
T4	3.2 mm and above

From this Table, it is possible to determine the value of the bow or twist. For example, a 2 per cent bow and twist will mean $0.02 \times$ thickness (millimeters).

14.6.4.2   Conductor Pattern Integrity

Conductor pattern integrity indicates the quality or the state of completeness of the conductors. It is used to determine if conductor widths are within tolerance and annular rings and contours are within drawing requirements. The easiest method to determine conductor pattern integrity is by using positive or negative film one lays on the finished printed circuit board.

Conductor Width: Conductor width has a direct effect on the electrical characteristics of the conductor as a decrease in conductor width increases electrical resistance and decreases the current-carrying capacity. Similarly, minimum clearance between adjacent conductors is ensured only if the maximum width is not exceeded where it may cause problems.

The conductor width is measured under a microscope or an optical comparator with a resolution of 25 microns. The conductor width is normally measured as the widest point on the conductor as shown in Figure 14.19. The conductor is viewed vertically from above.

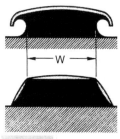

Fig. 14.19 Conductor width

While measuring the conductor width for PCB acceptance, the provisions of IPC-D-320 regarding process tolerances are followed, which are based on copper thickness upto and including 1 oz copper as 0.04 mm. For each ounce of additional copper, an additional 0.03 mm variation for conductor edge can be expected.

Conductor Spacing: The distance between adjacent edges (not between conductors centre line) of conductors is known as conductor spacing. Proper separation is necessary for sufficient insulation between conductors/circuits. A reduction in the spacing can bring down the electrical insulation between any two pairs of conductors. If clearance is critical, the board should be inspected under an optical comparator, and the electrical insulation can be measured using an insulation tester.

The cross-sectional width of conductors is usually non-uniform and therefore measurement should be taken at the closest point between the conductors and/or lands. The etch over-hang is taken into consideration while determining the conductor spacing.

Short-circuits, which are unwanted connections between conductors, result in insulation breakdown due to projections or unetched island of copper between conductors. They can normally be viewed visually or under an optical comparator. Short-circuit can be further confirmed using a multimeter to check for insulation. Normally, the boards with short-circuit are rejected according to most of the specifications. However, if the copper causing the short-circuit is small, the copper can be scraped with a scriber.

An open-circuit, which represents lack of continuity of a conductor is usually established with an ohm meter. Boards with open-circuit are normally rejected. Repaired boards with open-circuits by means of a simple solder bridge are not acceptable in professional electronics due to poor reliability of such connections.

Scratches (slight surface mask or cut): They are unacceptable if they cross one or more conductors.

Nicks: Local sharp reductions in the width of a conductor whose length does not exceed the average conductor width. They are usually acceptable if they are not more than one-third of the width.

Projections: Local sharp increases in the conductor width, whose length is less then the average conductor width.

Voids: The areas on the conductive pattern from which the copper has been wrongly etched away, or may even be missing on the original laminate, are called voids. They are largely acceptable if they are relatively small. This is shown in Figure 14.20.

If the voids are very small, they are called pits.

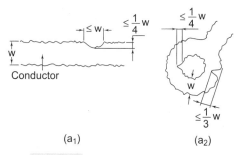

Fig. 14.20 Acceptance limit for voids

Nodules: Abnormal growth of the metallic coating, which appears as small metal balls. It is usually a plating defect. They are acceptable if they are solidly attached to the metal and exhibit no tendency to flake.

Edge Definition: Edge definition is the fidelity of reproduction of the pattern edge relative to the original master pattern. Irregular conductor edges can cause corona discharge in case of high voltage circuits. The edge definition is realized by measuring the distance from the crest to the trough as shown in Figure 14.21. An acceptable value of edge definition is 0.127 mm.

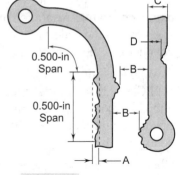

Fig. 14.21 Edge definition

Flush Conductor: A conductor, whose entire surface is in the same plane as the surface of the insulating material adjacent to the conductor, is called 'flush conductor'. It is used in rotary switches, commutators and potentiometers and makes use of wipers and brushes. Ideally, the conductors should be flush (Figure 14.22) for all such applications to reduce wiper vibration, wear and noisy signals. However, due to some mismatch, the length of the step allowed between pattern and base material would depend upon the relative wiper speed, the materials used and the design of tolerable electrical noise in the circuit. Commonly accepted height allowances are shown in Table 14.5.

Table 14.5 Wiper Speed

Wiper Speed(upto)	Height Allowance
50 r/min	2.032×10^{-4} to 5.08×10^{-3} mm
51–125 r/min	1.27×10^{-4} mm
126–500 r/min	0.762×10^{-4} mm

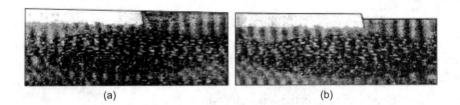

(a)   (b)

Fig. 14.22 Flush printed circuit (a) preferred (b) non-preferred conductor is not flush to the board surface and does not meet the specified tolerance (adopted from Coombs, 1988)

The degree of flushness is usually inspected with height gauges or visual inspection aided by microscope.

Current Breakdown Test: This test is performed to determine if sufficient plating is present within the plated through-hole to withstand a relatively high current. The test is performed as per IPC-TM-

650 which recommends a current of 10 A for 30 s. The test is performed either on an actual printed board or on a test pattern as shown in Figure 14.23.

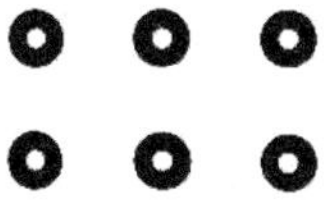

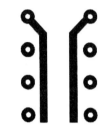

Fig. 14.23 Typical current break-down test pattern (IPC)

The test is conducted through the following simple steps:

- A load of approximate value is placed across the negative and positive terminals of a current regulated power supply.

- The supply current is adjusted to the desired value, say 10 A.

- One end of the resistor, from the positive supply terminal is removed.

- The desired plated through-hole is connected between the positive supply terminal and the disconnected end of the resistor.

- Observe whether or not the set current is maintained in the circuit.

Dielectric Withstanding Voltage: This test is performed to verify whether insulating materials and spacing in the component part are adequate. The test can be performed on either an actual board or on test pattern. Depending upon which application the board will be applied to, different test voltages, 500, 1000 and 5000 volts are applied, between virtually insulated parts of the specimen or between the insulated part and the ground. The voltage is steadily increased to the specified value where it is held for 30 seconds, and then gradually reduced. The specimen is visually examined for evidence of insulation breakdown or flash over between contacts. Figure 14.24 is a typical test pattern.

Fig. 14.24 Insulation resistance, dielectric withstanding voltage and moisture resistance test patterns

The test is performed basically to determine that the component part can operate safely at its rated voltage and withstand momentary over-potential due to surges, scotching operations, etc. The test can be destructive or non-destructive depending upon the degree of over-potential. Complete procedures for conducting the test are explained in IPC-ML-950 (Multi-layer PCBs) and MIL-STD-202 (electronic components method 301).

14.6.4.3 Holes

Hole Size Diameter Specifications: Hole size is the diameter of the finished PTH or unplated hole. Measurement of hole size is important to verify that the hole meets minimum and maximum fit requirements of a component lead, mounting hardware, etc, plus adequate clearance for solder. The hole diameter must be within specified limits, particularly where automatic assembly is to be used. A range of 0.1–0.15 mm is usually acceptable to most PCB manufacturers for holes upto 2.5 mm. The hole diameter is measured using a suitable plug gauge which should be of the go/no-go type or the optical method.

Hole Positions: The hole positions are usually specified with reference either to another hole or to the outline of the board. Strict tolerances between 50–150 microns are usually specified for boards in which components are assembled automatically.

Hole position can be measured with an electronic digital position gauge or with a microscope.

Annular Ring (Hole-to-land Registration): Annular ring is that position of the conductive material which completely surrounds the hole. It is basically a flange which provides an area for attachment of electronic component leads or wires.

For most holes, the hole centre is not exactly the land centre. Therefore, the maximum allowed deviation of the hole centre from the pad centre is usually specified for acceptance. Figure 14.25 shows the various terms associated with annular ring. An annular ring width of 0.254 mm is a standard requirement, which in special cases can be as small as 0.127 mm.

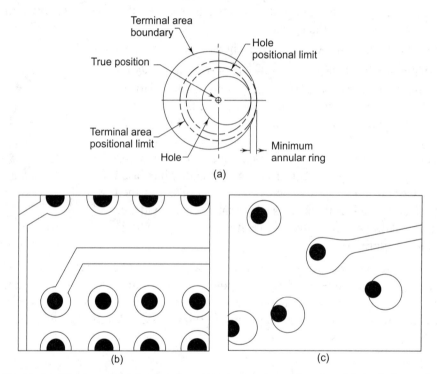

Fig. 14.25 Annular ring (a) area surrounding hole for making connections (b) preferred: holes in the centre in the land (c) not preferred: holes not in the centre (IPC)

Hole Metallization: This test is performed to check the quality of plated holes by examination of a metallographic cross-section of some holes. The test is done with the aid of a microscope and the following factors must be observed:

- Maximum thickness of all coatings;
- Minimum average of all coatings;
- Number and position of non-plated areas;
- Presence of nodules; and
- Presence of any other irregularity such as cracks.

These characteristics should be compared with the specifications and/or the agreed terms of acceptance with the supplier.

Resin Smear: During the drilling process, excessive heat may get generated which may soften the resin in holes and smear it over the exposed copper areas. This condition may result in loss of continuity between the internal land and the plated through-holes, thus leading to open-circuit as illustrated in Figure 14.26. Resin smear is inspected by viewing vertical and horizontal micro-section of plated through-holes.

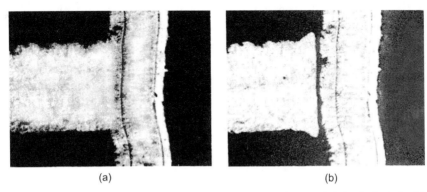

(a)   (b)

Fig. 14.26   Resin Smear (a) preferred: no evidence of resin smear between layer and plating in the hole; (b) not preferred: evidence of resin residue or resin between internal layer and plating in the hole (IPC)

Layer-to-layer Registration: The layer-to-layer registration test is performed to determine the degree of conformity of the position of a pattern with its intended position or with respect to any other conductor layer of a board. Misalignment of internal layers results in an increase of electrical resistance and in cases of severe misregistration, may even result in loss of continuity and open-circuit. Proper electrical connection between the plated through-holes and internal layers is ensured only if there is proper layer-to-layer registration.

The two popular methods of measuring layer-to-layer registration are listed below.

Micro-sectioning Method: It involves measuring each internal land area on a vertical plated through-hole cross-section and determining the centre line. The maximum variation between the centre lines is the maximum misregistration. This is shown as 'C' in Figure 14.27. This is a destructive method.

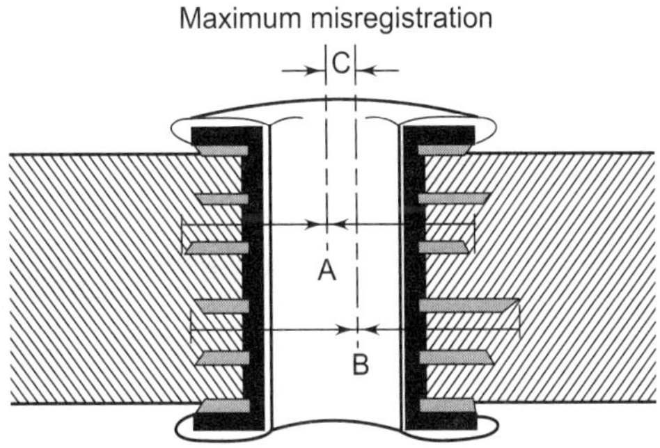

Inter-layer Registration: Accurate registration of internal layers is necessary to ensure good electrical

Fig. 14.27   Layer-to-layer registration micro-section method (redrwan after Coombs, 1998)

connection between the plated through-holes and the internal layers. Misregistration may lead to an increase in resistance, decrease in conductivity and in severe cases, creates an open circuit.

Layer-to-layer registration is usually measured by the X-ray method or the micro-sectioning method. X-ray method, which is non-destructive, utilizes an X-ray machine and a Polaroid film. If the annular ring is not properly visible, it shows severe misrepresentation on the film.

14.6.4.4 Etching Defects

- *Non-etched Copper*: Copper areas which have not been removed completely can generate webbing during automatic soldering process, and may therefore be rejected. Also, sometimes non-etched copper areas may cause short-circuit.

- *Pits*: Pits are small voids which may be present in the copper foils or inadequate protection of copper by the etch resist. They are normally accepted, except on contact tabs, where corrosion of the metal foil may degrade electrical contact problem.

- *Undercut*: Undercut is an etching defect for which acceptable limits should be specified. Undercut is the distance on one edge of a conductor measured parallel to the board surface from the outer edge of the conductor to the maximum point of indentation on the same edge. Referring to Figure 14.28, the

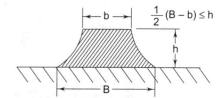

Fig. 14.28 Acceptance limits for undercut of conductors not plated or fused

$$\text{Undercut} \qquad u = \frac{B - b}{2}$$

Where B = Design width of the conductor

b = Minimum conductor width

U = undercut.

General methods of measuring linear distance may be used to determine undercut, if required.

- *Outgrowth:* Outgrowth is the increase in conductor width (Figure 14.29) at one side of the conductor. It is caused by plating build-up. Outgrowth length, approximately 1.5 times or less than the thickness of the outgrowth material is usually accepted. However, excessive outgrowth is likely to break off of the conductive pattern.

Fig. 14.29 Outgrowth extension is approximately ten times the thickness (Coombs, 1988)

- *Etch-back*: The process for the controlled removal of non-metallic materials from side walls of holes to a specified depth is called 'etch-back'. It is meant to remove resin smear and to expose additional internal conductor surfaces. Etch-back is usually specified as a minimum or maximum requirement which ranges from 0.00508 to 0.0762 mm. Figure 14.30 shows the etch-back configuration. The degree of etch-back is measured by micro-sectioning of multi-layer boards.

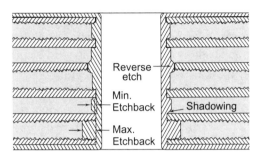

Fig. 14.30   Etch-back configurations (after Coombs, 1998)

14.6.4.5   Plating (Metallic Coatings)

- *Composition of Alloy*: The alloys which are popularly used in printed circuit manufacture are tin-lead or tin nickel. Their composition is usually specified. For example, when tin-lead is used, specifications usually require the tin content to be between 50 and 70 per cent. In case of tin-nickel, the percentage of tin is usually 65 per cent.

  The alloy composition on the plated printed board can be verified by wet analysis, atomic absorption and the beta backscatter. The beta backscatter allows the composition to be determined non-destructively.

- *Thickness*: The minimum thickness of all metallic coatings is usually specified. Although a number of methods are available for measuring the thickness of these coatings, the two popular methods are micro-sectioning followed by measurement with a microscope and beta backscatter.

- *Porosity*: Porosity is an important parameter to measure as corrosion of the base metal through the pores generates products which may have a detrimental effect on any electrical contact, especially on contact tabs. Several methods exist for measurement of porosity. They are electrochemical, electrographic and chemical. However, the most common methods are electrography and gaseous reagent tests.

- *Contact Resistance*: Contact resistance plays an important role, particularly on PCB contact tabs. A bad contact may generate electrical noise. Contact resistance is measured by recording the voltage across the contact at a specific current.

- *Wear Resistance*: Wear resistance provides important information concerning the failure mechanism of plating and may help in designing an improved coating. It is determined experimentally by mounting the specimen in contact with another part on a piece of equipment which provides relative movement of the two parts. The combination is kept working until failure occurs for a preset time. The test is usually conducted in a test chamber to simulate environmental dry corrosion conditions which may accelerate wear. The tested parts may be examined by the scanning electron microscope to determine the amount of material transferred from one contacting surface to another.

- *Adhesion*: A simple method to determine the plating adhesion is to carefully view the micro-sectioning process. In case of poor adhesion, the layers of plating will separate during preparation of the micro-section specimen. It will also indicate lack of adhesion at plating boundaries. Plating adhesion is also tested by tape test as per IPC-TM-650.

- *Solder Mask*: Solder mask is used to protect the untinned copper track from chemical and abrasive damage. This is also used to mask off selected areas of a pattern from the action of solder to prevent solder shorts, particularly during wave re-flow soldering. Such masks are screen printed and are usually 0.1 mm thick. The adequacy of solder mask is inspected for registration, wrinkles and de-lamination.

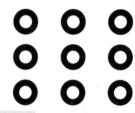

Fig. 14.31 Typical terminal pull test coupon

- *Bond Strength*: The bond strength test is conducted to test the plating adhesion to the laminate in the hole. Poor plating adhesion in the hole affects functionality. Plating adhesion is checked by micro-sectioning, whereas the terminal pull test is used to establish the condition. Figure 14.31 shows typical terminal pull test coupon.

14.6.4.6 Environmental Inspection

Environmental inspection is generally carried out as part of the acceptance procedure whenever high reliability performance is expected. This is to ensure that the printed board will perform its designed functions under the climatic condition to which it will be subjected during its normal use. Environmental tests are performed usually either on actual printed board or its coupon test patterns. The two commonly employed environmental tests are: thermal stress and thermal shock. These are detailed below.

- *Thermal stress*: Temperature-induced strain can lead to deformation of the printed board which can create serious problems during soldering such as plated through-hole degradation, separation of platings or conductors, or laminate de-lamination. Thermal stress inspection/test is generally carried out onto products to study the behaviour of the printed boards after soldering.

  Thermal stress is carried out by placing the printed board in a temperature controlled chamber at a temperature of 120°–150 °C for two hours to reduce moisture. It is then cooled by placing it in desiccators on a ceramic plate, fluxed and floated in a solder bath maintained at 287 °C ± 5 °C for ten seconds. Both visual and micro-sectioning tests for plated-length holes are performed to assess the effects of thermal stress.

- *Thermal shock*: The thermal shock test is useful in determining the behaviour of the printed board to exposures of extreme low and high temperatures. The thermal shock is induced by exposure to severe and rapid differences in temperature extremes, typically from + 125 °C to –60 °C within two minutes. Thermal shock effects may include cracking of plating in the holes and de-lamination. These effects manifest themselves as intermittent electrical connections, when continuous monitoring of electrical circuits is carried out during thermal shock cycling. A typical thermal shock test pattern is shown in Figure 14.32.

Fig.14.32 Typical thermal shock and moisture resistance test rpattern (IPC)

- *Moisture and Insulation Resistance*: This test is carried out to study the effect of high humidity and temperature on the performance of printed boards. The test conditions are usually specified as relative humidity of 90 to 98 per cent with temperature at 25 to 65 °C. After the required test cycles are completed, the insulation resistance is measured. After this test, the test specimen should not show any blistering, measling, warp or de-lamination.

14.6.4.7 Other Characteristics

Cleanliness: The board should be made free from dust, grease, oils, sweat or any other foreign material. Cleanliness can usually be established by visual inspection.

Solderability: The ability of the printed pattern to be wetted by solder for the joining of components to the board is known as 'solderability'. Solderability can be tested by several methods and IPC standards have been established for the same. Solderability of the PCB surface can be qualitatively assessed by the dip test. For PTH boards, the method involves lowering (keeping it horizontal) of the fluxed printed board onto the surface of the molten solder well, whose temperature is thermostatically controlled. With the board kept in contact with the solder, the time taken by the solder to rise through the hole to the top surface gives a measure of the solderability of the board.

Packaging: The board should be inspected for basic identification data, such as part number, assembly number, order number, etc. Each board should be separated from the others by a sheet of water leaf paper and should not include any foreign material such as metallic dips and dust, etc.

Batch Number: It is a good practice to ask for batch number meshed with a solvent resistant ink on the board. In case of a problem on the board, the batch number provides a useful tracking reference. The board can be visually inspected for the batch number and its reliability.

Approvals: If it is agreed upon to provide boards approved by a particular specified agency, such approvals should be indicated on the board. For example, the approval by underwrite Laboratories requires the boards to be marked as XXXXX 107, where XXXXX stands for PCB manufacturers code and 107 is an approved process.

 ## 14.7 Useful Standards

- *IPC-TM-650: Test Methods Manual*: Contains over 150 industry approved test techniques and procedures for chemical mechanical, electrical and environmental tests on all forms of printed boards and connectors.

- *IPC/EIA J/STD-001C: Requirements for Soldered Electrical and Electronic Assemblies*: Describes materials, methods and verification criteria for producing quality soldered interconnections and assemblies.

- *IPC-A-610C: Acceptability of Electronic Assemblies*: Illustrates industry consensus workmanship criteria for electronics assemblies through full-colour photographs and illustrations; including component orientation and soldering criteria for through-hole, SMT and discrete wiring assemblies, mechanical assembly, cleaning, marking, coating and laminate requirements.

- *IPC/WHMA-A-620: Requirements and Acceptance for Cable and Wire Harness Assemblies*: An Industry consensus standard for requirements and acceptance of cable and wire harness assemblies; covers criteria for wire prep, soldering to terminals, crimping of stamped and formed contacts and machined contacts, insulation displacement connectors, ultrasonic welding, splicing, connectors, molding, marking coax/twinax cables, wrapping/lacing, shielding, assembly and wire-wrap terminations.

- *IPC-DRM-SMT-C: Surface Mount Solder Joint Evaluation Desk Reference Manual*: Contains 3-D colour illustrations for chip component, gull wing and J-lead solder joints; also shows the minimum acceptable condition for each type of component misalignment and the minimum solder connections.

- *IPC-SM-785: Guidelines for Accelerated Reliability Testing of Surface Mount Attachments*: Covers guidelines for accelerated reliability testing of surface mount solder attachments useful while evaluating and extra-plating the results of these tests toward actual use environments of electronic assemblies.

- *IPC-9701: Performance Test Methods and Qualification Requirements for Surface Mount Solder Attachments*: Provides specific test methods to evaluate the performance and reliability of surface mount solder attachments of electronic assemblies; establishes levels of performance and reliability of the solder attachments of surface mount devices to rigid, flexible and rigid-flex circuit structures.

- *IPC-PCB-EVAL-CH: Printed Circuit Board Defect Evaluation Chart*: Identifies board defects as revealed by micro-sectioning such as nail heading plating, cracks, epoxy smear, negative etch-back, plating voids, insufficient plating thickness, de-lamination, plating folds.

- *IPC-6011: Generic Performance Specification for Printed Boards*: Defines general requirements and responsibilities for suppliers and users of printed boards, also describes mandatory quality and reliability assurance requirements.

- *IPC-6012-A-AM: Qualification and Performance Specification for Rigid Printed Boards*: Covers qualification and performance of rigid printed boards including single-sided, double-sided, with or without plated through-holes, multi-layer with or without blind/buried vias and metal core boards.

- *IPC-A-600F: Acceptability of Printed Boards*: Provides photographs and illustrations of the target, acceptable and non-conforming conditions observable on printed boards; covers measling and crazing of printed boards, annular ring requirements, plating voids for plated through-holes, flexible circuits, foil thickness for conductive patterns and flatness requirements.

- *IPC-QE-605A: Printed Board Quality Evaluation Handbook*: Contains photographic illustrations of various anomalies and characteristics of printed boards covering topics such as solder resist, plated through-holes, conductor characteristics and surface plating to aid the user in determining specific accept/non-conforming criteria for various anomalies.

- *IPC-TR-551: Quality Assessment of Printed Boards Used for Mounting and Interconnecting Electronic Components*: A compendium of technical methods and techniques used for evaluating the quality aspects of interconnection products and electronic assemblies; addresses base materials, conductor physical requirements, internal planes, construction, registration, plated through-holes, component mounting areas, cleaning evaluation, solder mask and printed board electrical requirements.

- *IPC-2524: PCB Fabrication Data Quality Rating System*: Describes PCB fabrication data quality rating system used by fabrication data quality rating system used by fabricators to evaluate incoming data package integrity; also includes information on conformance to both fabricator and customer design rules and can be used by printed board designers as an output quality check.

- *IPC-9151A: Printed Board Process, Capability, Quality and Relative Reliability (PCQR) Benchmark Test Standard and Database*: Describes the process for evaluating the manufacturing capability of key attributes specified in the design and acceptability standards controlled by IPC.

- *IPC-9191: General Guidelines for Implementation of Statistical Process Control (SPC)*: Outlines the SPC philosophy, implementation strategies, tools and techniques used for relating process control and capability to final product requirements.

- *IPC-9199: Statistical Process Control (SPC) Quality Rating*: The document is a tool for a customer or supplier organization's internal audit group to assess a statistical process control system against the requirements of IPC-9191.

- *IPC-9252: Guidelines and Requirements for Electrical Testing of Unpopulated Printed Boards*: Defines different levels of appropriate testing and assists in the selection of the test analyser, test parameters, test data and fixtures required to perform electrical test(s) on unpopulated board as inner layers.

- *IT-97061: PCB Hole and Land Misregistration: Causes and Reliability*: Discusses the root causes of misregistration, recommendations for proper constructions and the need for an intact annular ring with the overall intent of aiding in the manufacture of reliable, high density PCB products.

- *IT-98103: Reliability of Misregistrered and Landless Innerlayer Interconnects in Thick Panels*: Explains interconnect configuration, design and manufacturing relationships and their influence on interconnect reliability.

- *IPC-MS-810: Guidelines for High Volume Micro-section*: Discusses the many variables and problems associated with the process-from sample removal to micro-etch-and the variables common to high volume micro-section.

- *IPC-QL-653A: Certification of Facilities that Inspect/Test Printed Boards, Components and Materials*: This specification establishes the certification requirements for facilities that inspect/test printed boards, components and materials. This specification is intended to provide a minimum standardized basis for evaluating or auditing a technically oriented inspection/testing facility.

- *IPC-TR-486: Report on Round Robin Study to Correlate IST and Micro-sectioning Evaluations for Detecting the Presence of Inner-Layer Separation*: This technical report provides full detail of the round robin study charged with assessing the use of interconnect Stress Test (IST) as a test for incoming inspection in lieu of thermal stress in the detection of inner-layer separations in plated through-holes.

Environmental Concerns in PCB Industry

 ## 15.1 Pollution Control in PCB Industry

The PCB industry makes use of a large number of chemicals throughout the various stages in the process. A direct discharge of the solution containing hazardous chemicals and solutions containing contaminants into the sewer system is not permitted by regulations imposed by the local administration. This is because some chemicals cannot be mixed together as they produce hazardous reactions whereas others may generate toxic fumes or generate violent heat. The wastes that come out could be metal bearing or non-metal bearing. The high concentration of heavy metals such as Cu, Fe, Cr etc., in the effluent is very harmful to any biological process.

In order to protect an increasingly aware public, every civic authority adopts stringent regulations relating to the disposal of affected effluent and sludge (Spitz, 1990). These regulations are generally federally mandated and there are severe penalties for non-compliance and disregard for discharged requirements.

It is therefore essential that PCB manufacturers adopt such measures as are not only necessary but sufficient enough to neutralize the spent chemicals in such a manner so as to reduce the volume of toxic material to the minimum. For this purpose, detailed regulations issued by the Pollution Control Authorities should be thoroughly studied for compliance. Certification from such agencies may be required to the effect that: (i) volume and/or quantity and toxicity of the waste has been reduced to the maximum extent, and (ii) the method used to manage the waste minimizes risk to the extent practical.

Most regulations take concentration as the primary parameter to establish compliance with the regulations. However, in some cases, 'mass' is taken as the basis e.g. pounds of copper per day. The copper discharge per day can be calculated from the discharge concentration (in milligram per litre) and to the daily flow (in millions of gallons per day). The relationship is given by

Cu per day (lb/day) = concentration $\times$ 8.34 $\times$ the daily flow

If the expected flow from a facility per day and the allowable concentration is known, the mass of a specific pollutant to be discharged can be easily worked out.

 ## 15.2 Polluting Agents

Fabricating printed circuit boards involves a variety of different processes that are broadly divided into two categories:

- Dry fabrication; and
- Wet fabrication.

Dry Fabrication is composed of drilling, routing and imaging.

Wet Fabrication consists of scrubbing, developing, plating and etching. Almost all of the chemicals involved in the wet processing are hazardous to some degree (Breitengross, 1993). These hazards revolve around the properties of the chemicals. For example, strong acids are extremely corrosive and the vapours readily degrade most metallic parts on contact. Inhalation of these vapours can cause a range of problems such as slight respiratory irritation and pulmonary Oedema, and could be even fatal. All these hazards can be reduced or even eliminated from the shop by introducing engineering and administrative controls. Engineering controls, in a simple way, may include proper ventilation and material substitution. Administrative controls would include properly framed SOP (Standard Operating Procedures) and training.

Basically, the hazardous materials used in the wet fabrications processes, can be categorized into the following groups:

- Corrosive Materials: These include acids, bases and halogens. These materials could cause corrosion on substances with which they come into contact. Examples are strong mineral acids.
- Oxidizers: They are materials that readily oxidize substances with which they come into contact. A typical example is that of chlorine gas, which is used to regenerate an etching system.
- Toxins: They are materials that are known to have an adverse effect on humans. These materials are toxic, irritants and carcinogens. The examples are chrome trioxide, methylene chloride and xylene-based materials.

The most important responsibility of the management in the PCB industry is to protect personnel from the hazardous materials. This can be done by education, substitution and procedure development, particularly for waste water treatment and to reserve it to minimize or eliminate chemical discharge into the ponds.

15.3 Recycling of Water

In the PCB industry, there are many stages in the process sequence which can give out a lot of contaminants. The wastes that come out could be metal bearing or non-metal bearing. If the maximum copper content or any other ions in a waste stream is found to be less than the specified, they could be considered as low non-metal bearing wastes.

One of the best ways of reducing this is to limit the usage of water. The usage of water in a rinsing station could be limited by having a number of counter flow rinsing stages instead of only one stage rinsing. It has been found that by this method, the waste water will be reduced to about one-third of the output with only one stage rinsing and this could be effectively achieved by making use of water flow restrictors.

Apart from this, one should make use of the correct grade of filters in deburrers and the scrubbing stages which will help in reducing the copper content going out with the water. Most of the heavier metals could be filtered out by this method and after removing these heavy particles, the water is generally found to be suitable for recycling.

One of the most contaminating stages are the rinsers after plating tanks. Here, the drag out of the plating solutions into the rinsing stages could be reduced by holding the panels in air over the plating tanks thereby allowing the solution to drip back into the tank itself. The amount of time could be decreased by the introduction of an air knife which will help in faster removal of these excess solutions from the surface of the panels. These air knives have been found to reduce the drag-out by nearly 50 per cent. One of the widely used purifying treatments for rinsing water in a small set-up is to collect the water and subject it to batch treatment. Where the continuous process is used, the heavier metals have to be precipitated before being subjected to treatment.

15.4 Recovery Techniques

Earlier the printed circuit board industry had not considered the need to adopt metal recovery processes because the lower priced alternative i.e. offsite waste disposal, appeared to be more attractive. However, with disposal costs rising and land-fill sites becoming increasingly less accessible, metal recovery is gaining wider acceptance.

The amount of liquid effluent from a plating shop is very high. The major pollutants are copper, tin, lead, fluorine, phosphorous and metal. Their concentration in the effluent will vary from plant to plant depending upon the processes and the type of equipment used.

The waste from the etching area contains a considerable amount of copper, while other pollutants vary according to the etchant used. Chromic acid etchants are the most difficult to treat, even though they are highly toxic and tend to pollute the water bed.

Waste waters are received from the rinsing stages and process baths which are changed quite often. The major metal content that is found in waste water from the PCB industry is undoubtedly copper. The major sources of copper are deburrers, board scrubbers, and the bath and the rinser following micro-etching. The methods used for its recovery are:

15.4.1 Filtration

Filtration systems are used to remove particles from wash water from deburrers and scrubbing stages which will help in reducing the copper content going out with water. Filtering can be carried out by any of four systems, namely, a centrifuge, a paper filter, a sand filter, gravity setting and the filtration technique. Each of these methods has its own advantages and disadvantages. Most of the heavier metals can be removed by filtration and the water is generally found to be 100 per cent suitable for recycling.

15.4.2 Water Use Reduction Technique

One of the best ways of reducing the copper in the effluent is to limit the usage of water counter flow flare method. The most effective technique to reduce the water consumed in any plant is to use a number of counter-flow sousing stages than having only one stage rinsing. It is found that by this method, the waste water will be reduced to about one-third of the output with only one stage rinsing. The desired flow rate of rinse water is best maintained by using flow restrictors. Flow restrictors are of the flexible diaphragm type, which are rated for a specific flow rate in a specific pipe size. Usually, pulsating spray rinses are used to reduce the total water demand for a facility. However, the effectiveness of spray rinsing depends upon the geometry of the board, the water quality and the degree of rinsing required.

Figure 15.1 shows the principle of counter-current or cascade rinsing. The water flow is opposite to the direction of board movement. It may be observed that the last rinsing dip is in the almost fresh water. Several methods are available for purifying treatment for rinsing water. The method could be a batch treatment for smaller PCB shops, whereas for larger plants, a continuous treatment is necessary. For heavily contaminated rinsing water, the heavy metal precipitation process is used.

Most recovery systems are sized by knowing the volume of drag-out and concentration of metals in the drag-out. The drag-out of the plating solutions into the rinsing stages can be reduced by holding the panels in air even the plating tanks allowing the solution to drip back into the tank itself. The amount of time can be reduced by the introduction of air knife on the hoist.

Prior to international awareness and concern relating to ozone depletion, ozone-depleting chemicals were widely used in the electronics industry (Smith, 1993b). In response to pollution prevention initiatives, various alternative cleaning process operations have been developed. Wilk (1994) illustrates some of the new and alternative cleaning processes for recycling of waste water.

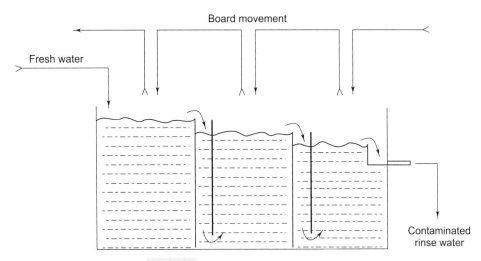

Fig. 15.1 Cascade rinsing process

Miller (1997) states that a closed-loop or zero discharge system is achievable by minimization of water usage through highly efficient rinsing, water re-use for facility requirements; develop, etch, strip solution re-use, and heavy metal removal and reuse. Integrated designs for metal recovery, water recycling and waste minimization through a combination of ion exchange, membrane and electrolytic metal recovery technologies have been suggested by Harnsberger and Saloka (1994).

Fulker (1992) looks at the waste discharge considerations as they apply to aqueous and semi-aqueous cleaning systems for PCB assemblies. The economics of these two types of cleaning are also examined in the publication.

15.4.3 Ion Exchange System

One of the highly recommended treatments for rinsing water is to use the resin purification process, which is a well known process for the production of de-ionized water. Special resins are available which selectively fix one or a few ions of the percolating liquid and replace them with H^+ or other non-polluting ions. On the other hand, most resins can be regenerated when exhausted.

The principle of the ion exchange system is very simple. The rinsing water is passed through two de-ionizing beds for cations and anions. It makes use of the fact that most of the water soluble chemicals used in PCB production are ionized in water, forming anions and cations. For example:

$$CuSO_4 \longrightarrow \underset{\text{(cation)}}{Cu^{+2}} + \underset{\text{(anion)}}{SO_4^{-2}}$$

where the rinsing water passes through the ion exchange system, the following changes take place:

- Cations replace hydrogen ions in the cation resin; and
- Anions replace OH ions in the anion resin (R = resin).

$$RH^+ + Cu^{+2} \longrightarrow 2H^+ + RCu^{+2}$$

$$ROH^- + SO_4^{-2} \longrightarrow 2\,OH^- + RSO_4^{-2}$$

Thereafter, H^+ (hydrogen ions) and OH^- (hydroxyl ions) combine together to form water

$$H^+ + OH \longrightarrow H_2O$$

A sample de-ionizing arrangement is shown in Figure 15.2.

After a certain time, the resins will get saturated, which can be detected by the rising conductivity of the de-ionized water. The resin regeneration process involves adding a strong acid to the cation resin and a strong base to the anion base. This results in the reverse reaction which returns the resin to the original state. In the regeneration process, the metals and ions are worked out in the concentrated form, which can then be subjected to the heavy metal precipitation process for purification. Figure 15.3 shows the scheme for regeneration of the de-ionizer.

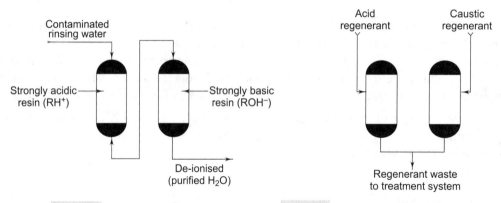

Fig. 15.2 De-ionizing method   **Fig. 15.3** Ion exchange system: regeneration system

Ion exchange is an expensive method and is therefore, usually performed as a final purification step on a limited amount of the waste which contains pollutants that cannot be kept within limits by chemical precipitation.

Spitz (1990) explains the use of the ion exchange technique for the recovery of metals and is shown in Figure 15.4. The low level metal streams from process rinse operations are first filtered through an active carbon filtration unit. This step removes traces of organic impurities such as proprietary bath additives. Next, the filtered stream passes through a collector ion exchange column which undergoes regeneration. The regeneration process involves washing the column with a reagent that displaces the metal from the ion exchange resin. The resulting metal concentrate solution can range from 1 to 30 grams/liter, and may be recycled directly to the process bath. Alternatively, the metals from the concentrate can be removed electrolytically, in a separate step, and recovered as metallic sheets.

Production (parts) flow

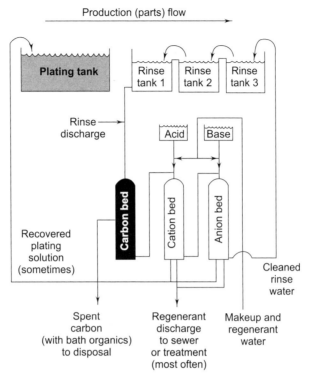

Fig. 15.4   Ion exchange system—the method is based on differences in concentrations between metal and exchanging ions (after Spitz, 1990)

15.4.4   Reverse Osmosis

The membrane separation technology uses permeable or semi-permeable membranes to separate metallic and other toxic components of a chemical waste stream from the solvent. Reverse osmosis is one of the more promising techniques in this technology.

In this process, the salts dissolved in the water are removed under high pressure. Figure 15.5 shows a general scheme for reverse osmosis: the rinsing water (feed) is made to flow across a semi-permeable membrane under a high pressure that may be of the order of 28-35 kg/cm^2 (400-500 psi). The membrane is permeable to water and impermeable to ions and the permeate, which is the water that passed through the membrane and the concentrate, which does not pass through the membrane. The permeate with a very

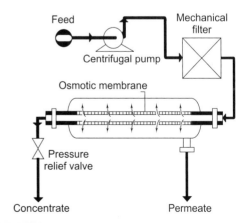

Fig. 15.5   Reverse osmosis uses pressure to drive dilute, ionic metal across an ion exchange membrane (after Leonida, 1989)

low concentration of all the substances (Figure 15.6) in the original solution can either be simply sent to sewage or can be used as de-ionized water for rinses. The concentrate, with a higher concentration of all the components of the plating bath, can be returned to the bath to compensate for the drag-out losses to some extent.

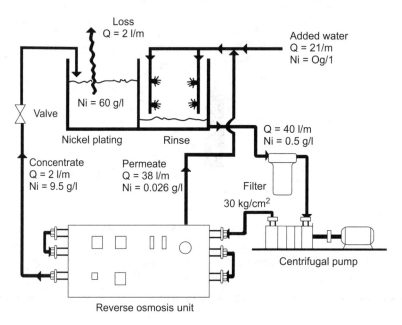

Fig. 15.6 Reverse osmosis: example of application to the recovery of nickel salts and water from a nickel plating cell (redrawn after Leonida, 1981)

The reverse osmosis technique has several advantages. It is found to be more economical to apply this technique directly to the plating tank, where the pollutant is generated. The main disadvantage is the cost of equipment and the running cost of the membranes. The technique is particularly useful for chromium and nickel plating bath. Broomfield (1992) examines the various issues related with the selection of water treatment plant for aqueous cleaners.

15.4.5 Evaporative Recovery

This technique makes use of heat to concentrate contaminants. Water molecules get evaporated as steam, which are then cooled to produce contaminant-free water, which is suitable for re-use. The concentrate can be treated to recover the salts or metals. The technique is used mostly to recover chromic acid in de-smear operations from the rinse water following that bath. The rinse waters are re-circulated. Impurities (trivalent chrome) are removed from the bath by using a cation exchange.

15.4.6 Precipitation of Heavy Metals

In order to neutralize the spent chemicals, it is necessary to reduce the volume of toxic material and to get a minimum quantity of highly concentrated and water insoluble sludge. The basic process involves heavy metal precipitation for which the batch diagram is shown in Figure 15.7.

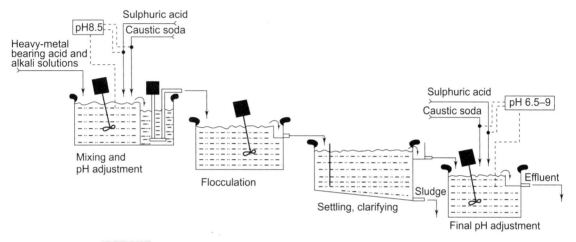

Fig. 15.7   Principle of heavy metal precipitation (adopted from Bosshart, 1983)

In principle, the pH value of the solution is raised to a certain level where the heavy metals like Cu, Ni, Pb, Fe or Cr get precipitated as water insoluble hydroxides. The sludge from the salting tank can be further concentrated by filtration. The hydroxide formation or actual neutralization takes place immediately after the critical pH value is reached. However, the flocculation and setting of the hydroxides may require several hours. It may be noted that the pH range in which precipitation occurs is not the same for all metals.

The sludge so formed can be processed to separate some chemicals which can be used again. This approach is gaining popularity because of the possibility of recovering heavy metals and other materials.

Alternatively, if it is not economical to follow the metal recovery route, the highly concentrated sludge can be buried in specially selected dumping sites where no ground water can be affected. However, the dumping site has to be selected carefully in consultation with the local civic authorities and care should be taken not to dump the sludge anywhere near the drinking water sources. This procedure is to be taken as a worst case treatment. Another option is to burn the sludge.

15.4.7 Electrolytic Recovery

Figure 15.8 shows the electrolytic process, which is commonly referred to as electrowinning, employs electrode-position technology. The reaction cell, which is the heart of the system, consists of a number of anodes and cathodes wired in parallel. As the solution to be treated is pumped into the reaction cell, direct electric current causes the metal ions in the solution to be reduced or plated out on the cathode surface. The metal-depleted solution is then circulated back into its original tank. Cathode deposits attain a thickness of 0.125 inch to 0.25 inch (3.175 to 6.35 mm) in each side of the cathode. Metal deposits, in the form of sheets or flakes, are separated manually from the cathodes, which have been removed from the cell. Re-installed, the cathodes are ready for the next cycle (Spitz, 1990).

Gemmell (2003) states that the key to pollution prevention in printed circuit board manufacturing is to minimize chemical drag-out; minimize the amount of water used for rinsing, and the recovery, re-use, and recycle of copper.

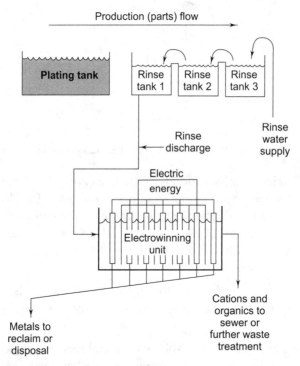

Fig. 15.8 Electrolytic techniques use electrical energy to reduce metal ions into their elemental state (redrawn after Spitz, 1990)

15.5 Air Pollution

There are several sources of air pollution in the PCB industry. Primarily, the pollution is of two kinds: dust and fumes.

15.5.1 Dust

Tiny glass epoxy particles in the form of fine dust, which are produced due to remains from mechanical operations such as drilling, cutting, sanding, routing, bevelling, slotting and milling operations, are a major source of the air pollution. The dust is harmful for both the workers and machinery. This type of pollution can be checked by using some kind of dust collectors, such as:

- Suction-based collectors: They are based on the principle of vacuum cleaner in which dust is collected in a bag, that can be cleaned periodically.
- Industrial Dust Collectors: These are based on the cyclone separation principle. The incoming air stream is brought into a double vortex where the stream starts rotating. The dust particles get separated by initial force. In the process, they settle down at the bottom of the equipment from where they can be cleaned. The air stream leaves the vortex from the top.

15.5.2 Fumes

There are several stages in the PCB fabrication process, which gives rise to a lot of fumes that are found to pollute the air around. For example, the process of etching and plating gives rise to a lot of chemical fumes which, though not really toxic, may harm the personnel due to continuous exposure. Hence providing a few exhaust fans on the walls may not really serve the purpose. There are several types of fume extractors that could be fitted on the individual machines or through a centralized system.

Acid fumes from acid cleaning and organic vapours from vapour de-greasing are usually not contaminated with other materials, and are therefore often kept separate for subsequent treatment. The exhaust facilities generally do not incorporate measures for treatment of such air before it is released into the atmosphere. This situation is mostly not acceptable for meeting the requirements of air pollution control agencies and the air must be cleaned before it is discharged into the atmosphere. This is essential particularly in those cases where large quantities of such fumes are produced. The method of cleaning the toxic fumes related pollutants are delineated below.

- Wet scrubbing: The fumes contaminated with acid ammonia or chloride are brought into close contact with water or aqueous solution wherein the contaminants get absorbed, before the air is released into the atmosphere. Various techniques are used to increase the surface contact area for better efficiency of absorption of the contaminants. These include passing air stream through a high velocity liquid spray or fog.

Various techniques are used to perform wet scrubbing. The contaminated air is either pushed or pulled through the wetted cleaning column, which contains small Teflon bath or rings, that help in increasing the surface area.

- The fluid recommended for acidic type of air is a caustic soda solution or water, while ammonia fumes are always treated with water.

- For treatment of chlorinated solvents, such as chlorinated hydrocarbons (such as trichloroethylene, trichloroethane and methylene chloride), active carbon absorption plant is used. This technique also helps enables to reclaim the chlorinated hydrocarbons for rc-use.

15.5.3   Clean Environment in Assembly Rooms

Cleanliness in an assembly environment is essential for the production of quality products. However, the level of cleanliness depends upon the criticality of the product, i.e. the quality and reliability standards which are expected to be satisfied. For example, normal air conditioning standards are adequate for most of the consumer products, whereas the equipment used in avionics and medical field requires the highest limits of cleanliness.

Clean room standards have been developed and classified into three types depending upon the number of particles of half a micron size or larger per cubic foot (per litre), under controlled temperature and humidity conditions. These are:

100,000 (3500) ; 10, 000 (350) and 1000 (3.5)

Figure 15.9 shows particle size distributions for the clean room standards classification as per the American Society for Testing and Materials. The number of particles in a clean room is Measured by using a light scattering mirror. The particles are counted in a known volume of air. A histogram showing the distribution of particles of various sizes can also be obtained by passing the air through various size filters.

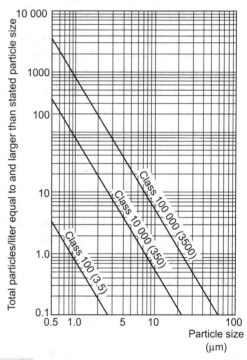

Fig. 15.9   Particle size distribution for three clean room standard classifications (redrawn after Haskard, 1998)

The major source of particulate contamination in air is the human operator. The particles are shed through movements and from the garments. It is usually advisable to provide some protective garment made of lint-free material to cover the head, face and feet.

Most printed circuit board manufacturers carry out processes in clean room environments to achieve acceptable yield rates, especially for fine line circuitry. Nevertheless, despite considerable investment in clean rooms, many companies still find significant amounts of contamination-related defects. Hamilton (1996) points out that the currently accepted methods of assessing contamination fail to deal fully with thc problem. He suggests a *Contamination Audit* to provide a wide ranging evaluation of the sources of contamination.

15.6   Recycling of Printed Circuit Boards

In recent years, there has been an increasing concern about the growing volume of end-of-life electronics. The problem is already acute in developed countries and is likely to become so in the developing countries in the very near future. For example, the European Commission (2001) estimates that electronics contribute 4 per cent of the municipal waste stream, and it is growing at the rate of 3 to 5 per cent per annum, three times the rate of growth of other wastes. Over 90 per cent of this waste is land-filled. Four hundred million cell phones are produced a year, and perhaps 500 million personal computers (PCs) will become obsolete by 2007.

In Europe, the precedent for product take-back and recycling has been established with the ELV (vehicle end-of-life) directive, which places the responsibility for disposal firmly on the importer or manufacturer. The directive also sets restrictions on a range of hazardous materials in manufacturing, a list that is being reportedly revised to include electronics assemblies (Rae, 2003). The WEEE (Waste Electrical and Electronic Equipment) directive and RoHS (Reduction of Hazardous Substances), adopted on May 15, 2001, follow this pattern, setting recycling targets and limiting lead, mercury, cadmium, hexavalent chromium and some brominated flame retardants.

Some legislation at the national levels is already in place. Sweden's recycling law was enacted in 2001. Japan's electronic appliance recycling law, enacted in April 2001, covers larger domestic appliances and is expected to be extended progressively to computers and other areas. A user fee of $30 is charged, and recycling plants have been set up by major electronics manufacturers. The awareness about environmental protection is fast developing in India also and the day is not far when a similar legislation would be introduced.

In the light of an eventual take-back legislation for electronic products, which is expected to come into force in the industrialized countries the world over, disposal processes for electronic products are obviously expected to receive special attention (Legarth, *et al.*, 1995). Electronic products are often defined as complex products with a content of printed boards. PCBs today constitute an environmentally problematic fraction in disposal, on the one hand. On the other hand, PCBs contain most of the elements in the periodic table and may thus be seen as a source of some rare and valuable resources.

15.6.1   Present Approach to PCB Scrap Disposal

The PCB scrap is generated at various sources such as PCB manufacturers, OEMs (Original Equipment Manufacturers), individuals, corporate and equipment dismantlers. The scrap from these sources can be directly sent for recycling, recovery operations or for land-fill. On an average, about 85 per cent of all the PCB scrap board waste is subject to land-fill and only 15 per cent is currently subjected to any form of recycling, (Goosey and kellner; 2002).

Scrap PCBs can be categorized into three grades depending upon the inherent precious metal content. These are:

- Low Grade Material: This comprises power supply units and television boards having ferrite transformers, large aluminium heat sink assemblies and laminate offcuts.
- Medium Grade Scrap: This contains precious metal content, generally from pin and edge connectors used in high reliability equipment.
- High Grade Material: This comprises high precious metal content boards, gold containing integrated circuits, discrete components, opto-electronic devices, gold and palladium pin boards, etc.

These grading materials help to determine the economics of applying recovery operations. However, it is possible to regrade the material from low to medium category through selective manual disassembly of high percentage mass ferrous and aluminium components.

Recycling involves the disassembly of scrap PCBs followed by sorting, grading and shredding operations. Iron and aluminium metals are removed from the final ground product by using magnetic and eddy current separation. The output from the recycler is either sent for land-fill or to a smelter. However, only those boards which contain sufficient gold or precious metal content are subject to *smelting*, otherwise all non-precious metal bearing board scrap is consigned to land-fill. About one per cent of the scrap is subject to specialized recycling operations solely for the purpose of precious metal recovery. Figure 15.10 shows a general scheme for PCB scrap disposal/treatment.

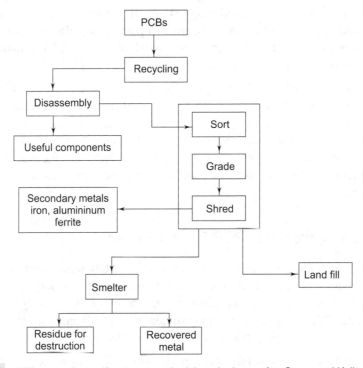

Fig. 15.10 PCB scrap disposal/treatment methodology (redrawn after Goosy and Kellnr, 2002)

15.6.2   Characteristics of PCB Scrap

PCB scrap, particularly populated scrap board, is highly heterogeneous and complex. These boards contain diverse levels of inorganics with relatively low levels of precious metals in conjunction with copper, solders, various alloy compositions, and non-ferrous and ferrous metals (Feldman and Scheller, 1994). The materials and components present in the scrap PCBs have widely differing intrinsic physical and chemical properties. The recycling techniques primarily depend upon various characteristics, which are detailed below.

15.6.2.1   Density Differences
The typical values of the specific gravity of materials contained within scrap PCBs are:

Materials	Specific Gravity range (g/cm^3)
Gold, Platinum group, tungsten	19.3–21.4
Lead, silver, molybdenum	10.2–22.3
Magnesium, aluminium, titanium	1.7–4.5
Copper, nickel, iron, zinc	7.0–9.0
Non-metallic Materials	1.8–2.0

It is evident that various materials can be separated by density-based separation systems, normally employed in the process industry (Barsky, *et al.*, 1991).

15.6.2.2   Magnetic and Electrical Conductivity Differences
The application of low intensity magnetic separators is well developed in the minerals processing industry and can be used to separate ferrous materials from PCB scrap. On the other hand, non-ferrous metals may be separated by means of electrostatic and eddy current separators which are well developed within the recycling industry (Iji and Yokoyama, 1997).

15.6.3   Disassembly of Equipment

In most of the operations for the recycling of printed circuit boards, disassembly is an essential part. The disassembly of components facilitates a selective and profitable recovery of metals and noble metals. In addition, the concentration of valuable materials could reduce the re-processing cost. Another advantage of disassembly is the isolation of hazardous components in order to prevent them from contaminating the shredded waste. Electronic components can be re-used after disassembly for economic and ecological reasons. However, the cost of disassembling, testing and selling the re-usable electronic components has to be seen in relation to the cost of a new product (Keimeier, 1994).

The disassembly of scrap is mostly carried out manually by using simple hand tools, which itself sometimes places limits on all such operations due to the costs involved. Disassembly is considered to be an area of great importance in case of recovery of low cost components and is considered to have a great impact on the overall future recycling strategies. In order to ensure safety during disassembly and take into consideration the cost factors, mechanical dismantling, and automated and robotic dismantling techniques have been practised (Feldmann and Scheller, 1994). The automated component disassembly operation basically involves scanning the board to read all component identification data; reading stored component database to determine their value; determining how the identified components are soldered or mounted; if mounted, disassembling is done via robot and if soldered, de-soldering is done by using laser or infra-red energy.

Yokoyama and IJi (1995) describe a recycling process of printed circuit boards with electronic components, which is shown in Figure 15.11. A practical process for pulverizing the PCB waste and separating the resulting powder into copper rich powder and glass fibre-resin powder was developed. With this process, up to 94 per cent of the copper was recovered in the pulverized PCBs of 100-300 microns average particle size. The recovered glass fibre-resin powder improved the mechanical strength and thermal expansion properties for epoxy resin type paints and adhesives. For the next step, the recycling of PCBs with electronic components by disassembling the components from the PCB, and by applying the PCB waste recycling process to the remaining board is being investigated.

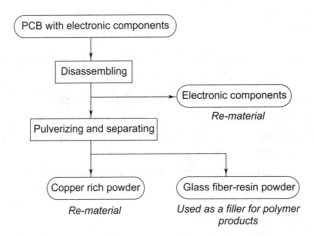

Fig. 15.11   Recycling process of a PCB with electronic components (after Yokoyama and Iji, 1995)

15.6.4   Technologies of Recycling of PCBs

Two approaches are emerging as the potential techniques for separation of materials in the recycling process. These are based on mechanical and hydrometallurgical methods.

15.6.4.1   Mechanical Methods

Mechanical systems for the treatment of a wide range of electronic scrap materials including populated and non-populated PCBs are commercially available. A practical process for pulverizing PCB waste and separating the resulting powder into copper rich powder and powder consisting of glass fibre and resin is described by Yokoyama and Iji (1995). The process, which involves pulverizing and separating, is shown in Figure 15.12.

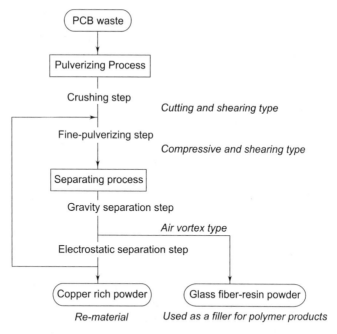

Fig. 15.12 Pulverizing and separating Process for the PCB wastes (adapted from Yokoyama and Iji, 1995)

In this process, the PCB waste is pulverized in a process consisting of a crushing step which uses cutting and shearing forces, and a fine pulverizing step, which uses compressive and shearing forces. This pulverizing process is highly effective, and is shown to have very high abrasion resistivity. The copper rich powder and the glass fibre-resin powder are recovered with a separating process consisting of a gravity separating step using an air vortex, and an electrostatic separation step. The effective average particle size for the pulverized waste was found to range from 100 to 300 μm. Up to 94 per cent of the copper was recovered from the PCB waste at this size. While the copper content in the PCB waste was 7 per cent, the content in the copper rich powder obtained from the gravity separation step was more than 20 per cent. Figure 15.13 shows the size distributions of each component.

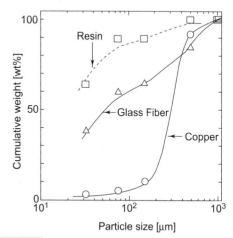

Fig. 15.13 Size distribution of pulverized PCB components (redrawn after Yokoyama and Iji, 1995)

15.6.4.2 Hydrometallurgical Methods

A number of hydrometallurgical methods have been developed which indicate the potential for the recovery of all materials from the PCB scrap. One such method is the recovery of gold from pins and edge connectors, which have been manually separated from the scrap board via the use of air knives, etc. In USA, a methodology based on solvolysis has been developed to enable both the more efficient recovery of metals and the recovery of plastic materials such as epoxides at high quality, with the additional benefit of the capability to extract both halogens and brominated hydrocarbon derivatives. (www.recyclers-info.de/de/bigat/prasengl.htm).

In addition, various studies have been undertaken to assess the viability of utilizing dilute mineral acids in conjunction with subsequent metal recovery techniques based on concentration and separation, such as solvent extraction, ion exchange, adsorption and cementation (Saito,1994).

 ## 15.7 Environmental Standards

The PCB industry worldwide is now taking up the parallel principles of quality control and waste management, and firmly believe that a more comprehensive approach towards environmental management is needed. Just as companies and nations have developed standards for everything from food packaging to the transmission of video and voice over fibre optic cables, international standards for environmental management and protection system called ISO-14000 have been developed by the International Organization for Standardization (ISO). This standard approaches environmental issues with a focus not on meeting limited static standards, but rather on driving continuous improvement across a broader spectrum (Bruhn, 1997).

ISO 14000 requires companies to identify and prioritize the significant environmental impacts of their operations, quantify and develop performance-based environmental objectives, define the structure and responsibilities of personnel and upper management involvement, develop internal auditing and corrective action programmes, and institute effective training and communication systems. This standard can be viewed as an extension of the total quality management approach to business management wherein the goal is to continually strive to achieve excellence, and constant improvement is the key objective. Perhaps the greatest benefit of the standard is that it encourages companies to be environmentally proactive and stay ahead of the regulatory curve. It stimulates the development method to eliminate materials, processes and wastes that make a company vulnerable to regulations.

There is a shift in handling the environmental issues from end-of-pipe solutions such as waste disposal and remediation to new emphasis on "design for environment" where the focus is to reduce energy and consumables in the manufacturing of products, and to design products to minimize environmental impact, so that they can be disassembled, re-used and/or recycled. This trend may increasingly affect the manufacturing and design of products, their technology development, plant locations and marketing strategy (Shaw, *et al.*, 1997). The participation of IPC in the EPA's (Environmental Protection Agency of USA) Design for the Environment Project is an example of how the effective participation of suppliers, manufacturers and customers can expedite the development of new environmentally beneficial technologies.

15.8 Safety Precautions for the Personnel

The use of different kinds of highly concentrated chemicals in a PCB fabrication facility involves a lot of health hazards and risks, if these chemicals are not properly handled. It is therefore mandatory for the personnel handling these chemicals to be educated about their usage and precautions. The following precautions should be followed:

- All protective devices such as goggles, face mask, gas mask, gloves, etc. must be available immediately to the operator for his use.
- Highly concentrated chemicals are used in the PCB industry. They need to be handled carefully, stored properly in tightly closed bottles and containers, and kept in well ventilated rooms. No smoking and eating is advised in the area where these chemicals are kept.
- Personal safely should be accorded highest priority. It is advisable to wear impermeable and acid-proof aprons, gloves, face masks with filters and goggles.
- All care should be taken to prevent accidents. Always observe and act upon the known and recommended safety rules. In case of accident involving personnel, call the doctor immediately.

15.9 Toxic Chemicals in PCB Fabrication

Several chemicals are used in the PCB fabrication process. Many of these chemicals have specific dangers and require careful handling. The following is a list of toxic chemicals used in the PCB industry:

- Chemicals which cause heavy burns and etchings;
  - HNO_3, nitric acid (20-70 % concentration);
  - HNO_3, nitric acid (>70 % concentration); with danger of fire
  - H_3PO_4, phosphoric acid (> 25 % concentration);
  - H_2SO_4, sulphuric acid (> 20 % concentration);
  - KOH, potassium hydroxide (anhydrous and solutions with > 10% KOH);
  - NaOH, sodium hydroxide (caustic soda);
  - HCOOH, formic acid (> 25 % concentration);
  - $HClO_4$, perchloric acid (10-50 % concentration);
  - HCl, hydrochloric acid (> 25 % concentration);
  - CH_3COOH, acetic acid (> 25 % concentration); and
  - CrO_3, chromic anhydride. inflammable

These chemicals should be stored in tightly closed bottles. Do not breathe in vapours. Avoid contact with skin, eyes or clothes.

- Chemicals with serious toxic danger if swallowed:
  - $Me^I CN$, $Me^{II} (CN)_2$, cyanides;
  - $Me^I F$, $Me^{II} F_2$, fluorides (soluble);
  - HF, hydrofluoric acid; causes heavy burns
  - HCHO, formaldehyde. causing burns and etchings

These chemicals should also be stored in tightly closed bottles, and contact with skin and eyes should be avoided.

- Chemicals with toxic danger that attack breathing path:
  - NH_3, ammonia (gas); attacks skin, eyes
  - NH_4OH, ammonia (10-35 % NH_3 concentration); attacks skin, eyes
  - NH_4OH, ammonia (> 35 % NH_3 concentration); causes burns, etching
  - NO_2, N_2O_4, nitric oxide; attacks skin and eyes
  - Cl_2, chlorine. attacks skin, eyes

These chemicals also attack the skin and eyes, and cause burns and etchings. They should be stored in tightly closed bottles in a cool and ventilated place. Do not breathe in the gas/vapours. Rinse splashes on skin or eyes with plenty of water.

- Chemicals which are inflammable:
  - C_6H_6, benzene; highly toxic
  - CH_3OH, methyl alcohol; toxic
  - C_2H_5OH, ethyl alcohol.

In all these cases, no smoking is allowed in the working areas. Bottles should be kept tightly closed in cool and well ventilated places. Keep away from open fires, heat sources and sparks.

- Chemicals with health affecting vapours:
  - $CCl_2{:}CCl_2$,, perchloroethylene; and
  - $CHCl{:}CCl_2$, irichloroethylene.

Do not breathe in the vapours and ensure that there is sufficient air circulation in the area and ensure that they are stored only in tightly closed bottles.

 ## 15.10   Lead-Free Soldering

A serious issue in the electronic manufacturing industry is the elimination of lead, which is the primary constituent of traditional eutectic solder alloys of Tin/Lead (Sn/Pb). Its good performance in terms of electrical and thermal conductivity as well as of the low eutectic point of 183 °C for the alloy Sn 63 % and Pb 37 % is because of the presence of lead.

The use of lead in electronics assembly operations has come under scrutiny due to health and environmental concerns associated with lead exposure (Melton and Fuerhaupter, 1997). Lead affects the nervous system, the kidneys and the reproductive system, if it enters the body. Generally, eating, drinking, smoking, or the application of cosmetics at workstations is prohibited in an industry where lead solder or soldered parts are handled. However, hand contamination, to some extent, is probably unavoidable, since lead is so soft that it is easily rubbed off onto the fingers in the handling of boards that have been coated with tin/lead or soldered, whether with solder paste, or on a wave-soldering machine or from manual soldering. Molten lead does not evaporate into the air, but lead oxide, the so-called dross, is a loose powder that can become airborne. The handling of dross should therefore be carefully done during the skimming operation at the hot air levelling (HAL) station or the soldering station as well as by shaking into another container.

Altogether, the control of lead absorption in PCB fabrication and PCB assembling shops is not so difficult since the exposure intensities are quite low. By paying proper attention, shops that use lead-containing solder can be kept as safe and healthy as any workplace with no lead being used in production processes. However, there is another important driver to ban the lead out of the electronic items: i.e. the huge growth of electronic waste! This waste can affect the groundwater.

The maximum level of lead permitted in drinking water in India as well as in the USA is 0.015 mg per litre. The residence of lead in our body is extremely low and on an average is as follows:

- Residence of lead in blood   : about 25 days
- Residence of lead in soft tissue   : about 40 days
- Residence of lead in no labile bones: > 25 years

Severe lead poisoning can be caused by the level > 1 mg per litre for adults and @> 0.7 mg/l for children. Severe toxicity can lead to colic (severe abdominal cramps), changes in consciousness, coma and death. Manko (1994) lists the safety aspects and precautions to be observed while carrying out lead-based soldering.

Any circuit board scrap containing tin/lead solder will fail a toxicity characteristic leaching procedure (TCLP) Test for lead. In countries like Germany and the USA, electronic waste containing lead is listed in the so-called banned list and is banned for normal waste disposal. Tin/lead soldered boards exceed the specified value of characteristic of hazardous waste. These characteristics include ignitability, corrosivity reacting or toxicity. This means that electronic waste needs to be managed as a potentially hazardous material if disposed off improperly.

Another driver, besides environmental regulations and take-back regulations, is the marketing aspect. Most of the major electronic companies in Japan have announced their lead replacement timetable and identified their lead-free products using environmental symbols. It appears that worldwide, the real pressure to move to lead-free manufacturing will come from the customers. Several studies have shown the market advantage of being perceived as a "green" company. In case of products with price and features being equal, consumers tend to choose a product that is labelled as environmentally safe.

The European Union has nearly finalized its legislation and is likely to ban the following materials from all electronic items with effect from July 1, 2006:

- Lead (exceptions);
- Mercury;
- Cadmium;
- Hexavalent (six-valued) chromium;
- Halogenated flame retardant.

Ban on products containing lead in Japan, Denmark and the European Union (EU) will have an effect in the global marketplace, which will force many US manufacturers to eliminate lead as effectively as a US legislative ban would.

European manufacturers (Nimmo, 2003) have indicated plans to make lead-free products available in early 2003 and all new products lead-free in 2004. An exemption will be granted only if the "use of lead is technically or scientifically unavoidable or where the negative environmental and/or health impacts caused by substitution are likely to outweigh the environmental benefits thereof".

15.10.1   Substitutes for Tin/Lead Solders

A substitute for a lead/tin solder should have the following desirable properties:

- **General**
  - Less toxic;
  - Better environmental properties;
  - Melting point at about 185 °C;
  - Electrical conductivity as Sn/Pb or better; and
  - Thermal conductivity as Sn/Pb or better.
- **Soldering**
  - Performing an intermolecular bond to all base metals;
  - Low surface tension;
  - Compatible with laminates/components;
  - Compatible with Flux Formulations
- **Physical**
  - Improved mechanical properties in
  - Shear strength
  - Creep strength
  - Thermal stability

The front running alloys as substitutes for tin/lead alloy in the electronic industry are tin-copper for wave soldering and tin-copper-silver for re-flow soldering. Other alloys such as tin-silver or tin-antimony are used in automotive and industrial applications. Antimony and Bismuth containing alloys are favoured strongly for consumer applications where in their poor high temperature strength above 100 °C is no disadvantage. Tin-Bismuth, with a low melting temperature of 139 °C, is very useful for heat-sensitive systems. Tin-zinc alloys have a very attractive temperature range as well as cost but may be subject to oxidation and corrosion before and after processing. Many more alloys(Slezak, 1994) such as tin-copper-gold or alloys containing indium or germanium are under test and consideration. However, it would be advantageous if the number of different alloys get shrunk to a minimum in the near future.

Table 15.2 gives a list of alloys and their melting points which are being considered as possible substitutes of tin/lead solder alloys.

Table 15.2   Possible Substitutes for Lead (Courtesy Braun, 2003)

Alloy Group	Composition	Liquidus	Characteristics	Re-flow solder	Wave solder
Sn-Pb	Sn-37 Pb	183 °C	Standard and target	***	***
	Sn-36Pb-2Ag	179 °C			
Sn-Cu	Sn-0.7Cu	227 °C	Moderate higher cost		***
Sn-Ag	Sn-3.5Ag	221 °C	Higher cost and high	***	***
	Sn-3.5Ag-0.7Cu	217 °C	temperature		
Sn-Ag-Bi	Sn-2.8Ag-1Bi	220 °C	)		***
	Sn-2.0Ag-3Bi	215 °C	)	***	
	Sn-3.5Ag-3Bi	210 °C	) + Cu, In	***	
	Sn-3.5Ag-6Bi	206 °C	)	***	
Sn-Zn	Sn-9Zn	199 °C	Zn oxidation problems	***	
	Sn-8Zn-3Bi	190 °C		***	
Sn-Bi	Sn-58Bi	138 °C	Low melting point	***	

15.11   Useful Standards

- *IPC-WP/TR-584: IPC White Paper and Technical Report on Halogen-free Materials used for Printed Circuit Boards and Assemblies*: This document is a combination of an IPC White Paper and Technical Report summarizing both the IPC position as well as significant data on the subject of halogen-free materials for the electronics industry.

- *IPC-Environment: Environmental Best Practices Guide*: This book is published by the British PCB Association. It details the industries, environmental success and makes recommendations to minimize copper waste water discharges, necessary for steps on adopting environmental management systems. It also presents detailed steps to prevent pollution and minimize the generation of waste.

- *IPC-1331: Voluntary Safety Standard for Electrically Heated Process Equipment*: This voluntary standard establishes the minimum requirements for design, installation, operation and maintenance of electrically heated process equipment to minimize electrical hazards and prevent fires in combustible tanks, tank liners and drying equipment.

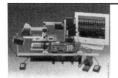

Glossary

Acceptance Test: The tests that determine the acceptability of boards as agreed to by purchaser/vendor.

Accepted Quality Level (AQL): The maximum number of defects per 100 units that can be considered satisfactory as a process average.

Access Holes: A series of holes in successive layers. Each set has a common centre or axis. The holes of a multi-layer printed board provide access to the surface of the land in one of the layers of the board.

Accuracy: The deviation of the measured or observed value from the accepted reference.

Acrylic Resin: A thermosetting, transparent resin.

Activating: A chemical treatment for conditioning the surface of non-conductive materials for improved adhesion.

Additive Process: A process for obtaining conductive patterns by the selective deposition of conductive material on a clad or unclad base material.

Adhesion Layer: The metal layer that adheres a barrier metal to a metal land on the surface of an integrated circuit.

Adhesive: A substance such as glue or cement used to fasten objects together. In surface mounting, an epoxy adhesive is used to adhere SMDs to the substrate.

Aging: The change of a property, e.g. solderability, with time.

Air Gap: The non-conductive air space between current carrying conductors such as traces, pads, ground planes, etc.

Algorithm: A procedure for solving a problem, usually mathematical.

Alkyd: A thermosetting resin with good electrical properties used for moulding the bodies of components.

Ambient: The surrounding environment coming into contact with the system or PCB in question.

Analog Circuit Simulator: A tool used to verify the design (or portions thereof) in the analog domain by applying virtual test signals to a virtual model of the design.

Analog Circuit: A circuit comprised primarily of individual (discrete) components, such as resistors, capacitors, diodes, transistors, etc. The circuit output is a continuous electrical signal that varies in frequency, amplitude, etc., as a function of the input. The magnitude is represented by physical variables such as voltages, current, resistance, rotation, etc.

Analog Functional Test: At the board level, various analog test signals are applied to a PCB through a switch to point out current outputs.

Analog In-circuit Test: A system measuring component values on a populated PCB before power is applied.

Anchoring Spur: An extension of a land (like one or two blind paths) on a flexible printed board that extends beneath the cover lay to assist in holding the land to the base material.

Angle of Attack: The angle between the face of the squeegee and the surface of the screen.

Annotation: Text or legend pertinent to a board design; text appears off the board areas and consists of lettering and symbols while legend appears on the boards.

Annular Ring: The width of the conductor surrounding a hole through a printed circuit pad.

ANSI: Acronym for "American National Standards Institute", an organization formed by industry and the US Government to develop trade and communication standards.

AOI: Automatic Optical Inspection.

Aperture List: List containing the shapes and dimensions of pads and tracks, etc., to expose on the film in a photoplotter.

AQL: Acceptable Quality Level. The maximum number of defectives likely to exist within a population lot that can be considered to be contractually tolerable.

Aqueous Cleaning: A water-based method that may include neutralizers, saponifiers, surfactants, dispersants and anti-foaming agents.

Arc Resistance: The resistance of a material to the effects of a high voltage, low-current, under prescribed conditions, passing across the surface of the material. The resistance is stated as a measure of the total elapsed time required to form a conductive path of the surface (material carbonized by the arc).

Artwork Master: An accurately scaled image of the conductive pattern of a PCB which is used to produce the 1:1 production master. The scale is chosen to provide the necessary degree of accuracy.

Artwork: An accurately scaled configuration used to produce a master pattern. It is generally prepared at an enlarged scale using various width tapes and special shapes to represent conductors.

ASCII: Acronym for "American Standard Code for Information Interchange"; a seven-bit code that assigns numeric values to letters of the alphabet, the ten decimal digits, punctuation marks and other characters.

Aspect Ratio: The ratio of the circuit board thickness to the smallest hole diameter.

Assembly Drawing: A drawing depicting the locations of components, with their reference designators (q.v.), on a printed circuit.

Assembly Drawing: The document that shows a printed board, components and any information necessary to describe joining them together in order to perform a specific function.

Assembly House: A manufacturing facility for attaching and soldering components to a printed circuit.

Assembly Language: A computer language of brief expressions for translation into a machine language.

ATE: (see Automatic Test Equipment).

Automated Component Insertion: Assembling discrete components to PCBs via electrically controlled equipment.

Automatic Test Equipment: Hardware that automatically analyses functional or static parameters to evaluate performance degradation. It also performs fault isolation.

AWG: American Wire Gauge. A method of specifying wire diameter; the higher the number, the smaller the diameter.

Axial Lead: A lead extending out the end and along the axis of a resistor, capacitor or other axial part rather than from the bottom.

Azeotrope: Two or more polar and non-polar solvents that behave when mixed as a single solvent to remove polar and non-polar contaminants with a boiling point lower than its components.

B-stage Material: Laminate impregnated with a resin and cured to an intermediate stage (B-stage resin). Normally designated as prepreg.

B&B: Blind and Buried via holes.

Backdriving: An in-circuit test procedure for digital circuitry.

Backpanel (Backplane): see "Mother Board".

Bare Board: A PCB having all lines, pads and layers intact but without components installed. An unassembled PCB.

Barrel: The cylinder formed by plating through a drilled hole.

Base Copper: The original, thin copper foil present on one or both sides of a copper clad laminate. During PCB manufacture, part of this base copper will be removed by etching. Conductors on the simplest PCBs consist of base copper only.

Base Laminate: The substrate material upon which the conductive wiring pattern may be formed.

Base Material Thickness: The thickness of the base material, excluding metal foil or material deposited on the surfaces.

Base Material: The insulating material (either rigid or flexible) as well as the copper foils bonded on one or both sides. It is a synonym for copper-clad laminate, i.e., the basic raw material for PCB manufacture. This also supports all components after assembly.

Base Solderability: The ease with which a metal or metal alloy surface can be wetted by molten solder under minimum realistic conditions.

Basic Wettability: The ease with which a metal or metal alloy can be wetted by molten solder.

BBT: Bare Board Test.

Bed-of-Nails: A method of PCB testing involving a fixture containing a field of spring-loaded pins that are co-ordinated with strategic points or nodes on the board to which they are brought into contact.

Bellows Contact: A connector contact which is a flat sprint folded to provide a uniform spring rate over the full tolerance range of the mating unit.

BGA: Ball Grid Array. Leadless array packaging technology in which solder balls are mounted to the underside of the package.

Biscuit Frame: An array of circuits on a larger 'mother' panel.

Bleeding: A condition in which liquid solder resist or rotation spreads larger than the defined apperture.

Blind Via: A via that reaches only one layer beneath the outer layer on one side of a multi-layer board.

Blister: De-lamination in the form of a localized swelling and separation between any of the layers of a lamination base material, or between base material and conductive foil or protective coating.

Blistering: A localized swelling and separation between any of the layers of a laminated base material, or between base material and conductive foil. It is a form of de-lamination. Also separation of solder mask layer and conductive pattern.

Blow Hole: A void in the solder fillet caused by outgassing from the barrel of a plated through hole. (See also outgassing).

Bluetooth: Bluetooth is a short-range (up to 10 m) 2.4 GHz wireless connectivity standard intended for such applications as wireless personal area networks (PANs). These PANs can be used to exchange data between devices such as cellphones, digital cameras, printers and household appliances at data rates of up to 721 Kbps (kilo bits per second).

Blutter Coat: An external layer of resin over the reinforcing structure of base material.

Board Density: A measure of the ratio of the area of the board used by parts to the total available area of the board. A board with less than 50 to 60 per cent of the available area should be able to be done single-sided, more than 75 to 80 per cent may have to go to multi-layer.

Board Thickness: The overall thickness of the base material and all conductive materials deposited thereon.

Body: The portion of an electronic component exclusive of its pins or leads.

Bond Interface: The common area between a lead and a land to which it has been terminated.

Bond Lift-off: A failure condition in which a lead is separated from its bonding surface.

Bond Strength: The force perpendicular to a board's surface required to separate two adjacent layers of the board, expressed as force per unit area.

Bonding Layer: An adhesive layer used in bonding together other discrete layers of a multi-layer printed board during lamination.

Bonding Time: The duration from hot-bar heat-up (contact with lead and pad) until the solder joint is completed.

Border Data: Patterns that appear in the border area, such as tooling features, test patterns and registration marks.

Boundary Scan: A self-test designed into components at the silicon level, permitting testing via a built-in, four-or-five-pin test bus accessing I/O pins.

Bow: The deviation from the flatness of a board characterized by a roughly cylindrical or spherical curvature. If the board is rectangular, its four corners are in the same plane (see also: "twist").

Branched Conductor: A conductor which connects electrically two or more leads on a printed board assembly. Some branched conductors, notably ground, support and re-set signal, connect many leads.

Brazing Alloy: A metal alloy (solder) which melts above 450°C but below metals being joined.

Breakdown Voltage: The voltage at which an insulator or dielectric ruptures or at which ionization and conduction take place in a gas or vapour.

Bridged Joint: Solder that spans across two or more conductors causing an electrical short-circuit

Bridging: A condition where excess solder builds up in the air gap between conductors and causes them to short together. Solder bridging generally occurs during the wave soldering process. Bridging is basically short-circuiting of a pad to an adjacent track or pad on a board.

B-stage Resin: A resin in an intermediate state of cure. The cure is normally completed during the laminating cycle.

B-stage: (prepreg) Partially cured resin (mostly reinforced with glass cloth) which will soften under a special range of temperature and which can be used to bond together cured laminate sheets to form a multi-layer board.

Bulge: A swelling of a printed board that is usually caused by internal de-lamination or separation of fibres.

Buried Via: A via that does not reach a surface layer on either side of a multi-layer board. The via transcends only inner layers of the board.

Burn-in: A method of testing devices via electrical stress vs temperature and/or time so that units prone to failure are eliminated.

Bus Bar: A conduit, such as a conductor on a printed board, for distributing electrical energy.

Bus: A heavy trace or conductive metal strip on the printed circuit board used to distribute voltage, grounds, etc., to smaller branch trances.

Bypass Capacitor: A capacitor which provides a comparatively low impedance ac (alternating current) around a circuit element.

CAD system: Computer-aided design that permits the interactive use of computers, programs and procedures in the design process. Decisions are made by the operator while the computer provides the data manipulation functions.

CAD/CAM system: Computer-aided design is the use of special software tools to formulate printed circuit patterns. Computer-aided manufacturing translates such designs into actual products.

Capacitive Coupling: The electrical interaction between two conductors caused by the capacitance between them.

Capillary Action: A phenomenon of force adhesion and cohesion that prompts liquids (molten solder) to flow against gravity between closely spaced solid surfaces, such as component leads and pads.

CARD: See "Printed Board".

Cast Adhesive: Special sheet adhesive material used for bonding polyimide multi-layer boards and flexi-rigids (similar to prepregs=B-Stages).

Catalyst: A chemical that speeds or changes the cure of a resin but does not become a part of the final product. Examples: hardeners, promoters, inhibitors.

Chamfer: A corner which has been rounded or shaped to eliminate an otherwise sharp edge.

Characteristic Impedance: The ratio of voltage to current in a propagating wave, i.e., the impedance which is offered to this wave at any point of the line. The characteristic impedance is expressed in ohms.

Check Plots: Pen plots that are suitable for checking only. Pads are represented as circles and thick traces as rectangular outlines instead of filled-in artwork. This technique is used to enhance the transparency of multiple layers.

Chemical Conversion Coating: A protective coating produced by the chemical reaction of a metal with a chemical solution.

Chemical Hole Clearing: The chemical process for cleaning conductive surfaces exposed within a hole (see also "Etch-back").

Chemically-deposited Printed Chip on board: In this technology integrated circuits are glued and wire-bonded directly to printed circuit boards instead of first being packaged.

Chip Testers: Large computer-based systems that test individual ICs, especially LSI and VLSI.

Circuit Tester: Generic term for volume tester of PCBs, such as bed-of-nails, footprint, guided probe, internal trace, loaded board, bare board and parts testing.

Circumferential Separation: A crack in the plating extending around the entire circumference of a plated through-hole.

Clad or Cladding: A relatively thin layer or sheet of metal foil which is bonded to a laminate core to form the base material for printed circuits.

Clad: (Adj). A condition of the base material to which a relatively thin layer or sheet of metal foil has been bonded at one or both of its sides, e.g. a metal clad base material.

Clamshell Fixture: An in-circuit test fixture to probe both sides of a PCB.

Clearance Hole: A hole in the conductive pattern larger than, but concentric with, a hole in the printed board base material.

Clearance: Metal to metal gaps on a board.

Clinched-wire through Connection: A connection made by a wire which is passed through a hole in a printed circuit board, and subsequently formed, or clinched, in contact with the conductive pattern on each side of the board, and soldered.

Coating: A thin layer of material, either conductive or insulating, applied over components or base materials.

COB: Chip on board. Component packaging technology in which bare integrated circuits are attached directly to the surface of a substrate and interconnected to the substrate most often by means of microscopic wires.

Coefficient of Expansion, Thermal-CTE: The fractional change in dimension of a material for a unit change in temperature expressed in ppm/C.

Cold Solder Connection: A solder connection that exhibits poor wetting and that is characterized by a greyish porous appearance.

Compatibility: In which materials can be mixed or brought into contact with no reaction or separation.

Compiler: A software module that analyses and converts programs from a high-level language to binary machine codes.

Component Density: The quantity of components on a unit area of printed board.

Component Hole: A hole in a PCB through which a component lead passes in order to be soldered or connected mechanically to the printed circuit and electrically to the conductive pattern. *Synonym:*

mounting hole. The hole is used for the attachment and electrical connection of component terminations, including pins and wires, to the printed board.

Component side: (Primary side) The surface layer of a board on which most of the components are placed. Component side is also referred to as the top side (layer one-counting downwards) of the board.

Component: Any of the basic parts used in building electronic equipment, such as a resistor, capacitor, DIP or connector, etc.

Conductive Adhesive: A material to which metal particles (usually silver) are added to increase electrical conductivity.

Conductive Foil: A thin sheet of metal that may cover one or both sides of the base material and is intended to form the conductive pattern.

Conductive Pattern: The configuration or design of the conductive material on the base laminate. Includes conductors, lands, and through connections.

Conductor Base Width: The conductor width at the plane of the surface of the base material. See also: Conductor width.

Conductor Layer: The total conductive pattern formed upon one side of a single layer of base material.

Conductor Pattern: See "Conductive Pattern".

Conductor Side: The side of a single-sided printed board containing the conductive pattern.

Conductor Spacing: The average or minimum (as specified) distance between the adjacent edges of conductors on the same layer of a printed board.

Conductor Thickness: The thickness of the conductor including all metallic coatings.

Conductor-to-hole Spacing: The distance between the edge of a conductor and the edge of a supported or unsupported hole.

Conductor Width: The observable width of the pertinent conductor at any point chosen at random on the printed board, normally viewed from vertically above unless otherwise specified.

Conductor: A single conductive path in conductive pattern. A PCB has at least one layer of conductors. *Synonyms:* path, trace.

Conformal Coating: A thin dielectric cover brushed, dipped or sprayed over parts and circuits of PCBs for environmental and mechanical protection.

Connection: One plug or receptacle which can be easily joined to or separated from its mate. Multiple-contact connectors join two or more conductors with others in one mechanical assembly.

Connector Tongue: A protrusion of the Printed Circuit Board edge that is manufactured to a configuration to mate with a receptacle that provides electrical and/or mechanical junction between the Printed Circuit Board and other circuitry.

Contact Area: The common area between a conductor and a connector through which the flow of electricity takes place.

Contact Resistance: The electrical resistance of the metallic surfaces at their interface in the contact area under specified conditions.

Contact Spacing: The distance between the centre lines of adjacent contact areas.

Contaminate/Contaminant: An impurity or foreign substance whose presence on printed wiring assemblies could electrolytically, chemically, or galvanically corrode the system.

Continuity Testing: A procedure in which voltage is applied to two interconnected lands to ascertain the presence or absence of current flow.

Copper Clad: A material, usually supplied in large sheets, consisting of a base material to one or both sides of which a thin copper foil is bonded. *Synonym:* laminate.

Copper Foil-See Foil: Quality electrolytic copper used to form conductive patterns on printed circuits. It is available in a number of weights (thickness) the traditional weights are 0.5, 1 and 2 ounces per square foot. (18, 35 and 70 μm thick).

Copper-mirror Test: A corrosivity test for fluxes in which the compound's reaction to a copper film vacuum-deposited on a glass plate is noted.

Corner Mark: Marks placed inside the edges of PCB corners to align and register the layers.

Corrosive Fluxes: Materials of inorganic acids and salts for surfaces of difficult solderability. Also called acid fluxes.

Cosmetic Defects: A smear or stain on the board representing flux residue after soldering and a variation from conventional appearance.

Coupon: One of the patterns of the quality conformance test circuitry area. (See: also Test Coupon).

Cover Lay, Cover Layer, Cover Coat: An outer layer(s) of insulating material applied over the conductive pattern on the surface of the printed board.

Cracking: A condition that makes breaks or separations in coatings that extends through to an underlying surface.

Crazing: A condition existing in the base material in the form of connected white spots or "crosses" on or below the surface of the base material, reflecting the separation of fibres in the glass cloth and connecting weave intersections.

Crimp Contact: A type of connector contact whose end is a hollow cylinder that can be crimped onto a wire inserted within it.

Cross-hatching: The breaking of large conductive areas by the use of a pattern of voids in the conductive material.

Cross-linking: The formation of chemical ties between reactive atoms in the molecular chain of a plastic.

Cross-talk: The undesirable interference caused by the coupling of energy between signal paths.

C-stage Material Laminate: The condition of a resin polymer when it is in the solid-state, with high molecular weight, being insoluble and infusible.

Cure: Change of physical properties of a material via chemical reaction or heat and catalysts, usually from a liquid to a solid.

Current-carrying Capacity: The maximum current which can be carried continuously, under specified conditions, by a conductor without causing degradation of electrical or mechanical properties of the printed circuit board.

Cut: Open circuit. An unwanted break in the continuity of an electrical circuit, that prevents current from flowing.

Cycle Rate: A component placement term measuring machine speed from pick-up to board location and return.

Database: A collection of inter-related data items stored together without unnecessary redundancy to serve one or more applications.

Datum: A defined point, line, or plane used to locate the pattern of layer for manufacturing or inspection, or for both purposes.

De-bugging: The process of locating and fixing problems (bugs) in the hardware and software portions of an electronic system.

De-lamination: A separation between any of the layers of a base material or between the laminate and the conductive foil, or both.

Dendrite: Metallic filaments growing between pads and traces resulting from electromigration.

Design Width of Conductor: The width of a conductor as delineated or noted on the master drawing. (See also: Conductor Base Width and Conductor Width).

Device: An individual electrical element, usually in an independent body, which cannot be further reduced without destroying its stated function.

De-wetting: A solder coating that has receded, leaving irregular deposits and indicating that the base metal has not been adequately de-oxidized.

Dezincification: Kind of galvanic corrosion, generally associated with two-phase brass alloys, in which the zinc-rich beta phase is selectively leached out of the brass. It occurs when brazed joints are exposed to salt or seawater.

DFSM: Dry Film Solder Mask.

Dice: Two or more dies.

Die: The uncased and normally leadless form of an electronic component that is either active or passive, discrete or integrated.

Dielectric Constant: The property of a dielectric which determines the electrostatic energy stored per unit volume for a potential gradient.

Dielectric Strength: The maximum voltage that a dielectric material can withstand, under specified conditions, without rupturing (usually expressed as volts/units thickness.)

Dielectric: An insulating medium which occupies the region between the conductors. It is also the distance between bonded inner layer conductors.

Digital Circuit: A circuit comprised mostly of integrated circuits and operates like a switch (i.e., it is either "ON" or "OFF").

Digital Logic Simulator: A tool used to verify the design (or portions thereof) in the digital domain by applying virtual test signals to a virtual model of the design.

Digitizing: Any method of reducing feature locations on a flat plane to digital representation of X-Y co-ordinates.

Dimensional Hole: A hole in a printed board where the means of determining location is by co-ordinate values not necessarily coinciding with the stated grid.

Dimensional Stability: A measure of dimensional change cause by such factors as temperature, humidity, chemical treatment, age, or stress (usually expressed as units/unit).

DIP Soldering: A process whereby printed boards are brought into contact with the surface of a static pool of molten solder for the purpose of soldering the entire exposed conductive pattern in one operation.

Discrete Component: A component which has been fabricated prior to its installation (i.e., resistors, capacitors, diodes, transistors, capacitors, diodes, transistors, etc). A single function component.

Dispersent: A chemical added to the cleaning solution to improve its particulate-removal ability.

Disturbed Solder Connection: A solder connection that is characterized by the appearance that there was motion between the metals being joined when the solder was solidifying (not accepted).

Documentation: Information on an assembly that explains the basic design concept, type and quantity of parts and materials, special manufacturing instructions and up-to-date revisions.

Double-sided Assembly: A PCB with components mounted on both sides.

Double-sided Board: A printed board with a conductive pattern on both sides.

DPF: Disc Plot Format.

Drag Soldering: The making of soldered terminations by dipping the solder side of a supported printed board with through-hole mounted components into the surface of a static pool of molten solder and moving it horizontal in one direction.

Drag-out: A measure of cleaning solution lost via board emergence after immersion.

DRC: Design Rule Check: Checks the integrity of print (PCB) so that problems like short-circuits, unrouted nets, etc. are recognized.

Drills: Solid, carbide cutting tools designed specifically for the fast removal of material in extremely abrasive, glass-epoxy materials.

Dross: Oxide and other contaminates which form on the surface of molten solder.

Dry Film Resists: Coating material in the form of laminated photo-sensitive foils specifically designed for use in the manufacture of printed circuit boards and chemically machined parts. They are resistant to various electroplating and etching processes.

Drying Time: Within the re-flow process after pre-heat and before peak re-flow temperature wherein volatile materials escape from the solder paste.

Dual In-line Package (DIP): A component which terminates in two straight rows of pins or lead wires.

Dual Solder Wave: In wave soldering, the first wave's multi-directional jet leaves solder on all contacted surfaces, followed by the second (flat) wave, which effects a finishing appearance by removing bridges and icicles. Intended for surface mount soldering.

Dummy Component: A mechanical package without the die, used to verify placement processes.

Durometer: A measurement of the hardness of a non-metal such as rubber, plastic, etc. Also the instrument for such measurement.

Edge Clearance: The distance of a pattern, components, or both, from the edges of the printed circuit board.

Edge Connector: A connector (can be gold-plated edge contacts or a series of parallel lines of holes).

Edge Spacing: The distance of a pattern, components, or both, from the edges of the printed board and intended for mating with an edge-board connector.

Edge-board Contacts: A series of contacts printed on or near any edge of a printed board and intended for mating with an edge-board connector.

Electroless Copper: A layer of copper plated on to an insulating or conductive surface of a PCB by chemical reduction, that is, without the use of applied electrical current.

Electroless Deposition: The deposition of conductive material from an auto-catalytic plating solution without application of electrical current.

Electron-beam Bonding: Terminations made by heating with a stream of electrons in a vacuum.

Electroplating: The electrodeposition of a metal coating on a conductive object. The object to be plated is placed in an electrolyte and connected to the relative terminal of a dc voltage source. The metal to be deposited is similarly immersed and connected to the positive terminal. Ions of the metal provide transfer to metal as they make up the current flow between the electrodes.

Embedded Component: A discrete component that is fabricated as an integral part of a printed board.

Emulsion Side: The side of the film on which the photographic image is defined.

Entrapment: The damaging admission and trapping of air, flux and fumes; it is caused by contamination and plating.

Epoxy Resins: Materials that form straight chain thermoplastic and thermosetting resins having good mechanical properties and dimensional stability.

Epoxy Smear: Epoxy resin which has been deposited onto the surface or edges of the conductive pattern during drilling. Also called resin smear.

ERBGF: Epoxy Resin Bonded Glass Fibre (See FR4).

ERC: Electrical Rule Check: checks the integrity of a circuit (SCH) so that problems like open inputs, shorted outputs and much more are recognized.

Escape Rate: The percentage of defects not defected vs. the total inspected.

ESD Sensitivity: (ESD susceptibility) The device's ability to dissipate the energy of the discharge or withstand the current levels involved.

ESD: Electro-Static-Discharge, the sudden transfer or discharge of electricity from one object to another.

Etch Factor: The ratio of the depth of etch (conductor thickness) to the amount of lateral etch (undercut).

ETCH Resist: An organic ink, lacquer, photo-resist, self-adhesive plastic tape, metal deposit or other material which will prevent specific areas of the metal on a panel from being attacked by an etchant.

Etchant: A solution used to remove, by chemical reaction, the unwanted portion of material from a printed board.

Etch-back: The controlled removal of all resins of base material on the side wall of holes in order to expose internal conductor areas.

Etched Printed Board: A board having a conductive pattern formed by the chemical removal of unwanted portions of the conductive foil.

Etching: Removal of metal from the surface of PCB by chemical dissolution. The process is normally carried out selectively by masking areas of metal which are to be left on the PCB.

Exotherm: The characteristic curve of a resin during cure showing reaction of temperature vs. time.

Extraneous Copper: Unwanted copper remaining on the base material after chemical processing.

Eyelet: A hollow tube inserted in a terminal or printed board to provide mechanical support for component leads or electrical connection.

Fault List: A listing of shorts and opens on a PCB to be repaired.

Feed-through: A plated through-hole in a printed circuit board that is used to provide electrical connection between a trace on one side of the printed circuit board to a trace on the other side. Since it is not used to mount component leads, it is generally a small hole and pad diameter.

Fibre Exposure: A condition in which broken reinforcing fibres of the base material are exposed and protrude in machined or abraded areas.

FICS: Flashscan Image Control Software: A DOS Program which sends Gerber files to the plotter.

Fiducial: A mark both in the artwork and etched with the circuit traces of the PCB. It is used to identify artwork orientation on the board. Global fiducials locate the overall pattern; local fiducials pinpoint component sites, typically fine-pitch.

Fillet: The concave formation of soldified solder between the land or pad and the component lead.

Fine Line Design: Printed circuit design permitting two, and nowadays even three, traces between adjacent dip pins. It entails the use of photo-imageable solder mask (PISM).

Fine Pitch: Refers to chip packages with lead pitches less than 1.25 mm or 50 mils.

Finger: A gold-plated terminal of a card-edge connector.

First Pass Yield: The per centage of finished assemblies to pass all tests without re-work.

Fixture: A device that enables interfacing a printed circuit board with a spring-contact probe test pattern.

Flat Cable: A cable with two or more parallel, round or flat, conductors in the same plane encapsulated by an insulating material.

Flexible Printed Wiring: A random arrangement of printed wiring utilizing flexible base material with or without flexible cover layers.

Flexure Failure: A conductor failure due to repeated flexing which is indicated by an increase of resistance to a specified value for a specified time.

Flood Bar: A device on a screen printer that drags solder paste back to the starting point after the squeegee has made a stroke. The return is for set-up of the next stroke as it does no printing on the backstroke.

Flow Soldering: Also called wave soldering. A method of soldering printed circuit boards by moving them over a flowing wave of molten solder in a solder bath.

Fluorocarbon: An organic compound having fluorine atoms in its structure to lend chemical and thermal stability to plastics.

Flush Conductor: A conductor whose outer surface is in the same plane as the surface of the insulating material adjacent to the conductor.

Flux: A substance used to promote or facilitate fusion, such as a material used to remove oxides from surfaces to be joined by soldering welding.

Flux Residue: A flux-related contaminant that is present on or near the surface after soldering and, if possible, should be washed away.

Flux, Activated Rosin Flux: A mixture of rosin and small amounts of organic-halide or organic-acid activator, which improves the ability of a flux to remove surface oxides from the surfaces being joined.

Flux-cored Solder: A wire of solder that contains one or more continuous flux-filled cavities along its length.

Foil: A thin sheet of metal, usually copper, used as the conductor for printed circuits.

Footprint: (See also Land Pattern). A set of properly sized and placed pads of a PCB on which a surface mounted component can be placed and soldered. Alternatively, the footprint is the board area occupied by a surface mounted component and its mounting pads.

Footprint: The pattern and space on a board taken up by a component.

Forced-air Convection: Convection, consisting of flow, rate, velocity and temperature, as heat transfer of fluid or gas over solder joints to be re-flowed.

FR4: Flame retardent laminate made from woven glass fibre material impregnated with epoxy resin.

Fully-additive Process: An additive process wherein the entire thickness of electrically isolated conductors is built-up by copper.

Functional Test: To check an assembly using equipment that tests for the functions intended and engaging inputs and outputs.

Fused Coating: A metallic coating, usually tin or solder alluvia, which has been melted and solidified forming a metallurgical bond to the base material.

Gas Blanket: A flowing inert gas atmosphere used to keep metal from oxidizing.

Gerber: Vector-based language, developed by Gerber Scientific Instrument Company, for sending commands to photoplotters.

GI: Laminate made from woven glass fibre material impregnated with polyimide resin.

Glass Epoxy: A material used to fabricate printed circuit boards. The base material (fibre-glass) is impregnated with an epoxy filler which then must have copper laminated to its outer surface to form the material required to manufacture printed circuit board.

Glass Transition Temperature: T_G-value, temperature at which resin ceases to act as a solid. Till this temperature, no separation will take place between resin/copper/reinforcement.

Glob Top: A coating process in which a set portion of resin is dispensed on the top of a chip or board. After spreading over the entire surface, it is cured to form a solid protective coating.

Go/No-Go Test: Procedure to yield only pass or fail.

Grid: An orthogonal network of two sets of parallel, equidistant lines used for locating points on a printed circuit board.

Ground Plane: A conductor layer, or portion of a conductor layer, used as a common reference point for circuit returns, shielding, or heat sinking. All those areas, not consumed by traces or pads, of the PCB which are left unetched and tied to the ground on the board.

Ground Plane Clearance: Removed portions of a ground plane that serves to isolate it form a hole in the base material to which the plane is attached.

Guided Probe Method: A technique for volume production of PCBs based on incoming inspection to catch the majority of device failures and inspection prior to populating, that will eliminate most manufacturing errors.

Gull Wing Lead: A surface mounted device lead which flares outward from the device body.

H.D.I.: High Density Interconnect.

Halides: Compounds containing fluorine, chlorine, bromine or iodine, which may be part of the activators in the flux system and must be cleaned.

Halogenated Polyester: A polyester resin modified with halogens to reduce flammability.

Haloing Mechanically: Induced fracturing or de-lamination on or below the surface of the base material, it is usually exhibited by a light area around holes, other machined areas, or both.

HASL: Hot Air Solder Level (See Solder Levelling).

Heat and Pull: A de-soldering method using a device that grasps, heats and pulls the leads to be removed.

Heat Sink: Any device that absorbs and draws off heat from a hot object, radiating it into the surrounding atmosphere.

Heel, Bonding: The part of a lead adjacent to a termination that has been deformed by the edge of the bonding tool.

Hipot Test: Wherein the assembly or component undergoes a high potential (ac) current.

Hole Breakout: A condition in which a hole is not completely surrounded by the land.

Hole Density: The quantity of holes in a printed circuit board per unit area.

Hole Location: The dimensional location of the centre of a hole.

Hole Pattern: The arrangement of all holes in a printed board.

Hole Pull Strength: The force necessary to rupture a plated through-hole when loaded or pulled in the direction of the axis of the hole.

Hole Void: A void in the metallic deposit of a plated through-hole exposing the base material.

Hot Zone: The section of a re-flow oven held at maximum temperature. Other zones include pre-heat and cooling.

Hygroscopic: The ability of a material to absorb and retain moisture from the air.

IC: Integrated Circuit.

Indentation: See "Pit".

Immersion Plating: The chemical deposition of a thin metallic coating over certain basis metals that is achieved by a partial displacement of the basis metal.

Impedance, Characteristic: The resistance of a parallel conductor structure to the flow of alternating current (ac), usually applied to high-speed circuits and normally consisting of a constant value over a wide range of frequencies.

In-circuit Test: A check of specific components(s) or circuits(s) within an assembly without their de-coupling from the primary circuit.

Inclusion: A foreign particle in the conductive layer, plating, or base material.

INDEX EDGE, INDEX EDGE MARKER, INDEXING HOLE, INDEXING NOTCH, INDEXING SLOT: See Locating Edge, Location Edge Marker, etc.

Initiating: See "Activating".

Inspection Lot: A collection of units of products bearing identification and treated as a unique entity from which a sample is to be drawn and inspected to determine conformance with the acceptability criteria.

Inspection Lot: Assemblies of a specific production run selected for inspection or test as a sample of the entire run.

Inspection Overlay: A positive or negative transparency made from the production master and used as an inspection aid.

Insulation Resistance: The electrical resistance of the insulating material (determined under specified conditions) between any pair of contacts, conductors, or grounding device in various combinations.

Insulation Resistance: The electrical resistance of the insulating material (determined under specified conditions) as measured between any pair of contacts or conductors.

Inter-facial Connection: See Through Connection.

Inter-layer Connection: An electrical connection between conductive patterns in different layers of a multi-layer printed board. (See also Through Connection.)

Internal Layer: A conductive pattern which is contained entirely within a multi-layer printed board.

Interstitial Via Hole: A plated through-hole connecting two or more conductor layers of a multi-layer printed board but not extending fully through all the layers of base material comprising the board.

IPC: The Institute for Interconnecting and Packaging Electronic Circuits, an American organization.

Isolation: The clearance around a pad, track, zone or via that defines the nearest approach allowed by conductors of another signal set.

J-Leads: The preferred surface mount lead form used on PLCCs, so named because the lead departs the package body near its Z-axis centre-line, is formed down the rolled under the package. Leads so formed are shaped like the letter "J".

Jumper: An electrical connection between two points on a printed board added after the intended conductive pattern is formed.

Just-in-time (JIT):   Minimization of inventory by supplying material and components to the production line directly before placement into the product.

Kapton: Du Pont trade name for polyimide film.

Key: A device designed to assure that the coupling of two components can occur in only one position.

Keying Slot or Polarizing Slot: A slot in a printed circuit board that polarizes it, thereby permitting it to be plugged into its mating receptacle with pins properly aligned, but preventing it from being reversed or plugged into any other receptacle.

Keyway: A slot used to assure the correct location in a mating connector.

Laminate: A product made by bonding together two or more layers of material.

Laminate Thickness: Thickness of the metal clad base material, single- or double-sided, prior to any subsequent processing.

Laminate Void: Absence of laminate material in an area which normally contains laminate material.

Laminating Presses: Multi-layer equipment that applies both pressure and heat to laminate and prepreg to make multi-layer boards.

Lamination: The process of manufacturing a laminate; also the process used for application of a dry film photo-resist.

Land Pattern: A combination of lands that is used to mount, interconnect and test a particular component.

Land: On a PCB, the conductive area(s) to which components are attached. Also called pad.

Landless Hole: A plated through-hole without a land(s).

Laser: Light Amplified by Stimulated Emission of Radiation

Layer-to-layer Spacing: The thickness of dielectric material between adjacent layers of conductive circuitry in a multi-layer printed circuit board.

Layer: One in a series of levels in a board on which tracks are arranged to connect components. Vias connect tracks and zones between layers.

LCCC: Leadless ceramic chip carrier.

Lead: (Pronounced "Leed")-A terminal on a component.

Lead Mounting Hole: See Component Hole.

Lead Projection: The distance that a component lead protrudes through the side of a board that is opposite from the one upon which the component is mounted.

Legend: A format of lettering or symbols on the printed board, e.g. part number, component locations, and patterns.

Lifted Land: A land that has fully or partially separated (lifted) from the base material, whether or not any resin is lifted with the land.

Line: See Conductor.

Liquation: If a solder alloy with a long melting range is heated too slowly, the phase with the lowest melting point begins to flow first. The material left behind has a changed composition and a higher melting point and will not flow readily. An unsound and unsightly joint is the usual result of liquation.

Liquids: The lowest temperature at which filler metal (solder) is completely liquid.

Load Test: A mass re-flow soldering system test for the capacity repeatedly to process boards regardless of their volume through the oven.

Locating Edge, Locating Hole, Locating Notch, Locating Slot: A physical feature in a panel or printed board used to position the board or mounted components accurately.

Locating Hole, Locating Notch, Locating Slot: A hole, notch or slot in the panel or printed board to enable it to be positioned accurately during manufacture and/or assembly. *Synonyms*: fabrication hole (or notch or slot), indexing hole, location hole, manufacturing hole, outrigger hole, tolling hole.

Logic Diagram: A drawing that depicts the multi-state device implementation of logic functions with logic symbols and supplementary notations, showing details of signal flow and control, but not necessarily the point-to-point wiring.

Major Defect: A defect that could result in failure or significantly reduce the usability of the part for its intended purpose.

Manhattan Distance: The orthogonal distance between two points.

MAR (Minimum Annular Ring):   The minimum metal width, at the narrowest point between the circumference of the hole and the outer circumference of the land. This measurement is made to the drilled hole on internal layers of multi-layer printed circuit board and to the edge of the plating on outside layers of multi-layer boards and double-sided boards.

Margin: The distance between the reference edge of a flat cable and the nearest edge of the first conductor. (See also Edge Spacing.)

Mask: A material applied to enable selective etching, plating or the application of solder to a printed circuit board.

Mass Soldering: Methods of soldering in which many joints are made in the same operation.

Master Artwork: A document showing dimensional limits and grid locations for all parts of an assembly to be fabricated. It includes the arrangement of conductors and non-conductive patterns and the size, type and location of holes.

Master Dot Pattern: See Hole Pattern.

Master Drawing: A document that shows the dimensional limits or grid locations applicable to any or all parts of a printed board (rigid or flexible), including the arrangement of conductive and non-conductive patterns or elements; size, type and location of holes.

Master Pattern: An accurately scaled pattern which is used to produce the printed circuit within the accuracy specified on the master drawing.

Maximum, Plated Through-hole Size: A hole size equal to the specified hole size before plating, plus the manufacturing tolerance, less twice the minimum plating thickness.

MCR: Moulded Carrier Ring. A type of fine-pitch chip package named for the method of supporting and protecting the leads. The leads are left straight; the ends of the leads are embedded in a strip of plastic, which is the Moulded Carrier Ring.

Mean Time between Failure (MTBF):   The statistical mean average time interval, usually in hours, that may be expected between failures of an operating unit. Results should be designated actual, predicted or calculated.

Measling: A condition existing in the base laminate in the form of discrete white sports or crosses below the surface of the base laminate, reflecting a separation of fibres in the glass cloth at the weave intersection.

MELF: A metal electrode leadless face surface mount component that is round with metallic cap terminations.

Melting Range: The temperature range over which the solder alloy melts. An alloy with a single melting point, rather than a melting range, is known as a eutectic alloy.

Meniscus: The contour or shape of molten solder as formed by surface tension forces in turn controlled by wetting.

Metal Clad Base Material: Base material covered with metal on one or both of its sides.

Metallization: A deposited or plated thin metallic film used for its protective or electrical properties.

Micro-sectioning: A destructive test procedure in which a section of a specimen it cut and removed for close examination.

Micro-strip: A type of transmission line configuration which consists of a conductor over a parallel ground plane, and separated by a dielectric.

Mil: One thousandth (0.001) of an inch . Not to be confused with Mil (short for millimetre) .1 Inch = 25.4 mm; 1mil = 0.0254 mm.

Minimum Electrical Spacing: The minimum allowable distance between adjacent conductors that is sufficient to prevent dielectric breakdown, between the conductors at any given voltage.

Minor Defect: A defect which is not likely to reduce the usability of the unit for its intended purpose. It may be a departure from established standards having no significant bearing on the effective use or operation of the unit.

Mis-registration: The lack of dimensional conformity between successively produced features or terns.

Mixed Component-Mounting Technology: A component mounting technology that uses both through-hole and surface-mounting technologies on the same packaging and interconnecting structure.

Modifier: A chemically inert substance added to a resin to change its properties.

Module: A separable unit in a packaging scheme displaying regularity of dimensions.

Mother Board: Also called back plane, or matrix board. A relatively large printed circuit board on which modules, connectors, sub-assemblies or other printed circuit boards are mounted and inter connections made by means of traces on the board.

Mounting Hole: A hole used for the mechanical mounting of a printed board or for the mechanical attachment of components to the printed board.

Muffle: An enclosure that is located between the heating elements and the parts being processed that contains the atmosphere required for the re-flow soldering process.

Multi-layer PCB: Printed circuit boards consisting of three or more conducting circuit planes separated by insulating material and bonded together with internal and external connections of the circuitry as required.

Multimeter: A portable test instrument which can be used to measure voltage, current and resistance.

Multiple-image Production Master: A production master used in the process of making two or more printed boards simultaneously.

Nail Heading: The flared condition of copper on the inner conductor layers of a multi-layer board usually caused by hole drilling.

Negative (Noun):   An artwork, artwork master, or production master in which the intended conductive pattern is transparent to light, and the areas to be free from conductive material are opaque.

Negative-acting Resist: A resist which is polymerized (hardened) by light and which, after exposure and development remains on the surface of a laminate in those areas which were under the transparent parts of a production master.

Netlist: A net is a junction of component nodes. A netlist is a collection of nets that define all the connections in a circuit. It is obtained automatically from a schematic capture program.

Neutralizer: An alkaline chemical added to water to improve its ability to dissolve flux residues.

NFP: Non-functional pad.

Node: A pin, lead or even junction which will have at least one wire connected to it.

Non-clean Solder: A process using specially formulated low-solid solder pastes whose residues require no cleaning.

Non-conductive Epoxy: An epoxy resin with or without a filler, which may be added to improve thermal conductivity.

Non-conductive Pattern: A configuration formed by functional non-conductive material of a printed circuit.

Non-functional Land: A land on internal or external layers, not connected to the conductive pattern on its layer.

Non-polar Compound: Material having electrical charges distributed over the surface of the molecule, thereby showing an electrical effect in solution.

Non-wetting: A condition whereby a surface has contacted molten solder, but has had none of the solder adhere to it.

NPTH: Non plated though-hole

Omegameter: A test instrument measuring ionic residues on PCBs via the drop of resistivity over a specific time.

One-sided Board: See Single-sided Board.

Open: An area of a bare PCB which, due to over-etching or fabrication problems, separates two electrically connected points.

Organic Activated (OA): A water-soluble flux using organic acids as activators.

OSP: Organic Solderable Preservative.

Outgassing: The gaseous emission from a laminate printed board or component when the board or the printed board assembly is exposed to heat or reduced air pressure or both.

Outgrowth: The increase in conductor width at one side of a conductor, caused by plating build-up, over that delineated on the production master.

Overhang: Increase in printed circuit conductor width caused by plating build-up or by undercutting during etching.

Packaging Density: Quantity of functions (components, interconnection devices, mechanical devices) per unit volume, usually expressed in qualitative terms, such as high, medium, or low.

Pad: A portion of the conductive area of which components, terminals, traces, etc., are mechanically attached. (Also called land).

Panel: The base material containing one or more circuit patterns that passes successively through the production sequence and from which printed circuit boards are extracted. See Backplanes/Backpanels.

Panel Plating: The plating of the entire surface of a panel (including holes).

Parylene: A polymer resin (polyparaxylense) that provides a thin, uniform coating on PCBs and components. It can be applied via vacuum for deposition on sharp edges and complex shapes.

Pattern Plating: Selective plating of a conducive pattern.

Pattern: The configuration of all conductive and/or non-conductive areas on a PCB. Letters and inscriptions may also be included. Pattern also denotes the circuit configuration on related tools, drawing and masters. *Synonym:* image.

PCB: Printed Circuit Board.

PEC: Printed Electronic Component.

Peel Strength: The force per unit width required to peel the conductor or foil from the base material.

Permanent Mask: A resist which is not removed after processing, e.g., plating resist used in the fully-additive process.

Photomaster: See Artwork Master.

Photopolymer: A polymer that changes characteristics when exposed to light of a specific frequency.

Photoplotter: A plotter that writes using light.

Physical Layer: A conductive board layer or artwork image representing a complete conductive layer.

Pick-and-place Machine: A programmable machine usually with a robot arm for picking components from a feeder. It moves the part for placement and/or insertion to a specific site on the board.

PIH Assembly: Pin-in-hole a printed board assembly made up of components with leads which pass through holes in the board and lands. *Synonyms:* traditional assembly, conventional assembly.

Pilot Hole: See Locating Hole.

Pin Density: The quantity of pins on a printed board per unit area.

Pin: A terminal on a component. A component lead that is not readily formable without being damaged.

Pinhole: A minute hole through a layer of pattern.

Pinholes: Small imperfections which penetrate entirely through the conductor and/or solder.

Pink Ring: Chemically-induced fracturing or de-lamination on or below the surface of the base material; it is usually exhibited by a light area around holes, other etched areas or both.

Pit: A depression in the conductive layer that does not penetrate entirely through it.

Pitting: Small holes or sharp edges on the surface of a solder joint generally caused by flux entrapment, oxidation or over-heating.

Plate Finish: Pertaining to Laminating. The finish present on the metallic surface of metal clad base material resulting from direct contact with the laminating press plates without modification by any subsequent finishing process.

Plated Through-hole: A hole with the deposition of metal (usually copper) on its sides to provide electrical connections between internal or external conductive patterns.

Plating Bar: The temporary conductive path interconnecting areas of a printed board to be electroplated, usually located on the panel outside the borders of such a board.

Plating Resists: Materials which, when deposited on conductive areas, prevent the plating of the covered areas. Resists are available both as screened-on materials and as dry-film photopolymer resists.

Plating Up: The process consisting of the electrochemical deposition of a conductive material on the base material (surface holes, etc.) after the base material has been made conductive.

Plating: A uniform coating of conductive metal upon the base material of the printed circuit board.

Plotting: The practice of mechanically converting X-Y positional information into a visual pattern, such as artwork.

Poise: A centimeter-gram-second unit of viscosity equal to that of a fluid requiring a shearing force of one dyne to move from a square centimeter area with a velocity of one centimeter per second (cps).

Polarization: A technique of eliminating symmetry within a plane so that parts can be engaged in only one way in order to minimize the possibility of electrical and mechanical damage or malfunction.

Polyester: (Mylar) low melting point plastic film used for cheap flexible circuits.

Polyimide Resins: High temperature thermoplastics used with glass to produce printed circuit laminates from multi-layer and other circuit applications requiring high temperature performance.

Polyimide: (Kapton) Higher melting point plastic film used as base for flexible portions of flexi-rigid boards as well as for many flexible circuits.

Populated PCB: A printed board on to which all passive and active components have been assembled. *Synonyms:* printed board assembly (PBA), card, assembled board.

Positive (Noun):   An artwork, artwork master, or production master in which the intended conductive pattern is opaque to light, and the areas intended to be free from conductive material are transparent.

Positive-acting Resist: A resist which is decomposed (softened) by light and which, after exposure and development, is removed from those areas which were under the transparent parts of a production master.

Pre-heat: The process portion of the re-flow heat curve in which the PCB is heated from ambient at a pre-set rate and prior to full liquidus at the solder joint areas.

Prepreg: Sheet material consisting of the base material impregnated with a synthetic resin, such as epoxy or polymide, partially cured to the B-stage.

Press-fit Contact: An electrical pin contact which can be pressed into a hole in a printed board to make immediate contact.

Primer: A coating applied before the application of an adhesive to improve the bond.

Printed Board Assembly: A printed board with electrical or mechanical components, other printed boards, or a combination of these, attached to it with all manufacturing processes, soldering, coating, etc. completed.

Printed Board: The general term for completely processed printed circuit or printed wiring configuration. It includes single, double, and multi-layer boards, both rigid and flexible.

Printed Circuit Assembly: A printed circuit board to which discrete components, hardware, and other electronic devices have been attached to form a complete operating unit.

Printed Circuit Board: An insulating material onto which an electronic circuit has been printed or etched.

Printed Circuit: Circuit where the interconnections between components, terminals, sub-assemblies, etc., are made by conductive strips (traces) that have been printed or etched onto an insulating board.

Printed Component: A component part, such as an inductor, resistor, capacitor, or transmission line, which is formed as part of the conductive pattern of the printed board.

Printed Contact: A portion of a conductive pattern formed by printing, serving as one part of a contact system.

Printed Wiring Assembly Drawing: A document that shows the printed board (rigid or flexible), the separately manufactured components which are to be added to the board, and any other information necessary to describe the joining of these parts to perform a specific function.

Printed Wiring Board: See Printed Board.

Printed Wiring Layout: A sketch that depicts the printed wiring substrate, the physical size and location of electronic and mechanical components, and the routing of conductors that electrically interconnect components, in sufficient details to allow the preparation of documentation and artwork.

Printed Wiring: A conductive pattern within or bonded to the surface of a base material intended for point-to-point connection of separate components and not containing printed components.

Probing Systems: Equipment for making electrical contact between the bare PCB, components or assemblies and the continuity tester. Probing devices range from manual units for low volumes to computer-controlled systems.

Process Indicator: A detectable anomaly, other than a defect, that is reflective of material, equipment, process and/or workmanship variations.

Production Master: A 1 to 1 (1:1) scale pattern which is used to produce one or more printed boards (rigid or flexible) within the accuracy specified on the master drawing.

PTH: Plated through-holes. Plating the holes on their internal wall.

Pull Strength: See Bond Strength.

PWB: Printed Wiring Board; same as PCB.

Radial Lead: A lead extending out of the side of a component, rather than from the end.

Reference Edge: The edge of cable or conductor from which measurements are made.

Re-flow Soldering: Joining components to substrates by placing the parts into solder paste and then melting the paste to achieve re-flow and the interconnection.

Re-flow Spike: The portion of the re-flow soldering process during which the temperature of the solder is raised to a value that is sufficient to cause the solder to melt.

Re-flowing: The melting of an electro-deposit followed by solidification.

Register Mark: A symbol used as a reference point to maintain registration.

Registration: The alignment of a pad on one side of the printed circuit board or layers of a multi-layer board to its mating pad on the opposite side.

Relative Humidity: The ratio of the quantity of water vapour present in the air to the quantity which would saturate the air at the given temperature.

Reliability: The probability that a component, device or assembly will function properly for a defined period of time under the influence of specific environmental and operational conditions.

Repair: The process of restoring the functional capability of a defective component or circuit.

Repeatability: The ability of a system to return to a specific parameter, said of equipment when evaluating its consistency of processing.

Residue: Any visual or measurable form of process-related contamination.

Resin Smear: Resin transferred from the base material onto the surface or edge of the conductive pattern normally caused by drilling. Sometimes called epoxy smear.

Resin: A high-molecular-weight organic material with no specific melting point. A polymer.

Resist: Coating material used to mask or to protect selected areas of a pattern from the action of an etchant, solder, or plating. Also see Dry Film, Resists, Plating Resists and Solder Resists.

Reverse Image: The film pattern on a printed circuit board enabling the exposure of conductive areas for subsequent plating.

Re-work: A manufacturing step or process that is repeated to bring a non-performing or non-conforming component or circuit to a functional condition.

Re-working: The act of repeating one or more manufacturing operations for the purpose of improving the yield of acceptable part.

Rheology: Science of flow, is the study of the flow and deformation of matter and is particularly important with regard to colloidal systems.

Ribbon Cable: A flat cable with round conductors.

Right-angle Edge Connector: A connector which terminates conductors at the edge of a printed board, while bringing the terminations out at right angles to the plane of the board conductors.

Roadmap: A printed pattern of non-conductive material by which the circuitry and components are delineated on a board to aid in service and repair of the board.

Rosin Flux: The mildest of solder fluxes and generally requiring added organic activating agents.

Rosin: A hard, natural resin (nowadays also synthetic), consisting of abietic and primaric acids and their isomers, some fatty acids and terpene hydrocarbons, that is extracted from pine trees and subsequently refined.

Saponifier: An aqueous organic or inorganic base solution with additives that promote the removal of flux.

SBU: Sequential Build-up.

Schematic Diagram: A drawing that shows, by means of graphic symbols, the electrical connections, components and functions of a specific electronic circuit arrangement.

Screen Printing: A process for transferring an image to a surface by forcing suitable media through a stencil screen with a squeegee. Also called silk screening.

Screen: A network of metal or fabric standards mounted tautly on a frame and upon which the PCB's circuit pattern is superimposed by photographic means.

Semi-additive Process: A process for obtaining conductive patterns by a combination of electroless metal deposition with etching and/or electroplating. A semi-additive process is used in conjunction with a metal clad base material.

Semi-aqueous Cleaning: A technique involving the use of a solvent followed by hot-water rinses and drying.

Sensitizing: See Activating.

Shadowing: A condition occurring during etching in which the dielectric material, in intimate contact with the foil, is incompletely removed though acceptable etching may have been achieved elsewhere. In re-flow soldering, a condition in which component bodies block the infra-red energy from certain areas of the board. In wave soldering, the solder fails to wet some parts leads due to other devices blocking the flow of solder.

Shielding, Electronic: A physical barrier, usually electrically conductive, designed to reduce the interaction of electric or magnetic fields upon devices, circuits or portions of circuits.

Short: Short-circuit. An abnormal connection of relatively low resistance between two points of a circuit or conductors from different nets either touch or come closer than the minimum spacing allowed for the design rules being used.

Signal Conductor: An individual conductor used to transmit an impressed signal.

Signal Plane: A conductor layer intended to carry signals, rather than serve as a ground or other fixed voltage function.

Signal: An electrical impulse of a pre-determined voltage, current, polarity and pulse width.

Silkscreen: Often used for legend print and means the printed reference designators on a printed wiring board.

Simulation: The process of creating a virtual representation (a computer model) of an electronic component, circuit board, or system and applying virtual test signals to the model to verify its functionality and possibly its timing.

Single-sided Board: A printed board with a conductive pattern on one side only.

Single-image Production Master: A production master used in the process of making a single printed board.

Slump: A spreading of the solder paste after printing but before re-flow soldering. If excessive, a loss of definition may result.

SMD: Surface Mounted Device. Any component or hardware element designed to be mounted to a printed circuit board (PCB) without penetrating the board.

Smear: Resin, smeared over the edge of an internal copper layer of a multi-layer board during drilling, which prevents the layer from making electrical contact with the barrel of the hole.

SMOBC: Solder Mask Over Bare Copper.

SMT: Surface Mount Technology. Defines the entire body of processes and components which create printed circuit assemblies without components with leads that pierce the board.

Snap-off Distance: The space between the top surface of the substrate and the underside of the stencil when the squeegee is not in play.

Snap-off: The return of a stencil to normal level after deflection by the pressure of the squeegee moving across the surface.

Soak: The period after pre-heat and before re-flow peak temperature where the internal temperature differences between parts allowed to equalize.

Solder (Soft): A metal alloy with a melting temperature that is below 450°C.

Solder Bridging: The unwanted formation of a conductive path of solder between conductors.

Solder Connection Pinhole: A small hole that penetrates from the surface of a solder connection to a void of indeterminate size within the solder connection (process indicator).

Solder Fillet: A preferable concave surface of solder that is at the interconnection of the metal surfaces of a solder connection.

Solder Levelling: The process of immersing printed circuit boards into hot liquids. Often referred to as HASL or HAL (Hot Air Levelling)

Solder Marks: A screening defect characterized by prints having jagged edges, the result of incorrect moving pressure.

Solder Mask, Solder Resist: Coating with mask and insulate areas of a circuit pattern where solder is not desired.

Solder Oil (Blanket): Liquid formulations that are used in intermix wave soldering and as coverings on static and wave soldering pots in order to eliminate dross and to reduce surface tension during the soldering process.

Solder Paste: Finely divided particles of solder, with additives to promote wetting and to control viscosity, tackiness, drying rate, etc.

Solder Plug: A core of solder in a plated through-hole.

Solder Projection: An undesirable protrusion of solder from a solidified solder joint or coating.

Solder Resist: An ink, lacquer, photo-resist or metal coating which is not wetted by molten solder.

It is applied to specific areas of a PCB to stop them from being solder-coated, usually when mass soldering.

Solder Resist: See Resist.

Solder Side: The side of a printed board which is opposite to the component side.

Solderability Testing: The evaluation of a metal to determine its ability to be wetted by solder.

Solderability: The ability of a metal to be wetted by molten solder.

Soldering Iron Tip: The portion of a soldering iron that is used for the application of the heat that melts the solder.

Soldering: A process of joining metallic surfaces with solder, without the melting of the base metals.

Solderless Wrap: A method of connecting a solid wire to a square, rectangular, or V-shaped terminal by tightly wrapping the wire around the terminal with a special tool.

Solidus: The highest temperature at which filler metal (solder) is completely solid.

Spurious Signal: See Cross-talk.

Stamped Printed Wiring: Wiring which is produced by die stamping and which is bonded to an insulating base.

Statistical Process Control (SPC):   The use of statistical techniques to analyse the outputs of processes with the results guiding actions taken to adjust and/or maintain a state of quality control.

Step Soldering: The making of solder connections by sequentially using solder alloys with successively lower melting temperatures.

Step-and-repeat: A method by which successive exposures of a single image are made to produce a multiple-image production master.

Straight-through Lead: A component lead that extends through a hole and is terminated without subsequent forming.

Strain: The deformation resulting from a stress.

Stripline: A type of transmission line configuration which consists of a single narrow conductor parallel and equidistant to two parallel ground planes.

Substrate: See Base Material.

Subtractive Process: A process for obtaining conductive patterns by selective removal of unwanted areas of conductive foil from a metal clad base material.

Supported Hole: A hole in a printed board that has its inside surface plated or otherwise reinforced.

Surface Insulation Resistance (SIR) Test: Test for the level of resistance of an insulating material, such as FR-4, between conducting members of a board (traces, contacts).

Surface Insulation Resistance (SIR):   A measure in ohms of an insulating material's (as in FR-4) electrical resistance between conductors.

Surface Leakage: The passage of current over the boundary surface of an insulator as distinguished from passage through its volume.

Surface Mounting: The electrical connection of components to the surface of a conductive pattern that does not utilize component holes.

Surface Tension: The natural, inward, molecular attraction force that inhibits the spread of a liquid at its interface with a solid material.

Surfactant: A chemical added to the cleaning solution to lower surface tension and to promote wetting.

Swaged Lead: A component lead wire that extends through a hole in a printed board and its lead extension is flattened (swaged) to secure the component to the board during manufacturing operations.

Tape-and-reel: A packaging method of housing surface-mount parts in their own tape cavities in a long continuous strip. The cavities are covered so that the tape can be wound around a reel for convenient handling and machine set-up.

Taped Components: Components attached to continuous tape for automatic assembly.

Teflon: Du Pont trade name as an inventor for PTFE (Polytetrafluoroethylene).

Temperature Profile: The depiction of the temperature that a selected point traverses as it passes through the re-flow process.

Tented Via: A via with solder mask completely covering both its pad and its plated through-hole. This completely insulates the via from foreign objects, thus protecting against accidental shorts, but it also renders the via unusable as a test point.

Tenting: A printed board fabrication method of covering over plated through-holes and the surrounding conductive pattern with a resist, usually dry film.

Terminal Area: A portion of a conductive pattern usually, but not exclusively, used for the connection and/or attachment of components.

Termination: The part of a component that makes contact with a pad on a substrate.

Terpenes: (Turpentine) A solvent used in cleaning electrical assemblies.

Test Board: A printed board suitable for determining acceptability of the board or of a batch of boards produced with the same process so as to be representative of the production board.

Test Coupon: A pattern as an integral part of the PCB on which electrical tests may be made to non-destructively evaluate process control. A portion of a circuit used exclusively to functionally test the circuit as a whole.

Test Fixture: A device that adapts a specific assembly under test to a test system via inter-connection.

Test Pattern: A pattern used for inspection or testing purposes.

Test Point: Special points of access to an electrical circuit, used for testing purposes.

Thermocouple: A device made of two dissimilar metals which, when heated, generate a voltage that is used to measure temperatures.

Thermoplastic: A plastic set into final shape by forcing the melted polymer into a cooled mould. The hardened form can be re-melted several times.

Thermoset: A plastic cured or hardened by heating into a permanent shape. Thermosets cannot be re-melted.

Through Connection: An electrical connection between conductive patterns on opposite sides of an insulating base, e.g. plated through-hole or clinched jumber wire.

Tinning: The application of molten solder to a basis metal in order to increase its solderability.

Tolling Hole: Also called fabrication hole, pilot hole, or manufacturing hole. These are used as PCB reference points upon which other dimensions are based.

Tombstoning: A soldering defect in which a chip component moves into a vertical position during solidification of the solder so that only one terminal is connected. It is caused by defective re-flow processing.

Trace: A single conductive path in a conductive pattern.

Traces: The metallic conductive strips that provide connections between components, terminals, etc., on printed circuits.

Transmission Cable: Two or more transmission lines.

Transmission Line: A signal-carrying circuit composed of conductors and dielectric material with controlled electrical characteristics used for the transmission of high-frequency or narrow-pulse type signals.

Triazine: Dielectric material with higher glass transition temperature and better thermal stability than epoxy resin, however, more expensive and not in common use.

Trim Lines: Lines which define the borders of a printed board.

Twist: The deformation of a rectangular sheet, panel or printed board, that occurs parallel to a diagonal across its surface in such a way that one of the corners of the sheet is not in the plane formed by the other three corners.

UL: Underwriter's Laboratories, Inc., a corporation for the purpose of establishing safety standards on types of equipment or components in USA and Canada.

Ultrasonic Soldering: Fluxless soldering wherein molten solder is vibrated at ultrasonic frequencies while making the joint.

Undercut: A groove or excavation at one edge of a conductor caused by etching.

Underwriters' Symbol: A logotype authorized for placement on a product which has been recognized (accepted) by Underwriters Laboratories, Inc. (UL).

Unsupported Hold: A hole containing no conductive material nor any other type of reinforcement.

Vacuum Pick-up: A component handling tool with a small vacuum cup for ease of pick-up and removal during de-soldering.

Vapour Phase: The solder re-flow process that uses a vaporized solvent as the source for heating the solder beyond its melting point, creating the component-to-board solder joint.

Via Hole: A plated through-hole whose only purpose is to connect a track on one layer or side of the board through to a track on another layer or side. In a via, there is no intention to insert a component lead or other reinforcing material.

Via: A plated through-hole used as an interlayer connection and not as a terminating point for a component lead. It may also be blind (incomplete penetration) or buried (non-surfacing).

Virtual Prototype: A virtual (computer model) representation of an electronic product that can be used to explore different design scenarios and then verify that the product will work as planned before building a physical implementation.

Viscosity: The property of a fluid that enables it to develop and maintain a level of shearing stress dependent upon the velocity of flow and then offer continued resistance to flow. The absolute unit of viscosity measurement is poise, or more commonly, centipoise.

Void: The absence of any substances in a localized area.

Voltage Plane: A conductor or portion of a conductor layer on or in a printed board which is maintained at other than ground potential. It an also be used as a common voltage source, for heat sinking, or for shielding.

Wave Exposure: A surface condition of base material in which the unbroken fibres of woven glass cloth are not completely covered by resin.

Wave Soldering: The technique of joining parts to a PCB by passing the assembly over a wave of molten solder so as to coat the pre-fluxed areas to be joined.

Weave Exposure: A condition of base material in which a weave pattern of glass cloth is appearing on the surface though the unbroken fibres of the woven cloth are completely covered with resin.

Wetting: The formation of a relatively uniform, smooth, unbroken and adherent film of solder to a base material.

Whisker: A needle-shaped metallic growth on a printed circuit board.

Wicking: Capillary absorption of liquid along the fibres of the base material.

Yield: The ratio of usable parts at the end of a manufacturing process to the number of components submitted for processing.

Z-Stroke: The movement of the head of a component placement machine in the vertical place for parts orientation and insertion.

References

Alan Roads, R. (1991) Automation SMD Assembly, *Electronic Production*, March 1991, p. 17.

Anderson, M.J. (1998) Improve PCB Gold-plating Yields Using DOE, *Electronics Engineer*, November 1998, contact: info@statease.com.

Andrus J.J. (1990) Successful PCB Water Cleaning Systems, *Electronic Packaging and Production*, October 1990, p. 64.

Banks, S. (1995) Reflow Soldering to Gold, *Electronic Packaging and Production*, June 1995, p. 69.

Barclay, B and Morrell, M. (2001) Laser Direct Imaging: A User's Perspective, Internet notes www.circuitree.coml

Barsky, LA, Schubov L, Bondar I (1991) Recovery of Non-ferrous Metals from Industrial and Urban Wastes Using Electrodynamic Separators. In: Reprints of *XVII Intl Mineral Processing Congress*, Vol. VII, Dresden, Germany, Bergakademie Freiberg, PP 45–51

Bauer B. (1994a) Guide to No-Clean Solder Pastes, *Electronic Packaging and Production*, March 1994, p. 65.

Bauer B. (1994b) Guide to No-Clean Solder Pastes, *Electronic Packaging and Production*, August 1994, p. 12.

Baumgartner, D (1996) Quick-Turn Design and DFM, *Printed Circuit Fabrication Asia*, May/June 1996, p. 14.

Bhardwaj, A. (2001) Design and Test Tools for Controlled Impedance on High Speed Boards, *National Conference on Emerging Trends in Electronics Design and Technology*, CEDTI, Mohali,

Biancini A.J , Supernova (1991) Advanced Surface Mount Design for Manufacturability, *Electronic Packaging and Production*, March 1991, p. 40

Boothroyd G. (1990) Designing for Assembly, *Electronic Production*, February 1990, p. 32.

Bosshart, W.C (1983) Printed Circuit Boards, *Tata McGraw-Hill Publishing Company Ltd.*, New Delhi.

Boyd, J. (1998) The Pulse-heated Solution, *SMT*, February 1998, p. 114

Braun, M.S.W. (2003) PCB Handbook for Assembling and ESD Control, *CEDTI*, January 2003, p. 25

Braun, M.S.W., (2002), PCB Design Hand Book, *CEDTI*, Mohali.

Breitengross R.A. (1993) A Case Study from the Navy on the Fabrication of Printed Wiring Boards, Internet Notes, http//es.epa.gov/techinfo/case/navy-cs1

Brist, G., Stewart, J. and Bird, S. (1997) Plasma-Etched Microvias, *Printed Circuit Fabrication Asia*, Vol.5, No.2 March/April 1997, p. 19.

Brock P. (1992) Protecting Populated PCBs against ESD Damage in the OEM's Plant, *Electronic Production*, August 1992, p. 15.

Brooks D (2002) Basic Transmission Lines, Why Use Them At All?, *Ultracad Design, Inc. and Mentor Graphics Corporation*, Internet Notes.

Brooks D. (1997a) Cross-talk, Part 1: The Conversation We Wish Would Stop!, *UltraCAD Design, Inc.*

Brooks D. (1997b) Cross-talk, Part 2: How Loud Is Your Cross-talk?, *UltraCAD Design, Inc.*

Broomfield N. (1992) Selecting Water Treatment Plant for Aqueous Cleaners, *Electronic Production*, March 1992, p. 31.

Bruhn, B. (1997) Environmental Excellence, *Printed Circuit Fabrication Asia* January/February 1997, p. 10.

Buckley, D. (1990a) Surface Mount Soldering, *Electronic Production*, September 1990, p. 43

Buckley, D. (1990b) Cleaning, Inspection, Re-work and Testing, *Electronic Production*, October 1990, p. 33.

Buckley D. (1990c) Components and Substrates, *Electronic Production*, July 1990, p. 29.

Buckley, D. (1992) Mechanical Drilling Eliminated in New PCB Technology, *Electronic Production*, September 1992, p. 11.

Cannon, M. (2001) The Micro Wave Revolution, *Technical Article*, ERSA GmbH, www.Solder Well_e.indd

Cavallaro, K. and Marchitto, M. (1991) Solder Paste Dispensing Versus Screen Printing, *Circuit Assembly*, October 1991, p. 40.

Cavallaro, K.J. (1994) Dispensing Adhesives for High-throughput SMT Assembly, *Electronic Packaging and Production*, August 1994, p. 16.

Chilton, C. and Gaugler, K. (1990) Solder Creams for Fine Pitch Assemblies, *Electronic Production*, May 1990, p. 19.

Coombs, C.F. (1988) Printed Circuit Handbook, *McGraw-Hill Book Company.*

Corrigan, K. (1992) Flex Circuit Properties Impact Packaging Design, *Electronic Packaging and Production*, May 1992, p. 86.

Creighton, M. (1996) Express Delivery on the Information Highway, *Printed Circuit Fabrication Asia*, May/June 1996, p. 22.

Cronin, R. (1995) Bare Board Test Program Generation, *Electronic Packaging and Production*, January 1995, p. 54.

Crum, S. (1993) Chipshooters Provide High Speed Placement, *Electronic Packaging and Production*, April 1993, p. 32.

Crum, S. (1994a) Flexible Laminate Materials Affect Circuit Performance, *Electronic Packaging and Production*, November 1994, p. 34.

Crum, S. (1994b) Bare Board Test Services Keep Pace with Test Technology, *Electronic Packaging and Production*, June 1994, p. 52.

Crum, S. (1995) Rapid Prototyping of Printed Circuit Boards. *Electronic Packaging and Production*, Feb.1995, p. 58.

Daniels, R. (1991) Solder: Past, Present, Future, *Circuits Assembly* April-1991, p. 7

Dolberg, S. and Kovarsky, M. (1997) Replacing Gerber Format, *Printed Circuit Fabrication Asia*, May/June 1997, p. 12.

Doyle K. (1991) AOI—The Way Forward, *Electronic Production*, April 1991, p. 17.

Duck, A. (1996) Dispensing SMT Adhesives, *Electronic Production*, May 1996, p. 21.

Dytrych, N.M. (1993a) Reviewing the Basics of Mass Re-Flow Soldering, *Electronic Packaging and Production*, July 1993, p. 34.

Dytrych, N. M.(1993b) Ensuring Bare Board Quality with Electrical Test, *Electronic Packaging and Production*, February 1993, p. 38.

Ehrler, S. (2002) A Review of Epoxy Materials and Reinforcements, Part 1, *PC Fab*, April 2002, p. 32.

Engelmaier, W. (1991) Component Reliability, *Circuits Assembly*, March 1991, p. 44.

Erdmann, G. (1991) Improved Solder Paste Stencilling Technique, *Circuits Assembly*, February 1991, p. 66.

Evans, D. (1991) Tackling the Test Problem, *Electronic Production Supplement*, October 1991, p. 35.

Falco, M. (1991) Inexpensive Microsectioning, *Circuits Assembly*, June 1991, p. 46.

Feldmann, K. and Scheller H. (1994) Disassembly of Electronic Products. In: Proc 1994 *IEEE Intl Symposium on Electronics and Environment*, IEEE, Piscataway NJ USA, pp. 81–86.

Ferrari, G. (1997) The Evolution of New Design Standards, *Circuits Assembly Asia*, May/ June 1997, p. 32.

Filleul, M. (1990) Clean Re-flow Soldering, *Electronic Production*, May 1990, p. 10.

Finstad, M. (2001) Designing for Flexibility and Reliability, *Minco Application Aid* 31, www.minco.com.

Fishburn, J. (2002) Seeing the Light of Full Colour Inspection, Internet notes, oeiwcsnts1.omron.com.

Fjelstad, J. (2001) Flexible Thinking, Internet Note, www.circuitree.com.

Fleck, I. (1994) Laser-cut Stencils Control Print Volumes for No-clean Solder Cream Printing, *Electronic Packaging and Production*, August 1991, p. 8.

Fulker, P. (1992) Aqueous and Semi-aqueous Cleaning: Discharges and Costs, *Electronic Production*, February 1992, p. 17.

Gaudion M (2000) Controlled Impedance Test, *The Board Authority*, September 2000, p. 56.

GE Electromaterials (2001) *The manufacture of Laminates*, www.geplastics.com/electromaterials.

Gemmell, A. (2003) Printed Circuit Board Manufacturing Pollution Prevention *Opportunities Checklist*, Internet Notes; http://es.epa.gov/techinfo/facts/cheklst7.html.

George, G. (1999) Typical Component Lead Sizes, Internet Notes, contact: glen@caltech.edu.

Ginsberg, L.G. (19992a) Printed Circuit CAD Systems and Software: Part 3 *Electronic Packaging and Production*, April 1992, p. 52.

Ginsberg, L.G.(1992b). Printed Circuit CAD Systems and Software: Part 1 *Electronic Packaging and Production*, January 1992, p. 40.

Ginsberg, L.G. (1992c) CAD in Concurrent Engineering: Part 4, *Electronic Packaging and Production*, May 1992, p. 60.

Goosey M. and Kellner R. (2002) End-of-life Printed Circuit Boards: A Scoping Study supported by Department of Trade and Industry, August 2002.

Goosey , M (2003) New Printed Circuit Board Laminate Material, www.intellectuk.org, April 2003.

Gothard A. (1991) Inert Atmosphere, *Electronic Production*, May 1991, p. 11.

Grant J. (1990) Low Cost Static Protection, *Electronic Production*, February 1990, p. 37.

Guiles, C.L (1998) Everything You Ever Wanted to Know About Laminates…. But Were Afraid To Ask, 7th Edition, Arion Inc. (Rancho Cucamonga, CA), 1998.

Gurain, M, , Ivory N. (1995), Performance Requirements of Primary Liquid Resists, *Electronic Packaging and Production*, March 1995, p. 49.

Hall, S. (1994) Screen Printer Requirements for Low Defect Process Capability, *Electronic Packaging and Production*, June 1994, p. 4.

Hallee, P.J. (1996) Testing High-Tech PCBs, *Printed Circuit Fabrication Asia*, Vol.4, No.4 July/August 1996, p. 18.

Hamilton, S. (1996) The Contamination Audit—A Vital Tool For Yield Improvement, Circuit World, Vol. 22, No.3, p. 24.

Harnsberger S and Saloka, T. (1994) Integrated Solutions for Metal Recovery and Water Recycling, *Electronic Packaging and Production*, September 1994, p. 42.

Harris, N. (1991) SM Adhesives-Just Strong Enough , *Electronic Production*, October 1991, p. 13.

Haskard, M.R (1998) Electronic Circuit Cards and Surface Mount Technology, *Technical Reference Publications Ltd.*

Herrmann, G., and Egerer, K. (1992), Handbook of Printed Circuit Technology, *Electrochemical Publications Limited.*

Hinton E.P. (1992) Solving the Problems of Internal Layer Registration, *Electronic Packaging and Production*, January 1992, p. 28.

Hodson, L.T. (1991) The Plating Process Optimizes PCB Performance, *Electronic Packaging and Production*, September 1991, p. 52.

Hodson L.T. (1993a) Selecting Pick and Place Equipment, *Electronic Packaging and Production*, June 1993, p. 32.

Hodson L.T. (1993b) Reworking the Surface Mount Assembly, *Electronic Packaging and Production*, October 1993, p. 46.

Hodson L.T. (1992) Solving Rework and Repair Problems, *Electronic Packaging and Production*, January 1992, p. 22.

Holden, H. (2003a) HDI's Beneficial Influence of High-Frequency Signal Integrity, Internet Notes, Mentor Graphics Technical paper at www.mentor.com/pcb/tech_papers.

Holden, H. (1997) PWB Build-up Technologies: Smaller, *Thinner and Lighter, Circuit World*, Vol. 23, No.2, p. 14.

Holden, H., Kenyon L. (1994) Framework-Based Electronic Assembly, *Electronic Packaging and Production*, November 1994, p. 44.

Holden, H. and Charbonneau R. (2000) Predicting HDI Design Density, *The Board Authority*, Vol.2, No.1, April 2000, pp. 28-31.

Holden, H. (2003 b) How to get started in HDI with Microvias, Technical Paper Series by Mentor Graphics at www.mentor.com/pcb/tech_papers.

Hudson, K. (2003) IDCT Announces 275, 000 RPM Multi-Head Drilling with Vision, Internet Notes www.pcbdriller.com.

Hunn, N. (1990) Getting Results from Hot Gas Rework, *Circuit Assembly*, November 1990, p. 21.

Iji. M, Yokoyama S (1997) Recycling of Printed Wiring Board Mounted Electronic Components, *Circuit World*, 23, No. 3, pp. 10-15.

Isaac, J. (1995) New Solutions for High Performance PCB Design, *Electronic Packaging and Production*, March 1995, p. 54.

Jeffery J.E (1997) Impedance Control of Conductors Acting as Transmission Lines in Printed Boards for High-frequency Digital Applications, *Circuit World*, Vol. 23, No. 2, 1997, p. 22.

Johnson, G. and Sparkman, O. (1996) The Future of Small Hole Drilling, *Printed Circuit Fabrication Asia*, May/June 1996, p. 26.

Judd, M. and Brindley, K. (1992) Soldering in Electronics Assembly, *Newnes*.

Justino, P (2002) An Overview of Hybrid Laser Drilling, Internet Notes, www.circuitree.com.

Kamat, P., Spilar L. and Yeager J. (1995) SIR System Design Ensures Test Accuracy, *Electronic Packaging and Production*, March 1995, p. 60.

Kawai, M. (2003) Motorola Ships Passive-Embedded PCB for Mobile Phones, *Nikkei Electronics Asia*, May 2003, p. 30.

Keeler, R. (1990a) Electroless Copper Technology in Transition, *Electronic Packaging and Production*, October 1990, p. 70.

Keeler, R. (1990b) Fine Pitch Soldering Options, *Electronic Packaging and Production*, October 1990, p. 41.

Keeping, S. (2000) Lasers Lead the Way for Microvias, *Electronic Production*, Issue 7, Vol. 29, p. 51.

Keimeier, S. (1994) Re-use before Recovering—Second Hand for Electronic Components, *OTTI-Technologie-Kolleg*, Regensburg.

Kelley, A. and Jones, S. (2002) Application of Laser Direct Imaging, Internet Notes contact: Kelley@circatex.com and Jones: stephenjones@tiscali.co.uk.

Lange, B. and Vollrath, K (www.lpkfusa.com) Highly Versatile Laser System for the Production of Printed Circuit Boards, Internet Notes.

Lasky, C.R., Primavera, A., Borgesen, P. and Lassen, © (1996) Critical Issues in Electronic Packaging, Part-III, *Circuits Assembly Asia*, September/ October 1996, p. 28.

Lea, T. (1990) Testing SMT Bare Boards, *Electronic Production*, January 1990, p. 29.

Legarth, J.B. and Alting L. et al (1995) A New Strategy in the Recycling of Printed Circuit Boards, *Circuit World*, Vol. 21, No.3, p. 10.

Leonida, G. (1989) Handbook of Printed Circuit Design, Manufacture, Components and Assembly, *Electro chemical Publications Ltd*.

Lexin, J. (1993) Cleaning and Handling of Flex Circuits, *Electronic Packaging and Production*, October 1993, p. 42.

Lideen, D.J., Dahl, A. (1995) Process Techniques for Pitch Screen Printing, *Electronic Packaging and Production*, July 1995, p. 30.

Lin, J. (2003) High Tg and Low Dk Laminate for Next Generation Printed Wiring Board, *Product Bulletin of Nan Ya Plastics Corporation*, www.npc.com.

Lindsey, D. (1985), Analog Printed Circuit Design and Drafting, *Bishop Graphics, Inc.*, California, USA.

Linman, D. (1990) Pre-heat Improves VPS Process, *Electronics Production*, February 1990, p. 24.

Lucas, L. G. (1993) Laminate Developments Enhance PCB Performance, *Electronic Packaging and Production*, April 1993, p. 42.

Lum, S. and Waddell P. (1996) An EMI/EMC Primer, *Circuits Assembly Asia*, March/April 1996, p. 26.

Manko, H.H.(1994). Lead Poison, Solder and Safety in the Workplace, *Electronic Packaging and Production*, February 1994, p. 93.

Mantay, M.K, Range, L.A and Schoenberg, L.N (1991) Optimizing Auto-routing Boosts PCB Manufacturability, *Electronic Packaging and Production*, June 1991, p. 58.

Markstein, H.W. (1993) Inspecting Assembled PCBs, *Electronic Packaging and Production*, September 1993, p. 70.

Markstein, W.H. (1995) Effective Shielding Defeats EMI, *Electronic Packaging and Production*, February 1995, p. 76.

Masaoka, K., Tanaka, Y. and Kobayshi, H. (1993) A Newly Developed System for Small Annular Ring Formation, May 1993 p. 22.

Maxfield, Clive (Max) and Wiens, David (2000) System Solutions, Re-defining Systems Design for the Electronics Community, *Technical Publication*, Mentor Graphic Corporation, September 2000, www.mentor.com/pcb p. 1.

Meier, J.D and Schmidt, H.S (2002) PCB Laser Technology for Rigid and Flex HDI-Via Formation, Structuring, Routing, *IPC Printed Circuit Expo*, Long Beach, CA.

Melton, C. and Fuerhaupter, H. (1997) Lead-free Tin Surface Finish for PCB Assembly, *Circuit World*, Vol. 23, No. 2, p. 30.

Mentor Graphics (2001) Board Systems Design and Verification, www.mentor.co,. p. 1.

Mentor Graphics (2002) Design Exchange: Seizing Control of the Design Process, *Technical Publication*.

Meyer, J., Werke, Kathrein, Rosenheim, K.G.(1991) Automated Package Design for High Speed Analog PCB, *Electronic Packaging and Production*, February 1991, p. 92.

Miller, M.B. (1997) Zero Wastewater Discharge, *Printed Circuit Board Fabrication Asia*, January/February 1997, p. 16.

Minco Application Aid 24 (2000), *Flex-Circuit Design Guide*, www.minco.com.

Montrose, (2003) EMC Suppression Concepts for Printed Circuit Boards, Internet Notes, www.ieee.org.

Morris, B. (1990) Reworking Fine Pitch SMCs, *Electronic Production*, Februray 1990, p. 12.

Muller, K. D (2000) Multi-layer Prototype and Series Production, *LPKF Laser and Electronics Application Report*, January 2000, p. 4.

Nakahara, H. (1991) Full Build Electroless Copper Plating is the Process of the Future, *Electronic Packaging and Production*, January 1991, p. 50.

Nargi—Toth, K. (1994) Additive Processing on the Upswing, *Electronic Packaging and Production*, December 1994 p. 38.

Nasta, M., and Peebles, H.C (1995) A Model of the Solder Flux Reaction; Reactions at the Metal/Metal Oxide/Electrolyte Solution Interface, *Circuit World* Vol. 21, No. 4, 1995, p. 10.

Neues, A (2002) Intelligent Visual Inspection Guidance, *Technical Article, ERSA GmbH*, www.ersa.de. info@ersa.de

Nimmo K. (2003) European Legislation on Lead in Electronics Circuits Moves Forward, *Solder and Assembly Technology*, No. 2, p. 2.

Noble, P., and Moore, R. (1992) Determining the Accuracy of Screen Printing Machines, *Electronic Production*, September 1992, p. 31.

Murray, G. (1996) Tearing Down the Wall, *Printed Circuit Fabrication Asia*, May/June 1996, p. 18.

Okubo, S. and Otsuki T. (2003) Mobile Phones Integrate OCR, Remote Controller Functions, *Nikkli Electronic Asia*, March 2003, p. 34.

Olney, B. (2003) EMC Design for High Speed PCB's. www.icd.com.au/ar.ticles/emc.html.

Oresjo, S (2003) Selecting the Optimal Test Strategy, *Circuits Assembly*, July 2003, p. 14.

Oresjo, S. (1990) Boundary Scan, *Circuits Assembly*, December 1990, p. 38.

Patterson, T.B. (1992) Additive and Subtractive Process Join Forces, *Electronic Packaging and Production*, March 1992, p. 40.

Peace, G. (1991) Towards Tomorrow's PCBs. Electronic Production, Oct. 1991, p. 24.

Polar Instruments (2001) Transmission Line Configurations Application Note 121.

Polar Instruments (2002) PCB Test—Locate Faulty Digital ICs Easily with In-circuit Function Test, Application Note 111, www.polarinstruments.com.

Polar Instruments (2003a) Introduction to Controlled Impedance, Application Note 120, www.polarinstruments.com.

Polar Instruments (2003b), Microstrip Transmission Line Structures, Application Note 122.

Polar Instruments (2003c) Single-ended Stripline Structures, Application Note 123.

Polhamus L.R. (1991) Cleaning Circuit Boards Populated with SMDs, *Electronic Packaging and Production*, February 1991, p. 84.

Pollack, H.W and Jacques R.C (1992) Adhesiveless Laminates Improve Flex Circuit Performance, *Electronic Packaging and Production*, May 1992, p. 74.

Price, D (1992) SMT Board Finishes-Going for Gold, *Electronic Production*, April 1992, p. 9.

Purdie, D. (1991) Repairing/Modifying Surface Mount PCBs, *Electronic Production*, February 1991, p. 11.

Qazi, J.M. and Calla, D. (1997) The Benefits of Machine Vision, *Circuits Assembly Asia*, January/February 1997, p. 26.

Betancourt, R. *et al.* (1996) Diseno y construccion de un amplificador de bajo ruido para la banda de 8-18 GHz, Memorias del Congreso de Instrumentation SOMI, XI, 636.

Raby, J. (1994) Assuring Solder Joint Reliability in Repair, *Electronic Packaging and Production*, September 1994, p. 54.

Rae, A. (2003) The Costs of Going Green, *Circuits Assembly*, July 2003, p. 22.

Raman, S. (2001) Laser Microvia Productivity: Dual Head Laser Drilling Systems, Internet Notes www.circuitree.com.

Raman, S., Davignon, J. and DiMarchoberardino, M. (www.esi.com) Implementation of Laser Technology in New Applications on PCB.

Rangel, R., Betancourt, R. and Chavez, R. (1997) Laser Drilling on Alumina-Based Printed Microwave Circuits, *Instrumentation and Development*, Vol. 3, No. 8, p. 53.

Reithinger, M. (1991), IR Vs VPS—A User's Evaluation, *Electronic Production*, March 1991, p. 27.

Robinson M. (1990) Shielding Against EMI, *Electronic Production*, March 1990, p. 43.

Ross, M. W. and Leonida, G. (1997) General Principles of Design and Layout (Printing Board Assemblies), *Circuit World*, Vol. No.23, 1997, p. 18.

Ross, M.W. and Leonida, G. (1996a) General Principles of Design and Layout (of Printed Board Assemblies) *Circuit World*, Vol. 23, No. 1, 1996, p. 25.

Ross, M. W. and Leonida, G. (1996b) General Principles of Design and Layout (of Printed Board Assemblies) Circuit World, Vol. 22, No. 4, p. 24.

Rubin, W. (1995) The Concept and Success of No-clean Technology, *Circuit World*, Vol. 21, No. 2, p. 23.

Saito, (1994) National Institute Resources and Environment, Tsukuba, Japan (1994) Recovery of Valuable Metals from PWB Wastes (2) Hydrometallurgical Treatment of PWB Wastes. Trans Meter Res Soc Jpn (1994), 18A (Ecomaterials), 211-14, CODEN: TMRJE3 Joumai written in Engilish. CAN 123-88995 AN 1995:700167.

Sallau, A, and Wiemers, A. (1999) Laser Directo Imaging, Publication ILFA GmbH, Edition 5.9 a. Internet Notes.

Saltzberg, M.A, Neller, A.L, Harvey, C.S, Borninski T.E and Gordon, R.J (1996) Using Polymer Thick Film for Cost-effective EMC Protection on PCBs for Automotive Applications Vol. 22, No. 3, 1996, p. 67.

Samsami, D. (1990a) Fine Pitch Soldering Defects, *Electronic Packaging and Production*, November 1990, p. 35.

Samsami, D. (1990) Enhancing the Manual Inspection Process, *Electronic Packaging and Production*, December 1990, p. 37.

Savage, R. (1992) Manufacturing Copper Foil For Flexible Circuits, *Electronic Packaging and Production*, May 1992, p. 80.

Scaminaci, Jr., J. (1994) Avoiding Signal Integrity Problems in Back Pains, *Electronic Packaging and Production*, July 1994, p. 40.

Shaw, J.M., *et al.* (1997) Big Blue Goes Green, *Printed Circuit Board Fabrication Asia*, January/February 1997, p. 27.

Shipley, C. (1991) Fixtureless Fine Line Board Testing, *Electronic Production*, April 1991, p. 29.

Siemens, A.G. (1991) IR vs VPS—A User's Evaluation, *Electronic Production*, March-1991, p. 27.

Slezak, E. (1994) Soldering Materials Trends, *Electronic Packaging and Production*, December 1994, p. 11.

Smith, J. (1993a) Equipment for Cleaning the Surface Mount Assembly, *Electronic Packaging and Production*, August 1993, p. 40.

Smith, J. (1993b) Countdown to CFC Phase-Out, *Electronic Packaging and Production*, January 1993.

Smith, K. (1991) The Case for Hot Bar Reflow, *Electronic Production*, January 1991, p. 13.

Spiak, R. and Valiquette, K. (1994) Trends in the Laminate Industry. *Electronic Packaging and Production*, p. 5.

Spitz, S.L (1990) The Case for Metal Recovery, *Electronic Packaging and Production*, July 1990. p. 44.

Stearns, T. (1992) Dielectrics Influence Circuit Performance and Laminate Processing, *Electronic Packaging and Production*, May 1992, p. 66.

Straw, J.J. (1992) Solutions to Fine Pitch Bare Board Electrical Test Challenges, *Electronic Packaging and Production*, March 1992, p. 28.

Taylor, S. (1991a) Complex Synthetic Fluxes Offer Enhanced No-clean Performance, *Electronic Production,* October 1991, p. 27.

Taylor, S. (1991b) Controlling the Wave Soldering Process for Synthetic Fluxes, *Electronic Production*, December 1991, p. 9.

Tennant, T. (1994) Solder Mask Options for the '90s, *Electronic Packaging and Production*, February 1994, p. 99.

Tong, P. (2003) Using Visual Inspection in Your PCB Test Strategy, Internet notes, www.eetasia.com.

Travi, C., Albertini, M. and Gemme, C. (1996) Effects of High Electrical Stress in PCBs, *Circuit World*, Vol. 22, No. 2, p. 16.

UltraCad Design (2000) Controlling Impedance, Internet Notes, www.omnigraph.com.

Vandervelde, H. (2001) PCB Handbook, *McGraw-Hill Publishing Company*.

Vaucher, C. and Jaquet, R. (2002) Laser Direct Imaging and Structuring: An Update, Internet notes, www.circuitree.com.

Vernon, D. (2003) Machine in the Electronics and PCB Inspection Industry, Internet Notes, www-prima.inrialpes.fr.

Wallig, L. (1992) Adhesives Bond Flex Circuit Materials, *Electronic Packaging and Production*, May 1992, p. 71.

Ward, J, (1992) Options in Generating High Quality Phototools, *Electronic Packaging and Production*, July 1992, p. 62.

Waryold, J and Lawrence, J. (1991) Conformal Coating, *Circuits Assembly*, June 1991, p. 56..

Waryold, J. et al (1998) A conformal Coatings Selection Guide. *Surface Mount Technology (SMT) Magazine*, Vol.12, No.2, p. 84.

Watts, N. (1993) Establishing a PCB Quality Assurance and Reliability Program, *Electronic Packaging and Production*, May 1993, p. 25.

Wiens, D (2000) Printed Circuit Board Routing at the Threshold, *Mentor Graphics*, May 2000.

Wilk, F.L. (1994) Treatment/Recycle of Wastewaters from Alternative Cleaning Processes, *Electronic Packaging and Production*, June 1994, p. 46.

Williamson, I. (1990) Front-End Automation, *Electronic Production*, February 1990, p. 19.

Willis, B. (1992) Repairing Damaged Plated Through Holes, *Electronic Production*, December 1992, p. 9.

Winstanely, A. (2003) The Basic Soldering and Desoldering Guide, Internet Notes, www.epemag.wimborne.co.uk.

www.thinktinkr.com Internet notes (2003) Multi-layer PCB Prototyping.

Yokoyama, S. and Iji, M. (1995) Recycling of Printed Wiring Board Waste. Proceedings of IEEE, p. 132.

Zarrow, P. (2000) Coolest Things Since Slice Bread, Internet Notes, www.ITM-SMT.com.

Zweig, D. (1996) Concurrent Growth, Multi-layer PCBs and X-ray Inspection, *Printed Circuit Fabrication Asia*, September/October 1996, p. 22.

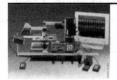

Index